Principles of Ecology

Principles of Ecology

Editor
Jennifer Heath

Salem Press
A Division of EBSCO Information Services, Inc.
Ipswich, Massachusetts

GREY HOUSE PUBLISHING

Cover photo: Butterfly kiss on a cosmos flower. By Jeffengeloutdoors.com (iStock).

Copyright © 2019, by Salem Press, A Division of EBSCO Information Services, Inc., and Grey House Publishing, Inc.

Principles of Ecology, published by Grey House Publishing, Inc., Amenia, NY, under exclusive license from EBSCO Information Services, Inc.

All rights reserved. No part of this work may be used or reproduced in any manner whatsoever or transmitted in any form or by any means, electronic or mechanical, including photocopy, recording, or any information storage and retrieval system, without written permission from the copyright owner. For information, contact Grey House Publishing/Salem Press, 4919 Route 22, PO Box 56, Amenia, NY 12501.

∞ The paper used in these volumes conforms to the American National Standard for Permanence of Paper for Printed Library Materials, Z39.48 1992 (R2009).

Publisher's Cataloging-In-Publication Data
(Prepared by The Donohue Group, Inc.)

Names: Heath, Jennifer, editor.
Title: Principles of ecology / editor, Jennifer Heath.
Description: [First edition]. | Ipswich, Massachusetts : Salem Press, a division of EBSCO Information Services, Inc. ; Amenia, NY : Grey House Publishing, [2019] | Series: Principles of | Includes bibliographical references and index.
Identifiers: ISBN 9781642650440
Subjects: LCSH: Ecology.
Classification: LCC QH541 .P75 2019 | DDC 577–dc23

FIRST PRINTING
PRINTED IN THE UNITED STATES OF AMERICA

Table of Contents

Publisher's Note vii
Editor's Introduction ix
Contributor List xiii

Accounting for nature 1
Agriculture 4
Air pollution and greenhouse gases 10
Alternative fuels 13
Amazon Deforestation 17
Amazon River 19
Amphibians 23
Anthropocentrism 24
Antienvironmentalism 25
Aquifers 29
Balance of nature 31
Beach erosion and sand mining 35
Bees and other pollinators 37
Biocentrism 40
Biodiversity 41
Biodiversity action plans 44
Biogeography 46
Biomagnification 50
Biomes 52
Biopesticides and bioassays 54
Bioprospecting and biopiracy 57
Bioregionalism 59
Bioremediation 61
Biosphere 64
Boreal and cloud forests 68
Brownfields 71
Carbon cycle and carbon footprint 73
Carbon dioxide air capture 74
Carbon trade 76
Carcinogens 78
Cattle 79
Clean Air Act 82
Clean Water Act 87
Climate Change 90
Climate change and human health 93
Climate change skeptics 96
Conference of the Parties 97
Coral reefs 100
Cultural ecology 102
Dams and reservoirs 105
Dead zones 108
Deep ecology 110
Deforestation 111

Desalination 114
Deserts and dunes 115
Dichloro-diphenyl-trichloroethane (DDT) 119
Drought 120
Dust Bowl disaster 124
Earth Day 127
Eat Local movement 130
Ecocentrism 132
Eco-fashion 133
Ecofeminism 135
Ecological economics 137
Ecological footprint 139
Ecology as concept and in history 141
Ecotourism 146
Electronic waste 150
Endangered species 152
Environmental determinism 157
Environmental ethics 158
Environmental justice and racism 161
Erosion 163
Estuaries 165
Fisheries 169
Floodplains 172
Fossil fuels 174
Genetic modification 179
Glacial melting 181
Global warming 183
Globalization 186
Grasslands 189
Green buildings 192
Groundwater pollution 194
Habitat destruction 197
Hazardous wastes 199
Hydraulic fracturing (fracking) 203
Indicator species 205
Intergenerational Environmental Justice 207
Intergovernmental Panel on Climate Change
  (IPCC) 209
International Nature Preservation Policies 212
Land-use policy 219
Light pollution 223
Marine life 225
National forests 229
National parks 232
Noise pollution 235
Nuclear accidents 237
Nuclear power industry 239

Ocean debris 245
Oceans and rising temperatures 248
Oil drilling 252
Oil shale and tar sands 254
Oil spills 255
Organic farming and gardening 259
Ozone layer 262
Population 269
Population control and social justice 270
Recycling 273
Renewable energy 276
Rising sea levels 280
Road systems 282
Severe and anomalous weather 285
Smog and heavy metals 288
Species loss and the Sixth Extinction 291
Tidal energy 297
Tragedy of the commons 298
Urban rain gardens and greenbelts 301
Urban sprawl and urban ecology 303
Water pollution 309
Watersheds 311
Wetlands 313
Wilderness areas 317
Wildlife refuges 319
Wind energy 322

Appendixes
U.S. federal laws concerning the environment ... 327
Major world national parks and protected areas 331
Environmental organizations 343
Directory of US National Parks 353
Subject index 357

■ Publisher's Note

Salem Press is pleased to add *Principles of Ecology* as the fourteenth title in the *Principles of* series that includes *Modern Agriculture, Physics, Astronomy, Computer Science, Physical Science, Biology, Scientific Research, Sustainability, Biotechnology, Programming & Coding, Climatology,* and *Robotics & Artificial Intelligence*. This new resource introduces students and researchers to the fundamentals of ecology using easy-to-understand language for a solid background and a deeper understanding and appreciation of this important and evolving subject. All of the 110 entries included in the volume are arranged in an A to Z order, making it easy to find the topic of interest.

Entries related to basic principles and concepts include the following:
- Fields of Study related to the topic;
- Principal Terms and definitions;
- A Summary that provides brief, concrete summary of the topic and how the entry is organized;
- Text that gives an explanation of the background and significance of the topic to modern agriculture by describing developments such as Bacterial resistance; Brownfields; Carrying capacity; Deep ecology; Geoengineering, and Superfund legislations;
- Illustrations and tables that clarify difficult concepts via models, diagrams, and charts of such key topics as the Glacial melting, Greenhouse gas emissions; Groundwater pollutions, and hazardous wastes; and
- Further Reading lists that relate to the entry.

This reference work begins with a comprehensive introduction to climatology, written by volume editor Jennifer Heath.

The book includes helpful appendixes as another valuable resource, including the following:
- Timeline of American Environmentalism, tracing the field back to the limits and restrictions on cutting timber and hunting to the current administration's proposed cuts to environmental programs
- Key Figures in the areas of ecology and environmentalism, offering biographical information and a brief description of their contributions
- U.S. federal laws concerning the environment
- Directory of U.S. National Parks
- Major world national parks and protected areas
- Environmental organizations
- Bibliography
- Subject Index

Salem Press and Grey House Publishing extend their appreciation to all involved in the development and production of this work. The entries have been written by experts in the field. Their names and affiliations follow the Editor's Introduction.

Principles of Ecology, as well as all Salem Press reference books, is available in print and as an e-book. Please visit www.salempress.com for more information.

■ Editor's Introduction

The word "ecology" was coined in 1873, by German zoologist Ernst Haeckel, as *oecology*, "a branch of science dealing with the relationship of livings things to their environments."

Ecology is a subsystem of biology and, in turn, of conservation biology. It is the study of interactions and speaks to the relationship between conservation biology and the activism whose objective is to conserve biodiversity. One absolute biological and ecological truism is that everything is connected. Biodiversity—an abbreviation of "biological diversity," a term that first appeared in 1968 in *A Different Kind of Country*, by scientist and conservationist Raymond F. Dasmann[1]—is at the crux of ecology. At the heart of biodiversity is variation. In popular usage, the word biodiversity generally refers to all the individuals and species living in a specific area or within an ecosystem.[2] Ecosystems consist of co-occurring organisms, living in conjunction with the nonliving components of their environment and interrelating as a community. One way to look at this ecological networking might be, as anthropologist Gregory Bateson put it, in a unrelated context, "There is a larger mind, of which the individual mind is only a subsystem."[3]

Ecology is a vast territory. A fundamental "Ten Principles of Ecology" has been devised by Oklahoma University biology professor, Michael Kaspari. He calls this his "working list":

1. Evolution organizes ecological systems into hierarchies;
2. The sun is the ultimate source of energy for most ecosystems;
3. Organisms are chemical machines that run on energy;
4. Chemical nutrients cycle repeatedly while energy flows through an ecosystem;
5. The rate that a population's abundance in a given area increases or decreases reflects the balance of its births, death and net migration into an area;
6. The rate that the diversity of species in an area changes reflects the balance of the number of new forms that arise, those that go extinct and those that migrate from the area;
7. Organisms interact in ways that influence their abundance;
8. Ecosystems are organized into webs of interactions;
9. Human populations have an outsized role in competing with, preying upon, and helping other organisms;
10. Ecosystems provide essential services to human populations; therefore, a key goal of ecology is to use principles 1-9 to preserve ecosystems.[4]

The study of ecology includes, among other areas, science, ethics, and history. Anthropology, sociology, engineering, geography, geology, oceanography, topology, mathematics, economics, and even religion are some of the many fields of study involved in the analysis of and efforts to save biodiversity. Across time, as illustrated in this encyclopedia, there has been a shift from the notion of preservation of wilderness to the conservation of biodiversity, that is, protection of the whole.

Principles of Ecology– whose contributors are all experts in their fields –strives to examine and define as many aspects and issues – principles – of ecology as possible, from bioregionalism, biogeography, and biocentrism to cultural ecology, environmental racism, social justice, globalization and wildlife refuges. Yet at the core of it all is the loss of biodiversity—the result of 3.5 billion years of evolution—caused by the destruction of ecosystems and habitats through deforestation, overpopulation, greenhouse gases, poisoning of soil and land from industrial, agricultural, and nuclear waste, mining, the depletion of natural resources and the ozone layer, the mass extinctions of plants and terrestrial and marine animals, and endangered species on land and in the seas. Here, readers will find reports of public health effects, of air and ocean pollution, as well as numerous informed articles, which, taken together, describe our complicated worldwide water crisis—including, for example, decreasing access to clean drinking water, the environmental impacts of desalination, accelerating drought (and desertification), glacial melting and rising sea levels, and the degradation of watersheds, aquifers, estuaries, and rivers.

There are—sometimes subtle—distinctions made throughout the book between "global warming" and "climate change." These are often used interchangeably, like "weather" and "climate." But weather refers to atmospheric conditions that occur locally over short periods of time, while climate is broader and refers to the long-term regional or even global average of temperature, humidity, and rainfall patterns over seasons, years, or decades. Global warming refers only to the Earth's rising surface temperature, while climate change is more expansive, including warming, as well as its side effects: melting glaciers, unusually heavy storms and anomalous weather, or progressively frequent drought. Global warming is a symptom of the much larger problem of human-caused climate change.

The main culprit is overuse of fossil fuels, which started in alarming earnest during the Industrial Revolution, which was initiated in Great Britain around 1750 and quickly spread throughout Europe and the United States. The Industrial Revolution moved manufacture from hand-production methods to machines, producing factories and, with them, the unprecedented use of coal, exacerbated by the invention of the efficient, highly polluting steam engine. Industrialization has certainly provided some benefits to humanity, but those have come at great cost, not only disrupting social systems, but putting ecosystems at risk more and more rapidly as our dependence on fossil fuels has compounded unrelentingly.

Readers will discover that the history of ecology had its genesis as long ago as ancient Greece and that the history of environmental degradation can be said to have begun with the invention of agriculture, with clearing of wild lands and forests, steel plows, and eventually monoculture – single crop – farming. The history of the Dust Bowl, which primarily impacted the Southern Plains region of the United States in the 1930s, is a major example of erosion from overcultivation. Land management helped rectify some of those problems, but modern practices also make heavy use of chemicals – pesticides and fertilizers that are deadly to soil, water, and human health. Genetic engineering, explained in this volume, affects the welfare of all sentient beings.

Much of the contents of this encyclopedia and the very real degeneration of Earth's environment may seem grim. Indeed, the prognoses aren't bright. But there is hope. There is movement, people pulling together to make change. In *Principles of Ecology*, readers can research laws, public policies, and action plans developed across time, nationally and internationally, to mitigate the destruction. Various articles also provide information that relates to what we ourselves can do in our daily lives to inspire a better future: organic farming and gardening, alternative fuels and renewable energy, recycling, and other endeavors that can be adopted at the grassroots, then trickle upward toward global solutions. "Think globally, act locally," the saying goes. At the back of this book, readers will find a selected list of organizations working toward environmental justice, biographies of key figures, a timeline of American environmentalism, a glossary, and a bibliography.

Here, too, readers can examine and weigh the points of view of climate-change disbelievers, from outright skeptics, opponents of the science, to anti-environmentalists who consider that the Earth and nature exist exclusively for human exploitation.

One essential first step in seeing ecology's larger picture is simply to look around. Everywhere, even in the ugliest most apparently hopeless places—sordid urban sprawl or beaches covered in medical waste and plastics – there is beauty or beauty waiting to be realized and reclaimed. All nature now exists in relation to human activity, so we might start by, as the poet William Blake wrote, seeing "the world in a grain of sand and heaven in a wildflower."[5] By looking and seeing, we can begin to value all things in nature and recognize, humbly, as indigenous peoples do worldwide, that humans are an inseparable part of nature. With that acceptance, we can connect, focus on the whole, and interact equitably with Earth's diverse bio-communities.

A second essential step is to understand time. Humans must appreciate time as larger and longer than our own short, individual lives. Biodiversity evolved across 3.5 billion years. Geologic time begins at the start of the Archean Eon (4.0 billion to 2.5 billion years ago) and continues to the present day. Native Americans speak of the "Seventh Generation Principle," that is, actions and decisions cannot be made without consideration of their descendants seven generations into the future.

An American environmental poet, Jack Collom, whose work is noted in *American Environmental Leaders*[6], wrote the following acrostic, "TIME," in 2014:

This is the problem. We destroy, pollute,
Incinerate, gobble up, and then all the
Mellow, philosophical thoughts in the world
can't recapture the lost
Eternity in every moment.[7]

The word *oecology* (basis of our word ecology) comes from the Greek, *oikos*, meaning "house, dwelling place, habitation." *Oikos*. Home. Our home. Our magnificent planet.

Jennifer Heath

Notes

1 Raymond F. Dasmann. *A Different Kind of Country*. New York: Collier MacMillan, 1968.
2 The word "biodiversity" came into common scientific usage in the 1980s, primarily at the 1986 National Forum on Biodiversity in Washington, D.C., where it was argued that the reduction of biological diversity is the most fundamental question of our time. The results of that forum appeared in a 1988 volume titled *Biodiversity*, published by the National Academy Press and edited by renowned biologist E.O. Wilson. It comprises sixty articles from leading authorities on the subject.
3 Gregory Bateson. *Steps to an Ecology of Mind: Collected Essays in Anthropology, Psychiatry, Evolution and Epistemology*. Northvale, N.J.: Jason Aronson, Inc., 1972.
4 The Kaspari Lab. https://michaelkaspari.org/2017/07/17/the-ten-principles-of-ecology/. Kaspari writes that in the first week of class his students "sample, expand on, and recombine [the list] throughout the rest of the semester." He adds that "every ecologist is different," and each may have a different approach to these principles."
5 William Blake, from "Auguries of Innocence," written approximately in 1803, available at the Poetry Foundation website, https://www.poetryfoundation.org/poems/43650/auguries-of-innocence. William Blake (1757-1827) was an English visionary poet and artist.
6 Anne Becher and Joseph Richey, eds., *American Environmental Leaders: From Colonial Times to the Present*. 3rd ed. Amenia, N.Y.: Grey House Publishing, 2018.
7 Unpublished, private collection, with permission from the Estate of Jack Collom (1931-2017).

■ Contributor List

Richard Adler

Emily Alward

Robin Attfield

Michael P. Auerbach

Elizabeth A. Barthelmes

Melissa A. Barton

Raymond D. Benge, Jr.

Margaret F. Boorstein

Richard G. Botzler

Pat Brereton

Jack Carter

Nader N. Chokr

Thomas Clarkin

Kathryn A. Cochran

Daniel J. Connell

Mark Coyne

Greg Cronin

Anna M. Cruse

Robert L. Cullers

Roy Darville

Joseph Dewey

Gordon Neal Diem

Stephen B. Dobrow

Frank N. Egerton

Jess W. Everett

C. R. de Freitas

Brian J. Gareau

James S. Godde

Lissy Goralnik

D. R. Gossett

Daniel G. Graetzer

Phillip A. Greenberg

Wendy Halpin Hallows

Wendy C. Hamblet

Clayton D. Harris

Gerald K. Harrison

Jennifer Heath

Thomas E. Hemmerly

Joseph W. Hinton

Robert M. Hordon

Solomon A. Isiorho

Bernard Jacobson

Bruce E. Johansen

Karen N. Kähler

Jamie Michael Kass

Julia Kendrick

Robert W. Kingsolver

Samuel V. A. Kisseadoo

Bill Kte'pi

Andrew Lambert

Timothy Lane

M. Lee

Thomas T. Lewis

Josué Njock Libii

Donald W. Lovejoy

R. C. Lutz

Marianne M. Madsen

Sergei A. Markov

Steven B. McBride

Roman Meinhold

Alice Myers

Nancy Farm Männikkö

Peter Neushul

Zaitao Pan

George R. Plitnik

Robert Powell

Donald F. Reaser

Claudia Reitinger

Raymond U. Roberts

Contributor List

James L. Robinson

Charles W. Rogers

Neil E. Salisbury

Robert M. Sanford

Martha A. Sherwood

Carlos Nunes Silva

Roger Smith

Courtney A. Smith

Toby Stewart

Dion Stewart

Toby Stewart,

Alexander R. Stine

Mary W. Stoertz

Hubert B. Stroud

Rena Christina Tabata

Julia Tanner

William R. Teska

John M. Theilmann

Oluseyi A. Vanderpuye

Megan E. Watson

Shawncey Webb

Robin L. Wulffson

A

■ Accounting for nature

FIELDS OF STUDY
Agriculture; Biology; Culture; Ecology; Economics; Environment; Environmentalism; Forestry; Wilderness; Nuclear Technology

SUMMARY
In comparisons of the costs and benefits of various actions related to the environment, the coordination of ecological and economic expertise can result in a balanced problem-solving approach.

PRINCIPAL TERMS
- **biological wealth:** the natural species of all living things responsible for the structure and maintenance of all ecosystems and that sustains human life and economic activity
- **contingent valuation:** survey method used to estimate the value of non-market resources
- **cost benefit:** economic analysis assigning a numerical value to the cost-effectiveness of an operation, procedure, or program
- **debt for nature swap:** concept to deal with developing-nation indebtedness and its consequent deleterious effect on the environment, so that ameliorating debt could simultaneously promote conservation
- **globalization:** the process of interaction and integration among people, companies, and governments worldwide.
- **Gross Domestic Product (GDP):** the broadest quantitative measure of a nation's total economic activity, representing the monetary value of all goods and services produced within a nation's geographic borders over a specified period
- **net economic welfare:** proposed national income measure that attempts to put a value on the costs of pollution, crime, congestion, and other negative spinoffs to find a better measure of true national income
- **polluter pay principle:** commonly accepted practice that those who produce pollution should bear the costs of managing it to prevent damage to human health or the environment

Photo of Dr. Paul R. Ehrlich, entomologist. By Ilka Hartmann (eBay).

Environmental decisions frequently set ecologists and economists on a collision course: Does society's demand for energy justify the environmental impacts of mining and burning coal? Is the increased economic efficiency of large corporate farms worth the loss of a rural lifestyle? How much should the public sacrifice to protect endangered birds or plants? These decisions are difficult because they force comparisons of "apples and oranges," pitting one set of values against another. Quantitative models (or accounting systems) that compare the costs and benefits of a course of action are frequently used to guide business and government decisions, but these have generally omitted environmental values. Accounting for nature in these models cannot produce completely objective solutions but can help coordinate ecological and economic expertise in a more balanced problem-solving approach.

Ecologist Edward O. Wilson, among those who popularized the term "biodiversity," has identified three kinds of national wealth: economic, cultural, and biological. He has observed that nations frequently create the illusion of a growing economy by consuming their biological or cultural "capital" to create short-term economic prosperity. For example, burning rain forests and replacing them with row crops may temporarily increase farm production in a developing nation, but tropical soils are often nutrient-poor and easily degraded by exposure to the sun and rain. If the biological basis of production is ignored, the population simply transfers wealth from one category to another, ensuring ecological disaster for its children in the process.

Developed nations have also neglected biological wealth in past cost-benefit analyses. Nations have justified the damming of wild rivers to make recreational lakes, for example, by counting the benefits of lumbering trees from the watershed but not the losses of aquatic and forest habitats. Recreational activities have been evaluated according to the money that people pay to participate in them; thus, hikers and canoeists in a wilderness area are given less consideration than water-skiers or drivers of recreational vehicles in a developed area because hikers and canoeists spend less money on equipment, fuel, and supplies.

Ecologists Eugene P. Odum and Howard T. Odum addressed this issue by calculating the value of ecosystem services provided by intact biological communities. Their approach was to measure the beneficial work performed by living systems and place a value on that service based on the time and energy required to replicate the service. A living tree, they reasoned, may provide a few hundred dollars in lumber if cut; if left alive, however, the oxygen it produces, carbon dioxide it absorbs, wildlife it feeds and shelters, soil it builds, evaporative cooling it yields, and flood protection it provides are worth far more on an annual basis.

National Wealth

In 1972, economists William Nordhaus and James Tobin refined the concept of national wealth by developing an index of net economic welfare (NEW) to replace the more familiar measurements of economic health such as gross domestic product (GDP). Their criticism of the GDP was that it counts any expenditure as a positive contribution to national wealth, whether or not the spending improves people's lives. A toxic waste dump, for example, contributes to the GDP when the pollutants are produced, again when millions of dollars are spent to clean it, and yet again if medical costs rise because of pollution-related illness. Nordhaus and Tobin's NEW index subtracts pollution abatement and other environmental costs from the value of goods and services that actually improve living standards.

Economist E. F. Schumacher subsequently argued that environmental costs should be "internalized," or charged to the industries that create them. This idea, also called the "polluter pays principle," not only generates funds for environmental cleanup but also encourages businesses to make environmentally sound decisions. The price of recycled paper, for example, would be more competitive if the public costs of deforestation and pollution from pulp mills were added to the price of virgin wood fiber. Proposals for internalizing environmental costs have ranged from centrally planned models, such as a carbon tax on fossil fuels, to free market trading of pollution credits. Debt-for-nature swaps, through which developing nations receive financial benefits for preserving natural ecosystems, represent environmental cost accounting on the asset side of the ledger.

A fundamental difference between economic and ecological worldviews is the time scale under consideration. Business strategies may look five years ahead, but ecological processes can take centuries. Thus, economic models that fail to take long-term issues into account are a frequent source of criticism by environmentalists. The U.S. decision to build nuclear fission reactors during the 1960s and 1970s is a case in point. Nuclear power appeared economically attractive over the thirty-five-year life span of a fission reactor, but the twenty-four-thousand-year half-life of radioactive plutonium 239 in spent fuel rods made skeptics wonder who would pay the costs of nuclear waste disposal for generations after the plants were closed.

Debates about growth are especially contentious. Traditional economists view the growth of populations, goods, and services as positive and necessary for economic progress and social stability. As early as 1798, however, economist Thomas Robert Malthus pointed out that on a finite Earth, an exponentially expanding human population would eventually run out of vital resources. In the closing decades of the twentieth century, Paul R. Ehrlich and Anne Ehrlich warned that unless population growth slowed soon, each person would have to consume less space, food, fuel, and other materials to avoid a global population crash. Whether one considers them economic pessimists or environmental realists, Malthus and the Ehrlichs demonstrate that taking a longer view is central to the task of accounting for nature. Sustainable development is the watchword for ecologists, economists, and political leaders attempting to create prosperity today while accounting for the welfare of future generations.

Contingent Valuation

Contingent valuation is used to assign value to nonmarket resources, such as renewable energy, open space, and sustainable development. While these resources provide utility, certain components of each do not have market prices; for example, renewable energy may reduce human-caused climate change or preserve fossil fuels for future generations. A contingent valuation survey is used to estimate a market price as a stated preference. The fundamental mechanism of the contingent valuation survey is asking people about their willingness to pay (WTP) to maintain an environmental feature or their willingness to accept (WTA) compensation for its loss.

Agricultural economist Siegfried von Ciriacy-Wantrup suggested the use of a direct interview method to measure the value of natural resources as early as 1947. Perhaps the first practical application was completed during the 1960s by economist Robert K. Davis, who measured the value of a specific wilderness area to hunters and recreationalists. Davis's contingent valuation results compared well with inferred value from cost associated with traveling to the wilderness area.

The method gained popularity during the 1970s in the United States as it was granted official recognition. Large numbers of studies were completed during the 1980s, with applications expanding to Europe and developing countries. However, criticism of the method also multiplied. Twenty-two expert economists on a panel convened in 1993 concluded that contingent valuation surveys must be carefully designed and controlled to ensure that valid results are obtained. They noted that individuals and organizations planning to employ contingent valuation should carefully review best practices before applying the method.

The panel offered specific recommendations concerning how contingent valuation surveys should be conducted, including the following. If possible, the survey interviews should be conducted in person; telephone surveys may be acceptable, but mail surveys should be avoided. A referendum format should be used in the questions; for example, "Would you be willing to contribute (or be taxed) D dollars to cover the cost of avoiding or repairing environmental damage X?" The results obtained from questions of this kind are more accurate than those gleaned from answers to open-ended questions, which are more likely to elicit strategic behavior, protest responses, biased answers, and incomplete consideration of personal income limits. The interviewers should ensure that respondents understand and accept the scenario they are asked to value. Respondents who do not accept the accuracy of the information concerning a scenario are in fact answering a question that is different from the one asked. Respondents should be reminded that their WTP for the specific scenario will reduce their ability to pay for other private or public goods.

Globalization

Globalization in the early twenty-first century rests on a free trade or neoliberal economic model that favors open markets and global competition among states and nonstate actors in the world economy. Intense competition among developing nations to secure investment and jobs from huge transnational corporations pushes ecological interests in those countries to the background of their political agendas. Corporate interests in the developed world tend to suppress movements for ecological reform that would cut into corporate profits.

Globalization is the subject of heated debate around the world, among politicians and economists as well as among scientists and environmental activists. From the standpoint of environmental ethics, globalizing trade practices have had devastating effects on the earth's natural environment. Regional neglect and the pollution of air, land, and ocean waters are driven by the "race to the bottom" phenomenon that pits developing countries against each other in efforts to lure global investors. Critics argue that the existing system is simply a broader-reaching, more profitable model of colonialism, a neocolonialism, whereby governments act as mere salespersons, promoting the profits of their corporations in a global marketplace.

Critics charge also that developing countries have no fighting chance in the global trade game, and so the rich get richer through the growing exploitation of the global poor and the devastation of the environment, in both the developed and the developing nations. Globalists, in contrast, assert that "free trade" promotes freedom and democracy, and that even as global inequality rises, poverty can be reduced through free trade. They argue that problems such as environmental degradation and global warming should be viewed as opportunities for entrepreneurial innovation and new economic ventures, and not as problems to be addressed through political intervention and legal restrictions.

Jess W. Everett, Wendy C. Hamblet, and Robert W. Kingsolver

Further Reading

Anderson, Terry L., Laura E. Huggins, and Thomas Michael Power, eds. *Accounting for Mother Nature: Changing Demands for Her Bounty.* Stanford, Calif.: Stanford University Press, 2008.

Baylis, John, ed., *The Globalization of World Politics: An Introduction to International Relations.* Oxford, U.K.: Oxford University Press, 2013.

Bowman, Troy, Jan Thompson, and Joe Colletti. "Valuation of Open Space and Conservation Features in Residential Subdivisions." *Journal of Environmental Management* 90 (2009): 321–30.

Brown, Peter and Peter Timmerman, eds. *Ecological Economics for the Anthropocene: An Emerging Paradigm.* New York: Columbia University Press, 2015.

Constanza, Robert, ed. *An Introduction to Ecological Economics.* Boca Raton, Fla.: St. Lucie Press, 1997.

Daly, Herman E. and Joshua Farley. *Ecological Economics: Principles and Applications.* Chicago: Island Press, 2010.

Kaplinsky, Raphael. *Globalization, Poverty, and Inequality: Between a Rock and Hard Place.* New York: John Wiley, 2013.

Pearce, Joseph. *Small Is Still Beautiful: Economics As If Families Mattered.* Wilmington, Del.: ISI Books, 2006.

Venkatachalam, L. "The Contingent Valuation Method: A Review." *Environmental Impact Assessment Review* 24 (2004): 89–124.

Wiser, Ryan H. "Using Contingent Valuation to Explore Willingness to Pay for Renewable Energy: A Comparison of Collective and Voluntary Payment Vehicles." *Ecological Economics* 62 (2007): 419–32.

■ Agriculture

FIELDS OF STUDY

Agriculture; Agronomy; Biology; Ecology; Environment; Environmental Engineering; Food Science; Genetically Modified Food Production; Genetically Modified Organisms; Genetic Engineering; Horticulture; Horticultural Science; Land-use Management; Soil Science

SUMMARY

Agriculture is the ability to produce enough food and fiber to feed and shelter the population. Modern agriculture has exacerbated ecological damage adding a host of issues critical to the environment. The beginnings of agriculture predate written history, when, some eight to ten thousand years ago, humans discovered that seeds from certain wild grasses could be collected and planted and the grasses later gathered for food.

PRINCIPAL TERMS

- **agricultural chemical industry:** suppliers and manufacturers of fertilizers, insecticides, anthelmintics, mineral and vitamin supplements, pharmaceuticals, growth promotants, feed additives, sanitation materials; producers of much of agriculture's research, maintaining the data

base of agricultural technical and scientific knowledge
- **assarting:** preparing fields by deforestation
- **genetically modified crops (GM):** plants used in agriculture, the DNA of which has been modified using genetic engineering methods generally in order to introduce a new trait to the plant not occurring naturally in the species. Also called biotech crops
- **Green Revolution:** term thought to have been introduced in 1968 describing the great increase in production of food grains due to the introduction of high-yielding varieties, to the use of pesticides, and to better management techniques
- **seminomadic farming:** seasonal migration when crops are cultivated during periods of settlement
- **slash-and-burn agriculture:** farming method involving the cutting and burning of plants in a forest or woodland to create a field called a swidden.
- **subsistence agriculture:** growing food crops for the survival of farmers and their families, with little or no surplus trade
- **swidden agriculture:** a technique of rotational farming in which land is cleared for cultivation, usually by fire, and left to regenerate. Also known as shifting cultivation

The earliest attempts to grow crops were primarily to supplement the food supply provided by hunting and gathering. However, as the ability to produce crops increased, people began to domesticate plants and animals, and their reliance on hunting and gathering decreased, allowing the development of permanent settlements in which humans could live. Six thousand years ago, agriculture was firmly established in Asia, India, Mesopotamia, Egypt, Mexico, Central America, and South America.

The first agricultural centers were located near large rivers that helped maintain soil fertility by the deposition of new topsoil with each annual flooding cycle. As agriculture moved into regions that lacked the annual flooding of the large rivers, people began to utilize a technique known as slash-and-burn agriculture, in which a farmer clears a field, burns the tress and brush, and farms the field. After a few years, soil nutrients become depleted, so the farmer must repeat the process at a new location. This type of agriculture is still practiced in some developing countries and is one reason tropical rain forests are disappearing at a fast rate.

In temperate climates, farmers often owned and lived on the land where they practiced swidden agriculture, clearing a portion of their land, burning the covering vegetation, growing grains for several years, then allowing that land to remain fallow, allowing the land and forests to recover. The slash-and-burn practice is not so much the problem as the length of a cycle. In some areas, land may require as little as five years to regain its maximum fertility; in others it may take one hundred years. Problems arise when growing populations pressure traditional farmers to return to fallow land too soon. Crops will be too small, leading to a vicious cycle in which the next strip of land is also farmed too soon, and each site yields less and less. As a result, more and more land must be cleared.

Another ancient form of agriculture, called nomadism, employs slash-and-burn techniques, but usually involves livestock rather than planting. Pressures generated by industrialized society are increasingly threatening traditional cultures of nomadic societies, such as the Bedouin of the Arabian Peninsula. Traditional grazing areas are being fenced off or developed for other purposes. Environmentalists are concerned about the ecological damage caused by nomadism. Wealth is measured by the size of a herd and herd animals eat increasingly large amounts of vegetation, which then has no opportunity to regenerate. Desertification may occur as a result.

Until the nineteenth century, most farms and ranches were family-owned, and most farmers practiced sustenance agriculture: Each farmer produced a variety of crops, enough to feed his or her family as well as a small excess sold for cash or bartered for other goods or services. Agricultural tools such as plows were made of wood, and almost all agricultural activities required human or animal labor. This situation placed a premium on large families to provide the help needed in the fields.

Land Clearance
Land clearance is the removal of plant life, stones, and other obstacles from a land surface to increase the area available for farming or construction of buildings. Although humans have modified land

surfaces for food production and habitation since prehistoric times, the process of land clearance escalated tremendously from the mid-twentieth century onward, causing dramatic changes in ecosystems around the world. These changes have raised environmental concerns regarding soil, water, and air pollution and the loss of animal habitats.

Land clearance increased significantly when human beings transitioned from hunter-gatherer to agricultural societies. As agriculture developed, extensive land areas were cleared for crop growth. With the advancement of technology, the process of land clearance escalated as humans using heavy equipment became capable of not only rapidly clearing but also reshaping the landscape. Often, areas of cleared land that have become less productive than they were previously are abandoned in favor of newly cleared land that is more supportive of agricultural endeavors. These deserted areas often become barren wastelands devoid of human and animal habitation.

Modern Agriculture

The arrival of the Industrial Revolution changed agriculture, just as it did almost all other industries. Eli Whitney invented the cotton gin in 1793. The mechanical reaper was invented by Cyrus McCormick, and John Lane and John Deere began the commercial manufacture of the steel plow in 1833 and 1837, respectively. Steel, engines and mass production resulted in equipment that led the way to the development of the many different types of agricultural machinery resulting in the mechanization of most farms and ranches.

With the steel plow, humans began reshaping the Earth's surface, destroying native plants and wildlife habitats, so that many species were driven to extinction. One notorious result of large-scale plowing was the Dust Bowl of the 1930s. Without the roots of native plants to hold the soil down, a drought turned the loose soil into dust that literally buried entire towns.

By the early part of the twentieth century, most agricultural enterprises in the United States were mechanized. American society was transformed from an agrarian into an urban society. People left farms to go to cities to work in factories. At the same time, there was no longer a need for large numbers of people to produce crops. As a result, fewer people were required to produce the growing amounts of agricultural products that supplied an increasing number of consumers.

As populations continued to grow, there was a need to select and produce crops with higher yields. The Green Revolution of the twentieth century helped to make these higher yields possible. Basic information supplied by biological scientists allowed agricultural scientists to develop new, higher-yielding varieties of numerous crops, particularly the seed grains which supply most of the calories necessary for maintenance of the world's population. These higher-yielding crop varieties, along with improved farming methods, resulted in tremendous increases in the world's food supply.

Monoculture

The new crop varieties also led to an increased reliance on monoculture, crop specialization emphasized in modern agriculture. Farmers, especially in industrialized regions, often grow a single crop on much of their land. Problems associated with this practice are exacerbated when a single variety or cultivar of a species is grown. Such a strategy allows the farmer to reduce costs, but it also makes the crop, and thus the farm and community, susceptible to widespread crop failure. The corn blight of 1970 devastated more than 15 percent of the North American corn crop. The corn was particularly susceptible to harmful organisms because 70 percent of the crop being grown was of the same high-yield variety. Chemical antidotes can fight pests, but they increase pollution. Maintaining species diversity or varietal diversity—growing several different crops instead of one or two—allows for crop failures without jeopardizing the entire economy of a farm or region that specializes in a monoculture, such as tobacco, coffee, or bananas.

Genetic Engineering

Growing genetically modified (GM) crops is one potential way to replace post-infestation chemical treatments. Recombinant technologies used to splice genes into varieties of rice or potatoes from other organisms are becoming increasingly common. The benefits of such GM crops include more pest-resistant plants and higher crop yields. However, environmentalists fear new genes could trigger unknown side effects with more serious, long-term environmental and economic consequences than the problems they were used to solve. GM plants designed to resist herbicide applications could potentially pass the resistant gene to closely related wild weed species that would then become "super weeds." Also, just as pests can develop resistance to pesticides, they may also become resistant to defenses engineered into GM plants.

Erosion
An age-old problem, soil loss from erosion occurs all over the world. As soil becomes unproductive or erodes away, more land is plowed. The newly plowed lands usually are considered marginal, meaning they are too steep, nonporous or too sandy, or deficient in some other way. When natural vegetative cover blankets these soils, it protects them from erosive agents: water, wind, ice, or gravity. Plant cover "catches" rainwater that seeps downward into the soil rather than running off into rivers. As marginal land is plowed or cleared to grow crops, erosion increases.

Expansion of land under cultivation has not been the only factor contributing to erosion. Fragile grasslands in dry areas have also been used more intensively. Grazing more livestock than these pastures can handle decreases the amount of grass in the pasture and exposes more of the soil to wind, the primary erosive agent in dry regions. Overgrazing can affect pastureland in tropical regions, too. Thousands of acres of tropical forest have been cleared to establish cattle-grazing ranges in Latin America. Tropical soils, although thick, are not very fertile. After one or two growing seasons, crops grown in these soils will yield substantially less than before.

Tropical fields require fallow periods of about ten years to restore the soil after it is depleted. Thus, tropical farmers using slash-and-burn agriculture move to new fields every few years in a cycle that returns them to the same place years later, after their lands have regenerated. Where there is heavy forest cover, soils are protected from exposure to the massive amounts of rainfall. Organic material for crops is present if the forest remains in place. When the forest is cleared, however, the resulting grassland cannot provide adequate protection, and erosion accelerates. Lands that are heavily grazed provide even less protection from heavy rains, and erosion accelerates even more.

The use of machines also promotes erosion, and modern agriculture relies on machinery such as tractors, harvesters, trucks, balers, and ditchers. Machinery is used intensely in industrialized nations, and its use has been on the rise in developing countries such as India, China, Mexico, and Indonesia, where traditional, nonmechanized farming methods were previously the norm. Farming machines, in gaining traction, loosen topsoil and inhibit vegetative cover growth, especially when farm implements designed to rid the soil of weeds are attached. The soil is then more prone to erosion.

Eco-fallow farming has become more popular in the United States and Europe to reduce erosion. This method of agriculture, which leaves the crop residue in place over the fallow (nongrowing) season, does not root the soil in place as well as living plants do, so some erosion continues. Additionally, eco-fallow methods require the heavy use of chemicals, such as herbicides, to "burn down" weed growth at the start of the growing season. This contributes to increased erosion and pollution.

Pollution and Silt
Besides causing resistance among harmful bacteria, insects, and weeds, pesticides inevitably wash into surface and groundwater supplies, contaminating them. Pesticides are potentially harmful to human health; there has been concern that their seepage into land and water is linked to cancer.

An increasingly heavy silt load has been choking the life out of streams and rivers. Accelerated erosion from water runoff carries silt particles into streams, where they remain suspended and inhibit the growth of many forms of plant and animal life. The silt load in American streams has become so heavy that the Mississippi River delta has been growing faster than it once did. Heavy silt loads, combined with chemical residues, have been creating an expanded dead zone. By taxing the capabilities of ecosystems around the delta, sediments have been filtered out slowly, plant absorption of nutrients has decreased, and salinity levels for aquatic life have been unable to be stabilized. Most of the world's population lives in coastal zones, and 80 percent of the world's fish catch comes from coastal waters over continental shelves that are most susceptible to this form of pollution.

Pesticide Resistance
With the onset of the Green Revolution of the mid-twentieth century, the use of herbicides, insecticides, and other pesticides increased dramatically all over

the world. An increasing awareness of problems caused by the overuse of pesticides followed, extending even to household antibacterial cleaning agents and other products. Mutations among the genes of bacteria and plants have allowed these organisms to resist the effects of chemicals that were toxic to their ancestors. The use of pesticides leads to a cycle wherein more or different combinations of chemicals are used, and more pests develop resistance to these toxins. Additionally, the development of herbicide-resistant crop plants enables greater use of herbicides to kill undesirable weeds on croplands.

Increasing interest in biopesticides (biological pesticides) may slow the cycle of pesticide resistance. Types of biopesticides include beneficial microbes, fungi, and insects such as ladybugs that can be released in infested areas to prey upon specific pests. Biopesticides may be naturally occurring or genetically modified organisms. Their use also avoids excessive reliance on chemical pesticides.

Fertilizers and Eutrophication

Increased use of fertilizers was another result of the Green Revolution. Particulate amounts of most fertilizers enter the hydrologic cycle through runoff. As a result, bodies of water become enriched by dissolved nutrients, such as nitrates and phosphates. The growth of aquatic plants in rivers and lakes is overstimulated, which results in the depletion of dissolved oxygen. This process of eutrophication can harm all aquatic life in these ecosystems.

Water Depletion

An increasing reliance on irrigation has contributed to the mismanagement and over tapping of groundwater resources. The rate of groundwater recharge is slow, usually between 0.1 and 0.3 percent per year. When the amount of water pumped out of the ground exceeds the recharge rate, it is referred to as aquifer overdraft. An aquifer is a water-bearing stratum of permeable rock, sand, or gravel.

In Tamil Nadu, India, groundwater levels dropped twenty-five to thirty meters during the 1970s due to excessive pumping for irrigation. In Tianjin, China, the groundwater level has declined 4.4 meters per year. In the United States, aquifer overdraft has averaged 25 percent over the replacement rate. The Ogallala aquifer—located under parts of South Dakota, Wyoming, Nebraska, Colorado, Kansas, Oklahoma, Texas, and New Mexico—represents an extreme example of overdraft: the rate of depletion has annually been three times faster than its rate of recharge. The capacity of the aquifer decreased by an estimated 33 percent between 1950 and 2004. At this rate, the Ogallala aquifer, which supplies water to countless communities and farms, has been projected to become nonproductive by 2030.

Soil Salinization

In addition, continued irrigation of arid regions can lead to soil problems. Soil salinization has been widespread in the small-grained soils of these regions, which have a high water absorption capacity and a low infiltration rate. Some irrigation practices add large amounts of salts into the soil, increasing its natural rate of salinization. This can also occur at the base of a hill slope. Soil salinization has been recognized as a major process of land degradation.

Although surface and groundwater resources cannot be enriched by technology, conservation and improved environmental management can make the use of precious freshwater more efficient. In agriculture, for example, drip irrigation can reduce water use by nearly 50 percent. In developing countries, though, equipment and installation costs often limit the availability of these more efficient technologies.

Urban Sprawl

With the increasing mechanization of farms, the need for farmers and farm workers has been drastically reduced. From a peak in 1935 of about 6.8 million farmers farming 1.1 billion acres in the United States, for example, the country at the end of the twentieth century counted fewer than two million farmers farming 950 million acres. In 2012, the number of farm operations was at 2.17 million, according to the U.S. Department of Agriculture, with only 914 million acres of land in use.

Urban sprawl converts a tremendous amount of cropland into parking lots, shopping malls, industrial parks, and suburban neighborhoods. If cities were located in marginal areas, then concern about the loss of farmland to commercial development would be nominal. However, the cities attracting the greatest numbers of people have too often replaced the best cropland. Taking the best cropland out of primary production imposes a severe economic penalty.

Energy and Technological Efficiency

The increasing mechanization of farms has led to major increases in the amounts of energy consumed by these farms, particularly those in industrialized nations. Farms use large quantities of energy for irrigation, to operate machinery, to heat and cool buildings, for food processing and shipment, to spray pesticides, and to fertilize crops. The latter two are products of fossil fuels. Raising livestock on grain also consumes large quantities of fossil fuel. Large-scale livestock farmers often feed their animals grains and protein byproducts rather than employing traditional methods of foraging and consuming crop waste. Grain and protein byproduct feeding require less land and allows for the animals to grow to market weight quickly. However, this method of feeding can be inefficient, as animals convert only a fraction of their food energy into growth; for example, it has been estimated that seven kilograms of grain are needed to produce only one kilogram of beef.

Practices such as conservation tilling, which requires less working of the soil, have helped reduce energy use on farms. Another practice is drip irrigation, in which water drips slowly to the roots of plants, thus saving on both water and fertilizer. Farmers have also begun to plant genetically modified crops that do not need pesticide, which itself has become more sophisticated and therefore used in smaller quantities than before. Other farms have opted to grow organic (pesticide-free) food and to raise animals that are not given growth hormones and that are free-range, or not always confined in tight quarters. While these practices can be expensive, they can also save on energy costs and alleviate consumer concerns about ingesting potentially harmful chemicals.

Another way farmers have learned to save money on energy is to use renewable energy sources such as wind power, solar power, and biomass products (also called biofuel). By having electric wind turbines built on their farms, farmers can produce their own energy. Wind power can be used to power an entire farm or to power a specific area of the farm, such as pumping water for cattle. Solar energy can power a farm's lighting and heating (e.g., in greenhouses), pump water, and produce electricity. It can even be used to dry crops faster and more evenly than crops left prone to damage in the fields. Many farmers grow corn to make ethanol. Other crops have begun to be used for fuel as well, since there is virtually no limit to the type of plant and organic waste that can be used to produce energy. Agriculture creates a lot of waste; there is the potential for taking that waste and converting it into energy, thus saving on energy production costs, disposal costs, and pollution. Crops grown specifically for biofuel—for farms and other consumers—can be produced in large quantities and thus become profitable when sold.

Other technologies that have made farming become more efficient—and that have saved and made farmers money—are broadband Internet access, smartphones, and Global Positioning System (GPS) technology. Use of the Internet has helped farmers quickly exchange important data with each other and has helped farmers connect directly to their markets and consumers. GPS technology has helped farmers navigate their fields in a fraction of the time it took before this invention; equipment can be guided through fields, with no overlap or gaps, so that seeds can be planted, and pesticides sprayed evenly. Smartphone applications (apps) for crop scouting can help farmers identify a problem and its specific source immediately, eliminating the need to apply pesticide, for example, to an entire field in the hopes of rectifying that one problem. There are numerous other mobile applications useful to farmers as well.

D.R. Gossett, James Knotwell, Denis Knotwell, Nancy Farm Männikö, Bryan Ness, Elizabeth Slocum

Further Reading

Acquaah, George. *Principles of Crop Production: Theory, Techniques, and Technology.* Englewood, N.J.: Prentice Hall, 2001.

Beckwith, Carol, and Marion Van Offelen. *Nomads of Niger.* New York: Academic Press, 1987.

Cockrall-King, Jennifer. *Food and the City: Urban Agriculture and the New Food Revolution.* Amherst, N.Y.: Prometheus Books, 2012.

Colfer, Carol J., with Nancy Peluso and Chin See Chung. *Beyond Slash and Burn: Building on Indigenous Management of Borneo's Tropical Rain Forest.* New York: New York Botanical Garden, 1994.

Goldstein, Melvyn C., and Cynthia M. Beall. *The Changing World of Mongolia's Nomads.* Berkeley: University of California Press, 1994.

Grigg, David. *An Introduction to Agricultural Geography.* London: Routledge, 2005.

_____. *An Introduction to Agricultural Geography.* London: Routledge, 2005.

Hoag, Dana L. *Agricultural Crisis in America: A Reference Handbook.* Santa Barbara, CA: ABC-CLIO, 1999.

Jackson, Wes. *New Roots for Agriculture.* 1980. Lincoln, NE: University of Nebraska Press, 1985.

Janick, Jules. *Horticultural Science.* 4th ed. New York: W. H. Freeman, 1986.

Lægreid, M., et al. *Agriculture, Fertilizers, and the Environment.* Cabi, 1999.

Magdoff, Fred, and Brian Tokar, eds. *Agriculture and Food in Crisis: Conflict, Resistance, and Renewal.* New York: Monthly Review Press, 2010.

McKee, Gregory J., et al., editors. *Pesticide Resistance, Population Dynamics, and Invasive Species Management.* Hauppauge, NY: Nova Science Publishers, 2010.

Mortimore, Michael. *Roots in the African Dust: Sustaining the Sub-Saharan Drylands.* New York: Cambridge University Press, 1998.

Paarlberg, Don, and Philip Paarlberg. *The Agricultural Revolution of the 20th Century.* Iowa City, IA: Iowa State University Press, 2000.

Pfeiffer, Dale A. *Eating Fossil Fuels: Oil, Food and the Coming Crisis in Agriculture.* Gabriola Island (BC: New Society Publishers, 2008.

Pimentel, David, and Anne Wilson. "World Population, Agriculture, and Malnutrition." *World Watch,* Sept.–Oct. 2004, pp. 22–25.

Rissler, Jane, and Margaret Mellon. *The Ecological Risks of Engineered Crops.* Cambridge, MA: The MIT Press, 1996.

Smith, Bruce D. *The Emergence of Agriculture.* New York: W. H. Freeman, 1994.

Zohary, Daniel, and Maria Hopf. *Domestication of Plants in the Old World: The Origin and Spread of Cultivated Plants in West Asia, Europe, and the Nile Valley.* New York: Oxford University Press, 2001.

Websites

Bloomberg
www.bloomberg.com/news/articles/2013-02-19/number-of-u-s-farms-fell-to-six-year-low-in-2012-usda-says-1

Union of Concerned Scientists, 2003
www.ucsusa.org/clean_energy/smart-energy-solutions/increase-renewables/renewable-energy-and.html.

■ Air pollution and greenhouse gases

FIELDS OF STUDY
Atmospheric Sciences; Atmospheric Structure and Dynamics; Chemistry; Climate Change; Climatology; Ecology; Elements; Emissions; Environment; Environmentalism; Hazardous Materials; Industries; Meteorology; Pollution; Substances; Toxic Waste

SUMMARY
Greenhouse gases raise the Earth's average temperature 33 degrees Celsius (59 degrees Fahrenheit) above what it would be if these gases were not present in the atmosphere. Most scientists agree that human activities are contributing to increased concentrations of greenhouse gases and thus to increases in the average surface temperature of the earth.

PRINCIPAL TERMS
- **carbon dioxide**: a colorless, odorless, incombustible gas, CO_2, present in the atmosphere and formed during respiration, usually obtained from coal, coke, or natural gas by combustion
- **fluorocarbons:** compounds characterized by great chemical stability, used chiefly as a lubricant, refrigerant, fire extinguishing agent, and in industrial and other applications in which chemical, electrical, flame, and heat resistance is essential, and banned as an aerosol propellant in the U.S. because of concern about ozone layer depletion.
- **global warming:** an increase in the earth's average atmospheric temperature that causes corresponding changes in climate and that may result from the greenhouse effect
- **greenhouse gases:** atmospheric gases that allow sunlight to reach the earth's surface but at least partially block infrared from radiating back into space
- **Kyoto Protocol:** an international treaty extending the 1992 United Nations Framework Convention on Climate Change (UNFCCC) that commits state parties to reduce greenhouse gas emissions, based on the scientific consensus that global warming is occurring that it is extremely likely that human-made emissions have caused it

- **methane:** a colorless, odorless, flammable gas, CH4, the main constituent of marsh gas and the firedamp of coal mines, obtained commercially from natural gas
- **nitrous oxide:** a toxic reddish-brown gas NO_2, a strong oxidizing agent produced by combustion, as of fossil fuels, and is an atmospheric pollutant, as in smog
- **ozone:** a form of oxygen, O3, with a peculiar odor suggesting weak chlorine produced when an electric spark or ultra violet light is passed through air

Airborne particulate matter represents a complex mixture of organic and inorganic substances and varies in size, composition, and origin. Some are known as primary particles, emitted directly from sources such as construction sites, unpaved roads, fields, smokestacks, and fire. Secondary particles are formed by reactions of gases, such as sulfur dioxide and nitrogen oxides, that are emitted from power plants, industrial plants, and automobiles, as well as dust, dirt, soot, and smoke, large enough to be visible to the naked eye, although other forms are so small, they require electron microscopes for detection. The inhalation of microscopic particles can have serious adverse effects on human respiratory and cardiovascular health.

Most scientists accept that a small global warming has taken place—that is, the earth's surface temperature has warmed about 0.7 degree Celsius (1.3 degrees Fahrenheit)—since the end of the nineteenth century. At least half of this temperature rise is attributed to the release of greenhouse gases into the atmosphere by human beings. It is thought that if greenhouse gases were returned to their 1990 levels, the temperature would still rise another 0.5 degree Celsius (0.9-degree Fahrenheit). Since greenhouse gases are transparent to visible light, sunlight passes through the atmosphere and warms the earth's surface. The warmed surface radiates infrared into the sky, where greenhouse gases absorb infrared and then reradiate it. They radiate about half of the infrared upward into space and half of it back down to the earth's surface. Since the earth absorbs more energy than it radiates back into space, it heats up until the energies entering and leaving are equal. Balance is possible because a hotter earth radiates with greater intensity and at shorter wavelengths where greenhouse gases allow more infrared to escape into space.

Water vapor strongly absorbs infrared of about 3 microns wavelength (3,000 nanometers), and so does carbon dioxide. Adding more carbon dioxide will not change the amount of energy passing out into space, since water vapor absorbs all of the energy near that wavelength. However, more carbon dioxide would make a difference at 4.5 microns wavelength (4,500 nanometers) because water vapor does not absorb at that wavelength. This means that doubling the amount of carbon dioxide in the atmosphere does not necessarily double the effect of carbon dioxide. In fact, most climate scientists believe that increasing carbon dioxide will increase the surface temperature, which will cause more water to evaporate. More water vapor in the atmosphere should further warm the earth's surface, but it will also cause more clouds that reflect sunlight back into space. More clouds will cool the earth. Under these competing processes it is believed that temperature will increase until a new equilibrium is reached.

National Ambient Air Quality Standards for Criteria Pollutants

Pollutant	Averaging Time	Pollutant Level	Effects on Health
Carbon monoxide: colorless, odorless, tasteless gas; it is primarily the result of incomplete combustion; in urban areas the major sources are motor vehicle emissions and wood burning.	1-hour	35 ppm	The body is deprived of oxygen; central nervous system affected; decreased exercise capacity; headaches; individuals suffering from angina, other cardiovascular disease; those with pulmonary disease, anemic persons, pregnant women and their unborn children are especially susceptible.
	8-hour	9 ppm	
Ozone: highly reactive gas, the main component of smog.	1-hour	0.120 ppm	Impaired mechanical function of the lungs; may induce respiratory symptoms in individuals with asthma, emphysema, or reduced lung function; decreased athletic performance; headache; potentially reduced immune system capacity; irritant to mucous membranes of eyes and throat.
	8-hour	0.080 ppm	
Particulate matter < 10 microns (PM10): tiny particles of solid or semisolid material found in the atmosphere.	24-hour	150 $\mu g/m^3$	Reduced lung function; aggravation of respiratory ailments; long-term risk of increased cancer rates or development of respiratory problems.
	Annual arithmetic mean	50 $\mu g/m^3$	
Particulate matter < 2.5 microns (PM2.5): fine particles of solid or semisolid material found in the atmosphere.	24-hour	65 $\mu g/m^3$	Same as PM10 above.
	Annual arithmetic mean	15 $\mu g/m^3$	
Lead: attached to inhalable particulate matter; primary source is motor vehicles that burn unleaded gasoline and re-entrainment of contaminated soil.	Calendar quarter	1.5 $\mu g/m^3$	Impaired production of hemoglobin; intestinal cramps; peripheral nerve paralysis; anemia; severe fatigue.
Sulfur dioxide: colorless gas with a pungent odor.	3-hour	0.5 ppm	Aggravation of respiratory tract and impairment of pulmonary functions; increased risk of asthma attacks.
	24-hour	0.14 ppm	
	Annual arithmetic mean	0.03 ppm	
Nitrogen dioxide: gas contributing to photochemical smog production and emitted from combustion sources.	Annual arithmetic mean	0.053 ppm	Increased respiratory problems; mild symptomatic effects in asthmatics; increased susceptibility to respiratory infections.

Note: ppm equals parts per million and $\mu g/m^3$ equals micrograms per cubic meter.
Source: United States Environmental Protection Agency (EPA); URL http://www.epa.gov.

Earth's Major Greenhouse Gases

Water vapor, carbon dioxide, methane, nitrous oxide, fluorocarbons, and ozone are the major greenhouse gases. The effect of a particular gas depends on which other gases are present, how much of the gas is present, and how likely a gas molecule is to absorb infrared radiation.

Water vapor is the most important greenhouse gas, but human activities have no direct effect on the global average amount of water vapor, which is fixed mainly by evaporation from the earth's oceans. Other greenhouse gases are significantly affected by human activities. Burning fossil fuels and deforestation in the Tropics increase the amount of carbon dioxide in the atmosphere. Rice paddy farming and the digestive processes of livestock produce large amounts of methane. Fertilizers used in farming produce nitrous oxide, and refrigeration systems and some manufacturing processes release chlorofluorocarbons, other perfluorocarbons, and sulfur hexafluoride.

Since the mid-nineteenth century, carbon dioxide concentration in the air has increased from 280 parts per million to almost 380 parts per million. Normal carbon is carbon 12 (6 protons and 6 neutrons in the nucleus), but about 1 percent of carbon is carbon 13 (6 protons and 7 neutrons). Plants prefer carbon 12, so plant carbon (and fossil fuels from plants) has a smaller ratio of carbon 13 to carbon 12 than the atmosphere does. Analysis of air bubbles in ice cores shows that the ratio of carbon 13 to carbon 12 in the atmosphere has been decreasing since the mid-nineteenth century, presumably because of increased burning of fossil fuels and the practice of setting fire to forests to clear land.

The Clean Air Act

The Clean Air Act, passed by the U.S. Congress in 1963, laid the foundation for what some consider to be the most progressive, wide-reaching, and complicated environmental cleanup legislation in the world. When the Clean Air Act and other early federal, state, and local clean air laws proved to be relatively ineffective, several sweeping amendments to the laws were enacted. In 1970, groundbreaking improvements resulted in emissions standards for automobiles and new industries in addition to establishing air-quality standards for urban areas and stimulating many states to pass regional and local air-pollution legislation, with some areas eventually passing laws that later proved to be even more stringent than federally established guidelines.

During this period, the newly created U.S. Environmental Protection Agency (EPA) began strongly suggesting the tightening of rules regulating the amount of lead that could be added to gasoline, a significant source of lead poisoning in urban children and young adults, thus laying the groundwork for the future elimination of all leaded gasolines. Many sectors of the business community challenged the wording of some of the 1970 amendments, arguing that the language was vague and required clarification, particularly regarding the deterioration of air quality in areas that were already meeting federal standards.

In December 2009, the EPA released a finding that greenhouse gases threaten the public health and welfare of the American people. The finding allowed the EPA to regulate greenhouse gas emissions in the United States if the Congress fails to pass legislation to do so.

International Politics

A 1990 report by the Intergovernmental Panel on Climate Change (IPCC) found that the continued increase of greenhouse gas concentrations in the earth's atmosphere would lead to increases in global average temperature. In response to this report, the United Nations facilitated negotiations among nations to determine international efforts to address the threat of global warming. The resulting treaty, the United Nations Framework Convention on Climate Change (UNFCCC), was signed by 192 countries in 1992 and established goals of quantifying and reducing greenhouse gas emissions, while also differentiating between the responsibilities of developed and developing nations in responding to the global warming.

Under the sponsorship of the United Nations, the Kyoto Protocol was adopted in December 1997, and went into force in February 2005. Seeing the protocol as flawed because it places no limits on China (the world's largest polluter) or India, the United States opted out of the protocol. Under the protocol, industrialized nations set goals for greenhouse gas reductions, and developing nations negotiated for money from those nations to help them industrialize with fewer greenhouse emissions. By 2007 only Germany, Norway, France, and the United Kingdom were meeting their goals. Several nations that had been part of the former Soviet Union reduced emissions, in large part because their economies floundered.

In November 2009, many e-mails were stolen from the Climatic Research Unit at the University of East Anglia, United Kingdom. Some climate change skeptics alleged that the e-mails provided evidence that scientists had doctored data on global warming. Subsequent investigations found evidence of frustration on the part of the climate researchers and a lack of willingness among them to share raw data, but no evidence of wrongdoing. This incident nevertheless cast some doubt on the East Anglia data on global warming, even though the findings of the Climatic Research Unit are supported by a great deal of data from other sources. This doubt may have contributed to the weakness of the agreements reached regarding greenhouse gas emissions at the United Nations Climate Change Conference held in Copenhagen, Denmark, in December 2009.

Daniel G. Graetze, Bernard Jacobson, Karen N. Kähler and Charles W. Rogers

Further Reading

Blau, Judith R. *The Paris Agreement: Climate Change, Solidarity, and Human Rights.* Cham, Switzerland: Palgrave Macmillan, 2017.

Houghton, John. *Global Warming: The Complete Briefing.* New York: Cambridge University Press, 2009.

Singer, S. Fred, and Dennis T. Avery. *Unstoppable Global Warming: Every 1,500 Years.* Updated ed. Lanham, Md.: Rowman & Littlefield, 2008.

U.S. Environmental Protection Agency. *The Plain English Guide to the Clean Air Act.* Research Triangle Park, N.C.: Office of Air Quality Planning and Standards, 2007

Weart, Spencer W. *The Discovery of Global Warming.* Rev. ed. Cambridge, Mass.: Harvard University Press, 2008.

Websites

Environmental Protection Agency
www.epa.gov/pm-pollution

■ Alternative fuels

FIELDS OF STUDY
Air Pollution; Biofuels and Synthetic Fuels; Chemistry; Ecology; Environmental Engineering; Electric Automobile Technology; Hybrid Vehicle Technologies; Mechanical Engineering; Nuclear Science; Steam Energy Technology; Transportation Engineering.

SUMMARY
The development of alternatives to fossil fuels (gasoline, diesel, natural gas, and coal) has been spurred by growing awareness of the environmental damage associated with the burning of fossil fuels, as well as by the knowledge that at some time in the future the earth's supply of fossil fuels will be exhausted.

PRINCIPAL TERMS
- **ambient heat:** heat available for free from environmental sources like the sun, able to be stored or conserved with insulation methods
- **biofuels:** fuels taken from organic matter, including ethanol, which can be derived from corn and sugarcane, and biodiesel, which can be formed from vegetable oils
- **butanol:** butyl alcohol made from fossil fuels or certain plants of algae carbon dioxide
- **fossil fuel:** fuel (such as coal, oil, or natural gas) formed in the earth from plant or animal remains
- **greenhouse effect:** exchange of excessive incoming and outgoing radiation that warms the Earth and heats the oceans resulting in global warming
- **greenhouse gases:** any gases whose absorption of solar radiation is responsible for the greenhouse effect, including carbon dioxide, methane, ozone, and the fluorocarbons
- **methane:** colorless odorless flammable gaseous hydrocarbon, a product of decomposition of organic matter and carbonization of coal
- **natural gas:** combustible mixture of gaseous hydrocarbons that accumulates in porous sedimentary rocks, especially those yielding petroleum

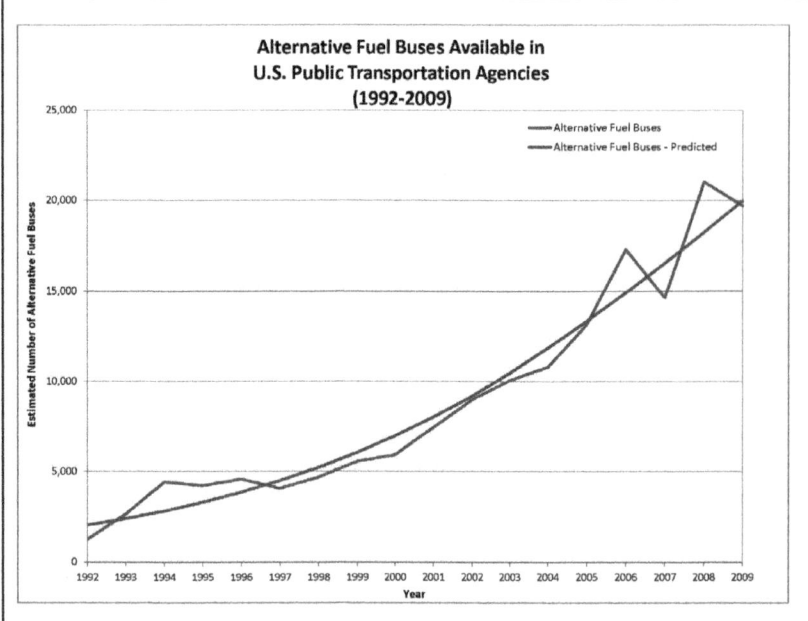

Predicted and Actual Alternative Fuel Bus Graph (APTA): 1992-2009. By Nesse069 (Own work).

Except for nuclear-powered seagoing vessels, most vehicles are powered by internal combustion engines that use either gasoline or diesel fuel. Gasoline and diesel release significant amounts of greenhouse gases into the atmosphere when burned; these gases include water vapor, carbon dioxide, ozone, nitrous oxide, and methane. These gases absorb and emit radiation in the infrared range; thus, they increase the earth's temperature. In addition to the fact that the internal combustion engine burns an environment-polluting fossil fuel, it also is an inefficient method for transferring the energy stored in the fuel into propulsion. Most of the stored energy is lost in heat, which escapes through the exhaust pipe. In addition, the pistons within the engine accelerate up, stop, accelerate down, and stop with each revolution. This rapid cycle of acceleration and deceleration wastes energy. Many of the alternative fuels available are used to power internal combustion engines and thus have the same limitations as fossil fuels in this regard. Using an alternative fuel such as stored electricity does not have these limitations because it does not produce heat and it propels an electric motor, which rotates (no starting and stopping with each cycle).

Comparisons of the costs and levels of pollutant production of nonfossil fuel sources must take into account the costs associated with production of the fuels. An example is the use of corn for the production of ethanol. Raising the crop requires energy for production, such as fuel for tractors. The corn must then be fermented (yeast converts the sugar in the corn into ethanol), and the fermented product must be distilled (boiled to release the alcohol), which requires fuel to heat the still. The process increases the cost of ethanol and produces pollution. The electricity that charges an electric vehicle may have been produced by a fossil-fuel source such as a diesel generator. Another problem with alternative fuels lies in the difficulty consumers may have in replenishing their supplies. Facilities distributing gasoline and diesel are prevalent throughout most developed nations; in contrast, sources of alternative fuels such as hydrogen and ammonia are not readily available. The ideal alternative fuel is one that is nonpolluting, cheap to produce, and easy to replenish.

Biofuels

Biofuels are derived from plant sources such as corn, sugarcane, and sugar beets; in some cases, they are blended with a fossil fuel, usually gasoline. Alcohol, methanol, butanol, biodiesel, biogas, and wood gas are all examples of biofuels.

Alcohol was initially used as a fuel in the Ford Model T automobile, which was first produced in 1908. The carburetor (a device that mixes fuel with air prior to entry into the engine) of the Model T could be adjusted to burn gasoline, ethanol, or a mixture. Many modern-day vehicles can run on a mixture of 10–15 percent ethanol and gasoline (E10, or gasohol, is 10 percent alcohol). The fuel known as E85 is a mixture of 85 percent ethanol and 15 percent gasoline; this fuel can be used only in flexible-fuel vehicles (FFVs). FFVs are designed to run on gasoline, E85, or any other gasoline-ethanol mixture. A disadvantage of ethanol is that it has approximately

34 percent less energy per volume than gasoline. Because ethanol has a high-octane rating, ethanol-only engines may have relatively high compression ratios, which increases efficiency. In developed nations such as the United States, ethanol blends are available in many areas. Critics of ethanol as an alternative fuel note that it requires a large amount of agricultural land, which is diverted from producing crops used for food; also, the use of crops such as corn for ethanol production drives up food prices.

Methanol can be used as an alternative fuel, but automakers have not yet produced any vehicles that can run on it. Butanol is more similar to gasoline than ethanol and can be used in some engines designed for use with gasoline without modification.

Biodiesel can be manufactured from vegetable oils and animal fats, including recycled restaurant grease. It is slightly more expensive than diesel; however, it is a safe, biodegradable fuel that produces fewer pollutants than diesel. Diesel engines are more efficient than gasoline engines (44 percent versus 25–30 percent efficiency); thus they have better fuel economy than gasoline engines. Some diesel engines can run on 100 percent biodiesel with only minor modifications. Biodiesel can be combined with regular diesel in various concentrations (for example, B2 is 2 percent biodiesel, B5 is 5 percent biodiesel, and B20 is 20 percent biodiesel).

Biogas is produced by the biological breakdown of organic materials—for example, rotting vegetables, plant wastes, and manure produce biogas—and the energy produced varies depending on the source. Biogas can replace compressed natural gas for fueling internal combustion engines. Wood gas is another biofuel that can power an internal combustion engine. It is produced by the incomplete burning of sawdust, wood chips, coal, charcoal, or rubber. Depending on the source, the gas produced varies in energy content and contaminants. Contaminants in wood gas can foul an engine.

Electric Vehicles

At the beginning of the twentieth century, automobiles powered by steam, gasoline, and electricity were available. Electric vehicles were popular into the early 1920s, when the automotive industry became dominated by gasoline-powered vehicles. The decline in electric-powered vehicles occurred for several reasons: road improvement allowed travel over longer distances, and the range of electric vehicles was limited; fossil fuels became cheap and plentiful; the electric starter replaced the hand crank on gasoline engines, which greatly simplified starting such engines; and mass production of automobiles by Henry Ford's company made gasoline-powered vehicles much less expensive than electric-powered vehicles ($650 versus $1,750 average price at that time). By the end of the twentieth century, a growing emphasis on environmentally friendly energy sources encouraged the reemergence of electric vehicles and the development of hybrid vehicles powered by both gasoline (or diesel) and electricity.

A hybrid vehicle contains an electric motor that can both propel the vehicle and recharge the battery. Hybrid vehicles have achieved greater popularity than electric-only vehicles, as electric-only vehicles continue to have some of the same basic problems as earlier electric cars: limited range and higher cost than gasoline-powered or hybrid vehicles. Public recharging facilities for electric vehicles remain few and far between; furthermore, recharging takes time. The latest electric and hybrid vehicles use lithium-ion batteries rather than the lead-acid batteries used by earlier versions (and still used in gasoline and diesel vehicles). Lithium-ion batteries are much lighter than lead-acid batteries and can be molded into a variety of shapes to fit available areas. One criticism of electric vehicles, including hybrids, is that many are small and lightweight and thus less safe for passengers, in the case of collisions, than are larger gasoline-powered vehicles.

Other Fuels Derived from Nonfossil Sources

Ammonia has been evaluated for use as an alternative fuel. It can run in either a spark-ignited engine (that is, a gasoline engine) or a diesel engine in which the fuel-air mixture ignites upon compression in the cylinder. Modern gasoline and diesel engines can be readily converted to run on ammonia. Although ammonia is a toxic substance, it is considered no more dangerous than gasoline or liquefied petroleum gas (LPG). Ammonia can be produced by electrical energy and has half the density of gasoline or diesel; thus it can be placed in a vehicle fuel tank in sufficient quantities to allow the vehicle to travel reasonable distances. Another advantage of ammonia is that it produces no harmful emissions; upon combustion, it produces nitrogen and water.

Compressed-air engines are piston engines that use compressed air as fuel. Air-engine-powered vehicles have been produced that have a range comparable to gasoline-powered vehicles. Compressed air is much less expensive than fossil fuels. Ambient heat (normal heat in the environment) naturally warms the cold compressed air upon the air's release from the storage tank, increasing its efficiency. The only exhaust is cold air, which can be used to cool the interior of the vehicle.

Hydrogen vehicles can be powered by the combustion of hydrogen in the engine much as the typical gasoline engine operates. Fuel cell conversion is another method of using hydrogen; in this type of vehicle, the hydrogen is converted to electricity. The most efficient use of hydrogen to power motor vehicles involves the use of fuel cells and electric motors. Hydrogen reacts with oxygen inside the fuel cells, which produces electricity to power the motors. With either method no harmful emissions are produced, as the spent hydrogen produces only water. Hydrogen is much more expensive than fossil fuels, and it contains significantly less energy on a per-volume basis, meaning that the vehicle's range is reduced. Experimental fuel cell vehicles have been produced, but such vehicles remain far too expensive for the average consumer.

Liquid nitrogen (LN_2) contains stored energy. Energy is used to liquefy air, then LN_2 is produced by evaporation. When LN_2 warms, nitrogen gas is produced; this gas can power a piston or turbine engine. Nitrogen-powered vehicles have been produced that have ranges comparable to gasoline-powered vehicles; these vehicles can be refueled in a matter of minutes. Nitrogen is an inert gas and makes up about 80 percent of air. It is virtually nonpolluting. Furthermore, it produces more energy than does compressed air.

Oxyhydrogen is a mixture of hydrogen and oxygen gases, usually in a 2:1 ratio, the same proportion as water. Oxyhydrogen can fuel internal combustion engines, and, as in hydrogen-fueled engines, no harmful emissions are produced.

Steam was a common method of propulsion for vehicles during the early twentieth century, but, like electricity, it fell into disfavor with the advent of the electric starter, cheap gasoline, and mass production of Ford automobiles. A disadvantage of steam-powered vehicles is the time required to produce the steam. A steam engine is an external combustion engine—that is, the power is produced outside rather than inside the engine. Steam engines are less energy-efficient than gasoline engines. Fuel for steam engines can be derived from fossil fuels or from nonfossil fuel sources.

Alternative Fossil Fuels

Some fossil fuels are less polluting than gasoline or diesel, and some are in plentiful supply. Natural gas vehicles use Compressed natural gas (CNG) or, less commonly, liquefied natural gas (LNG). Internal combustion engines can be readily converted to burn natural gas. Natural gas is 60-90 percent less polluting than gasoline or diesel and produces 30-40 percent less greenhouse gases. Furthermore, it is less expensive than gasoline. Limitations of natural gas vehicles include a lack of available fueling stations and limited space for fuel, given that natural gas must be stored in cylinders, which are commonly located in the vehicle's trunk.

Liquefied petroleum gas is suitable for fueling internal combustion engines. Like natural gas, LPG is less polluting than gasoline, with 20 percent less carbon dioxide emissions; it is also less expensive. LPG is added to a vehicle's fuel tank through the use of a specialized filling apparatus; a limitation to LPG is the lack of fueling stations.

Robin L. Wulffson

Further Reading

DeGunther, Rik. *Alternative Energy for Dummies.* Hoboken, N.J.: John Wiley & Sons, 2009.

Gibilisco, Stan. *Alternative Energy Demystified.* New York: McGraw-Hill, 2007.

Hordeski, Michael F. *Alternative Fuels: The Future of Hydrogen.* 2d ed. Lilburn, Ga.: Fairmont Press, 2008.

Lee, Sunggyu, James G. Speight, and Sudarshan K. Loyalka. *Handbook of Alternative Fuel Technologies.* Boca Raton, Fla.: CRC Press, 2007.

Nersesian, Roy L. *Energy for the Twenty-first Century: A Comprehensive Guide to Conventional and Alternative Sources.* New York: M. E. Sharpe, 2007.

Rhodes, Richard. *Energy: A Human History.* New York: Simon & Schuster, 2018.

Smil, Vaclav. Energy and Civilization: A History. Cambridge, Mass.: The MIT Press, 2017

Amazon Deforestation

FIELDS OF STUDY
Biology; Conservation Biology; Ecology; Environment; Forest Ecology; Silvology

PRINCIPAL TERMS
- **carbon sink**: an entity that absorbs and stores carbon, thereby removing CO_2 from the atmosphere
- **deforestation**: to divest or clear forests of trees
- **global warming**: increase in Earth's average atmospheric temperature causing corresponding changes in climate and resulting from the greenhouse effect
- **rainforest**: a tropical area dominated by evergreen trees whose leaves form a continuous canopy and that receives at least 254 centimeters (100 inches) of rain per year
- **savanna**: grassland with scattered trees, characteristic of tropical areas with seasonal rainfall on the order of 50 centimeters (19.68 inches) per year

SUMMARY
The Amazon rainforest, sometimes called "the lungs of the world," plays a key role in global climate and supports a diverse population of species, many of which exist nowhere else on Earth. Loss of Amazonian forest lands through both human clearing and drought has significant effects upon climate regulation and Earth's biodiversity.

The Amazon rainforest occupies more than 2,300,000 square kilometers (1,429,153.7 square miles) of land in South America, as of 2015. About 40 percent is in Brazil, and the remaining 60 percent includes parts of seven other South American countries: Bolivia, Peru, Ecuador, Colombia, Venezuela, Guyana, and Suriname, as well as French Guiana. Nearly untouched as late as 1970, the region underwent rapid development in the last quarter of the twentieth century. Annual rates of clearing in Brazil peaked at 3,341,908 square kilometers (2,076,565.36 square miles) in 2013. Between 1970 and 2013, the total area of rain forest in Brazil shrank from 4,100,000 square kilometers (2,547,621.9 square miles) to 3,341,908 square kilometers (2,076, 565.36 square miles), with only 81.5 percent of forest coverage remaining, leading to predictions of total annihilation of the Amazon within a century. As a result of international pressure and domestic conservation efforts, the annual clearing rate declined to 13,100 square kilometers (8,139.9626 square miles) in 2006, but it rose again in 2008 as increased world demand for soybeans and ethanol encouraged expansion of Brazilian agriculture. In 2019, with the inauguration of Brazil's president Jair Bolsonaro, the Amazon is once again threatened with further deforestation forestation for agriculture and commercial exploration.

The Forest and Deforestation
Terrestrial ecosystems in the region include the moist broadleaf forests of the Tio Negro Campinarana, Iquitos Varzea, Guryupa Varzea, Marajo Varzea, Purus Varzea and Monte Alegre Varzea, all of which are classified a threatened or endangered. The forests are full of plants that have significant commercial value, including mahogany (*Swietenia macrophylla*), balsam (*Myroxylon balsamum*), rubber (*Hevea brasiliensis*), strychnine (*Strychnos asperula*), and tagua nut (*Phytelephas microcarpa*). The south has the greatest number of palm species.

The Amazonian rubber boom contributed substantially to deforestation in the late 19th century and throughout the 20th century, due to the introduction of the pneumatic tire and the many commercial uses for rubber.

A forest of this magnitude affects world climate in numerous ways. On a regional level, dense vegetation supports higher temperatures, higher rainfall, less runoff, and lower daily and seasonal temperature fluctuations. In the long term, high global temperatures favor forests. On a geologic time scale, the warmest periods have coincided with the greatest extent of rain forest, whereas much of the area later occupied by the Amazon rainforest was savanna during the height of the last Pleistocene glaciation.

Plants extract carbon dioxide (CO_2) from the atmosphere via photosynthesis. An expanding forest acts as a carbon sink, removing CO_2 from the air and sequestering carbon in its woody parts. A mature forest is in equilibrium, emitting as much carbon through animal consumption and decomposition as it fixes through photosynthesis. Clearing or burning forests releases CO_2 into the atmosphere; however, if tree trunks are converted to lumber and the land is subsequently used to grow crops, the net carbon release may be relatively small.

Global Warming and Amazon Deforestation

Global climate change can affect a forest profoundly. Although warm temperatures in general favor forests, shifts in patterns of prevailing winds brought about by small changes in oceanic temperatures may bring drought to regions accustomed to high rainfall and flooding to formerly arid regions. Although such perturbations are common in the geologic record and the Earth's biota has repeatedly shown a rapid response, the rate of recovery is slow on a human time scale.

Increasing atmospheric CO_2 may stimulate forest growth in the tropics. High CO_2 levels favor rapid growth of trees, which tend to crowd out understory species, leaving fewer niches for animal species, particularly insects dependent on specific food plants. In the short term, such highly productive forests may be commercially desirable for lumber production, but ecological diversity and sustainability suffer.

Cycles of the El Niño-Southern Oscillation (ENSO) cause large natural fluctuations in rainfall in the Amazon basin. During the unusually severe drought of 2005-2006, some scientists predicted that tree species would die off and natural fires would destroy significant areas of forest, creating a climate feedback loop that would turn much of the Amazon into savanna. The forest appears to be unexpectedly resilient, however. During a drought year, deep-rooted trees remain green, and they even grow faster than normal, owing to the absence of cloud cover.

Some efforts to address environmental problems elsewhere in the world contribute to Amazon deforestation. Strenuous conservation efforts in the developed world, unaccompanied by reduction in wood-product consumption, increase logging pressure in places like Brazil. The United States' drive to produce and deploy corn-based ethanol opened the way for rapid growth in Brazil's soybean production. Brazil is also a leading producer and exporter of ethanol derived from sugarcane. These crops are rarely planted directly on cleared jungle land, but cattle ranchers displaced by soybeans and sugarcane migrate to the Amazon.

Finally, decreasing levels of sulfur dioxide resulting from more stringent pollution controls in Europe and North America were implicated in Brazil's devastating 2005 drought. This effect, the subject of a May 2008, article in Nature, was the first firm scientific evidence of the importance of sulfur dioxide emissions in canceling the greenhouse effect of CO_2.

Contribution of Deforestation to Global Warming

The effects of Amazonian deforestation on world species diversity eclipse its large-scale climatic effects, as loss of the forest may lead to the extinction of thousands of species. Nonetheless, even if only CO_2 emissions are considered, the deforestation's climatic effect is not negligible. CO_2 from slash burning following logging may account for as much as half of Brazil's carbon contribution to the atmosphere, estimated at 90 million tons in 2004; however, according to November 2012 article in Scientific American, Brazil is on track to achieve an 80 percent reduction by 2020. Although Brazil ranked fourth in terms of its contribution to world CO_2 pollution in a global warming study in January 2014, it accounts for only a little over 2 percent of the world total and has a very low per capita level of fossil fuel consumption because Brazilians rely on hydroelectric power and ethanol.

Some of the released carbon is recaptured when land is used for crops or pastureland. However, indiscriminate logging combined with a drier and more uncertain climate due to global warming may ultimately convert large areas of the Amazon to semiarid grassland of minimal value as a carbon sink. This scenario, which appeared to many to be imminent during the 2005-2006 drought, is now thought to be avoidable through feasible management schemes, some of which are already being implemented.

Context

It is tempting to view environmental threats en masse and to assume that a policy that ameliorates one ecological disaster will have a correspondingly benign effect on others. The interactions between global warming and the deforestation of the Amazon rainforest demonstrate that this is not always the case. The forest has shown itself to be more resilient to drought than scientists anticipated. The principal immediate, global-warming-related threat to the Amazon rainforest appears to be the rapid expansion of Brazilian agriculture in response to rising world demand for biofuels. Models for controlling this expansion so as to encourage efficient land use and sustainability favor large agricultural businesses over individual farmers and pay inadequate attention to preserving biodiversity.

No discussion of a global-warming issue is complete without mention of population issues. The populations of Brazil and other countries bordering on the Amazon are growing at a very rapid rate. Despite the overall low level of energy consumption in the area, this population growth increases human impact on the environment exponentially. The Amazon ecosystem is apparently robust enough to withstand present levels of global CO_2 emissions, but unless exponential trends are reversed, the grim scenario of degradation to savanna looms in the future.

Martha A. Sherwood and Bill Kte'pi

Further Reading

Alverson, Keith D., Raymond S. Bradley, and Thomas Pedersen, eds. *Paleoclimate, Global Change, and the Future.* Berlin, Germany: Springer Verlag, 2003.

Butler, Rhett. "Calculating Deforestation Figures for the Amazon." *Mongabay.com.* Mongaybay.com, 30 Oct. 2014. Web. 23 Mar. 2015.

Cox, Peter M., et al. "Increasing Risk of Amazonian Drought Due to Decreasing Aerosol Pollution." *Nature* 453 (May 8, 2008): 212-215.

Gash, John H. C. *Amazonian Deforestation and Climate.* New York: John Wiley, 1996.

Intergovernmental Panel on Climate Change. *Climate Change, 2007—Impacts, Adaptation, and Vulnerability: Contribution of Working Group II to the Fourth Assessment Report of the Intergovernmental Panel on Climate Change.* Edited by Martin Parry et al. New York: Cambridge University Press, 2007.

Krebs, Charles J. *Why Ecology Matters.* University of Chicago Press, 2016. An overview of twelve basic principles underlying ecology. A useful resource for non-specialists trying to understand ecological concerns.

London, Mark. *The Last Forest: The Amazon in the Age of Globalization.* New York: Random House, 2007.

Lorenz, R., A. J. Pitman, and S. A. Sisson. "Does Amazonian deforestation cause global effects; can we be sure?" *Journal of Geophysical Research: Atmospheres* (2016). Research into models of the effects of deforestation shows that the results depend critically on the assumptions of the models, and that field trothing is essential for accurate modeling.

Scheer, Roddy, and Doug Moss. "Deforestation and Its Extreme Effect on Global Warming." Earth Talk. *Scientific American*, 13 Nov. 2012. Web. 23 Mar. 2015.

■ Amazon River

FIELDS OF STUDY

Biology; Conservation Biology; Ecology; Environment; Forest Ecology; Hydrology; Inland Aquatic Biomes; Silvology

SUMMARY

The world's largest river is a diverse ecosystem providing a home to numerous species found nowhere else. The Amazon's size leads to habitats found nowhere else in the world. Amazonian deforestation has been a concern for centuries. It affects the greater ecosystem by impacts on the water cycle, which, in turn threatens the Earth's health. The conversion of forest to pasture or farmland leads to water running off into the river without being recycled through the trees; deforestation contributes to drought and wildfires and reduces rainfall critical to the water supply of the heavily populated parts of Brazil and Argentina. These direct effects of deforestation are seen in turn to contribute to global warming

PRINCIPAL TERMS

- **carbon dioxide:** colorless, odorless, incombustible gas present in the atmosphere, usually obtained from coal, coke, or natural gas by combustion
- **deforestation:** to divest or clear forests of trees
- **drainage basin:** the area drained by a river and all its tributaries
- **photosynthesis:** process by which carbon dioxide, water, and certain inorganic salts are converted into carbohydrates by green plants, algae, and certain bacteria, using energy from the sun and chlorophyll
- **rainforest:** a tropical area dominated by evergreen trees whose leaves form a continuous canopy and that receives at least 254 centimeters (19.6 inches) of rain per year

- **tributary (also affluent):** a stream or river that flows into a larger stream or main stem (or parent) river or a lake but does not flow directly into a sea or ocean.

The Amazon is the largest river by flow volume in the world, accounting for one-fifth of the world's total river flow. Accordingly, it has the largest drainage basin—about 2.7 million square miles (6,992,968 square kilometers) in area. It is so large that if accounted independently, two of its 1,000 tributaries—the Rio Negro and the Madeira River—would be among the ten largest rivers in the world. Nearly one-sixth of all the freshwater that drains into an ocean passes through the Amazon. Its width varies from one to six miles (two to 10 kilometers) when the river is low, expanding beyond 30 miles (48 kilometers) in the wet season.

The Amazon begins in the Andes Mountains and flows for about 4,000 miles (6,437 kilometers) through South America before entering the Atlantic Ocean in an estuary some 150 miles (241 kilometers) wide. At no point is the Amazon crossed by a bridge—not because of its width, which has been traversable by modern engineering for a century, but because so much of the Amazon passes through rainforest that there is little demand for a crossing. The river is about 11 million years old and has had its current shape for about 2.4 million years, according to a 2009 study of sediment columns.

The Amazon's size leads to habitats found nowhere else. The Piramutaba catfish, for example, one of the larger catfishes in the Amazon, migrates more than 2,000 miles (3,219 kilometers) from its nursery in the Guianan–Amazon mangroves to its upper-Amazon spawning grounds. The Amazon Basin is home to as many as 25 percent of the world's terrestrial species, and its flora accounts for about 15 percent of the world's land-based photosynthesis activity. Massive numbers of species remain unidentified. For a long time, the idea persisted that the white-water rivers of the Amazon were plentiful with fish, whereas the darker waters like the Rio Negro were "hunger rivers," void of most life. It has become clear that this is an oversimplification and that the black rivers are home to significant turtle populations, in addition to supporting fisheries.

Temperature, rainfall, and climate vary throughout the large region, but it is generally warm and humid. Because the wave and tidal energy of the massive river is enough to carry most of its sediments to sea, it doesn't form a true delta. Instead, it empties directly into the turbulent Atlantic Ocean, which rapidly carries the silt away.

The Amazon is joined to the Orinoco River basin by the Casiquiare canal. The Casiquiare is a distributary of the Orinoco and flows into the Rio Negro, one of the Amazon's tributaries.

Surrounding Life

Dramatic topographical variations have led to a rich diversity of Amazonian life, and for a long time, public discussion of biodiversity and conservation implicitly or explicitly centered on the Amazon. In the south, a dearth of human settlement and roads have kept the habitat intact. In the north, logging, mining, and increased settlement have posed hazards. The south has the greatest endemic richness, with 11 endemic mammal species (of about 250) and 17 endemic bird species (of nearly 800).

Widespread Amazonian fauna include tapirs (*Tapirus terrestris*), jaguars (*Panthera onca*), capybaras (*Hydrochoeris hydrochaeris*), and kinkajous (*Potos flavus*), as well the endangered woolly monkey (*Lagothrix lagotricha*), giant otter (*Pteronura brasiliensis*), giant anteater (*Myrmecophaga tridactyla*), and ocelot (*Leopardus pardalis*). The giant otter was once common throughout most of South America, but its population has declined to as few as 2,000, primarily in the protected areas of the Amazon Basin, as a result of being hunted for its fur.

Today, cattle ranchers are responsible for much of the deforestation, as demand for beef began a swift rise in the mid-20th century (principally due to the growing worldwide middle class—an important customer base—and to improvements in refrigeration and transport that expanded beef markets) and has stayed steady ever since.

The increasing demand for biofuel and other soy products has also increased deforestation by soybean farmers; in some areas, other livestock farmers and miners are the primary threats to the forest. Highway construction and land speculation, especially in the Brazilian stretch of the Amazon, are certainly factors, but outside Brazil, the problem of deforestation is thornier in its social and economic implications, because the laborers responsible for it are so often poor and deeply dependent on whatever work has displaced the forest. Each of these deforestation vectors has seen acceleration in recent decades—yielding an increasing stress level on the essential biology of the Amazon rainforest and other riverine ecosystems.

Even apart from the destruction of forest habitats and the various pollutants introduced by the purpose to which the deforested land is put, deforestation affects the greater ecosystem by impacts on the water cycle. The conversion of forest to pasture or farmland leads to water running off into the river without being recycled through the trees; deforestation contributes to drought and wildfires and reduces the rainfall that is critical to the water supply of the heavily populated parts of Brazil and Argentina. These direct effects of deforestation are seen in turn to contribute to global warming—both by the increased emanation of greenhouse gases from the industrial activities and more fires, and by depleting the carbon-absorbing power of the forest base. The feedback loop is completed when higher global air temperature and dislocation of normal precipitation patterns contribute to drought and drives up the spread of wildfires.

Upper Amazon Rivers and Streams

The major aquatic habitats of the Amazon are located in the Upper Amazon Rivers and Streams ecoregion, consisting of the muddy white-water tributaries in the west of the basin and the more-nutrient-poor black rivers of the Guiana Highlands. Hundreds of endemic (uniquely evolved to fit a locally biome niche) fish species can be found in this zone of the nearly 1,500 total fish species found throughout the Amazon. They include many ostariophysan fishes, including species from the siluroid catfish (*Siluridae*), minnow (*Cyprinidae*), and charcin (*Characidae*) families. Ostariophysan fishes are noteworthy for the way their gas-filled swim bladders amplify sound waves in water and transmit them to interear vertebrae, enhancing their sense of hearing.

Of the twenty species of piranha in the Amazon River, only five are a threat to humans. The piranha is perhaps the most famous Amazonian fish, and the razor-sharp teeth responsible for that fame have been used as cutting tools by Amazonian tribes for millennia. Less famous is their importance to the ecosystem, as they eat both dead fish and dead animals whose decomposition would otherwise pollute the river.

The Napo River alone is home to more than 500 species of fish. Other fish species in the region include electric eels, and Loricariid catfish. The shallow Napo waters are home to the anaconda, one of the largest snake species, which remains low in the water, with just its nostrils above the surface.

The Upper Amazon is one of the most jeopardized regions of the Amazon. The introduction of hydroelectric dams, while providing a relatively green source of power, has modified river flow and altered the movements of migratory fish species. Deforestation during construction has increased erosion of riverbanks, and when the cleared land is converted to pasture or farmland, it may contribute contaminants to the water in the form of fertilizer, pesticide, animal feces, detergents, and other pollutants.

A more recent threat is the Andes gold rush. Once the base of the Andes, where the Amazon begins, was a pristine environment even as other parts of the rainforest were threatened by deforestation. Now humans have dug thousands of mines there in search of gold. Though ranching and logging account for a greater amount of deforestation, gold miners—who have destroyed perhaps 100,000 acres (40,469 hectares) of rainforest in Peru so far—burn the forest in which they work, including trees more than a millennium old, and strip away the Earth's surface to a depth of 50 feet (15 meters). This damage is much harder to recover from than that of slash-and-burn agriculture. The mercury used in recovering gold from silt leaches into the watershed and is eventually taken up by Amazonian fish.

Most of the gold mines in Peru (as many as 98 percent) are illegal, making regulation a difficult remedy to apply. On a typical day, let alone a lucky one, a barely skilled laborer–turned–gold miner can earn twice as much as he or she would make in a month at his or her old job. The skyrocketing cost of gold has led many people to leave their old jobs and join into informal bands of miners who divide their findings among themselves. Other mine workers are teenagers, sold into servitude by their impoverished parents. Legal, commercial operations are larger in scale, with a typical mine removing sixteen dump-truck loads of rock and soil every hour, 18 hours a day. Though legal operations are required to pay environmental remediation fees, the fees do nothing to stop the changes to, and destruction of, the local Amazonian ecosystem.

Amazonian Flooded Forests

The Amazon's cycle of flooding leads to the most extensive system of floodplain habitats in the world, as the river rises more than 30 feet (nine meters) during the wet season. The flooded forested areas of north-central South America encompass an area about twice the size of California. These areas are home to a large number of freshwater fishes, reptiles, and other aquatic fauna that migrate into the newly flooded area, including two species of the freshwater dolphins characteristic of the Amazon. Floodplain lakes and floating meadows are replenished, and freshwater fish feed on fruit dropping from the trees of the flooded forests. Indeed, the reproduction cycle of many floodplain trees depends on their fruit being eaten, and seeds dispersed, by these fish.

Floodplain fish include the tambaqui (a fruit eater), arawana, pirarucu, arapaima, dourada catfish, tucunare, and *Lepidosiren paradoxa*, one of the few lungfish to evade extinction. Four species of the threatened uakari monkey (genus *Cacajao*) are among the floodplain mammals, as well as the pink and gray river dolphins and the manatee. The black caiman (*Melanosuchus niger*) is a local member of the alligator family; once nearly extinct due to hunting, the species has partially recovered its population, and it is classified as Conservation Dependent. The caiman depends primarily on fish for its diet, but larger adults will feed on mammals like tapirs and deer that come to the riverbank to drink. Jaguars prey on the juveniles.

The endangered Arrau turtle (*Podocnemis expansa*), the largest side-neck turtle, also migrates into the flooded forests, feeding on plants. In the dry season, the turtles travel in large numbers to find nesting areas, laying clutches of eggs in sandbanks. Most of the young are consumed almost immediately by predators; the survivors migrate to the floodplains once more.

The annual action of waters flowing into and then out of the floodplains contributes some of the Amazon River's trace mineral content. A 2004 study of the mineral content of the river, lakes, and leaves of the Ilha de Marchantaria forest affirmed that concentrations of manganese and copper could not be explained by tributary mixing or in-stream processes and were likely the result of sediment–water and plant–water interfaces. Though the results were less clear, the same could be true for concentrations of iron, aluminum, and rubidium. It may be that some of the plant species endemic to the region in fact depend on or benefit from these trace minerals, with which they would not otherwise be provided.

Lateral exchange with the floodplain also has significant effects on the carbon abundance and oxidation of the mainstream. The rate of respiration in the river is high, supported by organic material. Carbon-processing exchanges are most active at the places where floodplain waters interact with the mainstream water, which have a carbon and water mass balance anomaly. Most respiratory carbon dioxide (CO_2) is lost to the atmosphere, as is common in riverine ecosystems.

The flooded forests are jeopardized by logging of the virola (also known as epena, patricia, and cumala), used by tribes in the Amazon and Orinoco Basins for its resin, which includes hallucinogenic alkaloids such as the most potent forms of DMT. Another threat is logging of the kapok (*Ceiba pentandra*), the seed pods of which produce large amounts of lignin- and cellulose-rich fiber with numerous commercial applications.

Bill Kte'pi

Further Reading

Bayley, Peter B. "Understanding Large River-Floodplain Ecosystems." *BioScience* 45, no. 3 (1995).

Eisenberg, J. F. and K. H. Redford, eds. *Mammals of the Neotropics: The Central Neotropics*. Chicago: University of Chicago Press, 1999.

Foley, Jonathan A., et al. "Amazonia Revealed: Forest Degradation and Loss of Ecosystem Goods and Services in the Amazon Basin." *Frontiers in Ecology and the Environment* 5, no. 1 (2007).

Godar, Javier, Emilio Jorge Tizado, and Benno Pokorny. "Who is Responsible for Deforestation in the Amazon? A Spatially Explicit Analysis Along the Transamazon Highway in Brazil." *Forest Ecology and Management* 267, no. 58 (2012).

Henderson, A. *The Palms of the Amazon*. New York: Oxford University Press, 1995.

Richey, Jeffrey E., et al. "Biogeochemistry of Carbon in the Amazon River." *Limnology and Oceanography* 35, no. 2 (1990).

Smith, Nigel J. J. *Amazon Sweet Sea: Land, Life, and Water at the River's Mouth*. Austin: University of Texas Press, 2002.

Thorp, James H. "The Riverine Productivity Model: An Heuristic View of Carbon Sources and Organic Processing in Large River Ecosystems." *Oikos* 70, no. 2 (1994).

Viers, Jerome, et al. "The Influence of the Amazonian Floodplain Ecosystems on the Trace Element Dynamics of the Amazon River Mainstem." *Science of the Total Environment* 339, no. 1–3 (2005).

Amphibians

FIELDS OF STUDY
Biology; Conservation Biology; Ecology; Entomology; Environment; Epidemiology; Herpetology; Zoology

SUMMARY
Nearly all amphibians lead double lives and are therefore they affected by habitat alterations and environmental contaminants in both aquatic and terrestrial environments. Declining amphibian populations are of concern because they are important components of many ecosystems their decline may signal the onset of deteriorating conditions that will affect other forms of life, including humans.

PRINCIPAL TERMS
- **amphibian:** class of vertebrate animals with thin, permeable skin and generally characterized by an aquatic larval stage and a terrestrial adult stage
- **biotic communities:** a group of interdependent organisms inhabiting the same region and interacting with each other
- **chytrid fungus:** a type of fungus (Phylum Chytridiomycota) of which there are and there are approximately 1,000 different chytrid species that live exclusively in water or moist environments and are among the oldest (most primitive) types of fungi, some of which are parasites living on plants or invertebrate
- **larvae:** the immature, wingless, feeding stage of an insect that undergoes complete metamorphosis
- **parasite:** an organism that lives on or in an organism of another species, known as the host, from which it obtains nutriment
- **synergy:** the interaction of elements that when combined produce a total effect that is greater than the sum of the individual elements

Amphibians include frogs and toads, salamanders, and caecilians (rarely encountered legless tropical forms). Because nearly all amphibians lead double lives, they are affected by habitat alterations and environmental contaminants in both aquatic environments (as larvae, such as tadpoles) and terrestrial environments (as adults). Also, because amphibians breathe primarily through their skin, they are very sensitive to chemicals and poor water quality. Like the proverbial canary in the coal mine, amphibians serve as environmental monitors. Consequently, declining amphibian populations are of concern not only because amphibians are important components of many ecosystems but also because the declines may signal the onset of deteriorating conditions that will soon affect other forms of life, including humans.

About 32 percent of the world's amphibian species are known to be threatened or extinct, and insufficient data are available to allow determination of the threat status of another 25 percent. At least thirty-eight amphibian species are known to be extinct and at least another 120 species have not been found in recent years and could be extinct. At least 42 percent of all amphibian species are declining, suggesting that the number of threatened species will rise in the future.

Most of the declines are attributable to habitat alterations and destruction. Especially in the tropics, deforestation has affected a number of species, and the cool, wet mountaintop habitats favored by many amphibians are particularly vulnerable to global climate change. Complicating the issue is the rapid worldwide spread of the chytrid fungus, which has devastated amphibian populations. The fungus invades the thin skin of amphibians and kills them by disrupting their ability to regulate the movement of water and oxygen in their bodies. The fungus spreads in water and through direct body contact. A few resistant species, such as bullfrogs, clawed frogs, and cane toads, serve as vectors, transmitting the disease to sensitive species. Chytrid thrives in the same cool, wet conditions favored by most amphibians.

Deformed frogs with missing or extra limbs were brought to the attention of the world when they were discovered by schoolchildren in Minnesota in 1995. At first, chemical pollutants were thought to be the cause, but it became apparent that parasitic flatworms (trematodes) were responsible. However, experts believe that, like chytrid, stress imposed by deteriorating habitats and a changing climate may well be aggravating the problem in a synergistic fashion.

Another environmental concern associated with amphibians is not related to declines but is, in fact, caused by too many frogs in the wrong places. A few very hardy species have been transported by humans to places far from their native ranges. These invasive species upset biotic communities by competing with or eating native species. Large and voracious cane toads, introduced widely to control insect pests in sugarcane fields, have been implicated in the decline or extinction of native frogs in places as far from one another as Florida and Australia. Equally large and voracious bullfrogs, intentionally imported around the world for food, are responsible for declines of native frogs in the western United States, Latin America, and even Europe. Coquis, small tree frogs native to Puerto Rico, pose a threat to unique Hawaiian insects. Cuban tree frogs, introduced into Florida, have hitchhiked on tropical plants shipped around the world, and in their new habitats they eat other frogs and even small lizards.

Robert Powell

Further Reading

Collins, James P., and Martha L. Crump. *Extinction in Our Times: Global Amphibian Decline.* New York: Oxford University Press, 2009.

Heatwole, Harold, and John W. Wilkinson, eds. *Amphibian Decline: Diseases, Parasites, Maladies, and Pollution.* Chipping Norton, N.S.W.: Surrey Beatty & Sons, 2009.

Lannoo, Michael, ed. *Amphibian Declines: The Conservation Status of United States Species.* Berkeley: University of California Press, 2005.

Stuart, S. N., et al., eds. *Threatened Amphibians of the World.* Arlington, Va.: Conservation International, 2008.

■ Anthropocentrism

FIELDS OF STUDY
Anthropology; Biology; Controversies; Debates; Ecology; Environment; Environmentalism; Ethics; Public Policy

PRINCIPAL TERMS
- **anthropocentrism:** the view that human beings are of central importance in the universe
- **biodiversity:** the variability among all living organisms, plants and animals, and the ecological complexes of which they are a part, including diversity within species, between species, and of ecosystems, thus contributing to human well-being
- **speciesism:** prejudice or discrimination based on species, especially discrimination against animals
- **biocentrism:** ethical perspective holding that all life deserves equal moral consideration or has equal moral standing
- **sentientism:** the capacity to feel, perceive or experience subjectively central to the philosophy of animal rights
- **ecocentrism:** the belief that all living organisms and their natural environments have intrinsic value, regardless of their perceived usefulness or importance to human beings

SUMMARY
From the anthropocentric point of view, biodiversity should be preserved only if it is in the interest of humans to preserve it—any duties that human beings have to preserve biodiversity are owed to other humans, not to any other species. Environmentalists believe that anthropocentric attitudes are largely responsible for actions that have led to environmental calamities such as air and water pollution, species extinction, and global climate change.

For anthropocentrists human lives have greater value than do the lives of any other species. Anthropocentrists claim that humans are the only beings that possess certain capacities. They note that, unlike other animals, humans are typically intelligent, self-aware, autonomous, language users, and moral agents, engaging in play and creating art, among other complex cognitive tasks. For anthropocentrists only humans are intrinsically valuable. The rest of nature (including all plant and animal species) has only instrumental value, serving exclusively as a means to human ends. From the anthropocentric point of view, biodiversity should be preserved only if it is in the interest of humans to preserve it—any duties that human beings have to preserve biodiversity are owed to other humans, not to any other species.

Anthropocentrism is deeply rooted in most human cultures, but this viewpoint has come under increasing challenges by environmental activists, animal rights advocates, and others. Among the major arguments against anthropocentrism is that it is invidiously perfectionist—that is, logically, those humans who do not display all the characteristics that anthropocentrists assert are uniquely human (intelligence, self-awareness, autonomy, and so on) should be viewed as less valuable than those who do. Some environmental philosophers believe that humans must eradicate both anthropocentrism and the related viewpoint of speciesism, replacing them with deep ecology, biocentrism, sentientism (the view that all sentient beings have moral worth), and ecocentrism. The search for a convincing nonanthropocentric foundation for what is valuable in nature is at the center of environmental philosophy.

Julia Tanner

Further Reading

Crist, Eileen, and Helen Kopnina. "Unsettling Anthropocentrism" *Dialectical Anthropology* 38.4 (2014): 387–396. *Humanities International Complete*. Web. 14 Jan. 2015.

Liu, Alexander, and Sara Jill Unsworth. "Cross-Cultural Differences in Core Concepts of Humans as a Biological Species." *Jour. of Cognition & Culture* 14.3/4 (2014): 171–185. *Academic Search Complete*. Web. 14 Jan. 2015.

McIntyre-Mills, Janet Judy. "Anthropocentrism and Well-Being: A Way Out of the Lobster Pot?" *Systems Research & Behavioral Science* 30.2 (2013): 136–155. *Business Source Complete*. Web. 14 Jan. 2015.

Nolt, John. "Anthropocentrism and Egoism." *Environmental Values* 22.4 (2013): 441–459. *Environment Complete*. Web. 14 Jan. 2015.

Oriel, Elizabeth. "Whom Would Animals Designate as Persons? On Avoiding Anthropocentrism and Including Others." *Jour. of Evolution & Technology* 24.3 (2014): 44–49. *Academic Search Complete*. Web. 14 Jan. 2015.

Antienvironmentalism

FIELDS OF STUDY
Ethics; Advocacy; Policy; Protest; Ecology; Environment; Environmentalism; Public Policy; Debates; Controversies

SUMMARY
The long-standing debates that continue between anti-environmentalists and environmentalists have important influence on both legislators and policy makers as they address environment-related issues.

PRINCIPAL TERMS
- **anti-environmentalism**: philosophy that human beings' immediate economic and lifestyle needs are more important than concerns about the fates of other specifies and the general environment
- **Comprehensive Environmental Response, Compensation, and Liability Act (Superfund)**: allows EPA to clean up contaminated sites and forces parties responsible to either perform cleanups or reimburse the government for EPA-led cleanup work
- **Environmental Protection Agency (EPA)**: an agency of the United States federal government whose mission is to protect human and environmental health
- **environmentalism**: broad philosophy, ideology, and social movement for environmental protection and improvement of the health of the environment, seeking to incorporate the impact environmental changes on humans, animals, plants and non-living matter
- **global warming**: increase in the earth's atmospheric and oceanic temperatures due to an increase in the greenhouse effect resulting especially from pollution
- **ozone layer**: atmospheric layer normally characterized by high ozone content—a form of

oxygen—blocking most solar ultraviolet radiation from entry into the lower atmosphere
- **Sagebrush Rebellion:** a 1970s-1980s movement during seeking major changes to federal land control, use and disposal policy in the American West, including more state and local control and/or privatization
- **Wise-use movement:** a coalition of groups promoting the expansion of private property rights and reduction of government regulation of publicly held property

In the early twentieth century, environmentalism in the United States was largely fostered by wealthy sportsmen who saw the need to protect the outdoors in order to maintain satisfactory areas for their pursuits of hunting, fishing, and camping. The movement got a populist boost in 1962 from the publication of Rachel Carson's *Silent Spring*, which presents an easily understood account of the dangers of toxic substances in the environment. For the first time, the American public began to demand that laws be enacted to protect the environment and clean up land, water, and air that had already been polluted.

Growth of the Environmental Movement

For several years the environmental movement gathered strength as the public voted into office politicians with environmental orientations. Public outcry surged against polluting companies, leading to boycotts of products. Grassroots, citizen-led efforts such as recycling programs and litter patrols gained support as the public became more educated and concerned about environmental issues. Among the issues that pitted environmentalists against the government and industry were toxic waste incineration, habitat destruction by logging and mining companies, and use of public lands, including national parks.

Two oil crises during the 1970s served to focus awareness on energy conservation and the need to develop alternatives to energy derived from fossil fuels. Many feared that oil supplies were dwindling, while others wished to end US reliance on oil-exporting nations in the Middle East. One important result was a general reduction in the size of motor vehicles. This, along with other technological advances, helped lead to the development of cars that were more fuel-efficient. The research required to accomplish these changes, however, was costly to automakers.

The 1970s were also characterized by landmark legislation that imposed strict limits on pollution output and resource use and also provided for the remediation of polluted land and water. Earth Day was celebrated for the first time on April 22, 1970.

Thanks to the 1970s legislation, large fines were imposed for violations of the new laws, enforced by the newly formed US Environmental Protection Agency (EPA). One of the most important and pivotal developments was the passage of the Comprehensive Environmental Response, Compensation, and Liability Act, or Superfund, which provided vast sums of public money for the cleanup of designated industrial and military waste dumps and other degraded sites. Signed into law by the US Congress in 1980, Superfund's provisions allowed the government to bring lawsuits against the responsible parties, requiring them to help pay cleanup costs. In order to avoid fines, many industries were forced to develop and implement costly waste-processing technologies.

The political and economic situation began to change in the late 1970s as industry mounted a counteroffensive against environmental laws. Businesses contended with the burgeoning number of environmental regulations by finding and exploiting loopholes in legislation. A growing number of industries used stalling tactics and countersuits to delay or eliminate the need to implement required changes. Meanwhile, in the western United States, a coalition of loggers, miners, cattle ranchers, farmers, and developers demanded that the federal government transfer control of large tracts of federally owned land to individual states. Members of the so-called Sagebrush Rebellion felt that state ownership would give them more power to exploit the natural resources on the land.

Anti-environmentalist Backlash

A severe backlash against environmentalism began to occur when Ronald Reagan replaced Jimmy Carter as president of the United States in early 1981. Many environmental laws and regulations, which the new presidential administration viewed as barriers to economic progress, were weakened or abolished. Many federal judges who had started their careers during the 1960s retired, and they were replaced by politically conservative judges who began to interpret existing laws in favor of industry. The office of the EPA was weakened, and funding for environmental enforcement and remediation was slashed. Secretary of the Interior James Watt, who had been a leader of the Sagebrush Rebellion, promoted legislation to open previously protected areas to mining and oil exploration. The general public, experiencing growth and prosperity for the first time in many years, began to favor short-term economic gains and turned a blind eye to news of the weakening environmental movement.

The late 1980s saw the birth of the wise-use movement, which appeals to the pragmatic and optimistic aspects of human nature by asserting that some optimal balance of resource use and restoration is practicable and that technology, given time and funding, will develop workable solutions to existing environmental problems. This position assumes that human beings can understand the complex ecosystems involved well enough to know what these balances should be. Advocates of wise use believe that all public lands, including national parks, should be opened to mining and drilling. Like the Sagebrush Rebels, they also promote the strengthening of the rights of states and property owners to exploit resources with minimal federal regulation.

According to the tenets of wise use, the harvesting of timber from ancient forests would be followed by the planting of an equivalent acreage of saplings. Logging would be timed according to growth rates, and technology would produce fast-growing varieties of trees that would furnish adequate ecosystems for wildlife in the new forests. Environmentalists, in contrast, argue that ancient forests represent complex, irreplaceable ecosystems that cannot be substituted with new forests planted by logging companies. Similar disagreement exists regarding coastal wetlands, which provide vital habitat to numerous species and contain a high degree of biodiversity. Wetlands are frequently located in areas that are desired by real estate developers wishing to build vacation homes or resorts. Environmental protection laws based on the tenets of wise use mandate that destroyed wetlands must be replaced by new wetlands of equivalent size. Again, environmentalists worry that too many threatened species would be lost in the process of destroying and replacing wetland areas.

Scientific Controversy

Another significant trend in anti-environmentalism toward the end of the twentieth century and into the twenty-first involved public confusion over scientific debates concerning such topics as the hole in the ozone layer and global warming. Industry and government scientists often questioned and condemned dire predictions advanced by other scientists. The government frequently responded by requesting additional research before requiring vast, expensive reductions of known pollutants. Rather than believe the frightening scenarios painted by some scientists, many people sided with scientists who questioned the validity of these and other threats to the global environment.

Mainstream environmental organizations that had evolved from small groups of fervent individuals were now led by full-time professional lobbyists based in Washington, DC. Details of new legislation were negotiated among industry, government, and environmental leaders. National environmental organizations grew increasingly cumbersome and expensive to run. Many began accepting large donations from the same industries they were trying to monitor, which created serious conflicts of interest. Top industry executives became members of the boards of directors for environmental organizations. At the same time, these corporations also made large donations to elected officials and thus gained access and influence in government. As the 1990s progressed, membership in large environmental groups began to decline as "donor fatigue" set in, questions persisted about the true urgency of various environmental issues, and cynicism arose about the possibility of environmental progress under such circumstances.

A relaxation of concern about environmental problems came about in the late 1990s in the wake of encouraging news about improvements of environmental indicators such as air-pollution levels of certain gases in the aftermath of implementation of cleaner energy production. For example, atmospheric levels of sulfur dioxide, which leads to acid rain, decreased in the United States and Europe after cleaner coal- and oil-burning technologies were implemented. The air-quality goals of many cities were met through a combination of fuel efficiency and "scrubber" smokestacks. While environmentalists warned such progress did not indicate the end of environmental threats, the general public became less tuned to these issues once their direct negative effects were relieved.

A controversial strategy advanced by a coalition of industry, government, and environmental leaders involves tradable pollution permits. According to the plan, the government assigns utilities a certain number of pollution units per year. An especially clean-running plant will not need all of its units and will be able to sell them to plants that exceed their allotments. This system has been criticized by many environmentalists, who contend that some utilities are able to buy their way out of the need to reduce pollution. The position thought to be anti-environmentalist in this context would maintain that the plan is a realistic method of controlling overall levels of pollution without putting older utilities out of business while they endeavor to upgrade their performance.

The revelation that environmental degradation can, in certain cases, be reversed over a fairly short span of time led to the argument on the part of anti-environmentalists that nature is surprisingly resilient, and therefore environmental protection does not need to be so stringent, costly, and regressive. Environmentalists counter such arguments with a call to remain vigilant and to include the health of the environment in national and global visions of the future.

Political Considerations
The relative popularity and success of anti-environmentalism versus environmentalism often depends on the political climate. In the United States, political conservatism has tended to be more accepting of anti-environmentalism, elements of which came to often feature in Republican platforms. This alignment largely began during the Reagan administration, and grew stronger into the twenty-first century. For example, the Republican administration of President George W. Bush was noted for supporting policies seen as antienvironmental, such as oil drilling in the Arctic. However, having anti-environmentalists in power also helped motivate a surge in grassroots environmentalist organization, with greater public attention drawn to issues such as climate change in the 2000s.

The opposite effect was seen with the election of Democratic president Barack Obama in 2008. His administration was largely seen as pro-environmentalism, and strongly voiced support for climate change mitigation and other green policies, despite some environmentalists arguing it did not go far enough. In response, conservatives allied even more deeply with anti-environmentalist industries and other groups. The movement attacking mainstream climate science and other environmental science became more powerful, finding great success in sowing doubt in the general public.

A culmination of this attitude was seen in the election of business tycoon Donald Trump in the controversial 2016 presidential election, after he expressed radically anti-environmentalist views on the campaign trail. His cabinet eventually included several noted anti-environmentalists and climate change skeptics, bringing the movement an unprecedented level of power and visibility. The Trump administration took steps not only to block new environmental legislation, but also to undo existing protections and regulations and severely weaken the EPA. However, this rise in anti-environmentalism did appear to again incite a public backlash. Environmentalist activist groups mounted an increasingly organized resistance, staging events such as a March for Science in Washington, DC, on Earth Day 2017.

Wendy Halpin Hallows

Further Reading
Davis, Charles, ed. *Western Public Lands and Environmental Politics*. 2d ed. Boulder, Colo.: Westview Press, 2001.

Dowie, Mark. *Losing Ground: American Environmentalism at the Close of the Twentieth Century*. Cambridge, Mass.: MIT Press, 1995.

Easterbrook, Gregg. *A Moment on the Earth: The Coming Age of Environmental Optimism*. New York: Penguin Books, 1995.

Helvarg, David. *The War Against the Greens: The "Wise-Use" Movement, the New Right, and the Browning of America*. Rev. ed. Boulder, Colo.: Johnson Books, 2004.

Hirt, Paul W. *A Conspiracy of Optimism: Management of the National Forests Since World War Two*. Lincoln: University of Nebraska Press, 1994.

Jacques, Peter J. *Environmental Skepticism: Ecology, Power, and Public Life*. Burlington, Vt.: Ashgate, 2009.

McKibben, Bill. *The End of Nature*. 1989. Reprint. New York: Random House, 2006.

Shabecoff, Philip. *A Fierce Green Fire: The American Environmental Movement*. Rev. ed. Washington, D.C.: Island Press, 2003.

Young, John. *Sustaining the Earth: The Story of the Environmental Movement—Its Past Efforts and Future Challenges.* Cambridge, Mass.: Harvard University Press, 1990.

Websites
Vox, 7 June 2017
www.vox.com/2017/4/22/15377964/republicans-environmentalism

Aquifers

FIELDS OF STUDY
Ecology; Environment; Geology; Hazardous Waste; Hydrology

SUMMARY
Aquifers are important because groundwater supplies a substantial amount of the water available in many localities. The contamination of aquifers is thus a matter of concern, and so a variety of aquifer restoration techniques have been developed.

PRINCIPAL TERMS
- **anisotropic:** exhibiting properties with different values when measured in different directions
- **aquifers:** water-bearing geological formations that can store and transmit significant amounts of groundwater to wells and springs
- **artisan wells:** a well in which water is under pressure and where water flows to the surface naturally
- **hydraulic conductivity:** a physical property which measures the ability of the material to transmit fluid through pore spaces and fractures in the presence of an applied hydraulic gradient
- **hydraulic gradient:** a measure of the change in groundwater head over a given distance
- **igneous rock:** rock formed by solidification of a molten magma
- **isotropic:** an object or substance having a physical property with the same value when measured in different directions

All rocks found on or below the earth's surface can be categorized as either aquifers or confining beds. An aquifer is a rock unit that is sufficiently permeable to allow the transportation of water in usable amounts to a well or spring. (In geologic usage, the term "rock" also includes unconsolidated sediments such as sand, silt, and clay.) A confining bed is a rock unit that has such low hydraulic conductivity (or poor permeability) that it restricts the flow of groundwater into or out of nearby aquifers.

There are two major types of groundwater occurrence in aquifers. The first type includes those aquifers that are only partially filled with water. In those cases, the upper surface (or water table) of the saturated zone rises or declines in response to variations in precipitation, evaporation, and pumping from wells. The water in these formations is then classified as unconfined, and such aquifers are called unconfined or water-table aquifers. The second type occurs when water completely fills an aquifer that is located beneath a confining bed. In this case, the water is classified as confined, and the aquifers are called confined or artesian aquifers. In some fractured rock formations, such as those that occur in the west-central portions of New Jersey and eastern Pennsylvania, local geologic conditions result in semiconfined aquifers, which, as the name indicates, have hydrogeologic characteristics of both unconfined and confined aquifers.

Wells that are drilled into water-table aquifers are simply called water-table wells. The water level in these wells indicates the depth below the earth's surface of the water table, which is the top of the saturated zone. Wells that are drilled into confined aquifers are called artesian wells. The water level in artesian wells is generally located at a height above the top of the confined aquifer but not necessarily above the land surface. Flowing artesian wells occur when the water level stands above the land surface. The water level in tightly cased wells in artesian aquifers is called the potentiometric surface of the aquifer.

Water flows very slowly in aquifers, from recharge areas in interstream zones at higher elevations along watershed boundaries to discharge areas along streams and adjacent floodplains at lower elevations. Aquifers thus function as pipelines filled with various types of earth material. Darcy's law governing groundwater flow was developed in 1856 by Henry Darcy, a French engineer. In brief, Darcy's law states that the amount of water moving through an aquifer per unit of time is dependent on the hydraulic conductivity (or permeability) of the aquifer, the cross-sectional area (which is at a right angle to the direction of flow), and the hydraulic gradient. The hydraulic conductivity depends on the size and interconnectedness of the pores and fractures in an aquifer. It ranges through an astonishing twelve orders of magnitude. Very few other physical parameters exhibit such a wide range of values. For example, the hydraulic conductivity ranges from an extremely low 10^7 to 10^8 meters per day in unfractured igneous rock such as diabase and basalt to as much as 10^3 to 10^4 meters per day in cavernous limestone and coarse gravel. Typical low-permeability earth materials include unfractured shale, clay, and glacial till. High-permeability earth materials include lava flows, coarse sand, and gravel.

In addition to this wide range of values, hydraulic conductivity varies widely in place and directionality within the same aquifer. Aquifers are isotropic if the hydraulic conductivity is about the same in all directions and anisotropic if the hydraulic conductivity is different in different directions. As a result of all of these factors, groundwater yield is extremely variable both within the same aquifer and from one aquifer to another when they are composed of different rocks.

Because groundwater flows slowly in comparison with surface water, any contaminant that gets into the groundwater could be around for a long time, perhaps hundreds or thousands of years. It is thus simpler and much more cost-effective to prevent groundwater contamination than it is to try to correct a problem that has been in existence for years.

Restoration of a contaminated aquifer may be accomplished, albeit at a price, through one or more of the following procedures: inground treatment or containment, aboveground treatment, or removal or isolation of the source of contamination. The first approach involves natural treatment based on physical, chemical, or biological means, such as adding nutrients to existing subsurface bacteria to help them break down hazardous organic compounds into nonhazardous materials. The second approach uses engineered systems such as pumping wells or subsurface structures, which create hydraulic gradients that make the contaminated water stay in a specified location, facilitating removal for later treatment. Regardless of the restoration method selected, the source that is continuing to contaminate the aquifer must be removed, isolated, or treated.

Robert M. Hordon

Further Reading

Ahmed, Shakeel, R. Jayakumar, and Abdin Salih. *Groundwater Dynamics in Hard Rock Aquifers: Sustainable Management and Optimal Monitoring Network Design.* New York: Springer, 2008.

Fetter, Charles W. *Applied Hydrogeology.* 4th ed. Upper Saddle River, N.J.: Prentice Hall, 2001.

Kuo, Jeff. *Practical Design Calculations for Groundwater and Soil Remediation.* Boca Raton, Fla.: CRC Press, 1998.

Nonner, Johannes C. *Introduction to Hydrogeology.* 2d ed. Boca Raton, Fla.: CRC Press, 2010.

Todd, David K. *Groundwater Hydrology.* 3d ed. Hoboken, N.J.: John Wiley & Sons, 2005.

B

■ Balance of nature

FIELDS OF STUDY
Anthropology; Biology; Early Greek Science; Early Roman Science; Ecology; Environmental Chemistry; Environmental Microbiology; Geography; Golden Age of Islam; History of Medicine; History of Science

SUMMARY
The concept of the balance of nature has never been legitimated in science as either a hypothesis or a theory, but it persists as a designation for a healthy environment.

PRINCIPAL TERMS
- **balance of nature:** ecological concept that undisturbed nature achieves constant
- **equilibrium:** a state of rest or balance due to the equal action of opposing forces
- **biogeochemical cycles:** any of the natural pathways by which essential elements of living matter are circulated
- **biotic communities:** all living organisms which share a common environment
- **homeostasis:** the tendency of a system, especially the physiological system of higher animals, to maintain internal stability
- **hydrologic cycle:** the natural sequence through which water passes into the atmosphere as vapor, precipitates to earth in liquid or solid form, and returns to the atmosphere through evaporation
- **morphology:** branch of biology dealing with the form and structure of organisms
- **natural selection:** process resulting in the survival and reproductive success of individuals or groups best adjusted to their environment,

Balance of nature statue near VUDA Park in Visakhapatnam, Andhra Pradesh. By Adityamadhav83 (Own work).

leading to the perpetuation of genetic qualities best suited to that particular environment

Greek natural philosophers in the fifth and sixth centuries BCE attempted to explain naturalistically how nature works rather than depending upon myths. The atomistic theory of Leucippus and Democritus stated that matter can be transformed but is never created or destroyed. The Pythagoreans heard musical harmony in the universe. Hippocratic medicine taught that the balance of humors within the body produces health and an imbalance produces disease, and Greek physicians believed in the healing power of nature. Within this worldview, ecological balance would have been a compelling expectation.

Herodotus, known as the father of history, wrote about not only the human histories of Greece, Persia, and Egypt but also their geographies and natural histories. He was influenced by ideas in natural philosophy, and he was concerned with concrete examples that might illustrate generalities. In organizing his information on the lions, snakes, and hares of Arabia, he asked why predatory species do not eat up all of their prey. His answer was that a superintending "Providence" had created the different species with different capacities for reproduction. Predatory species, such as lions and snakes, produce fewer offspring than species that they eat, such as hares. From Egypt, Herodotus obtained a report about a mutually beneficial relationship between Nile crocodiles and plovers: Crocodiles allow plovers to sit on their teeth and eat the leeches that infest their mouths.

Plato lived after the natural philosophers and Herodotus, but he lacked their trust in sensory data, and therefore he explained nature with naturalistic myths. In a dialogue called *Pr tagoras* (c. 390 BCE; English translation, 1804), Plato asked Herodotus's question about why some species do not eat up the others, but he asked it more abstractly and not for its own sake. The point of this creation myth was to explain why humans do not have specialized traits such as wings or claws. The gods assigned to Epimetheus the task of creating each species with traits that would enable it to survive. He had given out all the specialized traits before he got around to creating humans, and his brother Prometheus had to save humanity by giving it reason and fire. The secondary point of the myth supplements what Herodotus had concluded about differences in reproduction with a conclusion about differential traits that ensure survival.

In the writings of Aristotle and his colleagues in Athens, full-fledged science emerged. These scholars realized the importance of both collecting data and ordering it so as to allow the drawing of conclusions. These scholars, however, focused on physiology and anatomy and neglected to look for what modern-day persons call ecological explanations of how nature works. For example, they explained that the greater number of offspring in hares than in lions is because of their size. Since hares are smaller, it is easier for more of them to grow within a female than for multiple larger lion cubs to grow in their mother. The balance-of-nature concept was not distinct enough to require either defending or refuting.

Scientific Revolution

The ancient Romans excelled in engineering, not science. They were mostly content with abbreviated Latin versions of Greek science. Roman writings are still worth mentioning, however, because of their influence on later European naturalists. Aelianus compiled a popular natural history book that, among other things, explained that jackdaws are friends to the farmer because jackdaws eat the eggs and young of locusts that would eat the farmer's produce. The philosopher Cicero wrote an influential book, *De natura deorum* (44 BCE; *On the Nature of the Gods*, 1683), in which he saw the work of Providence in endowing plants with the capacity to feed humans and animals and still be able to have seeds left over to ensure their own reproduction. Another philosopher, Plotinus, pondered the evil of suffering when predators kill animals for food. He decided that the existence of predation allows a greater diversity of life to exist than would be possible if all animals ate plants. These miscellaneous observations were insufficient for a theory, but they kept alive the notion of a balance of nature.

During the scientific revolution, fresh observations and conclusions appeared. Most significant, John Graunt, a merchant, analyzed London's baptismal and death records in 1662 and discovered the balance in the sex ratio and the regularity of most causes of death (excluding epidemics). England's chief justice, Sir Matthew Hale, was interested in Graunt's discoveries, but he nevertheless decided that the human population, in contrast to animal populations, must have steadily increased throughout history. He surveyed the known causes of animal mortality and in 1677 published the earliest explicit account of the balance of nature.

English scientist Robert Hooke studied fossils and in 1665 concluded that they represented the remains of plants and animals, some of which were probably extinct. John Ray, a clergyman and naturalist, argued in response that the extinction of species would contradict the wisdom of the ages, by which he seems to have meant the balance of nature. Ray also studied the hydrologic cycle, which is a kind of environmental balance of water. In the late seventeenth century, Antoni van Leeuwenhoek, one of the first investigators to make biological studies with a microscope, discovered that parasites are more prevalent than anyone had suspected and that they are often detrimental or even fatal to their hosts. Before that, it was commonly assumed that the relationship between host and parasite was mutually beneficial.

Richard Bradley, a botanist and popularizer of natural history, pointed out in 1718 that each species of plant has its own kind of insect and that there are even different insects that eat the leaves and bark of a tree. His book *A Philosophical Account of the Works of Nature* (1721) explored aspects of the balance of nature more thoroughly than had been done before. Ray's and Bradley's books may have inspired the comment in Alexander Pope's *An Essay on Man* (1733-1734) that all the species are so closely interdependent that the extinction of one would lead to the destruction of all living nature.

Toward a Science of Ecology
Swedish naturalist Carolus Linnaeus was an important proto-ecologist. In his essay *Oeconomia Naturae* (1749; *The Economy of Nature*, 1749) he attempted to organize the aspect of natural history dealing with the balance of nature, but he realized that one must study not only ways that plants and animals interact but also their habitats. He knew that while balance had to exist, there occurred over time a succession of plants, beginning with a bare field and ending with a forest. In *Politia Naturae* (1760; *Governing Nature*, 1760) he discussed the checks on populations that prevent some species from becoming so numerous that they eliminate others. He noticed the competition among different species of plants in a meadow and concluded that feeding insects kept them in check. French naturalist Comte de Buffon developed a dynamical perspective on the balance of nature from his studies of rodents and their predators. Rodents can increase in numbers to plague proportions, but then predators and climate reduce their numbers. Buffon also suspected that humans had exterminated some large mammals, such as mammoths and mastodons.

However, a later Frenchman, Jean-Baptiste Lamarck, published a book on evolution called *Philosophie zoologique* (1809; *Zoological Philosophy*, 1914), which cast doubt on extinction by arguing that fossils represent only early forms of living species: Mammoths and mastodons evolved into African and Indian elephants. In developing this idea, he minimized the importance of competition in nature. An English opponent, the geologist Charles Lyell, argued in 1833 that species do become extinct, primarily because of competition among species. Charles Darwin was inspired by his own investigations during a long voyage around the world and by his reading of the works of Linnaeus and Lyell. In his revolutionary book *On the Origin of Species by Means of Natural Selection: Or, The Preservation of Favoured Races in the Struggle for Life* (1859), Darwin argued an intermediate position between those of Lamarck and Lyell: Species do evolve into different species, but in the process, some species do indeed become extinct.

Darwin's theory of evolution might have brought an end to the balance-of-nature concept, but it did not. Instead, American zoologist Stephen A. Forbes developed an evolutionary concept of the balance of nature in his essay "The Lake as a Microcosm" (1887), in which he observes that although the reproductive rate of aquatic species is enormous and the struggle for existence among them is severe, "the little community secluded here is as prosperous as if its state were one of profound and perpetual peace." Forbes emphasized the stabilizing effects of natural selection.

Ecology
The science of ecology became formally organized in the period from the 1890s through the 1910s. One of its important organizing concepts was that of "biotic communities." Frederic E. Clements, an American plant ecologist, wrote a large monograph titled *Plant Succession* (1916), in which he drew a morphological and developmental analogy between organisms and plant communities. Both the individual and the

community have a life history during which each changes its anatomy and physiology. This super-organismic concept was an extreme version of the balance of nature that seemed plausible as long as one believed that a biotic community was a real entity rather than a convenient approximation of what one sees in a pond, a meadow, or a forest. However, the studies of Henry A. Gleason in 1917 and later indicated that plant species merely compete with one another in similar environments; he concluded that Clements's superorganism was poetry, not science.

While the balance-of-nature concept was giving way to ecological hypotheses and theories, Rachel Carson decided that she could not argue her case in *Silent Spring* (1962) without it. She admitted, "The balance of nature is not a *status quo*; it is fluid, ever shifting, in a constant state of adjustment." Nevertheless, for Carson the concept represented a healthy environment, which humans could upset. Her usage of the phrase has persisted within the environmental movement.

In 1972 English medical chemist James Lovelock developed a new balance-of-nature idea, which he called Gaia, named for a Greek earth goddess. His reasoning owed virtually nothing to previous balance-of-nature notions, which focused on the interactions of plants and animals. His concept emphasized the chemical cycles that flow from the earth to the waters, atmosphere, and living organisms. Lovelock soon had the assistance of a zoologist named Lynn Margulis, and the studies they conducted together convinced them that biogeochemical cycles are not random; rather, they exhibit homeostasis, just as some animals exhibit homeostasis in body heat and blood concentrations of various substances. Lovelock and Margulis argued that living beings, rather than inanimate forces, mainly control the earth's environment. In 1988 three scientific organizations sponsored a conference to evaluate their ideas; the conference was attended by 150 scientists from all over the world. Although science more or less understands how homeostasis works when a brain within an animal controls it, no one has succeeded in satisfactorily explaining how homeostasis can work in a world "system" that lacks a brain. The Gaia hypothesis is as untestable as were earlier balance-of-nature concepts.

Frank N. Egerton

Further Reading

Bauer, Susan Wise. *The Story of Western Science: From the Writings of Aristotle to the Big Bang Theory*. New York: W. W. Norton & Company, 2015.

Egerton, Frank N. "Changing Concepts of the Balance of Nature." *Quarterly Review of Biology* 48 (June 1973): 322-350.

_____. "The History and Present Entanglements of Some General Ecological Perspectives." In *Humans as Components of Ecosystems*, edited by Mark J. McDonnell and S. T. A. Pickett. New York: Springer, 1993.

Judd, Richard W. *The Untilled Garden: Natural History and the Spirit of Conservation in America, 1740-1840*. New York: Cambridge University Press, 2009.

Kirchner, James W. "The Gaia Hypotheses: Are They Testable? Are They Useful?" In *Scientists on Gaia*, edited by Stephen H. Schneider and Penelope J. Boston. Cambridge, Mass.: MIT Press, 1991.

Kricher, John C. *The Balance of Nature: Ecology's Enduring Myth*. Princeton, N.J.: Princeton University Press, 2009.

Milne, Lorus J., and Margery Milne. *The Balance of Nature*. 1960. Reprint. New York: Alfred A. Knopf, 1970.

Egerton, Frank N. "Changing Concepts of the Balance of Nature." *Quarterly Review of Biology* 48 (June 1973): 322-350.

Judd, Richard W. *The Untilled Garden: Natural History and the Spirit of Conservation in America, 1740-1840*. New York: Cambridge University Press, 2009.

Kirchner, James W. "The Gaia Hypotheses: Are They Testable? Are They Useful?" In *Scientists on Gaia*, edited by Stephen H. Schneider and Penelope J. Boston. Cambridge, Mass.: MIT Press, 1991.

Kricher, John C. *The Balance of Nature: Ecology's Enduring Myth*. Princeton, N.J.: Princeton University Press, 2009.

Milne, Lorus J., and Margery Milne. *The Balance of Nature*. New York: Alfred A. Knopf, 1970.

Beach erosion and sand mining

FIELDS OF STUDY
Conservation Biology; Ecology; Environment; Environmental Engineering; Environmental Science; Industrial Destruction; Oceanography

SUMMARY
The world faces a global sand crisis. Oceanographers calculate that 70 percent of the world's beaches are being cut back by erosion. Beaches worldwide are threatened by rises in sea level and industrial sand mining. Beaches are accumulations of loose sedimentary material found along the shorelines of bodies of water. They extend landward from the low-tide level to a place where there is a marked change in appearance or topography of the land surface, such as a cliff, a row of sand dunes, or human-made structures such as seawalls.

PRINCIPAL TERMS
- **beach:** sandy or rocky area on the shore of bodies of water
- **erosion:** the gradual destruction and removal of rock or soil in a particular area by rivers, the sea, or the weather
- **sand mining:** the extraction of sand from beaches or open pits for industrial purposes
- **seawalls:** a wall or embankment to protect the shore from erosion or to act as a breakwater
- **wrack:** seaweed or other vegetation cast on the shore

Beaches are accumulations of loose sedimentary material found along the shorelines of bodies of water. They extend landward from the low-tide level to a place where there is a marked change in appearance or topography of the land surface, such as a cliff, a row of sand dunes, or human-made structures such as seawalls. The two major subdivisions of a beach are the backshore, which extends seaward from the foot of a cliff, dunes, or seawall; and the foreshore, which usually has a steeper slope that continues down to water level. The boundary between the backshore and the foreshore is generally marked by a change in slope or by a small scarp excavated by wave activity. The seaweed and other debris that accumulate at this point are known as the wrack.

The backshore is the inactive part of a beach because there is little evidence for wave activity here. It is the part of a beach most used by humans.

The foreshore is the active part of the beach. It is covered daily by the rise and fall of the tides; no grasses or trailing vines are found here because the waves and tides would sweep them away immediately. All traces of human activity, such as footprints and lost toys, are removed by the waves as well. Upon going out, the tide leaves the surface of this part of the beach perfectly smooth, broken only at its lower end by a slight dip known as the low-tide step. This is a gentle trough excavated by the breakers at low tide. Waves grind up shells and pebbles here in a so-called wave mill, and the resulting fragments become new materials for the beach. They grow finer-grained as the wind and waves carry them upward toward the dunes.

The widths of beaches vary greatly around the world. Beaches at the foot of cliffs or seawalls may be just a few feet wide, whereas those on low-relief coasts can extend seaward for many miles.

Accelerating Beach Erosion

Oceanographers calculate that 70 percent of the world's beaches are being cut back by erosion, 10 percent are growing forward, and the rest remain unchanged. Erosion of a beach occurs when it loses more sediment than it gains from the sources that feed it. Losses may take place when beach materials are carried inland during storm surges, when they are carried seaward as big waves cut back the coast during severe storms, when they are carried along the shoreline by the longshore current or when the sand is mined for various industrial purposes. The longshore current is present along coasts whenever waves strike the shoreline at an acute angle rather than coming in head-on.

During severe storms the loss of beach materials can be rapid; evidence for this is seen in collapsed seawalls and cliff dunes on the backshore or in salt marsh deposits and tree stumps uncovered on the foreshore as the beach moves inland over a previously vegetated area. Aerial photographs are generally used to determine how much material has been removed from beaches, with recent photographs compared to photographs taken previously.

Beach erosion accelerated in the early twenty-first

Beach erosion coastal erosion. By Hillebrand Steve, US Fish and Wildlife Service.

century because of several factors: the active removal of materials by waves, wind, currents, rising sea level, which resulted in the migration of shorelines inland, increasing human modification of the coasts, including the mining of sand for industry. Human changes to coastlines that affect beach erosion include both the construction of seawalls, groins, offshore breakwaters, and jetties and the bringing in of new sand in a procedure known as beach nourishment. Seawalls are built at the backs of beaches to prevent erosion, but they may reflect the storm waves, causing them to carry sediment seaward. Groins, which are rock ribs extending seaward along the foreshore to trap sand on their up-drift sides, cause erosion on their down-drift sides, resulting in alternating scallops of erosion and deposition along the coast. Breakwaters, which are offshore structures built parallel to the coast, and jetties, which are walls extending seaward to stabilize inlets, both trap migrating sand, resulting in erosion on their down-drift sides. Even beach nourishment, a highly popular practice, has its negative effects. Because the sand that is added to beaches is dredged from offshore or trucked in from the land, it rarely has the same texture and composition of the original beach sand. The result is that this new sand usually erodes faster than the original sand that it replaces.

Sand Mining

Sand mining is the extraction of sand for industrial purposes, mainly through open pits, but also mined from beaches and inland dunes or dredged from oceans and riverbeds.

Little has been scientifically studied about this problem—considered a "hidden" global crisis—yet scientists are making a great effort to quantify how infrastructure systems such as roads and building affect the habitats that surround them, the impact of extracting construction materials such as sand from beaches to build those structures have, to date, been largely overlooked.

The call for sand, especially for construction, is voracious. The global urbanization boom devours huge amounts of sand, a key ingredient of concrete and asphalt for buildings and roads, as well as glass and electronics. Sand is also required in massive amounts for land reclamation projects, shale gas extraction and beach nourishment projects, In 2010, nations mined about 11 billion tons of sand for construction. In the United States alone, production and use of construction sand and gravel was valued at $8.9 billion in 2016. Most of the mining takes place in Texas and Illinoi. In Wisconsin, Minnesota, Illinois, Indiana, Iowa, and Florida, sand mining is used in hydrolytic fracturing, known as fracking. Because of sand mining, a beach just north of Monterey, California, has been losing eight acres a year of shore.

China, as of 2018, has used more cement than the United States exploited in the entire 20th century. Sand mining in Sierra Leone has contributed to the country's coastal erosion, which proceeds at approximately 6 meters (6.56 yards) a year.

Damages to waterways worldwide have caused profound environmental crises, not only erosion, but including the destruction of habitats of bottom-dwelling creatures and organisms. Churned sediment clouds the water, suffocating fish and blocking sunlight sustaining underwater vegetation. Disrupted riparian ecosystems throughout India have had fatal consequences for fish and other aquatic organisms, as well as for many bird species. Sand mining operations affect numerous animal species everywhere it is practiced, including fish, dolphins, crustaceans and crocodiles.

According to various government agencies, uneven record keeping in many countries hide real extraction rates. Official statistics widely underreport sand use and typically do not include non-construction purposes such as hydraulic fracturing and beach nourishment.

The supply of sand that can be mined sustainably is finite and its use is another example of unsustainable development and causes of beach erosion.

Donald W. Lovejoy

Further Reading

Bird, Eric. *Coastal Geomorphology: An Introduction.* 2d ed. Hoboken, N.J.: John Wiley & Sons, 2008.

Bird, Eric, and Nick Lewis. "Responses to Beach Erosion." *Beach Renourishment.* Heidelberg: Springer International, 2015. 29–39.

Davis, Richard A., Jr., and Duncan M. FitzGerald. *Beaches and Coasts.* Malden, Mass.: Blackwell, 2004.

Neal, William J., Orrin H. Pilkey, and Joseph T. Kelley. *Atlantic Coastal Beaches: A Guide to the Ripples, Dunes, and Other Natural Features of the Seashore.* Missoula, Mont.: Mountain Press, 2007.

Padmala, D. and K. Maya. *Sand Mining: Environmental Impacts and Selected Case Studies.* New York: Springer Publishing Company, 2014.

Pearson, Thomas W. *When the Hills Are Gone: Frac Sand Mining and the Struggle for Community.* Minneapolis, Minn.: University of Minnesota Press, 2017.

Pilkey, Orrin H., and J. Andrew G. Cooper. *The Last Beach.* Durham, N.C.: Duke University Press, 2014.

Websites

Al Jazeera
https://www.aljazeera.com/video/news/2017/06/farmers-sand-mining-destroying-environment-170626083142097.html

United States Geological Survey
http://coastal.er.usgs.gov/hurricanes/coastal-change/beach-erosion.php

Bees and other pollinators

FIELDS OF STUDY
Biology; Conservation Biology; Ecology; Entomology; Environment; Melittology

SUMMARY
Pollinating species are in decline worldwide, a trend that threatens biodiversity and food supplies. By the early twenty-first century, declines in pollinator populations around the world led environmental groups and government agencies to include pollinators in their conservation efforts. The introduction of killer bees was a failed experiment to produce a hybrid honeybee and an example of the potential for negative environmental impacts when nonnative organisms are released into the wild.

PRINCIPAL TERMS
- **biodiversity:** biological diversity in an environment as indicated by numbers of different species of plants and animals, including humans
- **colony collapse disorder (CCD):** the mass disappearance of worker honey bees from the hive resulting in a breakdown of the colony and insufficient workers to maintain it
- **ecosystem:** a system or group of interconnected elements formed by the interaction of a community of organisms with their environment
- **habitat:** the place or environment where a plant or animal naturally and normally lives and grows
- **killer bees:** aggressive genetic strain of honeybees
- **monoculture:** the practice of growing only one crop or keeping only one type of animal on an area of farm land
- **pollinators:** flying insects and other biological agents that facilitate plant reproduction by moving pollen among seed plants
- **symbiosis:** the living together relatively intimate association or close union of two dissimilar organisms

Bees are the best-known group of pollinators; however, other insects such as butterflies, wasps, beetles, and ants are pollinators, too, along with some species of bats and birds. Plants and their pollinators coevolved, forming symbiotic relationships in which plants provide food for the pollinators, which in turn aid the plants' reproduction. Many other species evolved over time to take advantage of this fundamental relationship between plant and pollinator, including humans and other animals that eat the results of the fertilized plants, such as nuts, seeds, and berries. Pollinators also contribute to the web of life by ensuring the continued survival of plants through the process of pollination; the loss of pollinators can affect the biodiversity and well-being of a whole ecosystem. By the early twenty-first century, declines in pollinator populations around the world led environmental groups and government agencies to include pollinators in their conservation efforts.

Bee. By Mark Winterbourne from Leeds, West Yorkshire, United Kingdom

Pollinator Decline

The decline of pollinating species has many causes, all for the most part related to human activity. One cause is the use (and misuse) of pesticides meant to kill other insects, such as mosquitoes; such pesticides also indiscriminately kill off other insect populations, many of them pollinator species. Another cause of pollinator decline is the large-scale transport of pollinators around the world for agricultural purposes; this practice also results in the large-scale transport of parasites and diseases that affect them, as well as the introduction of invasive species that can compete with or destroy native pollinator populations.

Human expansion into previously wild or relatively untouched spaces contributes to pollinator decline through habitat loss and degradation, as development brings with it pollution, hive destruction, and fragmentation of traditional "nectar corridors." Wild pollinators have fewer places to nest, mate, and roost safely, and their historic ranges have been destroyed or partitioned in such a way that it is far more difficult for them to find the nectar they need to survive. Many pollinators are already on endangered or threatened species lists, including more than one hundred avian species and dozens of mammals.

Climate change has also affected pollinators. A 2015 study found that warming temperatures had caused bumblebees' range to shrink by as much as 190 miles since the 1970s, with the bees staying farther and farther north as the climates in the southern parts of Europe and North America became less hospitable.

The Importance of Bees

The thousands of species of bees in the world are widely known for pollination and for their making of honey and beeswax. Bees are the primary pollinators of flowers around the world and are also responsible for the pollination of approximately one-third of the human food supply. Many agricultural crops, ranging from watermelons to cashews, rely on pollination by bees. European honeybees are extremely important to agriculture because of humans' ability to manage the species.

Pollination management, also called contract pollination, entails transporting honeybees in their hives to crops that are in need of pollination. Modern monoculture farming has created a situation in which the normal pollinators of some kinds of crops cannot pollinate them, either because of the huge size of the fields or because the pollinators have declined or died off. Honeybees, however, can be put in the fields in such large numbers that they make up for the fact that they are often less efficient pollinators than the native species that previously would have pollinated a particular crop. The economic viability of monoculture agriculture often depends entirely on pollination management, and most professional beekeepers in the twenty-first century focus on contract pollination rather than honey as their primary source of revenue.

The same factors that have caused declines in other pollinator populations since the late twentieth century have affected honeybees as well, causing a widespread honeybee population decline. In addition, across North America and Western Europe since at least 2006, honeybees have disappeared in what is called honeybee colony collapse disorder (CCD), the cause of which remains a mystery. CCD is characterized by the worker bees flying off and never returning to the hive or colony, leaving behind a healthy queen, an unhatched brood, and food supplies.

Evidence from around the world points to falling and increasingly unpredictable yields of insect-pollinated crops, particularly in the areas with the especially intensive farming. The most dramatic example comes from the apple and pear orchards of south west China, where wild bees have been eradicated by excessive pesticide use and a lack of natural habitat. In recent years, farmers have been forced to hand-pollinate their trees, carrying pots of pollen and paintbrushes with which to individually pollinate every flower, and using their children to climb up to the highest blossoms.

Given the importance of honeybees to agriculture around the world, scientists have investigated many possible causes for CCD; possibilities include a parasitic mite, insect diseases, and the use or misuse of insecticides. Some researchers have reported evidence that it may be viral. These possibilities are aggravated by other factors negatively affecting honeybee populations, such as poor diet related to monoculture, which can lead to malnutrition or impaired immune system, and migratory beekeeping, which spreads parasites and diseases among populations. One such disease, known as deformed wing virus after its chief symptom, became a serious threat to honeybee and bumblebee populations in the 2010s; a 2016 study found that the trade in pollinators was directly responsible for spreading the virus around the world.

Introduction and impact of Killer Bees

Killer bees—or, more properly, Africanized honeybees—are a genetic strain of honeybees introduced to Brazil in 1957 to hybridize the familiar European honeybee with an African subspecies. Although killer bees are no more venomous than ordinary honeybees, they exhibit lower average honey production, greater dispersal, and more aggressive behavior than their European counterparts. A person who disturbs a hive of Africanized honeybees may be pursued for distances of more than 90 meters (300 feet) by thousands of bees. Each bee stings only once, but an aggravated swarm can inflict a lethal number of stings. While the popular media have exaggerated their danger, killer bees can be as deadly as the name implies. They have caused hundreds of human casualties since 1957. Most of these victims were trapped or otherwise unable to run away from the bees. In South and Central America, wild bee colonies under bridges or near farm machinery are common sources of attacks. In the United States, where most honeybees are managed in hives, the Africanized bee problem has been less severe.

The problem of killer bees—the unintended consequence of a failed experiment to produce a hybrid honeybee—is an example of the potential for negative environmental impacts when nonnative organisms are released into the wild.

Eradication of the Africanized bee is unlikely, but steps can be taken to limit its impact. Beekeepers have responded by changing management practices and by maintaining tame varieties for breeding stock. Biologists continue to investigate honeybee genetics, and the public is learning to exercise caution around wild bee colonies.

Robert W. Kingsolver and Megan E. Watson

Further Reading

"Pollinators." *US Fish & Wildlife Service*. Dept. of the Interior, 1 Mar. 2016. Web. 2 Mar. 2016.

Buchmann, Stephen L., and Gary Paul Nabhan. *The Forgotten Pollinators*. Washington, DC: Island, 1996.

Buchmann, Stephen, with Banning Repplier. *Letters from the Hive: An Intimate History of Bees, Honey, and Humankind*. New York: Bantam Books, 2005.

Carl, A. Johansen and F. Mayer Daniel. *Pollinator Protections: A Bee and Pesticide Handbook*. Kalamazoo, Mich.: Wicwas Press, 2015.

Caron, Dewey M. *Africanized Honey Bees in the Americas*. Medina, Ohio: A. I. Root, 2001.

Carswell, Cally. "Bumblebees Being Crushed by Climate Change." *Science News*. AAAS, 9 July 2015.

Fischer, David, and Thomas Moriarty. *Pesticide Risk Assessment for Pollinators*. Chichester: Wiley, 2014.

Grimm, V., et al. "Ecological Modeling for Pesticide Risk Assessment for Honey Bees and Other Pollinators." *Pesticide Risk Assessment for Pollinators* (2014).

James, Rosalind R., and Theresa L. Pitts-Singer, eds. *Bee Pollination in Agricultural Ecosystems.* New York: Oxford University Press, 2008.

Kennedy, Merrit. "Report: More Pollinator Species in Jeopardy, Threatening World Food Supply." *NPR.* NPR, 26 Feb. 2016.

Miller-Struttmann, Nicole E., et al. "Functional Mismatch in a Bumble Bee Pollination Mutualism under Climate Change." *Science* 25 Sept. 2015: 1541–544.

St. Fleur, Nicholas. "Climate Change Is Shrinking Where Bumblebees Range, Research Finds." *New York Times.* New York Times, 9 July 2015.

Waser, Nickolas M., and Jeff Ollerton, eds. *Plant-Pollinator Interactions: From Specialization to Generalization.* Chicago: University of Chicago Press, 2006.

Wilfert, Lena, et al. "Deformed Wing Virus Is a Recent Global Epidemic in Honeybees Driven by *Varroa* Mites." *Science* 351.6,273 (2016): 594–7.

■ Biocentrism

FIELDS OF STUDY
Biology; Ecology; Environment; Environmental Ethics; Ethics

SUMMARY
Biocentrism has played a key role in the development of environmental ethics since that discipline was founded during the 1970's and had already been influential among important earlier thinkers, including Albert Schweitzer and Mahatma Gandhi.

PRINCIPAL TERMS
- **anthropocentric:** regarding the human being as the central fact of the universe
- **biocentrism:** life-centered stance that rejects the view that only human beings and their interests matter, while recognizing the moral standing of all living creatures
- **conservation:** planned management of a natural resource to prevent exploitation, destruction, or neglect
- **deontological ethics:** theories that place special emphasis on the relationship between duty and the morality of human actions
- **ecocentrism:** philosophy or perspective that places intrinsic value on all living organisms and their natural environments, regardless of their perceived usefulness or importance to human beings
- **global warming:** increase in the earth's average atmospheric temperature causing corresponding changes in climate believed to result from the greenhouse effect.
- **social justice:** concept of fair and just relations between the individual and society measured by the explicit and tacit terms for the distribution of wealth, opportunities for personal activity, and social privileges

Biocentrists hold that all living creatures have goods of their own and that their flourishing, or attaining their goods, is intrinsically valuable. This value should thus be taken into consideration whenever decisions affecting the flourishing of any creatures are being made. During the 1970s, Norwegian philosopher Arne Naess wanted to include living systems (such as habitats and ecosystems) within the scope of biocentrism as intrinsically valuable entities, but this view is now classified as "ecocentrism," and subsequent biocentrists have restricted moral standing to individual living creatures.

The most widespread kind of biocentrism was presented during the early 1980s by Paul Taylor, who advocated a life-centered ethic of respect for nature. In this ethic, the realization of the good of every living creature is intrinsically valuable and to be pursued for its own sake. In Taylor's version of biocentrism, not only is human superiority denied but also all living things are held to be equally worthy of respect, irrespective of differences of capacities, and to have the same moral importance. Taylor recommends defensible policies of conservation and social justice, but he has difficulty deriving them from these principles.

Another kind of biocentrism, not committed to equal respect for all living creatures, was proposed in 1981 by Robin Attfield and later by Gary E. Varner, who both defend the intrinsic value of the good of every creature, whether sentient or nonsentient. Unlike Taylor, however, Attfield upholds Peter Singer's principle of equal consideration for equal interests, applying it to the entire realm of life; thus, the satisfaction of greater interests takes priority over that of lesser ones when they conflict. (The capacities of living creatures for health and for being harmed are held to distinguish them from artifacts, which have no goods of their own.) Attfield integrates this account of intrinsic value into a consequentialist recognition of beneficial practices, general compliance with which makes actions right through generating or preserving greater value than would otherwise prevail.

More recently, James Sterba has defended a Taylor-like commitment to species equality through species-neutral principles authorizing resort to self-defense in certain circumstances (whether greater capacities are at stake or not). These principles are held to be defensible not by enhancing value but as part of a deontological ethical system. Simultaneously Sterba rejects the extension of biocentrism to ecosystems, holding that biotic communities have no clear goods of their own; individual living creatures constitute the limits of ethical egalitarianism.

Biocentrism in any form supports more radical policies to preserve habitats and curtail global warming than do anthropocentric stances. The biocentric viewpoint remains influential in debates about both normative principles and practical decision making.

Robin Attfield

Further Reading

Attfield, Robin. *Environmental Ethics.* Malden, Mass.: Blackwell, 2003.

Attfield, Robin, ed. *The Ethics of the Environment.* Burlington, Vt.: Ashgate, 2008.

Sterba, James. "A Biocentrist Strikes Back." *Environmental Ethics* 20, no. 4 (Winter, 1998): 361-376.

Taylor, Paul. *Respect for Nature: A Theory of Environmental Ethics.* Princeton, N.J.: Princeton University Press, 1986.

Varner, Gary E. "Biocentric Individualism." David Schmidtz and Elizabeth Willott, eds. *Environmental Ethics: What Really Matters, What Really Works.* New York: Oxford University Press, 2002.

■ Biodiversity

FIELDS OF STUDY
Animals; Biology; Botany; Conservation Biology; Ecology; Ecosystems; Environment; Forestry; Forests; Horticulture; Life Sciences; Marine Biology; Plants; Vegetation; Zoology

SUMMARY
Biodiversity—including diversity among species, among ecosystems, and among individuals within species—is generally assumed to enhance ecological stability. The concept recognizes that life on earth exists in great variety and at various levels of organization.

PRINCIPAL TERMS
- **biodiversity:** biological variety in a given environment, also called biological diversity
- **biogeography:** the study of the geographical distribution of living things.
- **biota:** the flora and fauna of a region
- **ecosystem:** a system, or a group of interconnected elements, formed by the interaction of a community of organisms with their environment
- **genetic diversity:** the vast numbers of different species as well as the diversity within a species, whose greater genetic diversity expands that species' chances of long-term survival
- **habitat degradation:** decline in species-specific habitat quality leading to reduced survival and reproductive success, related to changes in food availability or climate
- **species diversity:** a measurement of biological diversity to be found in a specific ecological community

In 1993 the Wildlife Society defined biodiversity – or biological diversity – as "the richness, abundance, and variability of plant and animal species and communities and the ecological processes that link them with one another and with soil, air, and water." Included in this concept is the recognition that life on earth exists in great variety and at various levels of organization. Many kinds of specialists—including organismic biologists, population and evolutionary biologists, geneticists, and ecologists—investigate biological processes that are encompassed by the concept of biodiversity. Conservation biologists are concerned with the totality of biodiversity, including the process of speciation that forms new species, the measurement of biodiversity, and factors involved in the extinction process. The primary thrust of their efforts, however, is the development of strategies to preserve biodiversity.

The biodiversity paradigm connects classical taxonomic and morphological studies of organisms with modern techniques employed by those working at the molecular level. It is generally accepted that biodiversity can be approached at three levels of organization, commonly identified as species diversity, ecosystem diversity, and genetic diversity. Some also recognize biological phenomena diversity.

Levels of organization for biodiversity include genetic diversity; organismal diversity; population diversity; species diversity; community diversity; ecosystem diversity; landscape and seascape diversity; and biogeographic diversity.

Species Diversity

No one knows how many species inhabit the earth. Estimates range from five million to several times that number. Each species consists of individuals that are somewhat similar and capable of interbreeding with other members of their species but are not usually able to interbreed with individuals of other species. The species that occupy a particular ecosystem are a subset of the species as a whole. Ecosystems are generally considered to be local units of nature; ponds, forests, and prairies are common examples.

Conservation biologists measure the species diversity of a given ecosystem by first conducting a careful, quantitative inventory. From such data, scientists may determine the "richness" of the ecosystem, which is simply a reflection of the number of species present. An island with three hundred species would thus be considered to be 50 percent richer than another with only two hundred species. Some ecosystems, especially tropical rain forests and coral reefs, are much richer than others. Among the least rich are tundra regions and deserts.

A second aspect of species diversity is "evenness," defined as the degree to which all the various elements are present in similar percentages of the total species. As an example, consider two forests, each of which has a total of twenty species of trees. Suppose that the first forest has a few tree species represented by rather high percentages and the remainder by low percentages. A second forest with species more evenly distributed would rate higher on a scale of evenness.

Species diversity, therefore, is a value that combines both species richness and species evenness measures. Values obtained from a diversity index are used in comparing species diversity among ecosystems of both the same and different types. They also have implications for the preservation of ecosystems; other things being equal, it would be preferable to preserve ecosystems with high diversity indexes, thus protecting a larger number of species.

Considerable effort has been expended to predict species diversity as determined by the nature of the area involved. For example, island biogeography theory suggests that islands that are larger, are nearer to other islands or continents, and have more heterogeneous landscapes would be expected to have higher species diversity than those possessing alternate traits. Such predictions apply not only to literal islands but also to other discontinuous ecosystems; examples would be alpine tundra of isolated mountaintops and ponds several miles apart.

The application of island biogeography theory to designing nature preserves was proposed by Jared Diamond in 1975. His suggestion began the "single larger or several smaller," or SLOSS, area controversy. Although island biogeography theory would, in many instances, suggest selecting one large area for a nature preserve, it is often the case that several smaller areas, if carefully selected, could preserve more species.

The species diversity of a particular ecosystem is subject to change over time. Pollution, deforestation, and other types of habitat degradation invariably reduce diversity. Conversely, during the extended process of ecological succession that follows disturbances, species diversity typically increases until a permanent, climax ecosystem with a large index of diversity results. Ecologists generally assume that more diverse ecosystems are more stable than those with less diversity. Certainly, the more species present, the greater the opportunity for various interactions, both with other species and with the environment. Examples of interspecific reactions include mutualism, predation, and parasitism. Such interactions apparently help to integrate a community into a whole, thus increasing its stability.

Ecosystem Diversity

Ecology can be defined as the study of ecosystems. From a conservation standpoint, ecosystems are important because they sustain their particular assemblages of living species. Conservation biologists also consider ecosystems to have intrinsic value beyond the species they harbor; therefore, it would be ideal if representative global ecosystems could be preserved. This is far from realization, however. Just deciding where to draw the line between interfacing ecosystems can be a problem. For example, the water level of a stream running through a forest is subject to seasonal fluctuation, causing a transitional zone characterized by the biota (that is, the flora and fauna) from both adjoining ecosystems. Such ubiquitous zones negate the view that ecosystems are discrete units with easily recognized boundaries.

The protection of diverse ecosystems is of utmost importance to the maintenance of biodiversity, but ecosystems throughout the world are threatened by global warming, air and water pollution, acid deposition, ozone depletion, and other destructive forces. At the local level, deforestation, thermal pollution, urbanization, and poor agricultural practices are among the problems affecting ecosystems and therefore reducing biodiversity. Both global and local environmental problems are amplified by rapidly increasing world population pressures.

In the process of determining which ecosystems are most in need of protection, many scientists have recognized that the creation of a system for naming and classifying ecosystems is highly desirable, if not imperative. Efforts are being made to establish a system similar to the hierarchical system applied to species that was developed by Swedish botanist Carolus Linnaeus during the eighteenth century. A comprehensive classification system for ecosystems is far from complete, but freshwater, marine, and terrestrial ecosystems are recognized as main categories, with each further divided into particular types. Though tentative, the system that has been established thus far has made possible the identification and preservation of a wide range of representative threatened ecosystems.

In 1995 conservation biologist Reed F. Noss of Oregon State University and his colleagues identified more than 126 types of ecosystems in the United States that are threatened or critically endangered. The following list illustrates their diversity: southern Appalachian spruce-fir forests; eastern grasslands, savannas, and barrens; California native grasslands; Hawaiian dry forests; caves and karst systems; old-growth forests of the Pacific Northwest; and southern forested wetlands.

Not all ecosystems can be saved. Establishing priorities involves many considerations, some of which are economic and political. Ideally, choices would be made on merit, taking into consideration the rarity, size, and number of endangered species that ecosystems include as well as other objective, scientific criteria.

Genetic and Biological Phenomena Diversity

Most of the variations among individuals of the same species are caused by the different genotypes (combinations of genes) the individuals possess. Such genetic diversity is readily apparent in cultivated or domesticated species such as cats, dogs, and corn, but it also exists, though usually to a lesser degree, in wild species. Genetic diversity can be measured only using exacting molecular laboratory procedures. Such tests detect the amount of variation in the deoxyribonucleic acid (DNA) or isoenzymes (chemically distinct enzymes) possessed by various individuals of the species in question.

A significant degree of genetic diversity within a population or species confers a great advantage. This diversity is the raw material that allows evolutionary processes to occur. When a local population becomes too small, it is subject to a serious decline in vigor from increased inbreeding. This leads in turn, to a downward, self-perpetuating spiral in genetic diversity and further reduction in population size. Extinction may be imminent. In the grand scheme of nature, this is a catastrophic event; never again will that particular genome (set of genes) exist anywhere on the earth. Extinction is the process by which global biodiversity is reduced.

The term "biological phenomena diversity" refers to the numerous unique biological events that occur in natural areas throughout the world. Examples include the congregation of thousands of monarch butterflies on tree limbs at Point Pelee in Ontario, Canada, as they await favorable conditions before continuing their migration, and the return of hundreds of loggerhead sea turtles each April to Padre Island in the Gulf of Mexico in order to lay their eggs.

Although biologists have been concerned with protecting plant and animal species for decades, only recently has conservation biology emerged as an identifiable discipline. Conceived in a perceived crisis of biological extinctions, conservation biology differs from related disciplines, such as ecology, in that it is advocative in nature, insisting that the maintenance of biodiversity is intrinsically good. Conservation biology is a value-laden science, and some critics consider it akin to a religion with an accepted dogma.

The prospect of preserving global ecosystems and the life processes they make possible, all necessary for maintaining global diversity, is not promising. Western culture does not give environmental concerns a high priority. Among those who do, greater concern is often expressed about the immediate health effects of environmental degradation than about the loss of biodiversity. The impetus necessary to save ecosystems and all their inhabitants—including humans—will develop only when education in basic biology and ecology at all levels is extended to include an awareness of the importance of biodiversity.

Why Preserve Biodiversity?

Biodiversity is defined as the total number of species within an ecosystem, and also as the resulting complexity of interactions among them. It measures the "richness" of an ecological community. Among an estimated 8 to 10 million unique and irreplaceable species existing on Earth, fewer than 1.6 million have been named. A tiny fraction of this number has been studied. Over thousands of years, organisms in a community have been molded by forces of natural selection exerted by other living species as well as by the nonliving environment that surrounds them. The result is a highly complex web of interdependent species whose interactions sustain one another and provide the basis for the very existence of human life as well.

Loss of biodiversity poses a serious challenge to the sustenance of many communities and ecosystems. For example, the destruction of tropical rain forests by clear-cut logging produces high rates of extinction of many species. Most of these species have never been named, and many never even discovered. As species are eliminated, the communities of which they were a part may change and become unstable and more vulnerable to damage by diseases or adverse environmental conditions. Aside from the disruption of natural food webs, potential sources of medicine, food, and raw materials for industry are also lost. As Harvard professor Edward O. Wilson once said, "The loss of species is the folly our descendants are least likely to forgive us."

Thomas E. Hemmerly

Further Reading

Chivian, Eric, and Aaron Bernstein, eds. *Sustaining Life: How Human Health Depends on Biodiversity.* New York: Oxford University Press, 2008.

Ehrlich, Paul R., and Anne H. Ehrlich. *Extinction: The Causes and Consequences of the Disappearance of Species.* New York: Ballantine Books, 1981.

Hunter, Malcolm L., Jr. *Fundamentals of Conservation Biology.* 3d ed. Hoboken, N.J.: Wiley-Blackwell, 2006.

Novacek, Michael J., ed. *The Biodiversity Crisis: Losing What Counts.* New York: New Press, 2001.

Primack, Richard B. *Essentials of Conservation Biology.* 5th ed. Sunderland, Mass.: Sinauer Associates, 2010.

Ray, Justina C., et al., eds. *Large Carnivores and the Conservation of Biodiversity.* Washington, D.C.: Island Press, 2005.

Shiva, Vandana. *Monocultures of the Mind: Perspectives on Biodiversity and Biotechnology.* 1993: Penang, Malaysia: Third World Network.

Wilson, Edward O. *Biophilia.* 1984. Reprint. Cambridge, Mass.: Harvard University Press, 2003.

_____. *The Diversity of Life.* New ed. New York: W. W. Norton, 1999.

Zeigler, David. *Understanding Biodiversity.* Westport, Conn.: Praeger, 2007.

■ Biodiversity action plans

FIELDS OF STUDY
Agreements; Animals; Biology; Ecology; Ecosystems; Environment; Environmentalism; Government; International Relations; Life Sciences; Marine Biology; Politics; Treaties; Zoology

PRINCIPAL TERMS
- **Biodiversity Action Plan (BAP):** an internationally recognized program addressing threatened species and habitats and is designed to protect and restore biological systems

- **biodiversity:** biological variety in an environment
- **Convention on Biological Diversity (CBD):** a multilateral treaty with three main goals including: the conservation of biological diversity (or biodiversity); the sustainable use of its components; and the fair and equitable sharing of benefits arising from genetic resources
- **vulnerable species:** any categorized by the International Union for Conservation of Nature as likely to become endangered unless the circumstances that are threatening its survival and reproduction improve

A Biodiversity Action Plan habitat in the United Kingdom. By Hugh Venables.

SUMMARY

Biodiversity action plans are important tools that governments use in protecting and restoring threatened ecosystems. These plans seek to implement the 1992 United Nations Convention on Biological Diversity, which demonstrates worldwide recognition that biodiversity is intrinsically valuable and deserves to be protected or restored through careful planning.

The schemes for protecting and restoring threatened species of plants and animals and their ecosystems known as biodiversity action plans (BAPs; also known as national biodiversity strategies and action plans, or NBSAPs) gained worldwide recognition with the signing of the United Nations Convention on Biological Diversity (CBD) at the Earth Summit in Rio de Janeiro, Brazil, in 1992. BAPs represent the steps that nations are taking to implement the provisions of the convention. However, although most of the world's nations have ratified the CBD, only a handful have developed substantive BAPs. Among those that have are the United States, Australia, the United Kingdom, Tanzania, and Uzbekistan.

A well-conceived BAP includes several components: plans for the carrying out of inventories and documentation of selected species and specific habitats, with particular emphasis placed on population distribution and conservation status within certain ecosystems; realistic targets or indicators for conservation and restoration; plans covering funding and time lines for achievement of specific goals; and plans for the establishment of partnerships among private and public institutions and agencies that will work together to achieve the goals set.

Obstacles and Criticisms

To implement a BAP effectively, a nation must overcome a number of obstacles; the difficulty of the process may explain in part why so few countries have attempted to develop such plans. In some parts of the world, for example, undertaking complete inventories of plant and animal species is not realistic. Scientists have estimated that only about 10 percent of the world's species have been characterized and documented; most of those still unknown include plants and lower animals such as insects. An ideal BAP includes the assessment of species population estimates over time so that the variability and degree of vulnerability of species can be determined; it also includes descriptions of species' ranges, habitats, behaviors, breeding practices, and interactions with other species. The collection of such fundamental information can be a daunting task. Another factor preventing some nations from developing BAPs is the cost involved. Depending on the size of the country, the cost of preparing a solid BAP can easily come to the equivalent of millions of U.S. dollars, with about 10 percent of the initial cost factored in for annual maintenance. It is therefore not surprising that the call for BAPs has been criticized by some developing countries.

In addition to the difficulty and expense involved in the implementation of BAPs, many developing countries are unwilling to create such plans because, they argue, the plans obviously favor the consideration of wildlife and plant protection over food production and industrial growth; in some cases, BAPs may even represent impediments to population growth. Most of the Middle Eastern countries and many African nations have shown little interest in participating in a substantive way in such plans. Others have simply opted to create pro forma plans that expend little on research and even less on the management of natural resources. In contrast, the European Union has chosen to divert the purpose of BAPs, instead implementing the CBD through a set of economic development policies while paying special attention to the protection of certain ecosystems.

It has become increasingly clear that what is at stake is the very definition of "biodiversity" itself. According to the CBD, biodiversity is the variation of life-forms within a given ecosystem; it is a combination of ecosystem structure and function as well as components (species, habitat, and genetic resources). The CBD states:

> In addressing the boundless complexity of biological diversity, it has become conventional to think in hierarchical terms, from the genetic material within individual cells, building through individual organisms, populations, species, and communities of species, to biosphere overall. . . . At the same time, in seeking to make management interventions as efficient as possible, it is essential to take a holistic view of biodiversity and address the interactions that species have with each other and their nonliving environment, i.e., to work from an ecological perspective.

At the World Summit on Sustainable Development in 2002, delegates adopted the objectives of the CBD and designated 2010 the Year of Biodiversity.

Nader N. Chokr

Further Reading

Chivian, Eric, and Aaron Bernstein, eds. *Sustaining Life: How Human Health Depends on Biodiversity.* New York: Oxford University Press, 2008.

O'Riordan, Tim, and Susanne Stoll-Kleemann, eds. *Biodiversity, Sustainability, and Human Communities: Protecting Beyond the Protected.* New York: Cambridge University Press, 2002.

Ray, Justina C., et al., eds. *Large Carnivores and the Conservation of Biodiversity.* Washington, D.C.: Island Press, 2005.

Wilson, Edward O. *The Diversity of Life.* New ed. New York: W. W. Norton, 1999.

Zeigler, David. *Understanding Biodiversity.* Westport, Conn.: Praeger, 2007.

■ Biogeography

FIELDS OF STUDY

Biology; Conservation; Environment; Evolutionary Theory; Geography; Paleontology; Statistics; Systems Ecology

SUMMARY

Biogeography is an interdisciplinary science that applies concepts and methodologies from biology, geography, systems ecology, evolutionary theory, conservation, statistics, and paleontology, among other fields, to study the features of the physical environment that affect the distribution of living organisms across space and time.

PRINCIPAL TERMS

- **alien:** a species of plant or animal not native but found to have successfully colonized, especially one that has been introduced as a result of human activity
- **allopatric:** of animals or plants occupying separate, non-overlapping distribution areas; of speciation, occurring as a result of such geographic separation
- **colonization:** long-term establishment of a successfully reproducing population of a plant or animal species in a habitat where it was not previously found
- **cosmopolitan:** animals or plants occupying distribution areas that span the entire or nearly entire globe

- **disjunct:** animals or plants with fragmented distribution areas including two or more geographically separate regions
- **ecogeographic rule:** principle describing correlations between the morphology of birds and mammals and the climatic or latitudinal gradients at which they are found
- **latitudinal gradient:** an increase in overall numbers of plants or animals found when moving from higher to lower latitudes
- **speciation:** the process by which new species arise or populations of the same species become reproductively isolated from each other, either by geographical separation or some other mechanism

The fundamental question that drives biogeographical study is: How is life distributed over the earth? This is a question that can be answered at many different levels. For example, any given taxon, or group of living organisms at a level of classification (such as a species, genus, family, or class), has a geographic range within which it is likely to be found. The plant *Yucca brevifolia*, a species of yucca also known as the Joshua tree, is only found growing in the dry desert soils, open plains, and mesas of the southwestern United States. This is what biogeographers refer to as the tree's distribution area. A species like *Yucca brevifolia* is said to have an endemic distribution area, meaning one that is restricted to a specific region of the world. In contrast, *Drosophila melanogaster* (the common fruit fly), which is found in virtually all types of habitats across almost the entire globe, is said to have a cosmopolitan distribution area.

Many taxons have what are known as disjunct distribution areas; this occurs when closely related species are found in widely geographically separated habitats. For example, many similar species of mosses, small mammals, and butterflies are found both in the mountain ranges of the far northerly Arctic, and in the more southerly mountain ranges of Europe, such as the Alps—but nowhere in between. There are several scientific hypotheses regarding historical reasons for the existence of disjunct biogeographic distribution areas.

Under the vicariance hypothesis, environmental events cause the range of a taxon to split into parts. For example, changes in the distribution of land can cause impassable fractures to form in what was once a continuous habitat. Plate tectonic shifts, continental drift, and the formation of mountains all have the potential to break apart a formerly intact habitat region. When this happens, allopatric speciation can occur. This is the evolution of new and distinct species of animals or plants as a result of a single species being separated into more than one non-overlapping distribution area. (Another form of vicariance occurs when individual members of a family that was once spread out over a large, continuous geographic range, go extinct. This, too, can cause fragmentation in the distribution area.) In contrast, some disjunct biogeographic distributions are believed to have been caused by dispersal, or the successful migration of species across an existing geographic barrier. For example, birds are capable of crossing oceans in order to colonize new habitats; other non-motile species, such as plants, may be transported along with them as seeds.

Besides studying how organisms range across the globe, biogeographers are interested in examining how individual organisms within a species are spatially arranged within the area where they are found. This is a concept known as species distribution. Three major types of species distribution are identified: clumped, random, and uniform. In a clumped species distribution, multiple groups of individuals tend to form; this is a common behavioral pattern found, for instance, in herding animals like deer or cattle. For such species, a clumped distribution is beneficial because it offers protection from predators. In a random species distribution, individuals are found in scattered locations with no organization or method—such as wildflowers whose seeds have been dispersed by the wind over a field, and which grow wherever they fall. In a uniform species distribution, individuals are spread out at a roughly even distance from each other. Many desert plants, for instance, are found in a uniform distribution because this is the pattern that enables each of them to absorb the most water from the arid environment.

Ecogeographic Rules

When biogeographic principles are used to look at the distribution of living organisms on a macroscopic scale, many striking principles emerge regarding the relative degree of biodiversity that can be found in different regions of the earth. One of the most significant of these "ecogeographic rules" is the latitudinal gradient in species diversity, or species richness.

This term describes a striking, though as yet not fully understood pattern in global biodiversity: the tremendous increase in the number of unique species that can be found when traveling from high latitudes (colder regions) to low latitudes (warmer regions). In other words, the equatorial tropics are home to a far greater diversity of life than the poles. This pattern applies across nearly every possible form of life, including birds, mammals, freshwater fish, marine invertebrates, insects, and flowering plants and trees. For example, more than two hundred different species of ants are known in Brazil; Alaska has fewer than ten. The 50,000 square kilometers (approximately 19,305 square miles) of land in the Malaysian peninsular is home to nearly 1,500 different species of trees and shrubs; the same area in the southeastern part of Russia holds fewer than 50 different species.

Paleobiogeographic data—data derived from analyzing the relationships between fossilized animals and plants and the locations upon the earth where those fossils were found—indicates that species richness latitude gradients are not a recent phenomenon. It has been shown, for instance, that during the Triassic (a period spanning about 245 million to 208 million years ago), ammonoids (a now-extinct group of mollusks) were more diverse at lower latitudes than at higher ones, and that the same was true of brachiopods, another group of marine invertebrates, during the Paleozoic ice age (about 570 million to 245 million years ago).

Dozens of competing hypotheses have been put forth regarding the potential mechanism or mechanisms responsible for the existence of species richness latitude gradients. Among them are the time and area hypothesis, the diversification rate hypothesis, and the hypothesis of effective evolutionary time. The time and area hypotheses argue that the tropical regions of the earth are both older and larger than the polar regions, giving them a longer span of evolutionary time in which speciation could have occurred and a greater amount of available habitat—both of which could have resulted in the higher degree of species richness that is observed at lower latitudes. (The time and area hypothesis does not account for certain anomalies, such as why the South East Asian sea—which is not as large as other bodies of water in the Indo-Pacific region—has the greatest degree of species richness.) The diversification rate hypothesis argues that speciation occurs faster and more frequently in tropical regions than in polar regions, either because of lower extinction rates or other factors, such as the increased likelihood of achieving reproductive isolation. The hypothesis of effective evolutionary time takes into consideration the longer amount of time ecosystems are believed to have existed in the tropics, but also gives weight to the effects of temperature on the rate at which evolution can occur. At higher temperatures, such as those associated with lower latitudes, genetic mutation rates, among other factors influencing speciation, are higher. To date, no single hypothesis about the cause of species richness latitude gradients has gained widespread acceptance within the scientific community.

Another important ecogeographic principle related to latitudes is Bergmann's rule. The nineteenth-century biologist Carl Bergmann noted that when the same species (or closely related species) of mammals and other warm-blooded animals were found in different parts of the earth, those that lived closer to the equator were smaller than those that lived at higher latitudes. For example, polar bears are much more massive than related bear species in the tropics. Bergmann's rule is often explained as a result of the direct relationship between body size and heat. Larger animals generate a greater amount of heat as a result of cell metabolic activities, and they also lose less heat to their surroundings, since the ratio of their surface area to overall mass is comparatively low. These adaptations make them well suited to living in colder, high-latitude environments.

Biogeographic Divisions

Biogeographers use a series of imaginary lines to divide the earth into six major divisions, or realms. A distinct biota, or group of plants, animals, and microbes, inhabits each zone. The most famous of these imaginary divides runs between the islands of Borneo and Sulawesi in the Indonesian archipelago, and is known as Wallace's line, after Alfred Russel Wallace. Wallace was a nineteenth-century British naturalist—a contemporary of Charles Darwin's who also developed a theory of evolution based on natural selection. It was Wallace who first noticed the sharp distinction between birds and animals on the western side of the line—which mostly came from taxons common in Asia—and those on the eastern side of the line—which mostly came from taxons common in Australia.

To the west of the Wallace line, for instance, one would find woodpeckers, pheasants, primates, squirrels, and big cats; to the east, cockatoos, bats, eucalyptus trees, and marsupials. The Wallace line is the best known of the world's biogeographical divisions not only because of the extreme contrasts between the zones on either side of it, but also because the entire region surrounding it is considered to be a very significant biodiversity hot spot. A biodiversity hot spot is a region characterized by a high density of species richness that is under threat of extinction due to human activity. Many of the species found in the region known as Wallacea have endemic distribution areas, including the caerulean paradise flycatcher (*Eutrichomyias rowleyi*) and the pig-like mammal known as the babirusa (*Babyrousa babyrussa*). Logging, mining, and agricultural development have all had a dramatic negative impact on the Wallacea ecosystem.

Biomes are another way in which biographers organize the earth's living communities. A biome is a large, naturally occurring ecosystem; the particular grouping of plants and animals that inhabit it are determined by its climate (especially temperature and precipitation) and geography. Different scientists classify biomes in different ways, but five of the most common terms used to describe biomes are aquatic (freshwater and marine habitats), desert (arid habitats), forest (densely wooded habitats), grassland (open plain habitats dominated by grasses), and tundra (open plain habitats with a permanently frozen layer of subsoil).

Theory of Island Biogeography

The theory of island biogeography was developed in the late 1960s by the ecologists E. O. Wilson, of Harvard University, and Robert MacArthur, of Princeton University, as a way of organizing and explaining the factors contributing to the relative richness of species biodiversity that occurs in the world's islands. In its most basic form, MacArthur and Wilson's theory states that the number of unique living species on any given island is primarily shaped by two competing forces: the rate at which new species move to the island and establish successful breeding populations there (the rate of immigration), and the rate at which existing populations of successfully breeding species disappear from the island (the rate of extinction).

Various other factors are responsible for determining the rate of immigration and extinction. For example, an island's rate of immigration is closely related to the distance of the island from other ecosystems from which potential colonizers could travel, such as other islands or a mainland. The theory predicts that the more geographically isolated an island is, the lower its rate of colonization will be; this pattern is known as the distance effect. Similarly, an island's rate of extinction is related to the size of the island is, the number of species that already inhabit it, and the availability of resources on it. According to the theory, a larger island will have more habitable area and more resources, like food, available—and will thus be able to support a greater number of species, lowering the extinction rate. As the number of existing species on an island increases, competition between these species for habitat and food will also cause its extinction rate to rise. By comparison, islands that have a variety of habitats (for example, mountainous regions as well as flat forested regions) will be able to support a larger number of different species.

Assuming these factors remain unchanged, MacArthur and Wilson's theory predicts that for any newly created island, the rate of immigration will start high and drop over time, while the rate of extinction will start at zero and rise over time. Eventually, the number of unique species that exist on a newly created island will become stable, as a balance, or equilibrium, is reached between the rate of immigration and the rate of extinction. (Even after equilibrium has been reached, the specific composition of the island's flora and fauna will be dynamic, rather than static, as different species continue to replace each other.) For example, the 1883 volcanic eruption on the tiny island of Krakatoa in Indonesia caused the total extinction of all its plant and animal species. Observations over the following hundred years showed that the rates of immigration and extinction on what was now effectively a "new" island followed the pattern predicted by MacArthur and Wallace: Between 1883 and 1934, thirty-four new species moved to and colonized the island, five of which went extinct. But between 1934 and 1985, an additional fourteen species began to successfully breed on the island (a lower immigration rate than in the previous period) and an additional eight became extinct (a higher extinction rate than in the previous period).

The theory of island biogeography has since been expanded in scope; it is now used to estimate the probable degree of species richness in any isolated habitat area, such as a river or lake surrounded by desert, or a wetland region surrounded by dry woodlands. It is a particularly useful tool for predicting the effects of habitat fragmentation due to human activity. For example, much of the once continuous deciduous forest of the eastern United States has been cut up into smaller, geographically separated, patches of trees, with roads, farms, and cities taking up the spaces in between. The distance effect predicts that these isolated patches of forest will have a relatively low rate of immigration, since it is difficult for plants and animals to move across the gaps between them. In addition, their small size will reduce the available habitat area and resources they contain, raising the rate of extinction. Overall, the theory of island biogeography predicts that habitat fragmentation will reduce the level of species richness in fragmented areas. Studies of avian populations in forest preserves surrounded by human development support this prediction.

M. Lee

Further Reading

Kruckeberg, Arthur R. *Geology and Plant Life: The Effects of Landforms and Rock Types on Plants.* Seattle, Wash.: University of Washington Press, 2004.

Lieberman, Bruce S. *Paleobiogeography: Using Fossils to Study Global Change, Plate Tectonics, and Evolution.* New York: Kluwer Academic, 2000.

Lomolino, Mark V., Dov F. Sax, and James H Brown. *Foundations of Biogeography: Classic Papers with Commentary.* Chicago: University of Chicago Press, 2004.

Losos, Jonathan B., Robert E. Ricklefs, and Robert H. MacArthur, eds. *The Theory of Island Geography Revisited.* Princeton, N.J.: Princeton University Press, 2010.

Morrone, Juan J. *Evolutionary Biogeography: An Integrative Approach with Case Studies.* New York: Columbia University Press, 2009.

Newton, Ian. *Speciation and Biogeography of Birds.* London: Academic Press, 2003.

Shugart, H. H. *How the Earthquake Bird Got Its Name and Other Tales of an Unbalanced Nature.* New Haven, Conn.: Yale University Press, 2007.

■ Biomagnification

FIELDS OF STUDY
Biology; Chemistry; Ecology; Environmental Biotechnology; Environmental Microbiology; Environmental Sciences

SUMMARY
Recognition of the detrimental environmental effects of biomagnification led to the adoption of several procedures to prevent the accumulation of toxic materials in higher organisms along the food chain. Some pesticides were banned outright, and others were modified to prevent their accumulation in the environment.

PRINCIPAL TERMS
- **bioaccumulation:** process by which some toxic substances accumulate and continue accumulating in living organisms, posing a threat to health, life and the environment
- **biomagnification:** accumulation of toxic contaminants in organisms as they move up through the food chain
- **Dichloro-diphenyl-trichloroethane (DDT):** non-water soluble, non-biodegradable chlorinated hydrocarbon used as an insecticide, primarily in agriculture to control plant pests, especially those affecting cotton and tobacco, causing widespread poisoning of birds and other small animals and plants
- **lipophilicity:** the ability of a chemical compound to dissolve in fats, oils, lipids, and non-polar solvents such as hexane or toluene
- **zooplankton:** a type of heterotrophic plankton ranging from microscopic organisms to large species, such as jellyfish, ecologically important organisms integral to the food chain, and found within large bodies of water, including oceans and freshwater systems.

As members of each level of the food chain are progressively eaten by those organisms found in higher levels of the chain, the potential grows for increased concentrations of toxic chemicals to accumulate within the tissues of the higher organisms. Not all chemicals, potentially toxic or not, are equally likely to undergo this accumulation, known as biomagnification. However, molecules susceptible to biomagnification have certain characteristics in common. They are resistant to natural microbial degradation and therefore persist in the environment. They are also lipophilic—that is, they tend to accumulate in the fatty tissue of organisms. In addition, chemicals must be biologically active in order to have effects on the organisms in which they are found. Such compounds are likely to be absorbed from food or water in the environment and stored within the membranes or fatty tissues.

The process of biomagnification usually begins with the spraying of pesticides for the purpose of controlling insect populations. Industrial contamination, including the release of heavy metals, can be an additional cause of such pollution. Biomagnification results when these chemicals contaminate the water supply and are absorbed into the lipid membranes of microbial organisms. This process, often referred to as bioaccumulation, results in the initial concentration of the chemical in the organisms in a form that is not naturally excreted with normal waste material. The levels of the chemical in the organisms may reach anywhere from one to three times the level found in the surrounding environment. Since the nature of the chemical is such that it is neither degraded nor excreted, it remains within the organisms.

As the organisms on the bottom of the food chain are eaten and digested by members of the next level in the chain, the concentration of the accumulated material significantly increases; at each subsequent level, the concentration may reach one order of magnitude (a tenfold increase) higher. Consequently, the concentrations of the pollutant at the top of the environmental food chain, such as in fish or carnivorous birds, and potentially even humans, may be as much as one million times as high as the original, presumably safe, levels in the environment. For example, studies of Dichloro-diphenyl-trichloroethane (DDT) levels in the 1960s found that zooplankton at the bottom of the food chain had accumulated nearly one thousand times the level of the pollutant in the surrounding water. Ingestion of the plankton by fish resulted in concentration by another factor of several hundred. By the time the fish were eaten by predatory birds, the level of DDT was concentrated by a factor of more than two hundred thousand.

Pollutants Subject to Biomagnification

DDT is characteristic of most pollutants subject to potential biomagnification. It is relatively stable in the environment, persisting for decades. It is soluble in lipids and readily incorporated into the membranes of organisms. While DDT represents the classic example of biomagnification of a toxic chemical, it is by no means the only representative of potential environmental pollutants. Other pesticides with similar characteristics include aldrin, chlordane, parathion, and toxaphene. In addition, cyanide, polychlorinated biphenyls (PCBs), and heavy metals—such as selenium, mercury, copper, lead, and zinc—have been found to concentrate within the food chain.

Some heavy metals are inherently toxic or may undergo microbial modification that increases their toxic potential. For example, mercury does not naturally accumulate in membranes and was therefore not originally viewed as a significant danger to the environment. However, some microorganisms are capable of adding a methyl group to the metal, producing methylmercury, a highly toxic material that does accumulate in fatty tissue and membranes.

Since pesticides are, by their nature, biologically active compounds, which reflects their ability to control insects, they are of particular concern if subject to biomagnification. DDT remains the classic example of how bioaccumulation and biomagnification may have effects on the environment. Initially introduced as a pesticide for control of insects and insect-borne disease, DDT was not thought to be particularly toxic. However, biomagnification of the chemical was found to result in the deaths of birds and other wildlife. In addition, DDT contamination was found to result in the formation of thin eggshells in birds, which greatly reduced the numbers of birds successfully hatched. Before the use of DDT was banned in the 1960s, the population levels of predatory birds such as eagles and falcons had fallen to a fraction of the levels found prior to use of the insecticide. Though it was unclear whether there was any direct effect on the human population in the United States, the discovery of elevated levels of DDT in human tissue contributed to the decision to ban the use of the chemical.

Several procedures have been adopted since the 1960s to prevent the biomagnification of toxic materials. In addition to outright bans, pesticides are often modified to prevent their accumulation in the environment. The chemical structures of most synthetic pesticides are easily degraded by microorganisms found in the environment. Ideally, a pesticide should survive no longer than a single growing season before being rendered harmless by the environmental flora. Often such chemical changes require only simple modification of a compound's basic structure.

Richard Adler

Further Reading

Atlas, Ronald, and Richard Bartha. *Microbial Ecology*. San Francisco: Benjamin Cummings, 1998.

Carson, Rachel. *Silent Spring*. Boston: Houghton Mifflin, 2002.

Colborn, Theo, Dianne Dumanoski, and John P. Myers. *Our Stolen Future: Are We Threatening Our Fertility, Intelligence, and Survival?* New York: Plume, 1996.

Karasov, William H., and Carlos Martínez del Rio. *Physiological Ecology: How Animals Process Energy, Nutrients, and Toxins*. Princeton, N.J.: Princeton University Press, 2007.

Primack, Richard B. *Essentials of Conservation Biology*. Sunderland, Mass.: Sinauer Associates, 2010.

Willmer, Pat, Graham Stone, and Ian Johnston. *Environmental Physiology of Animals*. Malden, Mass.: Blackwell Science, 2000.

■ Biomes

FIELDS OF STUDY

Botany; Conservation Biology; Ecology; Environment; Landscape Ecology; Land-use Management

SUMMARY

Biomes are major ecological zones shaped primarily by climate and characterized by the types of plant and animal life they support. The complexity of biomes exposes them to varying and sometimes cascading changes from both natural and human events. The consequences for the larger environment can range from minimal to catastrophic.

PRINCIPLE TERMS

- **anthropogenic:** caused or produced by humans
- **biomes:** complex biotic communities characterized by distinctive plant and animal species and maintained under the climatic conditions of the region
- **boreal forest:** a forest that grows in regions of the northern hemisphere with cold temperatures and comprised mostly of cold tolerant coniferous species such as spruce and fir; also known by the Russian word, *taiga*
- **chaparral:** dense growth of shrubs or small trees
- **rainforest:** a tropical forest, usually of tall, densely growing, broad- leaved evergreen trees in an area of high annual rainfall
- **savanna:** a plain with coarse grasses and scattered tree growth, especially on the margins of the tropics where rainfall is seasonal
- **tundra:** level or rolling treeless plain characteristic of arctic and subarctic regions, consisting of black mucky soil with a permanently frozen subsoil and dominant vegetation of mosses, lichens, herbs, and dwarf shrubs

Each of the earth's biomes has a typical ecological pattern, with its flora and fauna sharing similar characteristics, such as leaf forms and survival strategies. A study of any biome, on any continent, will show multiple ways its species carry out the essential tasks of life. Biomes are large and complex, with species forming intricate networks of life within subsets of the biomes.

Land-based major biomes include the following: tropical rain forest, tropical dry forest, savanna (tropical grassland), desert, temperate grassland, Mediterranean scrub or chaparral, temperate deciduous forest, temperate mixed and coniferous forest, boreal forest(taiga), and Arctic tundra. Marine biomes are basically characterized as oceanic or freshwater, although the lack of vegetation-climate links and the gradients between coastal and deep-sea habitats make the simplicity of this scheme problematic.

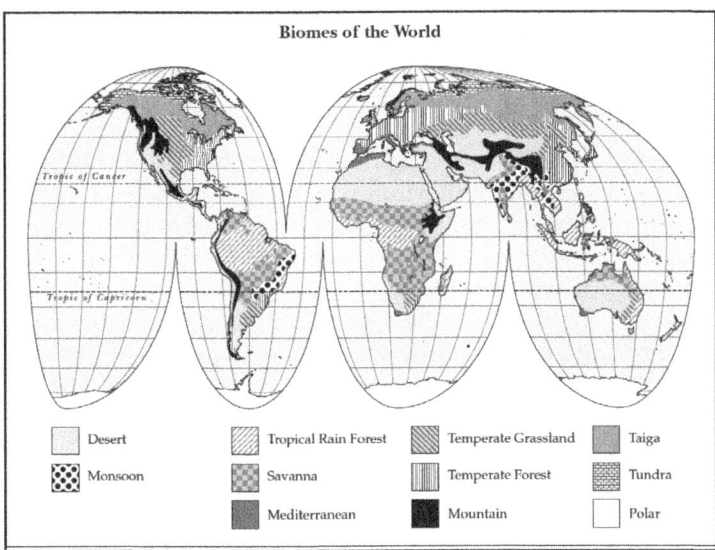

Finally, human activities have so transformed large portions of the earth that some biologists treat anthropogenic (human-dominated) biomes as a separate concept. These biomes range from dense settlement patterns in urban landscapes to cropland, rangeland, and forest.

Influences on Biomes

Biomes are sometimes viewed as giant ecosystems. Because of biomes' size and complexity, however, both scientists and lay observers find it more useful to view biomes as made up of many different ecosystems and habitats.

Land-based biomes are shaped largely by their climates. The most important climatic factors are latitude and humidity (total precipitation, seasonal rainfall patterns, ambient moisture); sometimes elevation is important also. Cross-cutting these climatic factors—though influenced by them—are such features as terrain, soil type and nutrient status, and prevailing winds. All these affect the kinds of vegetation that can flourish in a region, thus determining the types and amounts of animal life the plant cover will support.

Generally, biomes closer to the equator support more biodiversity, in the form of different species. An estimated 50 percent of all life-forms on earth are native to tropical forests. Tropical rain forests also have many layers of life systems, from the canopy ecosystems with their monkeys, marmosets, and tropical birds, such as macaws, down through several canopy-gap layers that shelter opossums and woodpeckers, to the forest floor, the base station for jaguars and frogs, among other jungle dwellers. What the tropical forests do not have is a rich nutrient base. They have poor soil quality, and most of the available nutrients are held within the trees themselves rather than fixed in the soil.

In contrast, temperate deciduous forests lack the great variety of species, and of levels, found in tropical rain forests. The life processes in temperate deciduous forests are highly regulated by seasonal temperature changes. As leaves fall to the ground, they carry calcium and other minerals to the forest floor; this litter turns into humus and becomes the nutrient-rich soil on which broad-leaved trees thrive.

Another mix of resources and life strategies is found in desert biomes. Most deserts do get some water and support some life, but anything growing there must adapt to little moisture and hot, unshaded land that is poor in nutrients. Desert plants store nutrients in underground organs such as their root systems, and cacti have expandable stems for storing water from sudden but infrequent downpours. Desert dwellers such as the kangaroo rat get their moisture from eating seeds. Their underground burrows retain three times as much humidity as exists aboveground.

Environmental Issues and Biomes

For the nonhuman organisms living in any given biome, the environmental features of that biome are essential to life. Human beings, in contrast, interact with biomes by finding ways to use the biomes' natural resources to sustain them. Human technologies and large numbers of humans, however, can upset a delicately balanced biome far beyond its ability to recover. Even in the ancient world, large-scale grazing and firewood collection apparently turned once-wooded coastal areas into the Mediterranean scrub that bears that region's name.

Throughout history, humans have cleared forests or modified grasslands for use in agriculture. Sometimes the environmental damage done by such processes is limited. For example, the prairie grasslands of the American Great Plains were originally roamed by bison, and the cattle herds that replaced the bison filled the same ecological niche. Other human-caused changes have brought near disaster to entire biomes.

Human activities have altered every one of the earth's biomes. Among the direst changes has been the destruction of tropical forests caused by timber harvesting and by the clearing of trees and other plant life to make way for ranching or industry. The damage caused when a biome is altered is not confined to the loss of diverse natural resources within that biome; rather, such change threatens the entire biosphere because it interferes with the role the biome plays in balancing the earth's systems. Loss of rain forest, for example, damages the atmosphere because of the accompanying loss of oxygen renewal that was provided by the trees.

The ocean biome is central to regulating heat and rainfall around the world, and, through its plankton, is the source of most photosynthesis on the earth. The crumbling of coral reefs and the huge trash vortex reported in the North Pacific in the late 1990's are early warning signals about the environmental threats to this biome's future. Freshwater bodies are susceptible to damage from industrial and fertilizer runoff. Other natural biomes face their own unique threats of rapid environmental degradation.

Anthropogenic biomes are by definition shaped by human activity, but humans can make conscious decisions that can reduce the negative impacts of such activity on the environment. Within cities, for example, humans can counter the effects of vast, heat-retaining stretches of pavement by creating rooftop gardens, rain gardens, and other green spaces.

Emily Alward

Further Reading

Butz, Stephen D. *Science of Earth Systems*. 2d ed. Clifton Park, N.Y.: Thomson Delmar Learning, 2008.

Whitfield, Philip, Peter D. Moore, and Barry Cox. *Biomes and Habitats*. New York: Macmillan, 2002.

Woodward, Susan L. *Biomes of Earth: Terrestrial, Aquatic, and Human-Dominated*. Westport, Conn.: Greenwood Press, 2003.

_____. *Introduction to Biomes*. Westport, Conn.: Greenwood Press, 2009.

■ Biopesticides and bioassays

FIELDS OF STUDY
Biochemistry Engineering; Chemistry; Ecology; Environment; Environmental Biotechnology; Environmental Engineering; Environmental Microbiology

SUMMARY
Biopesticides have significant advantages over commercial pesticides in that they appear to be environmentally safer, given that they do not accumulate in the food chain and they have only slight effects on ecological balances. Bioassays enable environmental scientists to evaluate the effects of the chemicals used in pesticides as well as the resistance of plants to pests.

PRINCIPAL TERMS

- **bioassays:** tests that use biological organisms to detect the presence of given chemical substances or determine the biological activity of known substances in certain environments
- **biopesticides:** biological agents that are used to control insect and weed pests
- **hyperparasitism:** an organism that is parasitic on or in another parasite
- **median lethal dose (LD_{50}):** the quantity of a lethal substance, as a poison or pathogen, or of ionizing radiation that will kill 50 percent of the organisms subjected to it in a specified time period
- **nematodes:** any unsegmented worm of the phylum Nematoda, having an elongated, cylindrical body
- **pathogenic fungi:** fungi that cause disease in humans or other organisms
- **protozoa:** a parasitic single-celled organism that can divide only within a host organism
- **saprophyte:** an organism which gets its energy from dead and decaying organic matter

Pests are any unwanted animals, plants, or microorganisms. When the environment has no natural resistance to a pest and when no natural antagonists are present, pests can run rampant. For example, the fungus *Endothia parasitica*, which entered New York State in 1904, caused the nearly complete destruction of the American chestnut tree because no natural control was present.

Biopesticides represent the biological, rather than the chemical, control of pests. Many plants and animals are protected from pests by passive means. For example, plant rotation is a traditional method of insect and disease protection in which the host plant is removed for a period long enough to reduce pathogen and pest populations.

Biopesticides have several significant advantages over commercial pesticides. They appear to be ecologically safer than commercial pesticides because they do not accumulate in the food chain. Some biopesticides also provide persistent control, because pests require more than a single mutation to adapt to them and because they can become an integral part of a pest's life cycle. In addition, biopesticides have only slight effects on ecological balances because they do not affect nontarget species. Finally, biopesticides are compatible with other control agents. The major drawbacks to using biopesticides are that, in comparison with chemical pesticides, biopesticides work less efficiently and take more time to kill their targets.

Biopesticides

Viruses, bacteria, fungi, protozoa, mites, and flowers have all been used as biopesticides. Viruses have been developed against insect pests such as *Lepidoptera*, *Hymenoptera*, and *Dipterans*. These viruses cause hyperparasitism. Gypsy moths and tent caterpillars, for example, periodically suffer from epidemic virus infestations.

Many saprophytic microorganisms that occur on plant roots and leaves can protect plants against microbial pests. *Bacillus cereus* has been used as an inoculum on soybean seeds to prevent infection by the fungal pathogen *Cercospora*. Some microorganisms used as biopesticides produce antibiotics, but the major mechanism for protection is probably competitive exclusion of a pest from sites on which the pest must grow. For example, *Agrobacterium radiobacter* antagonizes *Agrobacterium tumefaciens*, which causes crown gall disease. Two bacteria—*Bacillus* and *Streptomyces*—added as biopesticides to soil help control the damping off disease of cucumbers, peas, and lettuces caused by *Rhizoctonia solani*. *Bacillus subtilis* added to plant tissue also controls stem rot and wilt rot caused by the fungus *Fusarium*. *Mycobacteria* produce cellulose-degrading enzymes, and their addition to young seedlings helps control fungal infection by *Pythium*, *Rhizoctonia*, and *Fusarium*. *Bacillus* and *Pseudomonas* are bacteria that produce enzymes that dissolve fungal cell walls.

The best examples of microbial insecticides are *Bacillus thuringiensis* (*B.t.*) toxins, which were first used in 1901. They have had widespread commercial production and use since the 1960s and have been successfully tested on 140 insect species, including mosquitoes. *B.t.* produces insecticidal endotoxins during sporulation and also produces exotoxins contained in crystalline parasporal protein bodies. These protein crystals are insoluble in water but readily dissolve in an insect's gut. Once dissolved, the proteolytic enzymes paralyze the gut. *Bacillus* spores that have also been consumed germinate and kill the insect. *Bacillus popilliae* is a related bacterium that produces an insecticidal spore that has been used to control Japanese beetles, a pest of corn.

Biologists are seining for fish to be used in bioassays. EPA Gulf Breeze Laboratory

Comparison of the Properties of *Bacillus Thuringiensis* and *Bacillus Popilliae* as Microbial Biocontrol Agents

	BACILLUS THURINGIENSIS	BACILLUS POPILLIAE
Pest controlled	Lepidoptera (many)	Coleoptera (few)
Pathogenicity	low	high
Response time	immediate	slow
Formulation	spores and toxin crystals	spores
Production	in vitro	in vivo
Persistence	low	high
Resistance in pests	developing	reported

Source: Data adapted from J. W. Deacon, *Microbial Control of Plant Pests and Diseases* (1983).

versions of these naturally occurring compounds are found in products used to control head lice. Molecular genetics has also been used to insert the gene for the *B.t.* toxin into cotton and corn. *B.t.* cotton and *B.t.* corn both express the gene in their roots, which provides them with protection from root worms. Ecologists and environmentalists have expressed concern that constantly exposing pests to the toxin will cause insect resistance to develop rapidly and thus reduce the effectiveness of traditionally applied *B.t.*

Saprophytic fungi can compete with pathogenic fungi. Among the fungi used as biopesticides are *Gliocladium virens, Trichoderma hamatum, Trichoderma harzianum, Trichoderma viride,* and *Talaromyces flavus.* For example, *Trichoderma* competes with the pathogens *Verticillium* and *Fusarium*

Peniophora gigantea antagonizes the pine pathogen *Heterobasidion annosum* through three mechanisms: It prevents the pathogen from colonizing stumps and traveling down into the root zone, it prevents the pathogen from traveling between infected and uninfected trees along interconnected roots, and it prevents the pathogen from growing up to stump surfaces and sporulating.

Nematodes are pests that interfere with commercial button mushroom (*Agaricus bisporus*) production. Several types of nematode-trapping fungi can be used as biopesticides to trap, kill, and digest the nematode pests. The fungi produce structures such as constricting and non-constricting rings, sticky appendages, and spores, which attach to the nematodes. The most common nematode-trapping fungi are *Arthrobotrys oligospora, Arthrobotrys conoides, Dactylaria candida,* and *Meria coniospora.*

Protozoa have occasionally been used as biopesticide agents, but their use has suffered because of such difficulties as slow growth and complex culture conditions associated with their commercial production. Predaceous mites are used as a biopesticide to protect cotton from other insect pests such as the boll weevil.

Dalmatian and Persian insect powders contain pyrethrins, which are toxic insecticidal compounds produced in *Chrysanthemum* flowers. Synthetic

Bioassays

In many instances, a scientist may suspect that a certain chemical is present in a given environment but may not have access to a specific piece of equipment designed to measure the presence of the chemical. In some cases, an experimental protocol for the detection of the chemical may not exist. In either of these cases, the scientist may be able to detect the presence of the chemical by using a biological organism that responds in a specific manner when exposed to that particular chemical agent. At other times, a scientist may know that a certain chemical is present but not know how a particular organism will respond when exposed to the agent. In this case, the scientist will expose the test organism to the chemical and measure a particular physiological response.

Bioassays are utilized in many different areas of the biological sciences, including environmental studies. Some bioassay methods work better than others. A good bioassay meets two basic criteria. First, it is specific for a given physiological response. For example, if a given chemical is responsible for inhibiting the feeding response of an insect, then the bioassay for that chemical should measure only the inhibition of feeding of that insect and not some other physiological response to the chemical. Second, a good bioassay measures the same response in the laboratory that is observed in the field. Again, if a chemical inhibits the feeding response in the field, then the laboratory bioassay for that chemical should also inhibit feeding. An ongoing need exists for the development of accurate bioassay methods as well as the improvement of existing techniques.

Many different bioassays are used in environmental studies. One of the most common is the measure of the median lethal dose (LD_{50})—the concentration or dose of a chemical that will result in the deaths of one-half of a population of organisms—of a new pesticide on species of pest and non-pest organisms. To conduct this bioassay, the test species is exposed to a wide range of different concentrations of the chemical. The concentration of the pesticide that kills one-half of the test organisms represents the LD_{50}.

Another common environmental bioassay is the measure of resistance of plants to a particular insect pest. In order to reduce the dependence on chemical insecticides, plant breeders are continually trying to develop insect-resistant plants, either through traditional breeding programs or by using biotechnology to transfer resistance genes to susceptible crop strains. Bioassays are used to measure the degree of success of these attempts. In these bioassays, the same numbers of susceptible and resistant plants are subjected to infestation by equal numbers of the insect pest for which the breeder is trying to develop resistance. The two groups of plants are observed, and the degree of resistance, if any, is recorded.

Mark Coyne and D. R. Gossett

Further Reading

Churchill, B. W. *Biological Control of Weeds with Plant Pathogens*. Edited by R. Charudattan and H. Walker. New York: Wiley, 1982.

Deacon, J. W. *Microbial Control of Plant Pests and Diseases*. Research Triangle Park: Instrumentation Systems & Automation, 1983.

Metz, Matthew, ed. *Bacillus Thuringiensis: A Cornerstone of Modern Agriculture*. Binghamton, England: Haworth, 2003.

Ohkawa, H., H. Miyagawa, and P. W. Lee, eds. *Pesticide Chemistry: Crop Protection, Public Health, Environmental Safety*. New York: Wiley, 2007.

Rand, Gary M., ed. *Fundamentals of Aquatic Toxicology: Effects, Environmental Fate, and Risk Assessment*. 3d ed. Boca Raton, Fla.: CRC Press, 2008.

■ Bioprospecting and biopiracy

FIELDS OF STUDY
Botany; Chemistry; Ecology; Environment; Environmental Biotechnology; Environmental Engineering; Environmental Microbiology; Ethnobotany; Genetic Engineering

SUMMARY
The practice of extracting biological resources from regions of the world known for their great biological diversity, often carried out by scientists working for corporations or educational institutions, is the subject of ongoing debate. Many environmentalists and indigenous peoples see such resource extraction as a form of exploitation.

PRINCIPAL TERMS
- **actinobacteria:** a phylum of terrestrial or aquatic Gram-positive bacteria on which agriculture and forests depend for their contributions to soil systems, where they behave like fungi, helping to decompose organic matter of dead organisms
- **biological resources:** any substance or object in the environment required by an organism for normal growth, maintenance, and reproduction
- **bioprospecting/biopiracy:** extraction of biological resources from areas of diversity
- **genetic engineering:** development and application of scientific methods, procedures, and technologies that permit direct manipulation of genetic material in order to alter the hereditary traits of a cell, organism, or population
- **microorganisms:** any organism too small to be viewed by the unaided eye, such as bacteria, protozoa, and some fungi and algae
- **streptomycin:** an antibiotic organic base produced by a soil bacteria and used especially in the treatment of infections

With the signing of the United Nations Convention on Biological Diversity at the Earth Summit in Rio de Janeiro, Brazil, in 1992, participatory nations agreed to no longer consider biological resources the "common heritage of mankind" but conceded the rights to distribute such resources to the individual nations that housed them. Around the same time, the terms "biopiracy" and "bioprospecting" began to be used to describe the acquisition of these newly protected resources. The two terms refer to essentially the same thing, the extraction of biological resources from areas of biodiversity, but they have decidedly different tones, the former having been coined by opponents of such activity and the latter being preferred by the practitioners of this type of resource extraction.

Biological resources include whole organisms such as crops or livestock, chemical compounds that can be purified from specific organisms that produce them, and even the genetic material taken from organisms that can then be used to produce desired proteins, usually in conjunction with some form of genetic engineering. These resources hold value in that they can be used to improve agricultural yields, perform certain industrial processes, or serve various pharmaceutical applications. The debate over the appropriate acquisition of these resources led to the split in the terms used to describe the same activity. "Biopiracy" brings to mind a swashbuckler who pillages resources without regard to the victims; "bioprospecting," in contrast, conveys the image of a gold miner staking out a claim and then working it, with no guarantee of ultimate success.

Bioprospecting

The image of the gold-rush prospector is perhaps most appropriate for one resource collection: the biodiversity-driven, or random-collection, approach. Scientists taking this approach sample large amounts of organisms for a desired chemical activity or genetic attribute without prior knowledge of precisely where to look. The screened organisms are typically plants, microorganisms, insects, or marine invertebrates. This is called the biodiversity-driven approach because mass sampling is best done in areas with wide ranges of different organisms living in close proximity.

Black turmeric (Curcuma caesia) *is a kind of turmeric with bluish-black rhizome, famous for its unique medicinal properties. It is a perennial herb, belonging to Zingiberaceae (Ginger) family, native to North-East and Central India. Presently Black turmeric is on the verge of extinction due to biopiracy.* By Ramesh Raju (Own work).

Just as modern mining methods include scientific means for discovering deposits of minerals, however, bioprospecting often makes use of prior knowledge to narrow the pool of organisms being tested. This knowledge falls into three main categories: chemotaxonomic, ecological, and ethnobotanical/ethnopharmacological. The use of chemotaxonomic knowledge involves the sampling of organisms that belong to the same taxonomic class as an organism that is already known to have a desired property. An example would be screening several bacteria from the class Actinobacteria, the taxonomic group known to be responsible for the production of streptomycin, for antibiotic properties. Ecological knowledge is knowledge that can be gained from field observations of the interactions between organisms. Certain plants and animals, for example, produce chemical compounds called secondary metabolites that they use to defend themselves against predator attack. A scientist taking an ecological approach to bioprospecting may detect such interactions and choose species for further testing based on these observations.

The use of ethnobotanical/ethnopharmacological knowledge is the most controversial approach, as it seeks to capitalize on the medical practices of indigenous peoples who inhabit the areas of interest. Ethnobotanical knowledge focuses on plants that have traditionally been used for healing purposes by indigenous peoples, whereas ethnopharmacological knowledge is broader, encompassing all traditional drugs as well as their biological activities. Using such knowledge, scientists can screen specific organisms for desired properties with a much higher degree of success than is seen with randomly sampled collections.

Biopiracy

Much of the world's biodiversity lies in the tropical regions, often in developing countries that have historically experienced oppression by wealthier nations. It is not surprising, therefore, that indigenous peoples in these regions tend to be wary of the academic institutions and multinational corporations that engage in what these entities may view as simple bioprospecting. Often, indigenous peoples have concerns regarding the entire practice of treating biodiversity as a biological resource, including the patenting of living organisms and profiting from biological materials that for many years previously were exchanged freely among those who reaped the benefits. Even if these concerns are allayed, questions often remain about who should be compensated for traditional knowledge that leads to a "discovery," as well as what would constitute a fair level of compensation.

Although no entity has been prosecuted officially for biopiracy under the Convention on Biological Diversity, many allegations of biopiracy have been made, and several planned bioprospecting projects have been abandoned after information about them became public and protests ensued. It may be partially because of such controversies that bioprospecting activities actually decreased in the decades following the convention's signing, as many companies turned away from using natural resources and instead developed synthetic processes, such as combinatorial chemistry to produce lead compounds that could be screened for a desired activity.

James S. Godde

Further Reading

Godde, James S. "Genetic Resources." In *Encyclopedia of Global Resources*, Craig W. Allin, ed. Pasadena, Calif.: Salem Press, 2010.

Hamilton, Chris. "Biodiversity, Biopiracy, and Benefits: What Allegations of Biopiracy Tell Us About Intellectual Property." *Developing World Bioethics* 6 (2006): 158-173.

Paterson, Russell and Nelson Lima, eds. *Bioprospecting: Success, Potential and Constraints*. 2nd ed. New York: Springer, 2017

Shiva, Vandana. *Biopiracy: The Plunder of Nature and Knowledge*. Brooklyn, N.Y.: 1999.

Soejarto, D. D., et al. "Ethnobotany/Ethnopharmacology and Mass Bioprospecting: Issues on Intellectual Property and Benefit-Sharing." *Journal of Ethnopharmacology* 100, nos. 1-2 (August 2005): 15-22.

Tan, G., C. Gyllenhaal, and D. D. Soejarto. "Biodiversity as a Source of Anticancer Drugs." *Current Drug Targets* 7, no. 3 (March 2006): 265-277.

Tedlock, Barbara. "Indigenous Heritage and Biopiracy in the Age of Intellectual Property Rights." *Explore: The Journal of Science and Healing* 2, no. 3 (May 2006): 256-259.

■ Bioregionalism

FIELDS OF STUDY

Anthropology; Biology; Conservation; Ecology; Environment; Geography; Humanities; Landscape Ecology; Resource Management

SUMMARY

Bioregionalists are concerned with reversing the alienation from the land that is evident in the modern global economy, with protecting the environment from unsustainable human exploitation, and with ensuring that natural resources remain abundant and diverse for future generations.

PRINCIPAL TERMS
- **biodegradation:** the breakdown of organic matter by microorganisms, such as bacteria, fungi
- **bioregionalism:** environmental movement that holds that local populations should be self-sustaining based on the resources of their surrounding bioregions and largely self-governing
- **ecosystem:** a system or group of interconnected elements formed by the interaction of a community of organisms with their environment

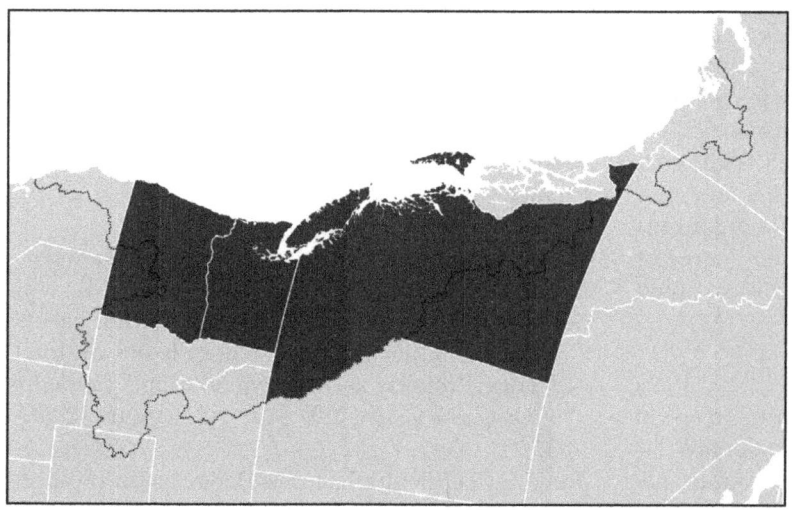

A map showing the two definitions of the proposed "Republic of Cascadia." Green shows the American states of Oregon and Washington; and the Canadian province of British Columbia (which make up the standard definition). The black-dotted line marks the border of the Cascadia bioregion (which is also mentioned as a border). By NuclearVacuum (Own work).

- **locavore:** one who tries to eat food grown, raised, or produced locally, usually within 100 miles of home
- **watershed:** region or area drained by a river, stream, or other water sources

Bioregionalism rests on two basic existential principles: that humanity is but one component of the "web of life" (that is, the ecosystem) and that humans best pursue their own welfare by living in balance with the local environment. Bioregionalism began as a philosophical offshoot of the environmental movement during the 1970s. Among its early proponents were environmental activist Peter Berg, ecologist Raymond F. Dasmann, and poet Gary Snyder. They particularly worried that technology, consumerism, postmodern culture, and global economics were producing a rootless population estranged from a sense of home and community. Bioregionalism is one of the "relocalization" movements seeking to reverse this trend.

A bioregion is an area defined by natural rather than political boundaries. It most commonly comprises a watershed, the plants, animals, climate, hydrology, and ecology of which give it a distinctive character. In 1978, Berg and Dasmann argued that people must concentrate on "living-in-place," which entails satisfying the necessities of life and enjoying life's pleasures as they are available in a particular area, as well as ensuring their long-term availability. The approach is pragmatic: It assumes that people who live long in one place come to know it thoroughly, come to care about it, and want to take care of it.

Bioregionalism is an eclectic movement. It fosters an awareness of local economic and cultural assets by applying lessons from physical geography, ecology, ecosystem management, sustainable agriculture, economics, literature, and political theory. For example, the bioregionalist approach to architecture involves the use of local materials and labor, as well as designs that reflect both regional traditions and the surrounding landscape while satisfying the requirements of present-day life. Bioregionalist political theory stresses participatory democracy and the resolution of social problems through the efforts of voluntary, nonprofit groups. Bioregionalist economics is centered on locally produced goods and services in place of imports and encourages recycling.

Bioregionalism is neither essentially hostile to technology nor divorced from global civilization. It acknowledges that bioregions are parts of larger economic, political, and cultural contexts but emphasizes local resources, both physical and intellectual. Moreover, bioregionalism is not doctrinaire. A variety of philosophical approaches and different kinds of activism—not all of them congruent—participate in or derive from bioregionalism. Among the organizations and movements it has fostered are grassroots environmental efforts to preserve natural features (such as Oregon's Friends of Trees), watershed conservancy, the "locavore" effort (a movement devoted to the promotion of eating only locally produced food), coordinated resource management plans, farmers' markets, Green political parties, community-based alternative energy projects, and educational and research institutions, such as the Bioregional Congress.

Critics contend that bioregionalism is utopian and impractical, at least for large populations.

Nonetheless, the movement attracted increasing interest during the late twentieth and early twenty-first centuries among persons interested in establishing self-sufficient economies, in preserving wildlife, and in safeguarding air and water against pollution.

Roger Smith

Further Reading

McGinnis, Michael Vincent. *Bioregionalism*. New York: Routledge, 1999.

Sale, Kirkpatrick. *Dwellers in the Land: The Bioregional Vision*. Athens, Ga.: University of Georgia Press, 2000.

Thayer, Robert L., Jr. *LifePlace: Bioregional Thought and Practice*. Berkeley, Calif.: University of California Press, 2003.

■ Bioremediation

FIELDS OF STUDY
Ecology; Environmental Biotechnology; Environmental Chemistry; Environmental Microbiology; Waste Management; Water Treatment

SUMMARY
The environmentally beneficial and inexpensive waste management strategy of bioremediation enables the degradation of toxic organic and inorganic compounds into environmentally harmless products.

PRINCIPAL TERMS
- **anaerobic:** living, active, occurring, or existing in the absence of free oxygen
- **bioreactor:** fermentation vat to produce living organisms, as bacteria or yeast, used in industrial processes such as waste recycling or in the manufacture of drugs or other products
- **bioremediation:** waste management technology that employs naturally occurring plants, microorganisms, and enzymes or genetically engineered organisms to clean contaminated environments
- **contaminant:** to make impure, unsuitable or harmful by contact or mixture with something unclean or by adding radioactive material
- **phytoremediation:** a process of decontaminating soil or water by using plants and trees to absorb or break down pollutants
- **toxic waste:** chemicals resulting from manufacturing or industry and poisonous to living things

Bioremediation uses biological agents to degrade or decompose toxic environmental compounds into less toxic forms. It is a beneficial and inexpensive strategy for waste management that is environmentally friendly in comparison with other remediation technologies. The products of waste decomposition are usually simple inorganic nutrients or gases.

Bioremediation works because, as a rule, all naturally occurring compounds in the environment are ultimately degraded by biological activity. Toxic and industrial wastes, and even some chemically synthesized compounds that do not naturally occur, can also be decomposed because parts of their structures resemble naturally occurring compounds that are sources of carbon and energy for biological systems. Wastes are either metabolized, in which case they are used as a source of carbon and energy, or co-metabolized, in which case they are simply modified so that they lose their toxicity or are bound to organic material in the environment and rendered unavailable.

Bioremediation can occur in situ (at the contaminated site) or ex situ, in which case contaminated soil or water is removed to a treatment facility where bioremediation takes place under controlled environmental conditions. Bioremediation can use organisms that naturally occur at a site, or it can be stimulated through the addition of organisms, sometimes genetically engineered organisms, to the contaminated site in a process known as "seeding." The first organism ever patented was a genetically engineered bacterium that had been designed to degrade the components of oil.

Techniques
Numerous approaches to bioremediation have been developed. One of the simplest is to fertilize a contaminated site to optimal nutrient levels and allow naturally occurring biodegrading populations to increase and become active. Organic contaminants have been mixed with decomposed and partially decomposed organic material and composted as a bioremediation process. In a method analogous to the

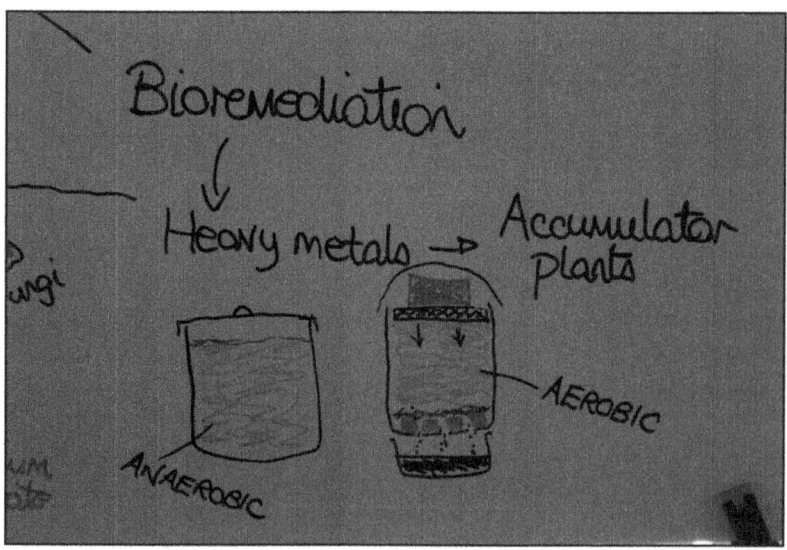

Lecture drawings by Aranya, from a Permaculture Design Course in Sweden, July 2011, showing bioremediation of heavy metals.

activated sludge process in wastewater treatment, contaminants are mixed in slurries and aerated to promote their decomposition. It is possible to obtain biosolids that are specially adapted for slurry systems because they have previously been exposed to similar organic wastes.

In situ restoration of contaminated groundwater is often accomplished through the injection of nutrients and oxygen into the aquifers to promote the population and activity of indigenous microorganisms. Trichloroethylene (TCE), for example, is cometabolized by methane-oxidizing bacteria and can be bioremediated through the injection of oxygen and methane into contaminated aquifers to stimulate the activity of these bacteria. Nitrate-contaminated aquifers have been successfully treated through the pumping of readily available carbon-containing methanol or ethanol into the aquifers to stimulate denitrifying bacteria, which subsequently convert the nitrate to harmless nitrogen gas.

Bioreactors have been used in which the contaminant is mixed with a solid carrier, or the organisms are immobilized to a solid surface and continuously exposed to the contaminant. This has been used with both bacteria and fungi. For example, *Phanerochaete chysosporium*, which produces an extracellular peroxidase and hydrogen peroxide (H_2O_2), has been used to cleave various organic contaminants such as dichloro-diphenyl-trichloroethane (DDT) in bioreactors.

Highly chlorinated organic contaminants such as TCE and polychlorinated biphenyls (PCBs) resist degradation aerobically, but the contaminants can be dechlorinated by anaerobic bacteria, which decreases their toxicity and makes them easier to decompose. High concentrations of PCBs in the Hudson River in New York have been dechlorinated to less toxic forms by anaerobic bacteria. Methanogens—anaerobic bacteria that produce methane—have been observed to dechlorinate TCE in anaerobic bioreactors.

One of the problems with some wastes is that they are mixed with radioactive materials that are highly toxic to living organisms. One solution to this problem has been the genetic engineering of radiation-resistant bacteria so that they also can bioremediate. For example, *Deinococcus radiodurans*, a bacterium that can survive in nuclear reactors, has been genetically engineered to contain genes for the metabolism of toluene, which will enable it to be used in the bioremediation of radiation- and organic waste-contaminated sites.

Phytoremediation

Phytoremediation is a special type of bioremediation in which plants—grasses, shrubs, trees, and algae—are used to biodegrade or immobilize environmental contaminants, usually metals. Types of phytoremediation include phytoextraction, in which the contaminant is extracted from soil by plant roots; phytostabilization, in which the contaminant is immobilized in the vicinity of plant roots; phytostimulation, in which the plant root exudates stimulate rhizosphere microorganisms that bioremediate the contaminant; phytovolatilization, in which the plant helps to volatilize the contaminant; and phytotransformation, in which the plant root and its enzymes actively transform the contaminant. For example, horseradish peroxidase is a plant enzyme that is used to oxidize and polymerize organic contaminants. The polymerized contaminants become insoluble and relatively unavailable.

Plants such as Indian mustard (*Brassica juncea*) and loco weed (*Astragalus*) are heavy metal accumulators and remove selenium and lead from soil. The aboveground plant parts are harvested to dispose of the metals. Algae are used to accumulate dissolved selenium in some treatments. Poplar trees have even been genetically engineered to contain a bacterial methyl reductase that lets them methylate and volatilize arsenic, mercury, and selenium absorbed by their roots.

Examples

A 1992 U.S. Environmental Protection Agency (EPA) survey indicated that of 132 well-documented bioremediation studies, 75 involved petroleum or related compounds, 13 involved wood preservatives such as creosote, 7 involved agricultural chemicals, 5 examined tars, 4 treated munitions such as trinitrotoluene (TNT), and the rest involved miscellaneous compounds. As this list suggests, bioremediation of oil spills has been the single best example of successful bioremediation in practice.

In March 1989, the *Exxon Valdez* oil tanker spilled millions of gallons of crude oil in Prince William Sound, Alaska. On many beaches, the EPA authorized the use of simple bioremediation techniques, such as stimulating the growth of indigenous oil-degrading bacteria by adding common inorganic fertilizers. Beaches cleaned by this method did as well as beaches cleaned by mechanical methods. In another instance of successful bioremediation, selenium-contaminated soil in the Kesterson National Wildlife Refuge in California was partially decontaminated in the 1980s through the method of supplying indigenous fungi with organic substrates such as casein and waste orange peels. This promoted as much as 60 percent selenium volatilization in less than two months.

A great deal of research into biodegradation and bioremediation of oil spills was conducted following the British Petroleum (BP) Deepwater Horizon oil spill in the Gulf of Mexico. Because the oil flowed from an offshore wellhead, affecting deep water, surface water, and coastal areas alike, the cleanup conditions differed from those experienced in previous spills. Moreover, oil-degrading bacteria were already prevalent at the spill location, feeding on naturally occurring oil seeps along the seafloor. Researchers proposed enhancing anaerobic degradation in affected marshes and introducing genetically modified bacteria. They also learned that the indigenous bacteria were not consuming polycyclic aromatic hydrocarbons (PAHs) and that certain microorganisms naturally inhabit water trapped within oil and feed off it, which may lead to improved methods of bioremediation in future.

Mark Coyne

Further Reading

Alexander, Martin. *Biodegradation and Bioremediation.* 2d ed. San Diego, Calif.: Academic Press, 1999.

Atlas, Ronald M., and Jim Philp, eds. *Bioremediation: Applied Microbial Solutions for Real-World Environmental Cleanup.* Washington, D.C.: ASM Press, 2005.

Boopathy, Raj, Sara Shields, and Siva Nunna. "Biodegradation of Crude Oil from the BP Oil Spill in the Marsh Sediments of Southeast Louisiana, USA." *Applied Biochemistry And Biotechnology* 167, no. 6 (July 2012): 1560–1568.

"A Citizen's Guide to Bioremediation." *EPA*, Sept. 2012. Accessed from https://clu-in.org/download/Citizens/a_citizens_guide_to_bioremediation.pdf.

Fingas, Merv. *The Basics of Oil Spill Cleanup.* 2d ed. Boca Raton, Fla.: CRC Press, 2001.

Frankenberger, William, and Sally Benson, eds. *Selenium in the Environment.* Boca Raton, Fla.: CRC Press, 1994.

Singh, V. P., and R. D. Stapleton, Jr., eds. *Biotransformations: Bioremediation Technology for Health and Environmental Protection.* New York: Elsevier Science, 2002.

Skipper, H. D., and R. F. Turco, eds. *Bioremediation: Science and Applications.* Madison, Wis.: American Society of Agronomy, 1995.

Stallard, Brian. "Oil Eaters: How Nature Cleans Up the Deepwater Horizon Spill." *Nature World News,* August 11, 2014.

Biosphere

FIELDS OF STUDY
Conservation Biology; Environmental Chemistry; Environmental Science; Geology; International Policy; Landscape Ecology; Land-use Management; Preservation; Public Policy

SUMMARY
The conceptualization of all life on earth as part of one large, integrated system, the biosphere, helps to encourage understanding of the interrelatedness of all ecosystems and biotic communities.

PRINCIPAL TERMS
- **biodiversity:** diversity among and within plant and animal species in an environment.
- **biomes:** a complex biotic community characterized by distinctive plant and animal species and maintained under the climatic conditions of the region
- **biosphere reserves:** UNESCO-designated sites where the preservation of natural resources is integrated with research and the sustainable management of those resources
- **biosphere:** the zone within which all life on Earth exists
- **carbon cycle:** the circulation of carbon atoms in the biosphere as a result of photosynthetic conversion of carbon dioxide into complex organic compounds by plants, then consumed by other organisms
- **lithosphere:** the crust and mantle or solid portion of Earth
- **water cycle:** continuous cycle whereby water changes from vapor in the atmosphere to liquid water through condensation and precipitation and then back to water vapor through evaporation, transpiration, and respiration

The concept of the biosphere was introduced in the nineteenth century by Austrian geologist Eduard Seuss. The biosphere is the zone, approximately 20 kilometers (12 miles) thick, that extends from the floors of the earth's oceans to the tops of the mountains, within which all life on the planet exists. It is thought to be more than 3.5 billion years old, and it supports nearly one dozen biomes, regions of similar climatic conditions within which distinct biotic communities reside.

Compounds of hydrogen, oxygen, carbon, nitrogen, potassium, and sulfur are cycled among the four major spheres—biosphere, lithosphere, hydrosphere, and atmosphere—to make the materials that are essential to the existence of life. The most critical of these compounds is water, and its movement between the spheres is called the water cycle. Dissolved water in the atmosphere condenses to form clouds, rain, and snow. The annual precipitation for a region is one of the major controlling factors in determining the terrestrial biome that can exist. The water cycle follows the precipitation through various paths leading to the formation of lakes and rivers. These flowing waters interact with the lithosphere to dissolve chemicals as they flow to the oceans, where about one-half of the biomes on earth occur. Evaporation of water from the oceans resupplies most of the moisture existing in the atmosphere. This cycle supplies water continuously for the needs of both the terrestrial and the oceanic biomes.

The biosphere is also dependent on the energy that is transferred from the various spheres. The incoming solar energy is the basis for all life. Light enters the life cycle as an essential ingredient in the

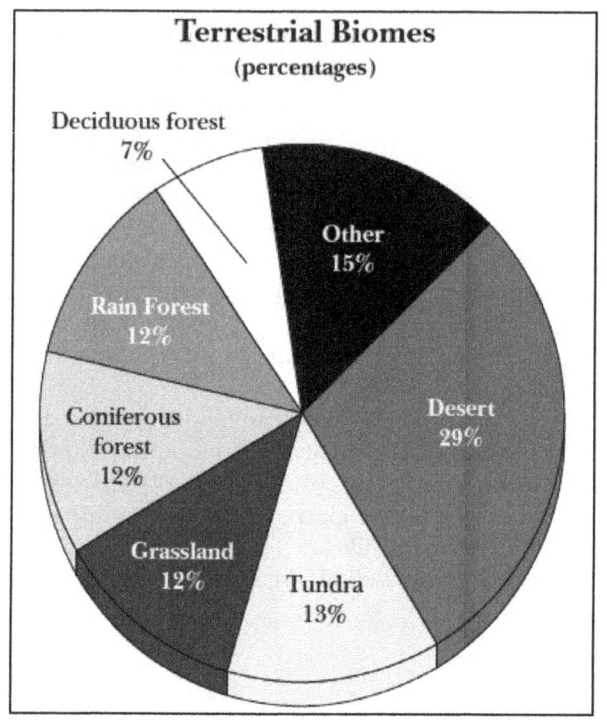

photosynthesis reaction. Plants take in carbon dioxide, water, and light energy, which is converted into chemical energy in the form of sugar, with oxygen generated as a by-product. Most animal life reverses this process during the respiration reaction, where chemical energy is released to do work by the oxidation of sugar to produce carbon dioxide and water.

The incoming solar energy also has a dramatic interaction with the water cycle and the worldwide distribution of biomes. Because of the earth's curvature, the equatorial regions receive a greater amount of solar heat than do the polar regions. Convective movements in the atmosphere (such as winds, high- and low-pressure systems, and weather fronts) and the hydrosphere (such as water currents) are generated during the redistribution of this heat. The weather patterns and general climates of earth are responses to these energy shifts. The seven types of climates are defined by mean annual temperature and mean annual precipitation, and there is a strong correspondence between the climate at a given location and the biome that will flourish.

Desert Biome
The major deserts of the world are located between 20 to 30 degrees latitude north and south of the equator. The annual precipitation in a desert biome is less than 25 centimeters (10 inches) per year. Deserts are located in northern and southwestern Africa, parts of the Middle East and Asia, Australia, the southwestern United States, and northern Mexico.

Deserts are characterized by life that is unique in its ability to capture and conserve water. Deserts show the greatest extreme in temperature fluctuations of all biomes: Daytime temperatures can exceed 49 degrees Celsius (120 degrees Fahrenheit), and night temperatures can drop to 0 degrees Celsius (32 degrees Fahrenheit). Most of the animals that live in desert biomes are active at night and retreat to underground burrows or crevices during the day to escape the heat. The water cycle in deserts rarely provides surface water, so plant life usually finds water through a wide distribution of shallow roots to capture the near-surface infiltration or a deep taproot system that finds groundwater located below the surface of dry streambeds. The plant life is characterized by scattered thorny bushes, shrubs, and occasional cacti. Animal life consists of an abundance of reptiles (mostly lizards and snakes), rodents, birds (many predatory types such as owls and hawks), and a wide variety of insects.

Deserts and semideserts cover approximately one-third of the land surface on earth. They continue to grow because of human influences such as deforestation and overgrazing.

Grassland Biome
Grasslands are found in a wide belt of latitudes higher than those in which desert biomes exist. Large grassland regions occur in central North America, central Russia and Siberia, subequatorial Africa and South America, northern India, and Australia. This biome flourishes in moderately dry conditions, having an annual rainfall between 25 and 150 centimeters (10 and 60 inches). Precipitation and solar heating are unevenly divided throughout the year, providing a wet, warm growing season and a cool, dry dormant season.

The animal life in grassland regions is characterized by large grazing mammals, such as wild horses, bison, antelopes, giraffes, zebras, and rhinoceroses, as well as smaller herbivores, such as rabbits, prairie dogs, mongooses, kangaroos, and warthogs. This abundance of herbivores allows for a large development of secondary and tertiary consumers in the food chain, such as lions, leopards, cheetahs, wolves, and coyotes. Grasslands have rich soils that provide the fertile growing conditions for a wide variety of tall and short grasses. Within a single square meter of this healthy soil, several hundred thousand living organisms can be found, from microbes to insects, beetles, and worms. The profusion of these smaller life-forms fosters an abundance of small birds.

Grasslands have been environmentally stressed as humans have converted them to farmland because of their rich soils and to rangeland because of the grass supply. It is estimated that only 25 percent of the world's original grasslands remain undisturbed by human development. Worldwide, overgrazing and mismanagement of rangeland have caused large tracts of fertile grassland to become desert or semidesert.

Tropical Rain-Forest Biome
Rain forests receive heavy rainfall almost daily, with an annual average of more than 240 centimeters (95 inches). Temperatures in these areas are fairly constant from day to day and season to season, with an annual mean value of about 28 degrees Celsius (82

degrees Fahrenheit). The combination of plentiful rain and high temperatures causes high humidity, allowing some plants to utilize the atmosphere for their water supply via "air roots." This biome is also unique because the chemical nutrients needed to sustain life within it are almost entirely contained in the lush vegetation of the biosphere itself and not in the upper layers of soil of the lithosphere. The soils are thin and poor in nutrients.

The tropical rain forests contain a wider diversity of plant and animal species per unit area than any other biome. It is estimated that nearly two-thirds of all the plants and insects found on earth are contained in tropical rain forests. This enormous biodiversity is accommodated in part because each form of plant or animal occupies a specialized niche based on its ability to thrive with a level of sunlight that corresponds to a given height above the forest floor within the forest canopy. Numerous exotic insects, amphibians, reptiles, birds, and small mammals can coexist within a single canopy level. The plants growing in tropical rain forests currently provide ingredients found in 25 percent of the world's prescription and nonprescription drugs. It is estimated that at least three thousand tropical plants contain cancer-fighting chemicals.

Tropical rain forests are being destroyed at a fast rate for farmland, timber operations, mining, and grazing. It has been estimated that most of the animal and plant diversity of the rain forests will be lost by the year 2050. Further, if the high rate of destruction continues, the release of carbon dioxide into the atmosphere from the burning of biomass and fossil fuels may no longer be offset by plant consumption, and the balance of the carbon cycle will shift toward a higher concentration of carbon dioxide in the atmosphere.

Temperate Forest Biome
Temperate forests exist in areas where temperatures change dramatically during the four distinct seasons. Temperatures fall below freezing during the winter, with warmer, more humid conditions during the summer. Rainfall averages between 75 and 200 centimeters (30 and 80 inches) per year. This biome is often divided into forests with broad-leaved deciduous (leaf-shedding) trees and those with coniferous (cone-bearing) trees. Deciduous forests develop in regions with higher precipitation values, whereas the needle-like evergreen leaves of conifers have scales and thick, waxy coatings that allow them to flourish at the lower end of the precipitation range.

Deciduous forests develop more solid canopies with widely branching trees such as elm, oak, maple, ash, beech, and other hardwood varieties. The forest floor often contains an abundance of ferns, shrubs, and mosses. Coniferous forests are usually dominated by pine, spruce, fir, cedar, and hemlock trees. Some deciduous trees, such as aspen and birch, often occur with the conifers. The coniferous forest floor is so acidic from decomposing evergreen needles that often only lichens and mosses can grow.

Temperate forests are home to diverse animal life. Common mammals of deciduous forests include squirrels, chipmunks, porcupines, raccoons, opossums, deer, foxes, black bears, and mice. Snakes, toads, frogs, and salamanders exist alongside a large bird population of thrushes, warblers, woodpeckers, owls, and hawks. The larger animals of the coniferous forests include moose, elk, wolves, lynx, grouse, jays, and migratory birds.

Forests have traditionally been viewed as a limitless timber resource by the lumber industry. The vast forests of Europe were cleared one thousand years ago. This liquidation of temperate forest continues in Siberia, where 25 percent of the world's timber reserves exist.

Tundra Biome
Tundras occur in areas near the Arctic ice cap and extend southward across the far northern parts of North America, Europe, and Asia. During the majority of the year, these largely treeless plains are covered with ice and snow and are battered by bitterly cold winds. Tundras are covered with thick mats of mosses, lichens, and sedges (grass-like plants). Because the winters are long and dark, tundra vegetation grows during the three months of summer, when there is almost constant sunlight.

Bogs, marshes, and ponds are common on the summer landscape because permafrost, a thick layer of ice that remains beneath the soil all year long, prevents drainage of melted waters. These wet areas provide perfect breeding grounds for mosquitoes, deerflies, and blackflies during the brief summer. These insects in turn serve as a source of food for migrating birds. Larger mammals, such as caribou, reindeer, musk ox, and mountain sheep, migrate in and out of the tundra. Some animals, such as lemmings, arctic

Composite image of the earth's biosphere shows the planet's heaviest vegetative biomass in the dark sections, known to be rain forests. (NASA)

hares, grizzly bears, and snowy owls, can be found in the tundra during all times of the year.

The tundra is the Earth's most fragile terrestrial biome. Vegetation disturbed by human activity can take decades to replenish itself. Roads and pipelines must be constructed on bedrock or layers of added gravel; otherwise, they will melt the upper layers of the permafrost.

Oceanic Biomes
Oceans cover 70 percent of the earth's surface and contain 97.6 percent of the water of the hydrosphere. They play a primary role in regulating the earth's distribution of heat, and they are central to the water cycle. Oceans are instrumental in the survival of all life on earth. In addition, oceans house more than 250,000 species of marine plants and animals that occur as six common biomes.

Coral reefs are coastal biomes that develop on continental shelves in regions of clear, tropical waters. Collectively, coastal biomes make up only 10 percent of the world's ocean area but contain 90 percent of all ocean species. These are the regions where most commercial fishing is done. The vast open oceans contain only 10 percent of all oceanic species. Vegetation is mostly limited to free-floating plankton. Exotic bottom fauna exists on deep hydrothermal vents. Animal life includes whales, dolphins, tuna, sharks, flying fish, and squids.

Several transitional biomes also exist at the ocean-land interface. The intertidal biome can be composed of sandy beaches or more rocky zones that are covered by water only during periods of high tide. A variety of crustaceans and mollusks are found on wet sandy beaches, whereas rock tidal pools contain kelp, Irish moss, and rockweed, all of which compete for space with snails, barnacles, sea urchins, and starfish.

Wilderness Issues
Through discussions that started in 1970, the United Nations Educational, Scientific, and Cultural Organization (UNESCO) initiated the Man and the Biosphere Programme to establish sites where the preservation of natural resources would be integrated with research and the sustainable management of those resources. The first biosphere reserve was designated in 1976, and by May 2009, 553 such sites existed across 107 countries. In 1995 UNESCO convened a conference in Spain and developed the Seville Strategy for Biosphere Reserves, which was designed to strengthen the international network and encourage the use of the sites for research, monitoring, education, and training. In the early years of the program, preservation was stressed. The adoption of the Seville Strategy by UNESCO emphasized the role of people in the use of their natural resources.

Each biosphere reserve contains a legally protected core area where there has been minimal disturbance by people. Only uses that are compatible with the preservation of biological diversity are permitted in the protected core. Surrounding the core is a managed-use or buffer zone; research and environmental education are examples of activities suitable for the buffer zone. Surrounding the buffer zone is a zone of cooperation or transition zone. The boundaries of the transition zone are loosely defined and often include local towns and communities. Economic activities such as farming, logging, mining, and recreation occur within the transition zone and are not restricted by the biosphere reserve.

Only the boundaries of the core area are legally defined. The designation of a biosphere reserve does not alter the legal ownership of the land or water that is included within its zones. UNESCO does not have jurisdiction over any nation's biosphere reserves. In many cases the areas within reserves reflect a mosaic of landownership, including federal, state, local, and private ownership. Even the core area may be privately owned, if it is managed for preservation.

In the United States, the core areas of some biosphere reserves are within national parks, such as Glacier and Yellowstone, whereas other biosphere reserves are composed of clusters of core areas, such as the ten units within the California Coastal Range Biosphere Reserve. The management and administration of a biosphere reserve often involves interested citizens, government agencies, and owners.

The worldwide network of biosphere reserves represents the only international network of protected areas that also emphasizes sustainable development and wise use of natural resources. Hence, they are sites where the objective of integrating conservation and development can be examined, demonstrated, and tested. Research at these sites serves to solve practical problems in resource management.

Toby Stewart, Dion Stewart and William R. Teska

Further Reading

Butz, Stephen D. *Science of Earth Systems*. 2nd ed. Clifton Park, N.Y.: Thomson, 2008. Print.

Ehrlich, Paul R., and Anne H. Ehrlich. *Extinction: The Causes and Consequences of the Disappearance of Species*. New York: Ballantine, 1981.

Hanna, Kevin S., Douglas A. Clark, and D. Scott Slocombe, eds. *Transforming Parks and Protected Areas: Policy and Governance in a Changing World*. New York: Routledge, 2008.

Krebs, Robert E. "Biosphere: Envelope of Life." In *The Basics of Earth Science*. Westport, Conn.: Greenwood, 2003.

McNeely, Jeffrey A., et al. *Conserving the World's Biological Diversity*. Washington, D.C.: Island, 1990.

Smil, Vaclav. *The Earth's Biosphere: Evolution, Dynamics, and Change*. Cambridge, Mass.: The MIT Press, 2002.

Sourd, Christine. *Explaining Biosphere Reserves*. Paris: UNESCO, 2004.

Weiner, Jonathan. *The Next One Hundred Years*. New York: Bantam, 1990.

Wilson, E. O., ed. *Biodiversity*. Washington, D.C.: National Academy, 1990.

Woodward, Susan L. *Introduction to Biomes*. Westport, Conn.: Greenwood, 2009.

■ Boreal and cloud forests

FIELDS OF STUDY
Conservation Biology; Ecology; Environment; Forestry; Landscape Ecology; Land-use Management; Silviculture; Urban Planning and Development

SUMMARY
Boreal forests play an important role in the cultural identities of people living in many parts of the world, particularly Canada and Siberia. These forests, which store large amounts of carbon and thus are a crucial element in the global carbon cycle, are threatened by resource development and climate change. Cloud forests are rich in biodiversity and act as important sources of water conservation. These valuable forest ecosystems are strongly threatened by human activity, particularly by deforestation, unsustainable agricultural use, and overharvesting of plants and animals.

PRINCIPAL TERMS
- **boreal forests:** coniferous forests, also called taiga, within the Subarctic land biome, existing in a nearly continuous band throughout the northern regions of North America, Europe, and Asia
- **cloud forests:** forests on moist tropical mountain slopes covered by tree-level clouds
- **deciduous:** trees and shrubs that shed leaves annually
- **deforestation:** the clearing of trees, transforming a forest into cleared land
- **epiphytes:** plants that grow above the ground, supported nonparasitically by other plants or objects, and deriving nutrients and water from rain, the air, or dust
- **temperate forests:** woodlands of usually mild climatic areas within the temperate zone that receives heavy rainfall, including numerous kinds of trees
- **tundra:** level or rolling treeless plain characteristic of arctic and subarctic regions, consisting of black mucky soil with a permanently frozen subsoil, and dominant vegetation of mosses, lichens, herbs, and dwarf shrubs
- **watershed:** the region or area drained by a river, stream

Boreal forests, or taiga, make up the world's largest land biome, covering vast parts of the Northern Hemisphere near 50 degrees of latitude and forming an ecologically sensitive habitat between the Arctic tundra and the temperate forest. Boreal forests extend through most of inland Canada, parts of Scandinavia, Russia (particularly Siberia), northern Mongolia and parts of central Asia, and northern Japan. They extend southward to parts of the Scottish highlands and the northern continental United States. Boreal forests are characterized by the dominance of conifers, such as firs, pines, spruces, hemlocks, and larches, and by a mean annual temperature between –5 and 5 degrees Celsius (23 and 41 degrees Fahrenheit).

Arctic bush along The Campbell Highway. By David Adamec (Own work)

Cloud forests cover humid tropical mountain slopes ranging from Central and South America to Africa, South and Southeast Asia, Papua New Guinea, and Pacific islands such as Hawaii. They are created when moist air meets mountain barriers and forms clouds covering the treetops. Because of prevailing wind directions, cloud forests generally cover eastern mountain slopes. They typically are found on mountains that range between 1,200 and 2,500 meters high (4,000 to 8,200 feet), but some cloud forests can be found on peaks as low as 300 meters (1,000 feet) on Pacific islands and as high as 3,500 meters (11,500 feet) in the Andes of South America and the Ruwenzori range of Uganda.

Boreal Forests

Boreal forests typically have short, cool summers and long, cold winters. Precipitation can vary from 20 to 200 centimeters (approximately 8 to 79 inches) per year and falls mostly as snow. Because there is little evaporation, the ground remains moist during the summer growing season. Owing to the high latitude, summer days are very long and winter days are very short.

In addition to conifers, mosses and lichens form an important part of the boreal forest ecosystem. Some broad-leaved deciduous trees, such as birch, aspen, and willow, are also present in boreal forests.

While low in biodiversity when compared with other biomes, boreal forests are home to a variety of animals as well as plants, many of them endangered. Animals found in boreal forests include reindeer or caribou; carnivores such as bear, wolverine, lynx, fox, weasel, and wolf; and more than three hundred species of birds.

Boreal forests rely on natural cycles of fire and insect damage followed by new tree growth; in North American boreal forests, these cycles are typically seventy to one hundred years long. Boreal forests store large amounts of carbon, perhaps more than the world's tropical and temperate forests combined.

The primary threats to boreal forests are logging, development of oil and natural gas reserves, and climate change. Large areas of Siberian and Canadian boreal forests have been logged, often by clear-cutting. Substantial reserves of oil and natural gas exist under boreal forests in Alaska, Canada, and Russia. As world oil demand and prices rise and technologies for working in very cold climates improve, these reserves become possible candidates for development. It is unclear whether slow-growing boreal forests would be able to recover from the environmental impacts that accompany oil and natural gas extraction.

Although boreal forests require fire and insect infestations for their natural cycle of death and renewal, the increased frequency of both that is associated with climate change may have negative effects. Also, as temperatures rise, deciduous broad-leaved trees can outcompete conifers in southern regions.

Several countries have enacted measures to protect boreal forests through improved management of resource development and conservation. For example, Canadian logging companies are generally required to replant or encourage natural forest renewal, and in 2010 the Canadian federal government enlarged the areas of boreal forests to be protected from development.

Cloud Forests

The higher the cloud forest, the smaller the trees tend to be, and the thicker and more gnarled are their stems and branches. The leaves of the trees are generally tough and small. The trees host a great variety of epiphytes (other plants that grow on them), such as bromeliads, orchids, lichens, mosses, and ferns. The soils are very moist and rich in organic material as they contain much humus and peat.

Cloud forests function as unique watersheds. The leaves on the top branches of the trees catch the moisture of the clouds driven there by the wind and let it drip to the forest floor. This process, scientifically called occult precipitation and popularly known as cloud stripping, accounts for doubling rainfall in dry seasons and still increasing it by about 10 percent in wet seasons when compared to areas outside the cloud forests. Cloud forests also act as water reservoirs, preventing runoff during rain and supplying a steady flow of water in dry times.

Cloud forests have come under severe threat by human activity. In 1974, cloud forests were estimated to cover 50 million hectares (124 million acres), or one-fourth of the hills and mountains of the Tropics. A 1990 survey by the United Nations Food and Agriculture Organization found that the annual deforestation rate in tropical mountain and highland forests was 1.1 percent for the decade of 1980 to 1990, and although the rate of deforestation subsequently slowed somewhat, it has been estimated that by the early years of the twenty-first century, as little as 35 million hectares (87 million acres) of cloud forests were in existence.

Initially, cloud forests enjoyed some natural protection because of their inaccessibility and the poor quality of their soil and timber for human uses. As farmers, ranchers, and loggers overexploited the tropical forests below cloud forests, however, they shifted their attentions upward into the cloud forests. The most severe threats to cloud forests are deforestation for cropland and tree cutting for fuelwood. In addition, unsustainable harvesting of plants such as rare orchids and ferns and the trapping and hunting of amphibians, birds, and mammals unique to the cloud forests threaten to destroy these ecosystems and lead to species extinction. In the Andes, the use of cloud forests for the planting of illegal crops such as coca (from which cocaine is derived) is another threat. Even tourism and recreation can threaten cloud forests if these are undertaken in an ecologically unsound manner.

Since the late twentieth century, environmentalists, policy makers, and the general public have grown increasingly aware that cloud forests need to be protected from degradation by humanity. The successful management of sustainable use of these forests depends on the participation of the forests' indigenous populations and on the creation of both economic incentives and legal sanctions supporting the forests' protection.

Melissa A. Barton and R. C. Lutz

Further Reading

Elliot-Fisk, Deborah L. "The Taiga and Boreal Forest." In *North American Terrestrial Vegetation*, edited by Michael G. Barbour and William Dwight Billings. New York: Cambridge University Press, 2000.

Gradstein, S. Robbert, Jürgen Homeier, and Dirk Gansert, eds. *The Tropical Mountain Forest: Patterns and Processes in a Biodiversity Hotspot.* Göttingen, Germany: Universitätsverlag Göttingen, 2008.

Haber, William. *An Introduction to Cloud Forest Trees.* 2d ed. Monteverde de Puntarenas, Costa Rica: Mountain Gem, 2000.

Hari, Pertti, and Liisa Kulmala, eds. *Boreal Forest and Climate Change.* Berlin: Springer, 2008.

Henry, J. David. *Canada's Boreal Forest.* Washington, D.C.: Smithsonian Institution Press, 2001.

Scatena, F. N., and L. S. Hamilton. *Tropical Montane Cloud Forests: Science for Conservation and Management.* New York: Cambridge University Press, 2010.

Brownfields

FIELDS OF STUDY
Ecology; Environment; Landscape Ecology; Land-use Management; Urban Planning and Development

SUMMARY
Governments have increasingly encouraged the redevelopment of brownfields, requiring that such redevelopment be preceded by the cleanup of any potentially hazardous substances in the soil and water on these abandoned industrial sites. In addition to the benefits of such cleanup for the environment, brownfield redevelopment reduces urban sprawl and improves property values.

PRINCIPAL TERMS
- **brownfields:** abandoned lots or other properties that were once used for commercial or industrial purposes and have the potential to contain harmful contaminants or pollutants
- **Environmental Protection Agency (EPA):** an agency of the United States government, established in 1970 for environmental protection
- **Housing and Urban Development (HUD):** agency of the United States government, established in 1965 to support community development, homeownership, improving affordable homeownership opportunities, increasing safe and affordable rental options, reducing chronic homelessness, fighting housing discrimination by ensuring equal opportunity in the rental and purchase markets, and supporting vulnerable populations.
- **Superfund:** a United States government program, established in 1980 as the Comprehensive Environmental Response, Compensation, and Liability Act and designed to fund the cleanup of sites contaminated with hazardous substances and pollutants
- **sustainability:** the quality of not being harmful to the environment or depleting natural resources, and thereby supporting long-term ecological balance

Brownfields are found in cities and towns throughout the world, usually located in industrial areas. The potential hazards and liability issues associated with such sites' past uses become important when these areas are targeted for redevelopment. In order for municipalities to redevelop brownfields, they must determine liability for any problems on the sites and take any cleanup measures required by law. In the United States, the Comprehensive Environmental Response, Compensation, and Liability Act, widely known as Superfund, mandates liability laws in regard to the cleanup and redevelopment of brownfields and also sets cleanup and development standards. The redevelopment of brownfields can be beneficial to cities, promoting urban and community revitalization through sustainable reuse of land, reducing sprawl, increasing local property taxes, promoting jobs and economic development, and protecting the environment from harmful substances and pollutants.

Barriers to Redevelopment
In the United States, when redevelopment of a brownfield is being considered, the U.S. Environmental Protection Agency (EPA), a state agency, or both must determine the kind and extent of cleanup required and help to determine the liability for the cleanup. Cleanup efforts for brownfield redevelopment are often costly, and it can be difficult for developers to obtain loans to develop potentially hazardous land. Because this is the case, federal, state, and municipal governments may offer some financial contributions under the EPA's Brownfields Initiative, the U.S. Department of Housing and Urban Development (HUD)'s Brownfields Economic Development Initiative, or state and municipal funding programs.

The potential presence of harmful substances and pollutants can increase the difficulty of assessing and cleaning up a brownfield site. Depending on their previous uses, brownfields may contain many hazardous substances, such as petroleum, lead, asbestos, hydrocarbon, pesticides, tributyltins, solvents, and diesel fuels; cleanup is thus imperative before such sites are developed. Sites that are determined to be harmful enough to be hazardous to individuals living and working around them are not considered brownfields; rather, these are designated as Superfund sites and are placed on either the Superfund National Priority List or a state priority list for cleanup.

Maine Central Railroad Depot, East Brownfield, Maine.

Laws and Regulations

The term "brownfields" was first introduced in 1992 during a congressional hearing. The first legislation concerning brownfields was passed in 1995 when the EPA created its Brownfields Program. Through this program, the EPA seeks to motivate communities, states, and organizations to work together to improve local brownfield sites and use them as resources for redevelopment. Since the 1990's U.S. states and municipalities, along with the federal government, have created programs to provide tax incentives and grants for developers, organizations, and governments to redevelop brownfield sites.

The Small Business Liability Relief and Brownfields Revitalization Act, enacted in 2002, provides support for small businesses to redevelop brownfields through financial assistance and state and local program promotion. HUD's Brownfields Economic Development Initiative provides grants to cities to promote economic and community development; this program is directed toward development in low- and moderate-income neighborhoods, with the aim of promoting economic stability, job creation or retention, business improvement, and increased property taxes.

Benefits of Brownfield Redevelopment

As land pressures and urban sprawl have become increasingly problematic, brownfield redevelopment has grown in importance. The redevelopment of brownfields allows developers to use existing properties within cities rather than create developments that sprawl outward from urban centers, possibly disrupting functioning ecosystems. Additionally, brownfield redevelopment projects often increase the property values in their surrounding areas because of the improvements made to the previously abandoned sites. Brownfield redevelopment also reduces the potential for harm to humans and other forms of life from environmental hazards and pollution, as cleanup of all hazardous substances is a requirement of redevelopment of any brownfield site.

Courtney A. Smith

Further Reading

Davis, Todd S. *Brownfields: A Comprehensive Guide to Redeveloping Contaminated Property.* 2d ed. Chicago: American Bar Association, 2002.

De Sousa, Christopher. *Brownfields Redevelopment and the Quest for Sustainability.* Boston: Elsevier, 2008.

Dixon, Tim, et al., eds. *Sustainable Brownfield Regeneration: Livable Places from Problem Spaces.* Malden, Mass.: Blackwell, 2007.

Witkin, James B. *Environmental Aspects of Real Estate and Commercial Transactions: From Brownfields to Green Buildings.* 3d ed. Chicago: American Bar Association, 2005.

C

Carbon cycle and carbon footprint

FIELDS OF STUDY
Biology; Chemicals; Chemicals; Chemistry; Compounds; Ecosystems; Elements; Geology; Geosciences; Life Sciences; Substances

SUMMARY
The balance of the carbon cycle determines the atmospheric concentration of carbon dioxide, which, through its role as a greenhouse gas, modulates the earth's temperature. Human activities that affect the carbon cycle thus influence the global climate. The total emissions caused by an individual, event, organization, or product, expressed as carbon dioxide equivalent is historically defined as a carbon footprint.

PRINCIPAL TERMS
- **carbon cycle:** pathways by which carbon moves through the environment
- **carbon dioxide:** a colorless, odorless, incombustible gas present in the atmosphere and formed during respiration, usually obtained from coal, coke, or natural gas by combustion
- **carbon footprint:** the amount of greenhouse gases, specifically carbon dioxide emitted by a person's activities or a product's manufacture and transport during a given period
- **ecological footprint:** measure used to quantify and assess the impact of human activities on ecosystems
- **greenhouse gas:** any of various gaseous compounds, such as carbon dioxide or methane, that absorb infrared radiation, trap heat in the atmosphere, and contribute to the greenhouse effect
- **photosynthesis:** the process by which carbon dioxide, water, and certain inorganic salts are converted into carbohydrates by green plants, algae, and certain bacteria, using energy from the sun and chlorophyll

Carbon is naturally exchanged between the atmosphere and the oceans, the terrestrial biosphere and soils, and the solid earth. The preindustrial balance of these carbon exchanges led to an atmospheric carbon dioxide concentration of 280 parts per million. Human industrial activities have added carbon dioxide to the atmosphere, increasing the atmospheric concentration to 380 parts per million.

Carbon is naturally removed from the atmosphere by photosynthesis on land and by dissolution of atmospheric carbon dioxide in the oceans. Carbon is introduced into the atmosphere by respiration and combustion of terrestrial organic matter, by outgassing of carbon dioxide from the oceans, and as a by-product of human industrial activities. Terrestrial plants remove about 15 percent of the atmosphere's carbon dioxide each year through photosynthesis. About half of this carbon is respired by these same plants as they release energy for their internal metabolic processes. The other half of the carbon removed from the atmosphere by terrestrial photosynthesis is primarily returned to the atmosphere through respiration of organic matter by decomposers. Most of this carbon is returned to the atmosphere as carbon dioxide through aerobic respiration, but in low-oxygen wetland environments carbon can be respired anaerobically and released to the atmosphere as methane, a much more potent greenhouse gas. This methane is then broken down to carbon dioxide in the atmosphere over a timescale of about eight years.

Approximately 12 percent of the atmosphere's carbon dioxide is exchanged with the oceans each year. Carbon is more soluble in cold water than in hot water, and so the oceans take up carbon at high northern latitudes and emit carbon in the Tropics. Photosynthesis in the surface ocean binds dissolved carbon into organic matter. Much of this carbon is returned to the surface ocean by breakdown of this organic matter, but some of this material is packaged into large enough clumps of organic matter that it sinks under its own weight and is re-dissolved in the deep ocean. This process, known as the biological pump, acts to move carbon (and nutrients) from the surface ocean to depth. In ocean upwelling zones, carbon-rich water from depth is brought up to the surface, producing a source of carbon to the atmosphere. Exchanges of carbon between the solid earth and the other reservoirs are important only on geologic timescales.

Human industrial activity introduces an amount of carbon to the atmosphere each year equivalent to about 1 percent of the total atmospheric carbon content. The rate of atmospheric carbon dioxide increase is, however, only half the rate at which industrial activity introduces carbon dioxide to the atmosphere. The other half is removed from the atmosphere by dissolution in the oceans and uptake by the land surfaces.

Carbon Footprint

The measurement of the environmental impact of an organization, event, person or product is called carbon footprint. It measures the total greenhouse gas emissions of a given activity or group of activities to provide a calculation of the extent to which these activities produce harmful emissions and therefore contribute to global warming. A carbon footprint is difficult to calculate. But once a carbon footprint has been established, efforts can be made reduce greenhouse gas emissions. These can include better process and product management, technological developments, carbon capture, and consumption modification strategies. Furthermore, projects such as solar and wind energy or reforestation can also be executed to reduce greenhouse gases (GHGs). Cloud computing is also considered an important step toward greener computing.

Carbon-cycle models predict that global warming will trigger processes that will alter the natural carbon cycle in a way that will decrease the ability of natural processes to remove industrial carbon from the atmosphere and accelerate the rate of increase of atmospheric greenhouse gases. Warming of the oceans decreases the solubility of carbon dioxide in water. Decomposition of organic material in soils to produce carbon dioxide proceeds more rapidly at higher temperatures. Permanently frozen soils at high latitudes (permafrost) globally hold more carbon than the atmosphere, and this carbon becomes exposed to decomposition when the permafrost thaws.

Alexander R. Stine

Further Reading

Field, Christopher B., and Michael R. Raupach, eds. *The Global Carbon Cycle: Integrating Humans, Climate, and the Natural World.* Washington, D.C.: Island Press, 2004.

Franchetti, Matthew John and Defne Apul. *Carbon Footprint Analysis: Concepts, Methods Implementation, and Case Studies.* Abingdon, UK: CRC Press/Taylor and Francis, 2012.

Houghton, R. A. "The Contemporary Carbon Cycle." *Treatise on Geochemistry* 8, no. 10 (2003): 473-513.

Wigley, T. M. L., and D. S. Schimel, eds. *The Carbon Cycle.* New York: Cambridge University Press, 2000.

■ Carbon dioxide air capture

FIELDS OF STUDY

Ecology; Emissions; Environmental Engineering; Hazardous Materials; Industries; Commercial Products; Inventions; Pollution; Technology and Applied Science; Toxic Waste

SUMMARY

Carbon dioxide capture and sequestration or storage is anticipated to play a bridging role between carbon dependence and a sustainable low-carbon energy future by serving as the critical enabling technology that will lead to a significant reduction in CO_2 release into the air while allowing industrial processes such as the burning of coal for power generation to continue to meet global energy needs.

PRINCIPAL TERMS

- **carbon capture and storage:** the process of capturing waste carbon dioxide from large point sources, such as fossil fuel power plants, transporting it to a storage site, and depositing it where it will not enter the atmosphere
- **carbon dioxide air capture:** trapping or elimination of carbon dioxide emitted from industrial or commercial sources before it can enter the atmosphere

- **carbon dioxide:** a colorless, odorless, incombustible gas present in the atmosphere and formed during respiration, usually obtained from coal, coke, or natural gas by combustion

The term "carbon dioxide air capture" is used to describe a set of technologies aimed at preventing the carbon dioxide (CO_2) emitted from industrial or commercial sources from entering the atmosphere. The processes involved are also commonly referred to as carbon dioxide capture and sequestration or carbon capture and storage (CCS). The geoengineering technique of scrubbing (absorbing) CO_2 from ambient air is also sometimes referred to as CSS, as are biological techniques that employ organisms (such as plankton) and organic matter to capture CO_2 from the air.

In CCS, the process of capturing CO_2 from a large emissions source is often coupled with the subsequent compression of the gas, storage of the gas (for example, by injecting it into deep underground geological formations or into deep ocean masses called saline aquifers, or by converting it into the form of mineral carbonate), and recycling of the gas to enhance industrial processes, such as that seen in CO_2-assisted enhanced oil recovery (EOR), in which CO_2 gas is injected into an oil-bearing stratum under high pressure to cause oil to be displaced upward.

The CCS process begins with the capture of CO_2 generated by a power station or other industrial facility, such as a cement factory, steelworks, or oil refinery. The CO_2 can be captured before, during, or after the source's combustion (burning) of fossil fuels. Pre-combustion capture involves the separation of fossil fuels into hydrogen and CO_2 before they are burned. For example, in the instance of coal, the process involves the conversion of coal into a synthetic gas (syngas) made up of carbon monoxide and hydrogen. The syngas is reacted further with steam to produce a CO_2-hydrogen mix. Further processing produces a mix with a high concentration of CO_2, which is separated out. The remaining hydrogen is then utilized as a CO_2-free energy source that produces only heat and water vapor when combusted. Pre-combustion capture technology has been widely implemented in the fertilizer industry as well as in natural gas forming.

In oxy-fuel combustion, burning of the fossil fuel in oxygen instead of air results in an exhaust gas that is CO_2-free. This technology is commonly used in the glass furnace industry. In post-combustion capture, the CO_2 is separated from flue (exhaust) gases after the combustion of fossil fuels. The CO_2 content is usually much lower than in the gas that is separated during oxy-fuel combustion or pre-combustion capture, with the volume of CO_2 in the range of 3 percent to 15 percent by volume.

Challenges to Implementation

One challenge facing CCS is the demonstration of its efficacy and safety on an industrial scale at competitive cost. While CCS is known to be safe and is well understood in terms of the fundamental science and technical requirements, no evidence has been gathered regarding the process's long-term impacts on the environment (for example, the safety of storing CO_2 in geological formations) or possible danger to humans (for instance, if CO_2 leaks from storage).

CCS applied to a modern power plant could reduce CO_2 emissions by up to 90 percent compared to an equivalent plant with no CCS devices, but the implementation of CO_2 capture is significantly more expensive than the use of traditional systems of emissions control. For example, capturing and compressing CO_2 increases the fuel requirement of a coal-fired plant by as much as 25-40 percent. It is estimated that the cost of energy produced by a new power plant with CCS is from 21 percent to 91 percent higher than that produced by a non-CCS power plant.

Aside from the cost and technical challenges of CCS, a regulatory framework needs to be established to support CCS and to clarify at regional, national, and international levels the long-term rights, liabilities, and technical requirements associated with the use of CCS technologies. Moreover, before investors, scientists, politicians, and industries can be persuaded that CCS is a worthwhile investment, agreements need to be reached on a price on carbon emissions and on whether a carbon tax, cap-and-trade regime, or other carbon-trading/taxation framework will be implemented.

Rena Christina Tabata

Further Reading

Hanjalić, K., R. van de Krol, and A. Lekić, eds. *Sustainable Energy Technologies: Options and Prospects.* London: Springer, 2008.

Kutz, Myer. *Environmentally Conscious Fossil Energy Production.* Hoboken, N.J.: John Wiley & Sons, 2009.

Rackley, Steve. *Carbon Capture and Storage.* Boston: Butterworth-Heinemann, 2009.

Rojey, Alexandre. *Energy and Climate: How to Achieve a Successful Energy Transition.* Chichester, England: John Wiley & Sons, 2009.

Shiosani, Fereidoon P. *Generating Electricity in a Carbon-Constrained World.* Burlington, Mass.: Academic Press, 2009.

■ Carbon trade

FIELDS OF STUDY
Economics; Capitalism; Climate Science; Climate Change; Government; International Policy

SUMMARY
Carbon trading emerged at the end of the 1990s to encourage reducing the emission of greenhouse gases through economic incentives. Carbon trading programs grant allowances, permits, or credits for participants to produce emissions up to a defined limit. These allowances can then additionally be traded, making it more profitable to reduce emissions and sell any surplus carbon credits. Carbon trading programs have been implemented around the world, although they continue to raise questions about whether market-based environmental policies are adequate solutions to the challenge of global climate change.

PRINCIPAL TERMS
- **cap and trade:** an emissions trading system in which a party receives a permit to emit greenhouse gases up to a set limit. Parties may additionally trade permits, selling them if their emission allowance exceeds their actual emissions or buying them if they emit more than they are permitted
- **carbon credits:** another term for the emissions permits that can be traded as part of emissions trading systems
- **Clean Development Mechanism:** lets a country earn emissions reduction credits per the terms of the Kyoto Protocol by establishing emission-reducing projects in developing countries
- **greenhouse gas:** the naturally occurring and anthropogenic gases that make up the atmosphere and absorb and emit radiation
- **Joint Implementation:** a process that allows developed countries to earn emissions reduction credits under the Kyoto Protocol from an emission-reduction project in another developed country
- **Kyoto Protocol:** established in 1997 as part of the United Nations Framework Convention on Climate Change, the Kyoto Protocol sets guidelines and suggests policies for nations to limit their emissions that contribute to global warming
- **United Nations Framework Convention on Climate Change:** an international treaty adopted in 1992 with the purpose of assessing global progress in combatting climate change

Carbon trading, also called emissions trading or cap and trade, emerged from the idea that the market could be put to use to reverse some of the pollution that the market itself had encouraged through the consumption of fossil fuels in the service of industrialization. Emissions trading was first introduced by the Kyoto Protocol, an international treaty negotiated in 1997 by members of the United Nations Framework Convention on Climate Change, and businesses, non-governmental organizations, or entire nations can participate in carbon trading systems.

The Kyoto Protocol identified a specific emission reduction schedule designed to decrease aggregate emissions by a certain percentage below a set baseline level over the 2008-2012 commitment period for signatory nations of the treaty. It also outlined three new policy mechanisms designed to help nations meet their reduction goals while also encouraging developing countries and the private sector to participate.

The first mechanism was emissions trading, or cap and trade. Under cap and trade systems, nations or companies are encouraged to decrease their emissions in order to sell allowances or credits to those who pollute more. The selling of these credits then created a carbon market, in which greenhouse gases are traded as a commodity. In these systems, the availability of carbon credits would decrease over time, thus mandating emissions reductions while maintaining the value of the carbon market through scarcity.

A second policy instrument in the Kyoto Protocol was the Clean Development Mechanism (CDM), which allows nations to earn emissions reduction credits to help them meet their reduction goals by establishing emission-reducing projects in developing countries. This mechanism is designed to give developed nations flexibility when it comes to meeting their reduction targets while also stimulating sustainable development around the world. CDM projects have included the Belgium investing in reforestation in Bolivia in 2009, or Sweden establishing a windfarm in China in 2015.

The third mechanism in the Kyoto Protocol is Joint Implementation (JI), which allows developed countries to earn emissions reduction credits from an emission-reduction project in another developed country. These projects are meant to help nations meet their emission reduction targets while allowing other nations to benefit from foreign investment. For example, New Zealand has installed a generator to capture electricity from a geothermal field with support from the Netherlands in a JI project in 2013.

Each of the mechanisms put forward in the Kyoto Protocol work together as a market-based approach to climate policy. The emissions reduction credits that parties can earn from JI or CDM projects thus enables them to have more carbon credits to sell on the carbon market. These mechanisms are therefore meant to work together to give nations an incentive to invest in projects that either produce lower emissions or remove them from the atmosphere, while also making economic sense.

While the Kyoto Protocol broadly introduced emissions trading, similar systems can also be established at the national or regional level. The European Union Emissions Trading Scheme (EU ETS) was established in 2005 and is the largest cap and trade system in the world, and accounts for more than 75% of the international carbon market. The EU ETS operates separately from the Kyoto Protocol, although many members of the EU ETS were also held to commitments under the treaty while it was in force. The United States, on the other hand, did not ratify the Kyoto Protocol and has not developed any other national system of carbon trading. But there have been several regional carbon trade programs in the United States, such as the Chicago Climate Exchange (CCX), which started trading carbon in 2003. Membership of the CCX was voluntary but legally binding and was made up of organizations that produced greenhouse gas emissions, as well as owners of qualifying emissions-offset projects that removed emissions from the atmosphere. While the CCX had several hundred members, it stopped trading in 2010 due to the low value of the United States' carbon market.

Proponents praise carbon trade programs for using market incentives to decrease emissions, but emissions tradinghas also been criticized for enabling continued pollution and encouraging profit over sustainable environmental action. From this perspective, market-based environmental agendas encourage pursuing the quickest method of saving small amounts of carbon in order to trade and profit in the short term, an impulse which may not be compatible with creating a sustainable future. The independent research group Carbon Trade Watch emerged in 2002 as a vocal critic of emissions trading. Instead of market-based environmental action, they promote environmental and climate justice movements that seek to detach human societies from their reliance on fossil fuels altogether.

Julia Kendrick

Further Reading

Chichilnisky, Graciela and Geoffrey Heal, eds. *Environmental Markets: Equity and Efficiency.* New York: Columbia University Press, 2000.

Hansjürgens, Bernd, editor. *Emissions Trading for Climate Policy: US and European Perspectives.* Cambridge: Cambridge University Press, 2005.

Hintermann, Beat and Marc Gronwald. *Emissions Trading as a Policy Instrument: Evaluation and Prospects.* Cambridge, Mass.: The MIT Press, 2015.

Newell, Peter. *Climate Capitalism: Global Warming and the Transformation of the Global Economy.* Cambridge: Cambridge University Press, 2010.

Stephan, Benjamin and Richard Lane, eds. *The Politics of Carbon Markets.* New York: Routledge, Taylor & Francis, 2015.

Websites

Carbon Trade Watch
http://www.carbontradewatch.org
European Commission
https://ec.europa.eu/clima/policies/ets_en

United Nations Framework Convention on Climate Change
https://unfccc.int/process/the-kyoto-protocol/mechanisms

World Resources Institute
https://www.wri.org/our-work/project/us-climate-initiative

■ Carcinogens

FIELDS OF STUDY

Air Quality Monitoring; Biology; Ecology; Environment; Environmental Microbiology; Medicine; Oncology

SUMMARY

The effects of human exposure to carcinogens in the environment may include the development of different types of illness, deaths, and economic obligations on a national and global scale.

PRINCIPAL TERMS

- **aflatoxins:** poisonous carcinogens that are produced by certain molds, which grow in soil, decaying vegetation, hay, and grains
- **asbestos:** a fibrous mineral, either amphibole or chrysotile, formerly used for making incombustible or fireproof articles
- **benzene:** a colorless, volatile, flammable, toxic, slightly water-soluble, liquid aromatic compound, obtained chiefly from coal tar
- **beryllium:** a toxic divalent metallic element, steel-gray in color, light and strong but brittle, used chiefly as a hardening agent in alloys
- **carcinogens:** substances or physical agents that cause or worsen cancer
- **dioxins:** any of several persistent toxic heterocyclic hydrocarbons occurring especially as by-products of industrial processes, such as pesticide manufacture and papermaking, and waste incineration
- **formaldehyde:** a colorless, toxic, potentially carcinogenic, water-soluble gas with a suffocating odor, usually derived from methyl alcohol and used chiefly in aqueous solutions, as disinfectants and preservatives, and in the manufacture of various resins and plastics.
- **vinyl chloride:** a colorless, easily liquefied, flammable, slightly water-soluble gas, with a pleasant, etherlike odor, used in the manufacture of plastics, as a refrigerant, and in the synthesis of polyvinyl chloride and other organic compounds

Cancer is a leading cause of death throughout the world. Environmentalists and others have raised concerns regarding the cancer-causing (carcinogenic) potential of exposure to a constantly growing number of both newly developed and long-existing chemicals in the environment. In addition, humans seem to be increasingly exposed to various sources of electromagnetic waves, such as microwaves, and some groups and individuals are concerned about the possible carcinogenicity of these physical phenomena.

Carcinogens can be categorized based on their origin as chemicals (naturally occurring or synthetic), physical agents, or infectious agents. Chemical carcinogens can be classed as compounds that occur naturally, such as aflatoxins, chromium 6 compounds, and arsenic compounds; and others that are

Fusarium ear rot, caused by the fungi Fusarium verticillioides and F. proliferatum, may typically be a more common ear rot of corn. By UIUC

largely synthetic in origin, such as benzene, formaldehyde, vinyl chloride, and dioxins. In addition, there are carcinogenic minerals, such as asbestos. Other carcinogenic chemicals are elements or substances such as radon (a radioactive gas), beryllium, and cadmium. Some carcinogens—such as tobacco smoke and alcoholic beverages—are mixtures of compounds. Physical agents that are carcinogens include solar radiation (primarily ultraviolet radiation), gamma rays, and x-rays.

Infectious disease agents that have been implicated as carcinogens include human papilloma virus (HPV), which can cause cervical cancer; *Helicobacter pylori* (*H. pylori*), a bacterium causally associated with stomach cancer; the hepatitis C and hepatitis B viruses, which can cause liver cancer; Epstein-Barr virus, which is associated with Burkitt's lymphoma; and human T-lymphotrophic virus type 1 (HTLV-1), which has been linked to leukemia in adults. The human immunodeficiency virus (HIV), which causes acquired immunodeficiency syndrome (AIDS), has been associated with Kaposi's sarcoma. Viruses have also been shown definitively to cause tumors in animals such as mice (mammary tumor virus), chickens (Rous sarcoma virus), and Tasmanian devils.

Exposure to carcinogens can be related to work environments, such as in the case of workers in the nuclear power and medical radioisotope industries. Sometimes carcinogenic agents happen to be concentrated in particular geographic regions; for example, widespread exposure to the carcinogen arsenic occurred in Bangladesh as the result of contaminated drinking water from a large number of wells that accessed groundwater in which arsenic was uncommonly abundant. Exposure to *H. pylori* is believed to occur through contaminated water supplies and thus is considered to be environmental in origin. Tobacco smoke, a major cause of lung cancer, is an environmental carcinogen to which people are widely exposed.

Carcinogens may lead to cancer by directly damaging deoxyribonucleic acid (DNA), as in the case of radiation; through conversion through metabolism; or through effects on metabolism. In the case of viruses, viral genetic material may be incorporated into host DNA at sites of oncogenes, which are genes whose altered function or disruption leads to cancer. Different carcinogens often lead to effects on different organs, thus *H. pylori* is associated mainly with stomach cancer, whereas tobacco smoking or use is associated with lung, oral, and laryngeal cancers (and also bladder, colon, and kidney cancers, among others), and asbestos is primarily linked to lung cancer. The time period and frequency of exposure to carcinogens as well as a person's genetic background can also influence the likelihood that a carcinogen causes cancer.

Oluseyi A. Vanderpuye

Further Reading

Hill, Marquita K. "Chemical Exposures and Risk Assessment." *Understanding Environmental Pollution*. 3rd ed. New York: Cambridge UP, 2010. Print.

"Known and Probable Human Carcinogens." *Cancer. org*. American Cancer Society, 12 Dec. 2014. Web. 29 Jan. 2015.

McKinnell, Robert. G., et al. *The Biological Basis of Cancer*. 2nd ed. New York: Cambridge UP, 2006.

National Institute of Environmental Health Sciences. "Report on Carcinogens." 12th ed. Research Triangle Park: US Dept. of Health and Human Services, Public Health Service, National Toxicology Program, 2011.

Ward, Elizabeth M. "Cancer." *Occupational and Environmental Health: Recognizing and Preventing Disease and Injury*. Ed. Barry S. Levy, et al. 6th ed. Philadelphia: Lippincott, 2011.

■ Cattle

FIELDS OF STUDY

Agriculture; Biology; Ecology; Environment; Environmentalism; Erosion Control; Land-Use Management

SUMMARY

The environmental impacts of the raising of cattle include the production of greenhouse gases that contribute to climate change, land degradation owing to increased erosion and overgrazing, and pollution of air and water caused by concentrated waste. These problems most often arise from the nature of the farming practices of cattle ranchers, and they can be alleviated somewhat by changes in management techniques.

PRINCIPAL TERMS
- **anaerobic:** living in the absence of air or free oxygen
- **cattle:** large domesticated mammals raised as livestock and as dairy animals
- **erosion:** the loss of topsoil through the actions of wind and water, and the efforts undertaken to mitigate such loss
- **greenhouse gas:** any of the gases whose absorption of solar radiation is responsible for the greenhouse effect, including carbon dioxide, methane, ozone, and the fluorocarbons
- **methane:** a colorless, odorless, flammable gas, CH_4, the main constituent of marsh gas and the firedamp of coal mines, obtained commercially from natural gas
- **methanogenesis:** the production of methane by bacteria or other living organisms
- **UN Food and Agriculture Organization (FAO):** a specialized agency of the United Nations that leads international efforts to defeat hunger

The collective species *Bos primigenius* includes three subspecies: *Bos primigenius taurus*, the European cattle; *Bos primigenius indicus*, the zebu; and the extinct *Bos primigenius primigenius*, the ancestor to modern domestic cattle and called the auroch. The term "cattle" is usually used to refer to domesticated *B. primigenius* that are bred for multiple uses, spanning the production of food to clothing to fuel, and to serve as work animals. Cattle are chief sources of food (from meat and milk), labor, clothing (from leather), and fuel (from dung) in many cultures worldwide.

According to a 2006 report from the United Nations Food and Agriculture Organization (FAO), the world's livestock, most of which are cattle, are among the major causes of serious global environmental problems, including land degradation, air and water pollution, and loss of biodiversity. Further, livestock are responsible for 18 percent of total global greenhouse gas emissions, which have been linked to climate change. The potential for those who raise livestock to contribute to solving these environmental problems is thus very large, and experts argue that major improvements can be achieved at reasonable costs.

The grazing of livestock occupies almost 30 percent of the earth's terrestrial surface, and the agricultural production of feed for livestock takes up about one-third of all arable land. In addition to affecting land that has already been converted to agriculture, the raising of cattle is responsible for the direct acceleration of deforestation in some parts of the world where ranchers burn forests to expand grazing land. This is a pressing problem in Latin America, where nearly 70 percent of previously forested land is used as pasture and much of the rest is used for growing livestock feed crops such as soybeans and corn. Tropical soils are especially susceptible to degradation by overgrazing, compaction, and erosion because of high precipitation and low nutrient loads. Ranching is thus usually not a sustainable practice in the Tropics, but it continues because it promises quick economic returns.

Concentrated animal feeding operations (CAFOs), in which livestock are fed in small confined areas for maximum profit, produce great amounts of highly concentrated waste that ultimately ends up in the water and as gases in the air. FAO has estimated that livestock manure is the largest sectoral source of water pollutants. In addition to manure, chemicals from pesticides and fertilizers used for feed crops and antibiotics administered to livestock also end up in groundwater and surface water, where they contribute to high nutrient loading and algal blooms, medicinal pollution that affects aquatic biology, and high sediment loads that reduce water quality.

Greenhouse Gas Emissions
Cattle produce methane through an anaerobic process in the gut called methanogenesis; the methane is released through belching and flatulence. Methane is an extremely effective greenhouse gas, having a warming effect 23 to 50 times greater than carbon dioxide; this causes concern as there are approximately 1.3 billion cattle worldwide, and the number only grows. According to FAO's 2006 report, the raising of cattle generates more greenhouse gases than do all forms of transportation. With increasing prosperity worldwide, people consume more meat and dairy products every year; the global production of both meat and dairy is expected to double by 2050.

Several possible ways of reducing methane production in cattle have been proposed or are under study. These include the administration of bovine medicines similar to the antacid Alka-Seltzer, the use of new varieties of feed grasses, and targeted breeding that selects for less gassy cattle. Any successes in this area will certainly be helpful, but scientists note that a global reduction in numbers of domestic cattle must also accompany these techniques if cattle's large-scale production of methane is to be effectively reduced.

Overgrazing
While grazing is of mutual benefit to plants and animals, overgrazing is ultimately detrimental to both the plant and animal populations and grassland ecosystems.

The Bureau of Land Management (BLM) manages livestock grazing on 64 million hectares (157 million acres) of the 99 million hectares (245 million acres) of public lands that it administers. The BLM administers roughly eighteen hundred permits and leases, which are held mostly by cattle and sheep ranchers. The U.S. Forest Service, which administers the 77 million hectares (191 million acres) of national forest system lands, manages some 39 million hectares (96 million acres) of rangelands. The Forest Service became the nation's first grazing control agency in the early 1900's. In 1934 the Department of the Interior's Division of Grazing Control (soon renamed the Division of Grazing) joined it; this division became the Grazing Service in 1939, which merged with the General Land Office in 1946 to form the BLM.

Both the Forest Service and the BLM implement a regulatory system of permits, rental fees, herd size limits, and grazing seasons. They must maintain a balance among several often-conflicting objectives: providing forage for grazing and browsing animals, ensuring the land's long-term health and productivity, protecting watersheds, managing wildlife habitat, administering permitted mineral and energy resource exploration and extraction, offering recreational opportunities, and preserving the land's distinctive character and aesthetic appeal. In order to meet the array of resource needs, rangeland management agencies inventory, classify, and monitor rangeland conditions. Where rangeland health needs improvement, they implement measures to restore ecosystem functions. Public land decision makers must consider a variety of factors that affect rangelands, including severe and extensive wildfires, invasive plant species, rural residential development driven by population increases, and global climate change.

Environmental Solutions
Humans certainly need not stop raising cattle entirely in order to ensure a healthy future for the environment, but some practices must change. Sustainable cattle ranching requires restoration of overgrazed and damaged land through soil conservation, the planting of trees, and protection of areas sensitive to erosion. Changing the ways cattle are fed to better reflect a natural diet can go a long way toward curbing greenhouse gas emissions. In addition, it has been argued that moving away from CAFOs to less intense feed operations can have positive effects on the cows, their environment, and the quality of all products that come from the cows. Increased use of processed manure as fertilizer can reduce waste pollution and reclaims a resource that should be valued and used instead of dumped into the water supply.

Many cattle ranchers and dairy farmers have become dedicated to reducing the negative environmental impacts of raising cattle because for them the advantages of sustainable management systems outweigh the pressures for higher profit. Persons living in subsistence communities often have little choice in how they raise cattle, however—they clear land and overgraze because they will starve otherwise. It has been proposed that developed nations should provide incentives to poorer countries to ensure that practicing deforestation to create pastureland is not the only option the people have for income.

Another solution to the negative environmental impacts of cattle raising that is sometimes proposed, primarily by animal rights and environmental conservation groups, is vegetarianism. If humans were to limit their beef consumption, this would certainly reduce demand and therefore reduce the number of cattle worldwide that are overgrazing, belching methane, and compacting soil. As critics of this approach have noted, however, expecting large numbers of people to change their diets radically is unrealistic. This is especially true in the United States, where the culture of beef eating is quite strong. For this reason, many environmentalists, scientists, and others have increasingly suggested that people could help reduce the negative environmental impacts of cattle raising simply by reducing the amount of beef in their diets.

Jamie Michael Kass and James L. Robinson, updated by Karen N. Kähler

Further Reading

Clutton-Brock, Juliet. *A Natural History of Domesticated Mammals.* 2d ed. New York: Cambridge University Press, 1999.

Du Toit, Johan, Richard Kock, and James Deutsch. *Wild Rangelands: Conserving Wildlife While Maintaining Livestock in Semi-Arid Ecosystems.* Chicester, England: Wiley, 2012.

Manske, Llewellyn, and Sheri A, Schneider. *Biologically Effective Management of Grazinglands.* Dickinson, N.D.: North Dakota State University, 2014.

Soliva, Carla Riccarda, Junichi Takahashi, and Michael Kreuzer, eds. *Greenhouse Gases and Animal Agriculture.* Boston: Elsevier, 2006.

Steinfeld, Henning, et al. *Livestock's Long Shadow: Environmental Issues and Options.* Rome, Italy: United Nations Food and Agriculture Organization, 2006.

■ Clean Air Act

FIELDS OF STUDY

Air Pollution Control; Air-Quality Management; Climate Engineering; Ecological Engineering; Environmental Sciences; Government; Waste Management;

SUMMARY

The Clean Air Act of 1963 and its amendments federalize the regulation of air pollution in the United States to a large degree. The act provides guidelines for minimum standards of air quality as well as maximum levels for the emissions of pollutants. It has served as a model for other federal environmental legislation.

PRINCIPAL TERMS

- **chlorofluorocarbons:** any of several simple gaseous compounds containing carbon, chlorine, fluorine, and sometimes hydrogen, and used as refrigerants, cleaning solvents, and aerosol propellants and in the manufacture of plastic foams, and believed to be a major cause of stratospheric ozone depletion
- **Clean Air Act:** U.S. federal laws that govern standards for air quality
- **Environmental Protection Agency (EPA):** an independent agency of the United States federal government established in 1970 for environmental protection
- **fluorocarbons:** any of a class of compounds produced by substituting fluorine for hydrogen in a hydrocarbon, and characterized by great chemical stability
- **greenhouse gases:** any of various gaseous compounds, such as carbon dioxide or methane, that absorb infrared radiation and trap heat in the atmosphere
- **hydrofluorocarbons:** any of several simple gaseous compounds that contain carbon, fluorine, and hydrogen
- **ozone:** a colorless gas and form of oxygen. There is a layer of ozone high above the Earth's surface, that protects us from harmful radiation from the sun.
- **sulfur hexafluoride:** an inorganic, colorless, odorless, non-flammable, extremely potent greenhouse gas

Since the 1880s state and local governments in the United States have put limits on smoke emissions and other forms of air pollution. Federal regulation of the problem, however, did not really begin until 1955, when the Air Pollution Control Act authorized the federal government to conduct research and aid state and local governments. This act included no national standards, and ceded responsibility for controlling air quality to the states.

Congress increased the federal role somewhat in the Clean Air Act of 1963. The secretary of the Department of Health, Education, and Welfare was authorized to call abatement conferences when air pollution from one state put citizens of another state in danger, but the Clean Air Act failed to include any sanctions for the enforcement of national standards. Meanwhile, evidence was accumulating that air pollution posed a serious threat to public health throughout the country. Incidents such as the November 1966, acute air-pollution episode in New York City, an event blamed for the deaths of some 168 people, served as a sobering example of how polluted America's air had become. President Lyndon

Time Line of U.S. Clean Air Laws and Policies

Year	Event
1955	Air Pollution Control Act, the first U.S. law to address air pollution and fund research into pollution prevention, is passed.
1963	Clean Air Act is the first U.S. law to provide for the monitoring and control of air pollution.
1967	Air Quality Act establishes enforcement provisions to reduce interstate air-pollution transport.
1970	Clean Air Act amendments establish the first comprehensive emission regulatory structure, including the National Ambient Air Quality Standards (NAAQS).
1977	Clean Air Act amendments provide for the prevention of deterioration in air quality in areas in compliance with the NAAQS.
1990	Clean Air Act amendments establish programs to control acid precipitation, as well as 189 specific toxic pollutants.
1995	Oil companies are required to sell reformulated gasoline in metropolitan regions, and gas stations are required to install vapor-retrieval devices on pumps.
2003	Proposed Clear Skies Bill is designed to amend the Clean Air Act with a cap-and-trade system.
2005	The EPA's Clean Air Interstate Rule (CAIR) begins a cap-and-trade program to keep air pollution generated in one state from rendering other states noncompliant with air-quality standards.
2008	A federal appeals court rules that CAIR exceeds the EPA's regulatory authority but later orders temporary reinstatement.

B. Johnson's Great Society looked to federal regulation as the only effective way to deal with such matters.

The 1967 Air Quality Act authorized the Department of Health, Education, and Welfare to consult with the states to determine air-quality standards in regions of particular concern, and the states were then given a year to formulate a plan to implement the guidelines. Environmentalists were disappointed that Congress still had not provided minimum standards of air quality or effective means for forcing the states to achieve their goals. The most significant aspect of the act was the authorization of some federal enforcement of vehicular emissions standards, with criminal fines of up to $1,000 for each violation of the standards. Relatively weak requirements based on grams of pollutants emitted per mile took effect for new automobiles in 1968.

1970 Amendments

Widespread support of the environmental movement was demonstrated by the enthusiastic response to Earth Day in 1970, and that same year the first report of the U.S. Council on Environmental Quality called on Congress to enact new laws to deal with several problems, including air pollution. Senator Edmund Muskie, a presidential hopeful, was the acknowledged congressional leader in the campaign for tough environmental reform, and he was the chief author of the 1970 clean air bill. President Richard Nixon also supported an aggressive bill. With this bipartisan support, Congress enacted a far-reaching amendment to the 1963 legislation, the Clean Air Act Extension of 1970, which initiated the federal government's regulation of air pollution.

Addressing perceived weaknesses in the existing law, the landmark 1970 amendments authorized the newly created Environmental Protection Agency (EPA) to establish standards that would be binding on states. Applying a command-and-control approach to regulation, the centerpiece of the legislation was a program for the EPA to determine National Ambient Air Quality Standards (NAAQS) that would define specific levels of air pollution considered harmful to public health. The EPA was also authorized to set emission limits on hazardous pollutants at levels allowing a sufficient margin of safety. Although states might exercise discretion in choosing how to meet the federal standards, they were required to develop state implementation plans (SIPs) that utilized appropriate measures to reach those standards. States could maintain the air-quality-control programs already in place for existing industrial plants while requiring new plants to meet stricter standards based on the best available technology that was economically feasible.

The 1970 legislation stunned American automobile manufacturers by requiring them to curtail their products' emissions of the "big three" pollutants—hydrocarbons, carbon oxides, and nitrogen oxides—by 90 percent within six years. The technology did not exist to allow them to meet the new standards, although it was hoped that new technology could be developed within the specified time. Most members of Congress, few of whom were willing to see the collapse of the automobile industry, understood that it might be necessary to extend the deadline. In fact, the deadlines for meeting the vehicular emission standards turned out to be excessively ambitious, and waivers for the standards were granted in 1971, 1973, 1974, and 1976.

1977 Amendments

With the enthusiastic support of President Jimmy Carter, Congress passed major revisions to the Clean Air Act on August 4, 1977. In addition to making NAAQS more stringent, the amendments required each state to designate "nonattainment" regions based on the NAAQS. Each state was then given the choice of either accepting statutory sanctions or revising its SIPs in order to meet the standards in a timely way. The amendments focused especially on coal-burning power plants, a significant source of sulfur dioxide in the atmosphere, which contributes to acid rain. Existing stationary sources of pollution were required to provide for "reasonably available control technology," and new or modified stationary sources were required to utilize technology meeting the "lowest achievable emission rate," which usually meant the use of expensive scrubbers.

In the case of clean air regions already in attainment, the amendments instituted the Prevention of Significant Deterioration (PSD) program, which was designed to prevent the EPA from allowing deterioration of air quality up to the national standards. Members of Congress from rural districts had unsuccessfully argued that such a program would unfairly restrict industrial growth in areas where air pollution was not a problem.

The 1977 amendments further extended the deadlines by which automobiles were required to achieve emissions-control standards. The stricter controls were scheduled for the 1980 model year. American automakers had insisted that they could not meet the requirements in the existing law, and they had threatened to shut down production lines. This marked the fifth relaxation of the vehicular deadlines.

1990 Amendments

Between 1977 and 1990 numerous efforts were undertaken both to strengthen and to weaken the Clean Air Act, but opposing interest groups prevented major changes in either direction. While there was widespread agreement that the Clean Air Act had been somewhat successful in improving air quality, the administration of President Ronald Reagan strongly opposed any expansion of environmental regulations. During the 1988 presidential election campaign, candidate George H. W. Bush pledged to be the "environmental president." When Bush entered the White House, the deadlines for compliance with most air-quality standards had passed, putting noncompliance regions in danger of losing many industrial jobs. In July, 1989, the Bush administration made a sweeping proposal that most Democrats and environmentalists could support, while conservative Republicans were divided on the issue. The resulting amendments were signed into law on November 15, 1990.

The 1990 amendments were extremely complex, requiring more than seven hundred pages. The major regulatory change, modeled on the Clean Water Act, was a requirement that all major sources of air pollution obtain state operating permits, with the EPA given the authority to veto such permits. The amendments provided additional regulations of emissions that were responsible for acid rain and established an allowance system based on a nationwide limit of 8.1 million metric tons (8.9 million U.S. tons) of sulfur dioxide per year. Other provisions included a phaseout program for chlorofluorocarbons (CFCs) and other ozone-depleting substances, as well as a requirement that industrial plants cut emissions of 189 toxic substances to the levels of the cleanest plants within their particular industries.

Because one important goal of the statute was to decrease urban smog, it included strict controls on automobile emissions and mandates for cleaner-burning fuels. Beginning with 1994 automobiles, tailpipe exhausts were required to contain 60 percent less nitrogen oxide, and emissions-control equipment was required to last ten years. The EPA was authorized to conduct a study to determine whether stricter standards were needed. A pilot program in California required an increasing number of cars and light trucks to run on batteries or non-gasoline fuels. Beginning in 1995, oil companies were required to sell only cleaner-burning reformulated gasoline in the smoggiest metropolitan regions, and gasoline stations were mandated to install devices to capture fumes during refueling.

The 1990 amendments considerably strengthened the enforcement provisions under the Clean Air Act. The EPA acquired new powers to issue administrative penalties of up to $25,000 per day, and individuals were empowered to take civil action against polluters. The EPA and the U.S. Department of Justice were given new authority to prosecute misdemeanor and felony violations of the act. The amendments increased maximum sentences for most violations from six months to two years and increased maximum fines from $25,000 to $500,000. An individual who released hazardous air pollutants into the air could henceforth be sentenced to fifteen years in prison and fined up to $250,000, and corporations could be fined up to $1 million.

At an estimated cost of $25 billion per year, the 1990 act is considered the most expensive piece of environmental legislation ever passed. It was expected that the costs would mostly be passed on to consumers in higher prices for cars, gasoline, electricity, and products containing chemicals. The EPA's estimated monetary value of the act, based on its public health and environmental benefits, offsets its cost.

The Clear Skies Initiative
In 2002, President George W. Bush announced the Clear Skies Initiative, a policy intended to incentivize innovation and cut costs through a market-based cap-and-trade program for reducing power plant emissions. Through such a program, polluters would have the right to emit a certain quantity of pollutants; a polluter that wishes to emit more than its allowance would have to purchase credits from one that emits less. The initiative, which prioritized economic growth, operated on the assumption that the market drives advances in pollution-control technologies and thereby hastens environmental progress. The initiative gave rise to the Clear Skies Bill of 2003, which would have amended the Clean Air Act with a cap-and-trade system. Critics—among them the Sierra Club and the Natural Resources Defense Council—argued that the proposed law was a propagandistically named reduction of air-quality protections that would allow increased toxic industrial emissions and hamper enforcement of pollution-control standards. The bill never moved beyond the Senate Environment and Public Works Committee and thus did not become law.

In 2005 the EPA introduced the Clean Air Interstate Rule (CAIR). A key measure of the deadlocked Clear Skies Bill, CAIR is a cap-and-trade program intended to ensure that air pollution generated in one state does not prevent another downwind state from meeting air-quality standards. CAIR, which was designed to reduce smog and soot pollution from power plants in the eastern United States, includes a permanent cap on the precursor pollutants sulfur dioxide and nitrogen oxides. In July, 2008, a federal appeals court vacated CAIR, citing several fundamental flaws in the rule. The EPA, the Environmental Defense Fund, and several states successfully appealed for a rehearing. The court determined that, despite the rule's shortcomings, the environmental and health benefits of CAIR are significant. (The EPA estimated that CAIR would prevent seventeen thousand deaths annually by 2015.) In December 2008, the court issued an order temporarily reinstating CAIR until the EPA could replace it with a rule that fully addresses CAIR's flaws.

Another measure of the Clear Skies Bill, the Clean Air Mercury Rule (CAMR), was also introduced in 2005. Environmental groups, several states, Native American tribes, and physicians' organizations opposed the rule, as its use of a cap-and-trade program in the case of a bio-accumulative, environmentally persistent material such as mercury would allow the development of toxic hot spots that would endanger human health. CAMR also removed oil- and coal-fired electric-utility steam-generating units from the list of hazardous air-pollutant sources. In 2008, a federal appeals court found CAMR to be in violation of the Clear Air Act and vacated the rule.

Subsequent Developments
In a 2007 case heard by the U.S. Supreme Court, twelve states, three major U.S. cities, a U.S. territory,

and several nongovernmental organizations sued the EPA for failing to regulate greenhouse gases (GHGs) as pollutants. The Court found that the EPA was again in violation of the Clean Air Act and charged the agency with determining whether GHG emissions from new vehicles are pollutants that endanger the public health or welfare. The EPA concluded that GHGs do in fact pose a danger to the public and submitted its endangerment finding to the White House. White House officials refused to read the EPA's report and took other measures to block the EPA's regulation of GHGs during George W. Bush's administration.

In 2009, under President Barack Obama's administration, the EPA issued its final findings regarding GHGs. It determined that current and projected concentrations of six GHGs in the atmosphere—carbon monoxide, methane, nitrous oxide, hydrofluorocarbons, perfluorocarbons, and sulfur hexafluoride—constitute a threat to the public health and welfare of current and future generations. It also found that new motor vehicles were contributing to GHG pollution. These findings may someday lead to more restrictive emissions limits for power plants, oil refineries, auto manufacturers, and other major GHG contributors.

Enforcement

Based on a command-and-control model, the Clean Air Act and its amendments provide a variety of strong mechanisms for enforcing their statutory and regulatory requirements. The EPA has primary responsibility for enforcement at the federal level, and the states share responsibility for regulating SIPs. Citizens are also given broad opportunities to participate in the enforcement process.

When the EPA finds evidence that a violation has occurred, a regional office of the agency issues a notice of violation to both the source and the state. Based on its investigations, the EPA has the discretion to determine whether further action is necessary. The agency may issue an administrative order requiring a person or institution to comply with the applicable statute or regulation. If the recipient of the order fails to comply, the EPA may enforce the order through a civil action. If there is probable cause that a crime has occurred, the EPA will initiate a criminal prosecution, but the Department of Justice usually takes charge of the legal actions.

In formulating its SIPs, each state is required to include a program of legal enforcement. The states are usually given the opportunity to lead in initiating enforcement action if they wish to do so. If a state does not do so, the EPA has the authority to proceed on its own. Any person, moreover, may bring a civil action against an individual or entity alleged to be in violation of the Clean Air Act. If a violation is proved, any monetary awards must be either turned over to the EPA's "penalty fund" or used for "beneficial mitigation projects."

When the landmark amendments of 1970 were passed, proponents of the act tended to be extremely optimistic about the prospects for achieving national air-quality standards without any serious economic costs. The act envisioned full attainment of the standards by 1975, but this expectation turned out to be unrealistic, especially in regard to ozone. In the 1977 amendments, Congress responded to the problem by explicitly recognizing noncompliance regions, which were thereafter required to improve incrementally. It was even more difficult to formulate vehicular emissions standards that were both meaningful and attainable, in part because no one could be certain about the prospects of technological improvement. When automotive technology improved, moreover, no one could be certain about whether Clean Air Act standards were a primary cause.

By the late 1990's, few people denied that the Clean Air Act had helped decrease air pollution and improve the public health. By its nature, however, such legislation does not completely satisfy everyone. Environmental organizations commonly argue that the EPA has not been aggressive enough in its enforcement efforts, while probusiness groups tend to blame the Clean Air Act for forcing American industries to close their doors and move to poor countries with weaker regulatory protections and a greater toleration for dirty air.

Thomas T. Lewis; updated by Karen N. Kähler

Further Reading

Bryner, Gary C. *Blue Skies, Green Politics: The Clean Air Act of 1990.* Rev. ed. Washington, D.C.: Congressional Quarterly Press, 1995.

Cohen, Richard E. *Washington at Work: Back Rooms and Clean Air.* 2d ed. Boston: Allyn & Bacon, 1995.

Griffin, Roger D. *Principles of Air Quality Management.* 2d ed. Boca Raton, Fla.: CRC Press, 2007.

Guyer, J. Paul. *An Introduction to Air Pollution Control.* Scottsvalley, Calif.: CreateSpace Independent Publishing Platform, 2016.

Lipton, James P., ed. *Clean Air Act: Interpretation and Analysis.* New York: Nova Science, 2006.

Rajan, Sudhir Chella. *The Enigma of Automobility: Democratic Politics and Pollution Control.* Pittsburgh: University of Pittsburgh Press, 1996.

U.S. Environmental Protection Agency. *The Plain English Guide to the Clean Air Act.* Research Triangle Park, N.C.: Office of Air Quality Planning and Standards, 2007.

■ Clean Water Act

FIELDS OF STUDY

Coastal Engineering; Ecological Engineering; Environmental Sciences; Government; Sanitary Engineering; Sewage Engineering; Waste Management; Water Pollution Control; Water-Quality Monitoring

SUMMARY

The legislation now called the Clean Water Act was largely shaped by the 1972 amendments to the Federal Water Quality Act of 1965, itself an amendment to 1948 legislation. The complex law was further strengthened by later amendments as the American public became increasingly aware of the importance of clean water supplies to the public health.

PRINCIPLE TERMS

- **ambient water:** water at room temperature
- **Environmental Protection Agency (EPA):** an inpendent agency of the United States federal government, established in 1970, for environmental protection
- **Federal Water pollution Control Act (FWPCA):** federal legislation enacted in 1972, designed to improve the quality of surface water throughout the United States
- **U.S. Army Corps of Engineers (USACE):** a nited States federal agency under the Department of Defense, comprising approximately 37,000 civilian and military personnel working in public engineering, design, and construction management agencies worldwide
- **wetlands:** areas where water covers the soil or is present at or near the soil all year or for varying periods of time

Before the mid-1960s government regulation of water pollution in the United States was mostly left up to individual states. The earliest U.S. federal environmental law was the Rivers and Harbors Act of 1899, which prohibited the dumping of debris into navigable waters. Although the law was intended to protect interstate navigation, it became an instrument for regulating water quality sixty years after its passage. The Oil Pollution Act of 1924 prohibited the discharge of oil into interstate waterways, with criminal sanctions for violations. The first Federal Water Pollution Control Act (FWPCA), passed in 1948, authorized the preparation of federal pollution-abatement plans, which the states could either accept or reject, and provided some financial assistance for state projects. Although the FWPCA was amended in 1956 and 1961, it still contained no effective mechanisms for the federal enforcement of standards.

By this period, however, many Americans were recognizing water pollution as a national problem that required a national solution. The Federal Water Quality Act of 1965 amended the 1948 legislation to introduce a policy of minimum water-quality standards that could be enforced in federal courts. The standards applied regardless of whether discharges could be proven to harm human health. The act also significantly increased federal funds for the construction of sewage plants. A 1966 amendment required the reporting of discharges into waterways, with civil penalties for failure to comply. Another amendment, the Water Quality Improvement Act of 1970, established federal licensing for the discharge of pollutants into navigable rivers and provided plans and funding for the detection and removal of oil spills.

The Advent of Modern Water-Protection Legislation

Congress and President Richard Nixon agreed that existing programs were ineffective in controlling water pollution. The resulting Federal Water Pollution Control Act Amendments of 1972 amended the Federal Water Quality Act to establish the basic

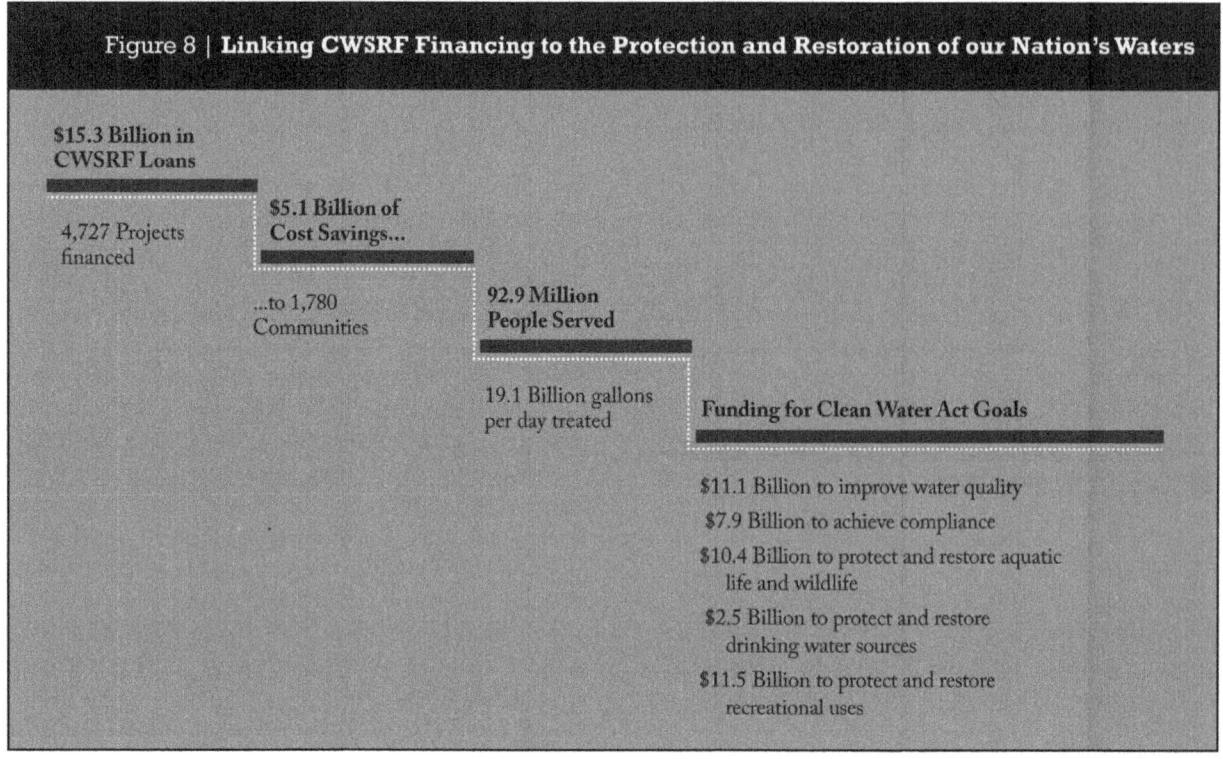

Clean Water State Revolving Funds provide loan assistance for water quality improvement projects. By Environmental Protection Agency (CWSRF Annual Report 2007).

framework for the Clean Water Act. The centerpiece of the landmark amendments was the National Pollutant Discharge Elimination System (NPDES), which utilizes the command-and-control methods earlier enacted in the Clean Air Act. The premise of the legislation was that polluting surface water is an unlawful activity, except for those exemptions specifically allowed in the act. The announced goal was to eliminate all pollutants discharged into U.S. surface waters by 1985.

In addition to standards of quality for ambient water, the amendments included technology-based standards. Industrial dischargers were given until 1977 to make use of the "best practicable technology" in their industries, and the standard was to be increased to the "best available technology" by 1983. The 1972 act also included stringent limitations on the release of toxic chemicals judged harmful to human health. For members of Congress, the most popular part of the act was the grant program for the construction of publicly owned treatment works (POTWs).

The U.S. Environmental Protection Agency (EPA), created just two years earlier, was assigned the primary responsibility for regulating and enforcing the legislation. The agency could issue five-year permits for the discharge of pollutants, and any discharge without a license or contrary to the terms of a license was punishable by either civil or criminal sanctions. When dealing with a discharge of oil or other hazardous substances, the EPA could go to court and seek a penalty of up to $50,000 per violation and up to $250,000 in the case of willful misconduct. In addition, a discharger might be assessed the costs of removal, up to $50 million. Because of the technical complexity of the law, the EPA for many years relied more on civil penalties than on criminal prosecutions.

The 1972 amendments prohibited the discharge of dredged or fill materials into navigable waters unless authorized by a permit issued by the U.S. Army Corps of Engineers (USACE). Based on the literal wording of the statute, the USACE at first regulated only actually, potentially, and historically navigable waters. In 1975, however, it revised its regulations to include jurisdiction over all coastal and freshwater wetlands, provided they were inundated often enough to support vegetation adapted for saturated soils. The Supreme Court endorsed the USACE's broad construction of the law. The USACE and the EPA later adopted a rule under which isolated waters that were actual or potential habitat for migratory birds that crossed state lines were subject to the provisions of the Clean Water Act.

The Clean Water Act amendments of 1977, which gave the legislation its current name, focused on a large variety of technical issues. They required industries to use the best available technology to remove toxic pollutants within six years. For conventional pollutants (such as ammonia, pathogens, phosphorus, and suspended solids), businesses could seek waivers from the technology requirements if the removal of the pollutants was not worth the cost. The act further required an environmental impact statement for any federal project involving wetlands, and it extended liability for oil-spill cleanups from 19 kilometers (12 miles) to 322 kilometers (200 miles) offshore.

Wastewater being additionally treated after primary treatment; prior to the Clean Water Act, wastewater was discharged into rivers after primary treatment. By 90.5 WESA (End of Primary Treatment).

Later Amendments

The Municipal Wastewater Treatment Construction Grants Amendments of 1981, an important piece of environmental public works legislation, streamlined the municipal construction grants process. This allowed for municipalities to improve their sewage treatment capabilities.

The amendments of 1987, entitled the Water Quality Act, were passed by Congress over President Ronald Reagan's veto. In addition to increasing the powers of the EPA, the act significantly raised the criminal penalties for acts of pollution. Individuals who knowingly discharged certain dangerous pollutants could receive a fine of up to $250,000 and imprisonment for up to fifteen years. The maximum prison term for making false statements or tampering with monitoring equipment was increased from six months to two years. The most controversial part of the act was its authorization of $18 billion for the construction of wastewater treatment plants. In addition, the 1987 amendments phased out the earlier construction grants program, replacing it with the State Water Pollution Control Revolving Fund. Also called the Clean Water State Revolving Fund, the new program relied on EPA-state partnerships.

The 1987 Water Quality Act also provided state funds for managing and controlling nonpoint source pollution, such as stormwater runoff from urban areas, forests, agricultural lands, and construction sites. Earlier legislation had focused more on pollution from discrete sources, such as industrial plants and municipal sewage facilities, that could be more easily identified and regulated. Roughly half of the nation's remaining water pollution stemmed from nonpoint sources.

In the wake of the 1989 *Exxon Valdez* oil spill, Congress passed the Oil Pollution Act of 1990. This legislation strengthened cleanup requirements and penalties for oil discharges.

Ongoing points of contention regarding the Clean Water Act have been its wetlands protection program, the loose interpretation of "navigable waters," and the EPA/USACE "migratory bird rule." In a 2001 case, *Solid Waste Agency of Northern Cook County v. Army Corps of Engineers*, the U.S. Supreme Court found that federal protection under the Clean Water Act did not apply in the case of isolated wetlands such as the area that Cook County, Illinois, planned to use as a landfill. In 2006 the Court also determined that the act was inapplicable in the related cases *Rapanos v. United States* and *Carabell v. Corps of Engineers*, which involved two Michigan landowners planning to develop on wetlands. In early 2010 a Clean Water Act amendment was proposed that would replace the phrase "navigable waters" with "waters of the United States."

Some of the worst causes of water pollution in the United States have been curtailed in the years since the Clean Water Act was overhauled in 1972, even though the act has manifestly failed to achieve its stated goals. The legislators who hoped to render all U.S. waters fishable and swimmable within a decade were clearly overly optimistic. It is probably inevitable that economic prosperity and population growth will mean that water in the United States will never be completely free of pollutants. Since 1972, nevertheless, the American public has become increasingly intolerant of dirty and unhealthful water, and Congress, reflecting public sentiment, has continued to strengthen the Clean Water Act.

Thomas T. Lewis; updated by Karen N. Kähler

Further Reading

Copeland, Claudia. *Clean Water Act: A Summary of the Law.* Washington, D.C.: Congressional Research Service, 2008.

Finkmoore, Richard J. *Environmental Law and the Values of Nature.* Durham, N.C.: Carolina Academic Press, 2010.

Freedman, Martin, and Bikki Jaggi. *Air and Water Pollution Regulation: Accomplishments and Economic Consequences.* Westport, Conn.: Quorum Books, 1993.

Lazarus, Richard J. *The Making of Environmental Law.* Chicago: University of Chicago Press, 2004.

Milazzo, Paul Charles. *Unlikely Environmentalists: Congress and Clean Water, 1945-1972.* Lawrence, Kansas: University Press of Kansas, 2006.

Ryan, Mark. *The Clean Water Act Handbook.* 2d ed. Chicago: American Bar Association, 2003.

WEBSITES

EPA
https://www.epa.gov/laws-regulations/history-clean-water-act.

Climate Change

FIELDS OF STUDY
Atmospheric sciences; Climate and Climate Change; Climatology; Atmospheric Structure and Dynamics; Ecology; Environment; Environmentalism; Land-Use Ecology; Meteorology

SUMMARY
While Earth's climate is constantly changing in various ways, the planet tends to experience long-term trends toward either warming or cooling. The potential or actual contribution of postindustrial human activity to climate change, the consequences of that contribution, and the proper response to those consequences remain matters of crucial importance and significant controversy.

PRINCIPAL TERMS
- **anthropogenic:** caused or produced by humans
- **climate change:** change of climate attributed directly or indirectly to human activity altering the composition of the global atmosphere and to natural climate variability observed over comparable time periods
- **climatology:** the science that deals with the phenomena of climates or climatic conditions
- **glaciation:** process or state of being covered by glaciers or ice sheets.
- **global warming:** increase in the average surface and ocean temperature of the Earth since 1850 and the projected persistence of the trend
- **Milankovi cycles:** the collective effects of changes in the Earth's movements on its climate over thousands of years, coined in the 1920s and named for Serbian geophysicist and astronomer Milutin Milankovi.
- **paleoclimate:** the climate of some former period of geologic time.

Climate is characterized by mean air temperature, humidity, winds, precipitation, and frequency of extreme weather events over a lengthy period, at least thirty years. Global warming is an example of climate change, and so are increases in the magnitude or frequency of floods and droughts experienced in many parts of the world during the past several decades.

Climate change includes both natural variability and anthropogenic changes.

Although climate changes on longer than millennial timescales are natural, the global warming of the past 150 years or so is likely anthropogenic, according to the Fourth Assessment Report of the Intergovernmental Panel on Climate Change (IPCC) of the United Nations. The United Nations is concerned primarily with anthropogenic climate change, both because it poses a threat to global security and because it can be altered by altering human and governmental behavior. For this reason, the United Nations Framework Convention on Climate Change (UNFCCC) defines climate change as a change of climate which is attributed directly or indirectly to human activity that alters the composition of the global atmosphere and which is in addition to natural climate variability observed over comparable time periods.

Climate Change Detection

Earth's atmosphere is chaotic, and weather can change dramatically in a matter of days or even hours. Temperature in some places may rise or fall by 20° Celsius or more in one day. On the other hand, climate, as the average of years of weather conditions, changes on a much smaller scale. For example, the global mean surface air temperature increased by only 0.6° Celsius during the twentieth century. By the same token, such a seemingly small increase can have extremely significant effects.

The climatic increase in mean surface air temperature is computed from tens of thousands of weather station records spanning decades. The difficulty of ensuring data continuity in time, uniformity in space, and constancy in observational methods poses serious challenges to climatologists. To discern slight trends amid diverging data, scientists use advanced mathematical tools to synchronize all observations, adjust discontinuities, and filter out local influences such as heat island effects.

Modern climate change has generally been observed with in situ thermometers and, later, with remote sensing devices. Paleoclimate change (change before about 1850) is inferred from proxy climate data. Tree rings can provide evidence of temperature and precipitation history for two to three thousand years, while tiny air bubbles trapped in the Antarctic ice deposits provide data on ice ages hundreds of thousands of years in the past. Pollen and

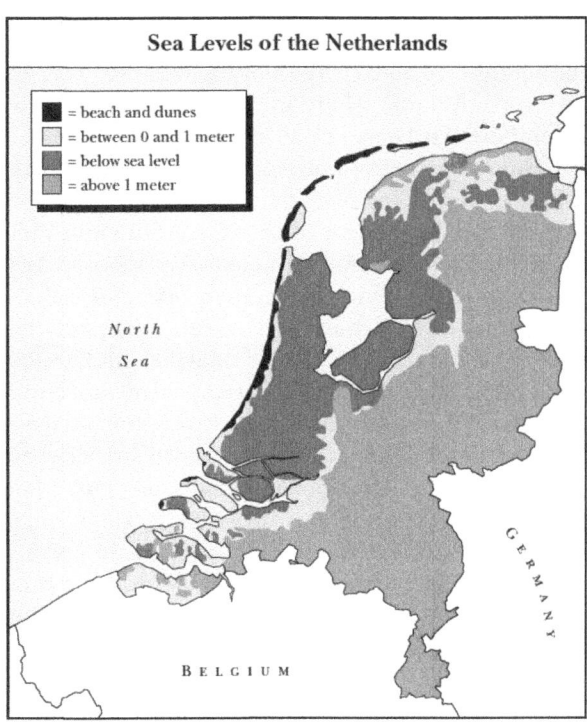

zooplankton cells in river and sea sediments also contain useful proxy climate data.

Detecting climate change depends on individual variables. Temperature change is the most reliable such variable, because its internal variability is small, and it is more widely observed than other variables. Long-term precipitation changes are more difficult to discern, because rain- and snowfall vary so greatly from one year to the next. The intensity and frequency of extreme weather events such as hundred-year floods are even more difficult to detect, because these events are rare, so a significant data set must cover many years.

Instrument records from land stations and ships indicate that the global annual mean surface air temperature rose during the twentieth century. The warming occurred more quickly in high latitudes than it did in the tropics. It was also faster over land than it was over the ocean and faster in the Northern Hemisphere than in the Southern Hemisphere. Winters warmed more than did summer, and nights warmed more than did days. Contemporary daily temperature ranges have narrowed, precisely because nights have warmed more than have days.

Extensive heat waves and intense floods have become more frequent in recent decades. Globally, the average number of tropical storms (about ninety per year) changed little during the twentieth century, although historical data are poor for some regions. In the North Atlantic, where the best records are available, there has been a clear increase in the number and intensity of tropical storms and major hurricanes. From 1997 to 2006, there were about fourteen tropical storms per year, including about eight hurricanes in the North Atlantic, compared to about ten storms and five hurricanes between 1850 and 1990.

On timescales of thousands of years or greater, the Earth's climate has been both warmer and much colder than it is today, although temperatures around the turn of the twenty-first century were the warmest in the past two thousand years. Based on ice-core proxy data, four major global glaciations occurred in past 450,000 years, about one every 100,000 years, correlating well with the cyclical variations in Earth's orbit known as the Milankovi cycles. Various ice ages occurred, with the most recent one ending about 11,500 years ago. Before that, much of North America was covered in permanent ice. Over the course of Earth's history, its temperature has swung more than 10° Celsius between cold and warm modes.

Climate Change Scenario
In the bed of what was once the Aral Sea, a Kazakh villager pulls water from a well sunk into the sandy, desert ground.

Future climate changes are predicted by climate models based on assumed greenhouse gas (GHG) emission scenarios. The scenarios range from high fossil fuel consumption, resulting in atmospheric carbon dioxide (CO_2) concentration of 800 parts per million, to low consumption, with CO_2 concentration reaching 550 parts per million. The reliability of these predictions depends on future global environmental, energy, and climate policy, as well as the accuracy of the models.

Most models project that climate change will accelerate during the twenty-first century and that the global average temperature will increase by between 1.8° Celsius and 4.0° Celsius by 2100. As in the past, warming will be more pronounced in the polar Northern Hemisphere during winter. Precipitation amounts are likely to increase in high latitudes and to decrease in most subtropical lands. Heat waves and heavy precipitation events will very likely increase in frequency. With warmer oceans, future tropical storms will become more intense, with greater peak wind speeds and heavier precipitation.

Context
Climate change may be attributed to natural processes or to human activity. Natural factors include the Earth's internal processes, such as volcanic eruptions, as well as external parameters, such as solar luminosity and Earth's orbital pattern around the Sun. Anthropogenic activity includes GHG and aerosol emission and, to a lesser degree, changes in land use. Separating natural and anthropogenic causes of climate change is challenging, if it is possible at all. Since no controlled laboratory setting exists in which to conduct climate change experiments, climate scientists have developed computer models based on the laws governing climate systems. By altering model settings, one can simulate natural and anthropogenic effects on climate, separately or in combination, thereby tracing the causes of climate change. In general, on scales of a decade to a century, climate change is attributable to atmosphere-ocean interaction and to human activity. On scales of millennia to hundreds of thousands of years, the variations in Earth's orbit directly controls the planet's climate. This orbit is described by the Milankovi cycles, which repeat every 20,000 to 100,000 years. Beyond the million-year timescale, tectonic drift is likely the main driver of climate change.

Zaitao Pan

Further Reading
Easterling, David R., et al. "Maximum and Minimum Temperature Trends for the Globe." *Science* 277 (July 18, 1997): 364-367.

Intergovernmental Panel on Climate Change. *Climate Change, 2007—The Physical Science Basis: Contribution of Working Group I to the Fourth Assessment Report of the Intergovernmental Panel on Climate Change.* Edited by Susan Solomon et al. New York: Cambridge University Press, 2007.

Klein, Naomi. *This Changes Everything: Capitalism vs. The Climate.* New York: Simon & Schuster, 2014.

Mann, Michael E., Raymond S. Bradley, and Malcolm K. Hughes. "Global-Scale Temperature Patterns and Climate Forcing over the Past Six Centuries." *Nature* 392 (April 23, 1998): 779-787.

Nakicenovic, N., et al., eds. *Special Report on Emission Scenarios: A Special Report of Working Group III of the*

Intergovernmental Panel on Climate Change. New York: Cambridge University Press, 2000.

National Research Council. *Abrupt Climate Change: Inevitable Surprises.* Washington, D.C.: National Academies Press, 2002.

Romm, Joseph. *Climate Change: What Everyone Needs to Know.* New York: Oxford University Press, 2018.

United Nations Framework Convention on Climate Change, Article 1: Definitions. Geneva, Switzerland: United Nations, 1992.

Climate change and human health

FIELDS OF STUDY
Atmospheric Sciences; Climate and Climate Change; Climatology; Disease; Medicine; Meteorology; Public Health; Weather

SUMMARY
Changes in climate can affect the numbers of people who die directly as the result of temperature extremes (either cold or hot) or violent weather and can also increase the ranges of certain diseases and other health problems, which can lead to lesser, but sometimes serious, health effects.

PRINCIPAL TERMS
- **climate change:** change of climate attributed directly or indirectly to human activity altering the composition of the global atmosphere and to natural climate variability observed over comparable time periods
- **global warming:** increase in the average surface and ocean temperature of the Earth since 1850 and the projected persistence of the trend
- **greenhouse gases:** any of the gases whose absorption of solar radiation is responsible for the greenhouse effect, including carbon dioxide, methane, ozone, and fluorocarbons.
- **drought:** a period of dry weather, especially a long one that is injurious to crop
- **World Health Organization (WHO):** a specialized agency of the United Nations, established in 1948, that is concerned with international public health

The threat of severe climate change, driven in large part by the emission of greenhouse gases (GHGs) into the atmosphere as a result of human activities, has aroused worldwide concern about the potential effects of such change on human health. The United Nations' Intergovernmental Panel on Climate Change (IPCC), in its 2014 report, assessed four potential scenarios for climate change throughout the twenty-first century based on future levels of GHG emissions, ranging from "stringent mitigation" of emissions to "very high GHG emissions." In all four scenarios, Earth's surface temperature was predicted to increase throughout the century, with the "stringent mitigation" scenario projected to result in a global surface temperature increase of between 0.3 and 1.7 degrees Celsius (relative to temperatures in 1986–2005) by the end of the twenty-first century, the "very high emissions" scenario projected to result in an increase between 2.6 and 4.8 degrees Celsius, and the other two scenarios falling somewhere in between.

Direct deaths from extremely cold or warm weather (e.g., hypothermia, heatstroke) are relatively rare in developed countries, but occasionally large numbers of people are killed or otherwise seriously affected by heart attacks caused by the weather (or by activity in response to the weather, such as people trying to shovel too much snow at once). The accelerated global warming trend that has been observed since the late twentieth century could affect the number of deaths directly related to weather, reducing the number harmed by severe cold but increasing the number harmed by extreme heat. Climate changes can also alter the numbers of people affected by flood and drought as well as the ranges of parasites and disease vectors.

In 2006, Anthony J. McMichael and colleagues published "Climate Change and Human Health: Present and Future Risks," a comprehensive summary and assessment of research performed to date on the direct and indirect health risks associated with climate change, including "thermal stress, extreme weather events, and infectious diseases, with some attention to estimates of future regional food yields and hunger prevalence." The evaluation notes some likely beneficial effects, including a reduction in cold-related deaths due to milder winters and a potential reduction in disease-carrying mosquito populations due to increased temperatures and decreased precipitation. However, these are outweighed by the projected adverse health effects, particularly in less developed regions that may lack the wealth or the technological resources to adapt to the stresses induced by climate change.

Direct Effects

Warmer temperatures tend to lead to more frequent deaths from excessive heat, although this is mitigated in the case of warming due to GHG emissions by the fact that the greatest such warming occurs at night and in winter, and especially in the Arctic and Antarctic regions, when the temperatures otherwise would be cooler. Warmer temperatures also reduce the numbers of deaths associated with extreme cold (including deaths due to cardiovascular and pulmonary diseases, such as influenza, that are worsened by exposure to cold).

The effects of rising temperatures, both beneficial and deleterious, are amplified in urban environments as a result of urban heat islands (UHIs), a fact that has become increasingly relevant as Earth's population has continued to migrate from rural areas to more densely concentrated suburbs and cities. Cities are warmer than surrounding countryside because traditional roofs and paving surfaces absorb more solar heat than do dirt and vegetation, and because significant heat is generated by industry, power plants, residential heating, and air conditioning. According to the United States Environmental Protection Agency (EPA), a city with a population of one million or more may have an annual mean air temperature of 1–3 degrees Celsius higher than the surrounding area, with the temperature difference increasing to as much as 12 degrees Celsius "on a clear, calm night."

It is difficult to accurately determine the number of deaths directly attributable to heat or cold, as many are not identified as such at the time of death. In addition, several such analyses—including that performed by Danish economist Bjørn Lomborg for his 2007 book *Cool It: The Skeptical Environmentalist's Guide to Global Warming*—fail to take into account normal seasonal variations in death rates, in which more people consistently die in winter than in summer for reasons unrelated to extreme cold. Nevertheless, studies have consistently shown that in areas that are prone to experiencing both extreme heat and extreme cold, there are more annual deaths from cold than from heat. A study of more than seventy-four million deaths in thirteen countries between 1985 and 2012, published by Antonio Gasparrini and colleagues in 2015, found that 7.29 percent of those deaths were attributable to cold versus 0.86 percent attributable to heat (although the study did not include any countries in Africa or the Middle East). Events such as the severe heat wave of summer 2003, which Jean-Marie Robine and colleagues found resulted in more than 70,000 deaths in Europe (calculated in part by comparing deaths in 2003 to mortality rates during the five years previous), are outliers in the overall trend. Still, as the IPCC's Working Group II noted in its contribution to the 2014 report, "the influence of seasonal factors other than temperature on winter mortality suggests that the impacts on health of more frequent heat extremes greatly outweigh benefits of fewer cold days." The group also pointed out that few studies have been conducted on temperature-related deaths in developing countries in tropical regions, where extreme cold is typically not a factor and where the health impact of extreme heat would likely be more significant than in more developed countries in temperate regions.

Climate change can also alter rainfall patterns, causing some areas to be more subject to floods (which inflict heavy damage and can also lead indirectly to other health problems) or droughts (which can lead to crop failures and water shortages). Rising sea levels caused by climate change also pose a flood threat to low-lying coastal areas, including many of the world's largest cities. Many scientists believe such changes are inevitable if glaciers, permafrost, and sea ice continue to melt at a rapid pace. In addition to the potential direct deaths and injuries due to floods, rising sea levels would displace many significant populations, exacerbating existing health concerns and introducing new ones.

Other potential direct health impacts due to climate change include landslides, greater exposure to both household and ambient air pollution—estimated by the World Health Organization (WHO) to be responsible for 7.3 million deaths per year—and increased exposure to ultraviolet radiation, potentially causing cancer and other adverse effects. Many experts believe climate change and its related extreme weather events are likely to cause an increase in mental health problems among the most directly affected populations.

Food and Water

Climate changes can also affect food crops. Judging from past experience during the Medieval Warm Period (MWP), which lasted from about 950 to 1250 CE, this can lead to greater production in some places (partly from longer growing seasons as well as the fertilizing effect of increased carbon dioxide) and shortages in others (caused by droughts and floods rather than the temperature changes, though this can change what crops are grown in particular areas). These shortages can lead to malnutrition, deficiency diseases, and even famine in less wealthy countries.

Drought also makes obtaining water supplies more difficult even as a population continues to grow. One consequence is that people are often forced to work hard (expending labor that would otherwise be available for other needs) for water that is often tainted, which leads to increasing outbreaks of diseases such as dysentery, typhoid fever, and cholera, as well as aquatic parasites such as guinea worms. When water is scarce, cleaning and other sanitation practices suffer. Unclean bodies (especially hands) help spread diseases, and unclean clothes can carry and spread parasites such as lice and fleas.

In 2016 the World Bank published the report *Shock Waves: Managing the Impacts of Climate Change on Poverty*, which highlighted the ways in which the economic effects of climate change can be linked to health issues. It warned that unchecked climate change could help cause the population of global poor to grow by one hundred thousand people by 2030. Impacts on agriculture would generate poverty, accompanied by health problems such as stunted growth, malaria, and chronic diarrhea.

Diseases

Climate changes can affect the ranges of various lifeforms in many ways. Warmer weather, particularly if it is also wetter, tends to increase the numbers of insects; the mild winters created by GHG warming are especially important for those insects that are otherwise susceptible to freezing temperatures. Many of these are disease vectors, spreading serious diseases such as malaria, dengue fever, yellow fever, typhus, and the plague. Not only may these insects cover larger areas (and also spread to higher elevations, as shown by a 1997 malaria outbreak in Papua New Guinea at an altitude of 2,100 meters, or 6,900 feet), but warmer temperatures can also enable them to be active for a longer portion of the year. This is especially crucial for mosquitoes that carry dengue fever, but malaria exposure may also increase. Although most malaria victims survive, the disease is persistent, with frequent relapses, and thus debilitating. On the other hand, it is estimated that warming will reduce the incidence of schistosomiasis, and possibly also the range of ticks that carry diseases such as Rocky Mountain spotted fever.

In areas that become significantly wetter, increased molds can lead to increases in hay fever and asthma, which can be fatal. Flooding can drive rodents, which help spread diseases such as the plague, from their burrows. When carbon dioxide increases, crop yields improve, but so does the growth of allergenic pollens such as ragweed.

Some skeptics, such as virologist Barry Beaty of Colorado State University, argue that the spread of diseases such as malaria seen in the late twentieth and early twenty-first centuries is the result primarily of non-climatic factors, such as resistance to drugs by the disease pathogens, resistance to pesticides by the disease vectors, and a collapse in public health measures in some areas. (More than 80 percent of the world population is theoretically vulnerable to malaria even without global warming.) Lomborg has argued that the most cost-effective way to deal with the various problems resulting from global warming is to fix the individual problems, such as improving public health and medical care, implementing desalination projects and improving infrastructure (pipes and faucets) to supply potable water, and improving the distribution of food.

However, mainstream scientists and organizations, such as the World Health Organization, increasingly connect climate change to the spread of certain diseases, in addition to more direct effects. WHO has listed climate change as the single biggest threat to overall global health interests for the twenty-first century.

Timothy Lane

Further Reading

Braasch, Gary. *Earth under Fire: How Global Warming Is Changing the World*. University of California Press, 2009.

Fagan, Brian. *The Great Warming: Climate Change and the Rise and Fall of Civilizations.* London: New Bloomsbury Press, 2008.

Gasparrini, Antonio, et al. "Mortality Risk Attributable to High and Low Ambient Temperature: A Multicountry Observational Study." *The Lancet*, vol. 386, no.

Field, Christopher B. Field, ed. *Assessment Report of the Intergovernmental Panel on Climate Change.* Cambridge, England: Cambridge University Press, 2014. IPCC:

Johansen, Bruce E. *The Global Warming Desk Reference.* Santa Barbara, Calif.: Greenwood Press, 2002.

Lomborg, Bjørn. *Cool It: The Skeptical Environmentalist's Guide to Global Warming.* New York: Alfred A. Knopf, 2007.

Mann, Michael E., and Lee R. Kump. *Dire Predictions: Understanding Climate Change.* London: DK Publishing, 2015.

Robine, Jean-Marie, et al. "Death Toll Exceeded 70,000 in Europe during the Summer of 2003." *Comptes Rendus Biologies*, vol. 331, no. 2, 2008, pp. 171–78.

Singer, S. Fred, and Dennis T. Avery. *Unstoppable Global Warming: Every 1,500 Years.* Lanham, Md.: Rowman & Littlefield Publishers, 2007.

Websites

Intergovernmental Panel on Climate Change,
www.ipcc.ch/report/ar5/wg2/.
IPCC: Intergovernmental Panel on Climate Change
www.ipcc.ch/report/ar5/wg2/.
IPCC: Intergovernmental Panel on Climate Change
www.ipcc.ch/report/ar5/syr/.
National Geographic,
search.ebscohost.com/login.aspx?direct=true&db=a9h&AN=48758786&site=ehost-live.
National Institute of Environmental Health Sciences,
The Lancet,
search.ebscohost.com/login.aspx?direct=true&db=a9h&AN=20031043&site=ehost-live.
United States Environmental Protection Agency,
www.epa.gov/heat-islands/learn-about-heat-islands.
World Bank eLibrary,
doi:10.1596/978-1-4648-0673-5.
World Health Organization
www.who.int/mediacentre/factsheets/fs266/en/.
www.niehs.nih.gov/research/programs/geh/climatechange/health_impacts/.

■ Climate change skeptics

FIELDS OF STUDY

Atmospheric Sciences; Atmospheric Structure and Dynamics; Climate and Climate Change; Climatology; Controversies; Debates; Meteorology; Public Policy

SUMMARY

Governments are under increasing pressure to adopt policies to mitigate the threat of climate change resulting from emissions of greenhouse gases, mainly carbon dioxide from the burning of fossil fuels. As industrialized countries are engaged in costly efforts to cut emissions, persons who are skeptical that climate change poses a danger to humankind debate the environmental and economic risks of global warming.

PRINCIPAL TERMS

- **anti-environmentalism**: philosophy that human beings' immediate economic and lifestyle needs are more important than concerns about the fates of other specifies and the general environment
- **carbon dioxide:** colorless, odorless, incombustible gas present in the atmosphere and formed during respiration, usually obtained from coal, coke, or natural gas by combustion
- **climate change:** change of climate attributed directly or indirectly to human activity altering the composition of the global atmosphere and to natural climate variability observed over comparable time periods
- **fossil fuels:** any combustible organic material, as oil, coal, or natural gas, derived from the remains of former life
- **global warming:** increase in the earth's atmospheric and oceanic temperatures due to an increase in the greenhouse effect resulting especially from pollution
- **greenhouse gases:** any of various gaseous compounds, such as carbon dioxide or methane, that absorb infrared radiation and trap heat in the atmosphere

A range of opinions is found among those persons described as climate change skeptics, who challenge the view that dangerous global climate change is under way as a result of the buildup of human-caused greenhouse gases in the atmosphere. Some believe it is impossible that global climate change could be caused by the burning of fossil fuels, perhaps because they simply do not understand the science underpinning the radiative effect of greenhouse gases on climate. Most climate change skeptics, however, in challenging those who believe humankind is in a planetary environmental emergency, focus their arguments on uncertainties in the answers to the following questions: Is global climate warming? If so, what part of that warming is caused by human activities? How good is the evidence? What are the risks?

Some skeptics accept that the rising atmospheric concentration of carbon dioxide can influence climate, but they argue that the net effect of this rise above twentieth century levels is minor compared to natural forces. They point out that no scientific evidence suggests anything tragic, and they maintain that this is because of the existence of natural feedback mechanisms within the global climate system, such as the increase in cloudiness, that suppress initial change imposed on the system by carbon dioxide increase (negative feedbacks). Skeptics challenge the claim that any climate change is likely to be dangerous on the grounds that there is no observational evidence to prove the hypothesis that small initial changes triggered by carbon dioxide will set in motion amplifying processes (positive feedbacks). They emphasize the fact that predictions based on climate models are not evidence of future climate because the models cannot adequately simulate climate.

The stance of the skeptics is reinforced by what they perceive as a range of factors that motivate climate change alarmism. They point out that the news media are drawn toward worst-case scenarios and argue that the public eventually comes to view these reported scenarios as reality if the stories are told often enough. Skeptics also assert that many others who share their views are reluctant to challenge the alarmist stance, as it is seen as "politically incorrect" to do so; such a challenge may be taken to imply a lack of concern for the environment. Some skeptics assert that politicians are drawn to the theme of climate change by the appeal of tackling something grand—a global environmental issue, as opposed to merely a local one. Another factor for politicians is the allure of the green vote. Skeptics have also argued that scientists promote alarmist speculation because heightened concern improves the chances that the funding of climate change research will be given high priority.

Many skeptics point out that the climate change issue is often confused with the separate matters of conservation of fossil fuels and the effects of burning such fuels on air quality. Skeptics further argue that the public in general relies too much on "authorities" to rule on scientific matters about which, they claim, there is a high degree of uncertainty.

C. R. de Freitas

Further Reading

Horner, Christopher C. *Red Hot Lies: How Global Warming Alarmists Use Threats, Fraud, and Deception to Keep You Misinformed.* Washington, D.C.: Regnery, 2008.

Michaels, Patrick J. *Meltdown: The Predictable Distortion of Global Warming by Scientists, Politicians, and the Media.* Washington, D.C.: Cato Institute, 2004.

_____, ed. *Shattered Consensus: The True State of Global Warming.* Lanham, Md.: Rowman & Littlefield, 2005.

Sinanian, Arek. *A Climate for Denial: Why Some People Still Reject Climate Change Science.* Sydney, Aus.: Longueville Media, 2017.

■ Conference of the Parties

FIELDS OF STUDY

Atmospheric Science; Climate Change; Global Warming; Government; International Policy; Meteorology; Organizations and Agencies

SUMMARY

The Conference of the Parties (COP), established in 1995, is the decision-making body of the United Nations Framework Convention on Climate Change (UNFCCC) and is the largest annual United Nations conference with an average attendance of 25,000 participants. Frequently hosted by member nations around the world, the Conference of the Parties annually reviews the implementation of any measures of the Convention and progress towards the Convention's goals of climate action.

PRINCIPAL TERMS
- **carbon offsets:** a decrease in emissions of carbon or other greenhouse gases in order to compensate for emissions produced elsewhere.
- **Clean Development Mechanism:** lets a country earn emissions reduction credits per the terms of the Kyoto Protocol by establishing emission-reducing projects in developing countries.
- **emissions trading:** a market-based method to decreasing greenhouse gas emissions in which countries receive permits to produce a specified amount of emissions, which they may then use or sell to other countries who may wish to emit more than their allotted quantity of greenhouse gases.
- **greenhouse gas:** the naturally-occurring and anthropogenic gases that make up the atmosphere and absorb and emit radiation.
- **Joint Implementation:** a process allowing developed countries to earn emissions reduction credits under the Kyoto Protocol from an emission-reduction project in another developed country.
- **Kyoto Protocol:** established in 1997 at the third Conference of the Parties, the Kyoto Protocol sets guidelines for nations to limit their emissions that contribute to global warming.
- **Paris Agreement:** adopted in 2015 at the twenty-first Conference of Parties, one of the primary goals of the agreement was to keep global warming below 2 °C.
- **United Nations Framework Convention on Climate Change:** an international treaty adopted in 1992 with the purpose of assessing global progress in combatting climate change.

A 1990 report by the Intergovernmental Panel on Climate Change (IPCC) found that the continued increase of greenhouse gas concentrations in the earth's atmosphere would lead to increases in global average temperature. In response to this report, the United Nations facilitated negotiations among nations to determine international efforts to address the threat of global warming. The resulting treaty, the United Nations Framework Convention on Climate Change(UNFCCC), was signed by 192 countries in 1992 and established goals of quantifying and reducing greenhouse gas emissions, while also differentiating between the responsibilities of developed and developing nations in responding to the global warming. The UNFCCC itself did not have any legally binding mechanism to enforce its goals, but the treaty included an annual meeting called the Conference of the Parties, at which member parties including the United States, Australia, Japan, Canada, nations in the European Union, and Russia would negotiate actions to meet specific goals.

The first Conference of the Parties (COP-1) met in Berlin in 1995. In what became known as the Berlin Mandate, negotiators set 1990 as a baseline year with which to measure changes in greenhouse gas emissions, and they established emissions targets for nations to work towards. By the time the third Conference of the Parties (COP-3) was held in Kyoto, Japan in 1997, many developed nations, like the United States and European Union member states, had failed to meet the greenhouse gas reduction targets set in Berlin three years earlier. COP-3 was therefore focused on determining new, individualized emissions targets for different nations, as well as policy options to achieve those targets. The Kyoto Protocol, as these policies were collectively known, involved environmental strategies that were well known at the end of the twentieth century, such as pursuing sustainable agriculture or renewable energy, but it also suggested three new mechanisms to help developed nations reduce greenhouse gas emissions in an economically viable way. These new tactics were the Clean Development Mechanism (CDM), Joint Implementation (JI), and emissions trading, and they each allowed nations to incentivize greenhouse gas emissions reductions for economic benefit.

In the first decades of the twenty-first century, there continued to be conflicting ideological approaches to climate action, as some nations favored more regulatory policies while others wanted free-market solutions. The Conference of the Parties has reflected this debate while maintaining focus on the broad goal of limiting greenhouse gas emissions and global warming. The twenty-first Conference of the Parties (COP-21) established the Paris Agreement in 2015, articulating its goal as limiting the average increase in global temperatures to 2 °C above levels at the time of the Industrial Revolution. Under the Paris Agreement, signatory nations must plan for itself how it will contribute to limiting global warming, with the terms of the agreement going into effect in 2020.

The Conference of the Parties has also continued to expand its goals for climate policy beyond its initial focus primary on economic strategies for opposing global warming. In 2015, members argued that climate action should take into account the inequalities in the ways that humans around the world are affected by global warming. Changes in climate have a greater impact on the world's poor, those who rely more directly on natural resources for their livings or those who are more vulnerable to extreme weather events. The UNFCCC has therefore emphasized the role that considerations of gender and indigenous peoples should play in its approach to climate action, and its strategies for adaptation to global warming thus reflect a growing recognition that responses to climate change must take place on multiple scales, from international treaties to individual communities.

Compliance
The UNFCCC was criticized from the beginning for lacking a mechanism to enforce international compliance with the treaty's terms, and the annual Conference of the Parties was a measure to address that issue of enforcement. But international treaties like those negotiated at the Conference of the Parties frequently struggle with how to how to compel ideologically diverse nations around the world to adhere to their agreements. In the lead up to COP-3 in Kyoto in 1998, media and lobbyist campaigns in the United States had worked to discredit global warming science in an effort to influence the nation's policy responses to the new Kyoto Protocol. The United States had signed the protocol at the fourth Conference of the Parties (COP-4) in Buenos Aires; however, any treaty must be ratified by the United States Senate for it to become binding. Because the political climate in the United States was so unreceptive to climate policy, the Clinton administration opted to postpone a ratification vote of the protocol until it a time when it might not be rejected outright. Alas, that time would not come, as George H. W. Bush rejected the Kyoto Protocol outright when he became President of the United States in 2001.

The Kyoto Protocol was in effect from 2008-2012, and while it was amended in 2012 to continue from 2013-2020, that amendment has not entered into force because not enough nations have accepted it. The Paris Agreement covers the period from 2020 on, and while it was adopted by all 197 member nations of the UNFCCC, it has also been subject to criticism. Some have argued that the Paris Agreement represented admirable goals, but it did not do enough to enforce compliance with its terms. Many people believe that measures such as a carbon tax that would actively make emissions economically damaging would be a more likely way to achieve the goals of greenhouse gas reduction, as the odds that many nations would voluntarily and drastically decrease their emissions can seem unlikely.

As if in demonstration of that weakness in the Paris Agreement's enforcement mechanisms, in June 2017, President Donald Trump announced that the United States planned to withdraw from the treaty. Trump offered his own critiques of the treaty, arguing that the Paris Agreement was unfairly punitive of the United States and threatened American sovereignty. Because the Paris Agreement stipulates terms for withdrawal, the earliest the United States could officially withdraw would be 2020.

The United States consistently represents a significant portion of global greenhouse gas emissions, and many environmentalists and politicians argue that the United States must be integral to any efforts to limit global warming. The international response to the United States' withdrawal from the Paris Agreement was therefore strongly negative, and several U.S. states formed the United States Climate Alliance with the intention of continuing to adhere to the objectives of the Paris Agreement in those states, even without the support of the federal government.

In another challenge to the goals of the UNFCCC, Brazil withdrew from hosting the twenty-fifth Conference of the Parties, which was planned to take place there in 2019. Brazil has historically played a significant leadership role in the United Nations' climate action, and the nation is also home to much of the Amazon, the world's largest rainforest which also absorbs substantial amounts of carbon as a carbon sink. But in 2018, following the election of Jair Bolsonaro as President of Brazil, who had discussed withdrawing from the Paris Agreement during his campaign, the Brazilian government announced that the country would no longer host the Conference of the Parties the following year because of budgeting constraints and the transition to a new presidential administration. Brazil's withdrawal from hosting the conference has been criticized as a failure of the new administration to live up to the leadership role that Brazil has historically played in international climate policy.

Julia Kendrick

Further Reading

Blau, Judith R. *The Paris Agreement: Climate Change, Solidarity, and Human Rights.* Cham, Switzerland: Palgrave Macmillan, 2017.

Glover, Leigh. *Postmodern Climate Change.* New York: Routledge, 2006.

Schneider, Stephen H., Armin Rosencranz, and John O. Niles, editors. *Climate Change Policy: A Survey.* Washington: Island Press, 2002.

van Ierland, Ekko C., Joyeeta Gupta, and Marcel T.J. Kok, eds. *Issues in International Climate Policy: Theory and Policy.* Cheltenham, UK: Edward Elgar Publishing Limited, 2003.

Vasser, Christophe P. *The Kyoto Protocol: Economic Assessments, Implementation Mechanisms, and Policy Implications.* Hauppauge, NY: Nova Science Publishers, 2009.

Websites

United Nations Framework Convention on Climate Change
https://unfccc.int/

United Nations.
https://www.un.org/sustainabledevelopment/climate-action/

■ Coral reefs

FIELDS OF STUDY

Aquatic Ecosystems; Climatology; Conservation; Ecology; Environment; Evolutionary Biology; Geoscience; Oceanography

SUMMARY

Coral reefs are being threatened in various ways, including a process known as coral bleaching, which kills the coral and destroys the reefs, thus also threatening all of the marine species dependent on the reefs for survival.

PRINCIPAL TERMS

- **coral reefs:** stony formations created by the depositing of exoskeletons by colonies of coral polyps
- **corals:** the hard, variously colored, calcareous skeleton secreted by certain marine polyps.
- **dead zones:** areas of deep-water oxygen depletion due to surface algal blooms or disruption of thermohaline circulation
- **El Niño–Southern Oscillation (ENSO):** periodic fluctuation of temperatures and currents in the Pacific Ocean on a four-, ten-, and ninety-year cycle
- **primary production:** production of fixed carbon through photosynthesis
- **thermohaline circulation:** the rising and sinking of water caused by differences in water density due to differences in temperature and salinity

Corals are living organisms, and approximately one thousand different species of coral are known to exist, some of which live in colonies. These colonies of coral polyps deposit their exoskeletons of calcium carbonate as they grow, and thus provide the durable formations that the polyps, along with other species of coral and many other animals, live on and around; these stony formations become coral reefs. Coral reefs are found in the oceans, most often in warm, shallow waters in the Tropics, particularly in the Pacific Ocean.

Reef-building corals need to be in shallow water in order to get energy from the sunlight that filters down through the water column. The corals themselves do not photosynthesize sunlight; rather, they have a symbiotic relationship with protozoa called zooxanthellae. These organisms live within the tissues of the corals and provide them with the by-products of photosynthesis, such as glucose and amino acids, that the corals use for energy. Reef-building corals are thus almost wholly dependent on the zooxanthellae and must live in shallow waters where the sunlight can reach them.

Coral reefs cover a total area equivalent to just 1 percent of the surface area of the oceans, yet they support one-fourth of marine life. For organisms such as phytoplankton and algae, which use corals for shelter and for the raw materials they need to produce energy, and larger animals such as fish and sea snakes, coral reefs offer sources of nutrients in addition to shelter. The diversity and sheer numbers of marine life make up a complex food web, with the corals and their symbionts at the center. It is because of their nutrient richness that most coral reefs are so important to so many different species of marine life. Given the variety of animals that live in reef communities, any threat to the corals is a threat to the entire ecosystem, and coral reefs are very fragile owing to their susceptibility to changes in water temperature and acidity.

Some significant threats to coral reefs come directly from human activities, such as the destructive fishing practices of overfishing, bottom trawling, and blast fishing. Coral bleaching, however, is the biggest threat facing reef ecosystems; bleaching events have been occurring with growing frequency across the globe since the 1980s. The term "coral bleaching" refers to the whitening of corals when the protozoa lose their pigmentation or the corals expel their symbionts in reaction to some form of outside stress, such as pollution.

The main causes of coral bleaching are related to climate change increases in water temperature and increased acidity of the oceans. Coral reefs continue to undergo bleaching even if the source of the stress is removed, and if the colonies survive the bleaching, the zooxanthellae do not immediately reappear, sometimes taking months to come back.

In 2014, for only the third time ever recorded, scientists announced the occurrence of a prodigious, worldwide instance of coral bleaching. Mainly caused by global warming and the increasing temperature of ocean waters, the bleaching was reported as impacting major bodies of water that included the Pacific and Indian Oceans. Exacerbated by a strong El Niño in 2015–16 and a subsequent La Niña, the event continued until May 2017. It was the longest bleaching event ever recorded and affected more coral reefs than any previous such event; as warm currents came and went, coral reefs in some parts of the Pacific experienced new rounds of bleaching every year while the event was ongoing. It also affected some reefs that had never undergone bleaching before, such as the northernmost part of the Great Barrier Reef off the coast of Australia. Scientists said that it would take ten to fifteen years for the affected reefs to recover but expressed concern that the reefs would not have that much time before waters warmed again.

The prospect of higher sea levels associated with climate change is also a threat to coral reefs, as corals need to be in shallow waters with a lot of sunlight to survive, and corals cannot grow fast enough to keep up with predicted sea-level rises. More than half of the world's coral reefs are threatened directly or indirectly by human activity, a situation that has led environmentalists increasingly to promote public awareness of the economic value of the reefs and the ecosystem services they offer, such as shoreline protection and well-stocked fisheries.

Great Barrier Reef

The Great Barrier Reef is a biodiverse ecosystem of more than three thousand coral reefs and seven hundred individual islands (some barely a few yards across, but twenty-seven of them large enough to have tourist resorts) that follows the Australian coast off the state of Queensland. With its tremendous size (at more than 337,000 square kilometers, or 130,000 square miles, it is visible from space) and its compelling beauty, the reef enthralls the imagination apart from its value as an intricate ecosystem. It is, in a

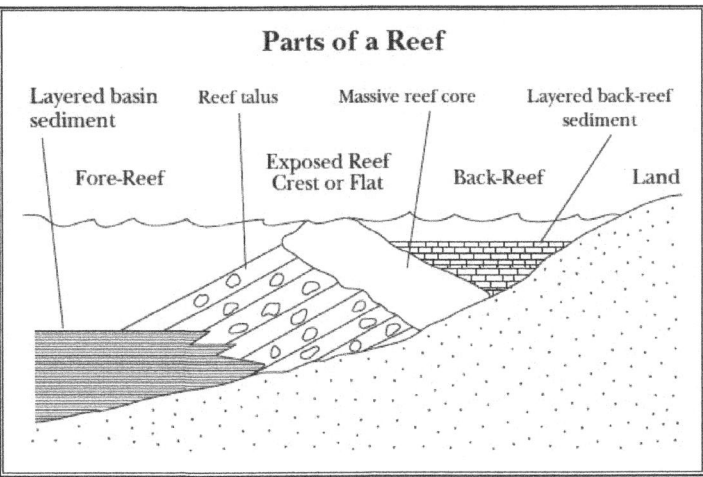

Parts of a Reef

sense, a single living organism—although more precisely it is a colony of millions of tiny coral polyps (living creatures inside colored hard shells of aragonite, a calcium derivative that shapes the familiar fan and branch shapes of coral) that live atop the dead, bleached remains of earlier generations, building slowly, steadily, century after century, into an incredibly dense superorganism.

Within this vast construction of accumulated coral structures (most of it just feet below the ocean's surface) thrives a diverse ecosystem in the pristine tropical waters that includes a wide variety of animal and plant species, among them green sea turtles, sharks, porpoises, whales, crocodiles, dugongs, and snakes, as well as more than one thousand species of fish and more than two hundred species of both land and marine birds. The Great Barrier Reef was designated a World Heritage Site in 1981 by the United Nations Educational, Scientific, and Cultural Organization (UNESCO), a recognition reserved for natural and cultural sites deemed an irreplaceable part of humanity's heritage.

Protecting the natural integrity and rich biodiversity of the massive reef is the special mission of the Great Barrier Reef Marine Park Authority, an oversight committee of the Queensland state government. In addition to natural threats—including cyclones, disease, and periodic infestations of crown-of-thorns starfish that attack the coral polyps—the most prominent threat measured since the 1990s has come from slowly rising ocean temperatures from the effects of El Niño weather conditions and, increasingly, global warming linked to humans' exponential burning of fossil fuels and subsequent release of greenhouse gases over the years.

Despite ongoing efforts to protect and preserve the reef, the survival of this intricate and delicate ecosystem is constantly threatened by both natural events and human activities.

Joseph Dewey and Megan E. Watson

Further Reading

"Coral Bleaching Crisis Spreads Worldwide—and It's Getting Worse." *CBS News*. CBS Interactive, 8 Oct. 2015.

Allsopp, Michelle, et al. *State of the World's Oceans*. New York: Springer, 2009. Print.

Mooney, Chris. "Scientists Say a Dramatic Worldwide Coral Bleaching Event is Now Underway." *Washington Post*. Washington Post, 8 Oct. 2015. Web. 2 Nov. 2015.

Peters, Robert L., and Thomas E. Lovejoy, eds. *Global Warming and Biological Diversity*. New Haven, Conn.: Yale University Press, 1992.

Reynolds, Colin S. *Ecology of Phytoplankton*. Cambridge, England: Cambridge University Press, 2006.

Sapp, Jan. *What Is Natural? Coral Reef Crisis*. New York: Oxford University Press, 1999.

Veron, John E. N. *A Reef in Time: The Great Barrier Reef from Beginning to End*. Cambridge, Mass.: Harvard University Press, 2010.

Websites

ARC Centre of Excellence for Coral Reef Studies.
www.coralcoe.org.au/media-releases/
    life-and-death-after-great-barrier-reef-bleaching.

The Guardian.
www.theguardian.com/environment/2017/jun/20/
    worst-global-coral-bleaching-event-eases-as-experts-await-next-one.

■ Cultural ecology

FIELDS OF STUDY
Agriculture; Anthropology; Art and Art History; Ecology; Environmentalism; Humanities; Political Science; Religious Studies

SUMMARY
While claims that environmental factors can wholly explain cultural traits and dynamics are no longer made, cultural ecology has become a useful approach for understanding a given society's customs, even those that initially make no sense to outsiders. Cultural ecology has proven to be a fruitful approach.

PRINCIPAL TERMS
- **carrying capacity:** the maximum equilibrium number of organisms of a species that can be supported indefinitely in a given environment
- **cultural ecology:** theory of anthropology that seeks to explain human cultures in terms of the

environmental conditions in their home territories
- **Web of Life:** All the living things in an ecosystem which depend on all the other things - living and non-living for continued survival - for food supplies and other needs

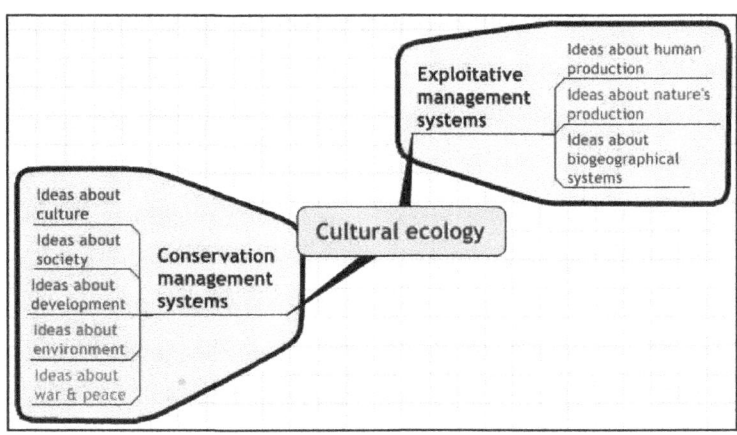

Cultural ecology was originally associated with anthropologist Julian Steward. He developed the theory during the 1950s, using it to study how cultural norms and customs are adaptive, given the environment that a group inhabits.

Early attempts to apply the theory to classic anthropological tropes, such as the frequent warfare and female infanticide of the Yanomami of South America and the custom of potlatch (competitive giving feasts) among indigenous peoples of the Pacific Northwest coast, had mixed results. For example, anthropologist Marvin Harris explained the fierce wars and infanticide of the Yanomami as methods of keeping their population levels within the carrying capacity of their jungle environment. Although war may be seen as one of the Four Horsemen of the Apocalypse, a consequence of population pressures, this is more a literary or religious concept than a sociological truth. There was little evidence that the Yanomami felt such pressures; other explanations fit their acknowledged cultural traits better. On the other hand, the potlatch took place within a culture region that usually had a surplus of food and other material resources. The gift-giving ceremonial not only built prestige for the sponsoring chief but also helped to counteract local scarcities caused by sudden natural events. Producing and storing food for a potlatch also gave a village a margin of safety against future natural disasters.

With worldwide movements for both increased food production and protection of the environment, insights from cultural ecology have become invaluable. For example, Western efforts to impose temperate-zone, mechanized farming methods on small-scale farmers in the tropics have often met with disaster. Indigenous patterns of shifting cultivation, growing several crops with different maturation dates together in the same small plot, and periodically using fire to put chemicals back into the soil quickly may actually reflect the optimum use of this land.

Early anthropologists, it has been noted, were more interested in recording a society's rain dances than in the rain itself and its place in the society's life. With longer terms of fieldwork and the tools of cultural ecology, observers can note not only the role of both in the culture but also the conditions under which imminent rain is felt to be essential to survival and thus must be called down.

In modern societies, the web of life support extends far beyond the immediate natural surroundings. Rain, or its lack, may be less immediately relevant to a society's present well-being than the infrastructure, which includes reservoirs, irrigation mechanisms, commercial and transportation arrangements for food distribution, and many other factors. Each society, however, no matter how complex, has its tipping point, and the social mechanisms for avoiding the tipping point are not always understood, much less practiced. Cultural ecology, while potentially useful, has only fitfully treated the complexity of the changing interfaces between human society and the environment in advanced societies. This remains a challenge for the future.

Emily Alward

Further Reading

Bennett, John E. *The Ecological Transition: Cultural Anthropology and Human Adaptation.* Piscataway, N.J.: Transaction, 2005.

Haeen, Nora, ed. *The Environment in Anthropology.* New York: New York University Press, 2005.

Netting, Robert M. *Cultural Ecology.* 2nd ed. Prospect Heights, Ill.: Waveland Press, 1986.

D

■ Dams and reservoirs

FIELDS OF STUDY
Agriculture; Conservation; Ecology; Electrical Engineering; Engineering; Environment; Geography; Hydraulic Engineering; Hydroelectric Power Plants

SUMMARY
Because dams obstruct the natural flow of water, they have significant effects on stream and river ecosystems. Although dams provide benefits such as flood control and hydroelectric power generation, they also have negative environmental impacts.

PRINCIPAL TERMS
- **barrages:** a low artificial obstruction in a watercourse to increase the depth of the water and facilitate irrigation
- **dams:** structures that obstruct the natural flow of water in rivers and streams, and the bodies of water created by the impoundment of water behind or upstream of such structures
- **hydroelectric power:** electricity produced from generators driven by turbines that convert the potential energy of falling or fast-flowing water into mechanical energy.
- **reservoirs:** a natural or artificial place where water is collected and stored for use, especially water for supplying a community, irrigating land, furnishing power
- **Tennessee Valley Authority:** a federally owned corporation in the United States, originally created in 1933, to provide navigation, flood control, electricity generation, fertilizer manufacturing, and economic development to the Tennessee Valley, a region particularly affected by the Great Depression
- **tributary:** a stream that flows to a larger stream or other body of water
- **watershed:** a region or area bounded peripherally by a divide and draining ultimately to a watercourse or body of water

Dams are designed for various purposes, including conservation, irrigation, flood control, hydroelectric power generation, navigation, and recreation. Not all dams create reservoirs of significant size. Low dams, or barrages, have been used to divert portions of stream flows into canals or aqueducts since human beings' first attempts at irrigation thousands of years ago. Canals, aqueducts, and pipelines are used to change the direction of water flow from a stream to

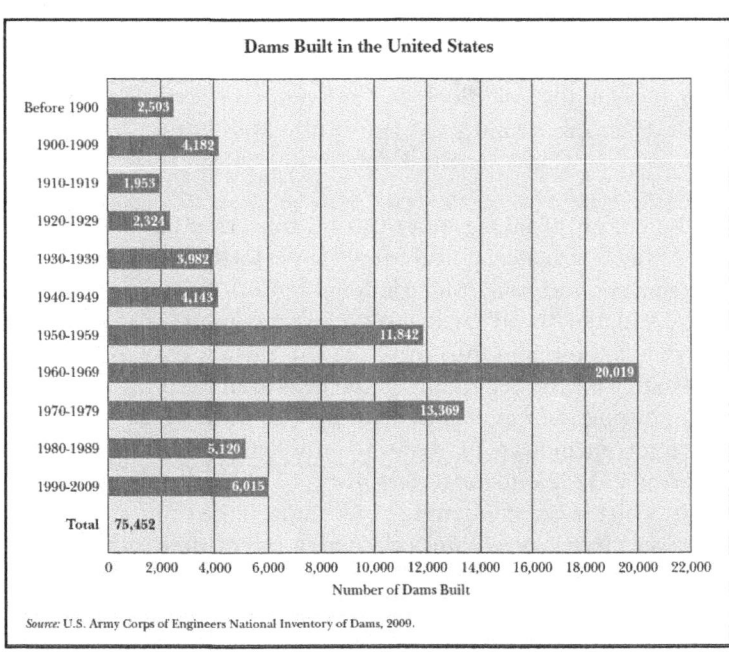

Source: U.S. Army Corps of Engineers National Inventory of Dams, 2009.

agricultural fields or areas with high population concentrations.

Dams and reservoirs provide the chief, and in most cases the sole, means of storing stream flow over time. Small dams and reservoirs are capable of storing water for weeks or months, allowing water use during local dry seasons. Large dams and reservoirs have the capacity to store water for several years. As urban populations in arid regions have grown and irrigation agriculture has dramatically expanded, dams and reservoirs have increased in size in response to demand. They are frequently located hundreds of kilometers from where the water is eventually used. The construction of larger dams and reservoirs has resulted in increasingly complex environmental and social problems that have affected large numbers of people. This has been particularly true in tropical and developing nations, where most of the large dam construction of the last three decades of the twentieth century was concentrated.

Sizes and Purposes of Dams
Early dams and their associated reservoirs were small, and dams remained small, for the most part, until the twentieth century. The first dams were simple barrages constructed across streams to divert water into irrigation canals. Water supply for humans and animals undoubtedly benefited from these diversions, but the storage capacity of most dams was small, reflecting the limited technology of the period. The earliest dams were constructed some five thousand years ago in the Middle East, and dams became common two thousand years ago in the Mediterranean region, China, Central America, and South Asia.

The energy of falling water can be converted by water wheels into mechanical energy to perform a variety of tasks, including the grinding of grain. Dams create a higher "head" or water level, increasing the potential energy, and thus served as the earliest energy source for the beginnings of the Industrial Revolution during the nineteenth century. The most significant contribution of dams to industrialization occurred in 1882 with the development of hydroelectricity, which permitted energy to be transferred to wherever electric power lines were built rather than being confined to river banks.

During the nineteenth century, large-scale settlement of the arid regions of western North America and Asia soon exhausted the meager local supplies of water and prompted demands for both exotic supplies from distant watersheds and storage for dry years. Big dams for storage and big projects for transportation of the water were thought to be the answer. Small dam projects could be financed locally; grander schemes required the assistance of federal or national governments. To justify expenditures on larger dams, promoters of these projects touted the multiple uses for reservoir water as benefits that would offset the projects' costs. Benefit-cost ratios thus became the tool by which potential projects were judged. To raise the ratio of benefits to costs, promoters placed increasing importance on intangible benefits—those to which it is difficult to assign universally agreeable currency values. While dams in arid regions were originally justified chiefly for irrigation, public water supply, and power, decisions to build dams in wetter areas were usually based on projected benefits from flood control, navigation, and recreation in addition to power generation and public water supply.

Complicating the equation is the fact that multiple uses are frequently conflicting uses. While all dams are built to even out the uneven flow of streams over time, flood control requires an empty reservoir to handle the largest floods; conversely, power generation requires a high level of water in the reservoir to provide the highest head. Public water supply and navigation benefit most from supplies that are manipulated in response to variable demand. Recreation, fishing, and the increasingly important factor of environmental concerns focus on in-stream uses of the water.

By the last two decades of the twentieth century, environmental costs and benefits and the issue of Native American water rights in the American West dominated decisions concerning dam projects in the United States, and few dams were constructed. Most of the best sites for the construction of large dams in the developed nations had been utilized, and the industry turned its attention to the developing nations. Most of the large dam projects of the last quarter of the twentieth century were constructed in or proposed for developing nations and the area of the former Soviet Union.

Human Impacts
Small dams have small impacts on the environment; they affect small watersheds and minor tributaries and usually have only a single purpose. Farm ponds

and tanks, as they are known in many parts of the world, generally cover a fraction of 1 hectare (2.47 acres) in area and are only a few meters in height. These tiny ponds are designed to store water for livestock and occasionally for human supply. They frequently serve recreational purposes as well, such as fishing. During dry spells they become stagnant and subject to contamination by algae and other noxious organisms, which can threaten the health of humans and livestock. Otherwise, they have little negative impact on the environment or on nearby people and animals.

Large dams and reservoirs are responsible for environmental and social impacts that often appear to be roughly related to their size: The larger the dam or reservoir, the greater the impact. Geographical location is also important in assessing a project's impact. Scenic areas in particular, or those with endangered species of plants or animals or irreplaceable cultural or archaeological features, raise more controversy and litigation if they are chosen as potential sites for dam and reservoir projects.

People in tropical regions suffer proportionately greater health-related impacts from dams than do those in corresponding non-tropical areas. The large numbers of workers required for the construction of big dams and associated irrigation projects can carry diseases into unprotected populations. Stagnant or slow-moving waters in reservoirs and irrigation canals, as well as fast-moving waters downstream of dams, are associated with particularly vicious tropical health risks. Snails in slow-moving water carry schistosomiasis, a parasitic disease that infects intestinal and urinary tracts, causing general listlessness and more serious consequences, including failure of internal organs and cancer. Estimates of the numbers of people infected range into the hundreds of millions. Malaria, lymphatic filariasis (including elephantiasis), and other diseases are carried by mosquitoes that breed in water; the incidence of such insect-borne diseases dramatically expands near irrigation projects and reservoirs. River blindness, which results from the bite of black flies, is associated with fast-flowing water downstream of dams and affects hundreds of thousands of humans.

The flooding of densely populated river valleys by reservoirs displaces greater numbers of people, with attendant health problems and social impacts, than similar projects in sparsely populated areas. Population displacement in developing countries, especially those in the tropics, causes greater health and social problems than in developed nations, where remedial measures and compensation are more likely to assuage the loss of homestead and community.

Environmental Impacts

Many of the environmental problems created by large dams are associated with rapid changes in water level below the dams or with the ponding of stream flow in the reservoirs, which replaces fast-flowing, oxygenated water with relatively stagnant conditions. Indigenous animal species, as well as some plant life, are adapted to seasonal changes in the natural stream flow and cannot adjust to the post-dam regime of the stream. Consequently, the survival of these species may be threatened. In 1973 the Tennessee Valley Authority—the worldwide model for builders of many large, integrated river basin projects—found that the potential demise of a small fish, the snail darter, stood in the way of the completion of the Tellico Dam. After considerable controversy and litigation, the dam was completed in 1979, but few large projects have been proposed in the United States since then, particularly in the humid East. Since the early 1970's, the arguments for abandoning dam projects have been more likely to be backed up by laws, regulations, and court decisions.

Construction of the Hetch Hetchy Dam in the Sierra Nevada mountain range of California in the early twentieth century sparked vigorous dissent, which is said to have led to the growth of the Sierra Club and organized environmental opposition to dam building. This opposition successfully challenged the construction of the Echo Park Dam on the Green River in Colorado in the early 1950's but was unsuccessful in stopping the construction of the Glen Canyon Dam on the Colorado River, which was completed in 1963. Glen Canyon, however, was the last of the big dams constructed in the American West. The preservationists, whose arguments chiefly concerned scenic and wilderness values, with attendant benefits to endangered species, lost the Glen Canyon battle but won the war against big dams. The controversy surrounding the Glen Canyon Dam continued for more than three decades after its completion, pitting wilderness and scenic preservationists against powerboat recreationists, who benefit from the access accorded by the dam's reservoir to the upstream canyonlands.

All reservoirs eventually fill with silt from upstream erosion; deltas form on their upstream ends. Heavier sediments, mainly sands, are trapped behind the dam and cannot progress downstream to the ocean. The Atlantic coastline of the southeastern United States suffers from beach erosion and retreat because the sands are no longer replenished by the natural flow of nearby rivers. The Aswan High Dam on the Nile River in Egypt has had a similar impact on the Nile Delta. Moreover, the natural flow of sediments downstream has historically replenished the fertility of floodplain soils during floods. To the extent that the flood-control function of a dam is successful, new fertile sediment never reaches downstream agricultural fields. While irrigation water provided by the dam may permit the expansion of cropland, this water in arid regions is often highly charged with salts, which then accumulate in the soils and eventually become toxic to plant life.

Reservoir waters release methane from decaying organic matter into the atmosphere. Methane is a greenhouse gas that promotes global warming, and some estimates suggest that the effect of large reservoirs is roughly equal to the greenhouse gas pollution of large thermal-powered electrical generation plants. The weight of the water in large reservoirs has also been implicated in causing earthquakes, which may lead to the failure of a dam. Dam failure may also occur because of poor construction or poor design, or because the builders had inadequate knowledge of the geology of the site. Tens of thousands of lives have been lost because of such failures.

Neil E. Salisbury

Further Reading

Berga, L., et al., eds. *Dams and Reservoirs, Societies, and Environment in the Twenty-first Century.* New York: Taylor & Francis, 2006.

Billington, David P., and Donald C. Jackson. *Big Dams of the New Deal Era: A Confluence of Engineering and Politics.* Norman, Ok.: University of Oklahoma Press, 2006.

Cech, Thomas V. "Dams." In *Principles of Water Resources: History, Development, Management, and Policy.* 3d ed. New York: John Wiley & Sons, 2010.

Goldsmith, Edward, and Nicholas Hildyard. *The Social and Environmental Effects of Large Dams.* New York: Random House, 1984.

Leslie, Jacques. *Deep Water: The Epic Struggle over Dams, Displaced People, and the Environment.* New York: Farrar, Straus and Giroux, 2005.

McCully, Patrick. *Silenced Rivers: The Ecology and Politics of Large Dams.* Enlarged ed. London: Zed Books, 2001.

Stevens, Joseph E. *Hoover Dam: An American Adventure.* Norman, Ok.: University of Oklahoma Press, 1988.

■ Dead zones

FIELDS OF STUDY

Agriculture; Aquatic Ecosystems; Conservation Biology; Ecology; Environmental Ethics; Hydrology; Public Policy; Waste Management

SUMMARY

The occurrence of dead zones in waters along heavily inhabited lakeshores and coastlines has increased drastically since the end of the twentieth century. Human nutrient inputs into these ecosystems from fertilizer use and pollution have contributed to more frequent eutrophication events, fueling this source of economic and ecological devastation.

PRINCIAL TERMS

- **algae:** plants with no stems or leaves that grow in water or on damp surfaces
- **dead zones:** aquatic environments incapable of sustaining life
- **eutrophication:** a body of water overly enriched with minerals and nutrients which induce excessive growth of plants and algae
- **nitrogen:** a colorless, odorless, gaseous element constituting about four-fifths of the atmosphere's volume and is present in combined form in animal and vegetable tissues, especially in proteins

In aquatic ecosystems, just as in all others, there are certain nutrients whose limited presence or availability prevents the unrestrained growth of certain species. In nearshore marine waters and lakes these so-called limiting nutrients are nitrogen and phosphorus. Their restricted availability keeps

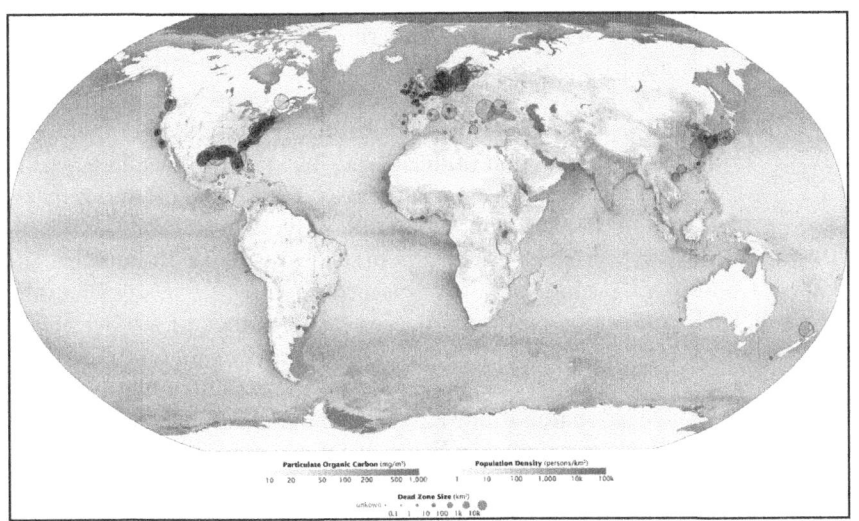

Data from Robert Diaz, Virginia Institute of Marine Science (dead zones); the GSFC Ocean Color team (particulate organic carbon); and the Socioeconomic Data and Applications Center (SEDAC) (population density). By Robert Simmon & Jesse Allen (NASA Earth Observatory).

microscopic organisms called algae from reaching unsustainable and destructive levels.

Normally these algae form the base of the food web, utilizing the sun's energy to convert carbon dioxide into oxygen and food. However, in the presence of excess nitrogen and phosphorus, the algae can reproduce uncontrollably in a process called eutrophication, resulting in an algal bloom. As the algae population peaks and begins its decline, microbes begin decomposing the dead algae, consuming oxygen in the process. Eventually the dissolved oxygen in the aquatic environment is depleted, resulting in the death of all organisms in the ecosystem. At this stage, the environment is deemed a hypoxic or dead zone.

In some cases, an influx of nitrogen and phosphorus into an aquatic ecosystem can be the result of natural processes. In certain areas on the western coasts of the continents, for example, a process called upwelling occurs, in which prevailing winds and currents bring nutrients up from the ocean depths and to the surface, where algae live. Humankind's interference with the natural cycling of nitrogen and phosphorus, however, has led to significant inputs of these nutrients into aquatic ecosystems as well. Nitrogen and phosphorus are mass-produced for use as fertilizers. After the fertilizers are sprayed on crop fields, rainwater and irrigation runoff carries these nutrients into rivers and streams, which in turn feed estuarine and marine habitats. Likewise, inadequately treated sewage and wastewater laden with nutrients from cities and towns is often pumped into aquatic habitats.

More than 140 dead zones have been documented around the world, and the number continues to rise. The largest known dead zone is in the Baltic Sea, where agricultural runoff and poorly treated sewage contribute to a dead zone that is tens of thousands of square kilometers in size. Similarly, in the United States a seasonal hypoxic zone the size of New Jersey forms during the summer months at the mouth of the Mississippi River in the Gulf of Mexico. Fishery yields there and in other locations affected by dead zones have sharply declined, sapping the lifeblood of coastal economies and cutting into marine food production.

While ominous, dead zones are not necessarily permanent. The infamous Black Sea dead zone was once the largest in the world, covering an area of approximately 40,000 square kilometers (15,000 square miles). When the Soviet Union collapsed in 1991, the industrialized agricultural economy of the region also collapsed, and the resulting drop in fertilizer use cut nitrogen and phosphorus input into the Black Sea by more than half. Within about a decade, the dead zone virtually disappeared. Although this example provides hope, the ecosystem of the Black Sea has yet to recover fully, as it supports limited marine life. To combat dead zones efficiently, political action through management of fertilizer use and pollution will be needed, not only to rein in current dead zones but also to prevent future outbreaks entirely.

Daniel J. Connell

Further Reading

Erisman, Jan Willem, et al. "Consequences of Human Modification of the Global Nitrogen Cycle." *Philosophical Transactions: Biological Sciences* 368.1621 (2013): 1–9.

Nassauer, Joan Iverson, Mary V. Santelmann, and Donald Scavia, eds. *From the Corn Belt to the Gulf: Societal and Environmental Implications of Alternative Agricultural Futures.* New York: Resources for the Future, 2007.

Sielen, Alan B. "The Devolution of the Seas." *Foreign Affairs* 92.6 (2013): 124–32.

Vernberg, F. John, and Winona B. Vernberg. *The Coastal Zone: Past, Present, and Future.* Columbia, S.C.: University of South Carolina Press, 2001.

■ Deep ecology

FIELDS OF STUDY
Ecofeminism; Ecology; Environmental Activism; Environmental Ethics; Environmental Philosophy; Ethics; Religious Studies

SUMMARY
The thinking of deep ecologists has attracted criticism from some quarters at the same time it has expanded approaches to environmental ethics.

PRINCIPAL TERMS
- **anthropocentric:** caused or produced by humans
- **deep ecology:** school of environmental philosophy based on environmental activism and ecological spirituality
- **ecocentrism:** a philosophy or perspective placing intrinsic value on all living organisms and their natural environments, regardless of their perceived usefulness or importance to human beings
- **ecofeminism:** a concept of gender that theorizes on the relationship between humans and the natural world
- **Gaia Hypothesis:** a model of the earth as a self-regulating organism, advanced as an alternative to a mechanistic model

The term "deep ecology" was first used by Norwegian philosopher Arne Naess in 1972 to suggest the need to go beyond the anthropocentric view that nature is merely a resource for human use. Since that time, the term has been used in three major ways. First, it has been used to refer to a commitment to deep questioning about environmental ethics and the causes of environmental problems; such questioning leads to critical reflection on the fundamental worldviews that underlie specific environmental ideas and practices. Second, the term has been used to refer to a platform of generally agreed-upon values that a variety of environmental activists share; these values include an affirmation of the intrinsic value of nature, the recognition of the importance of biodiversity, a call for a reduction of human impact on the natural world, greater concern with quality of life than with material affluence, and a commitment to changing economic policies and the dominant view of nature. Third, "deep ecology" has been used to refer to particular philosophies of nature that tend to emphasize the value of nature as a whole (ecocentrism), an

Deep ecology. By Xvegandettax (Own work)

identification of the self with the natural world, and an intuitive and sensuous communion with the earth.

Because of its emphasis on fundamental worldviews, deep ecology is often associated with non-Western spiritual traditions, such as Buddhism and Native American cultures, as well as with radical Western philosophers such as Baruch Spinoza and Martin Heidegger. It has also drawn on the nature writing of Henry David Thoreau, John Muir, Robinson Jeffers, and Gary Snyder. Deep ecology's holistic tendencies have led to associations with the Gaia hypothesis, and its emphasis on diversity and intimacy with nature has linked it to bioregionalism. Deep ecological views have also had a strong impact on environmental activism, including the Earth First! movement.

Deep ecologists have sometimes criticized the animal rights perspective for continuing the traditional Western emphasis on individuals while neglecting whole systems, as well as for a revised speciesism that still values certain parts of nature (animals) over others. Some deep ecologists have also been critical of mainstream environmental organizations such as the Sierra Club for not confronting the root causes of environmental degradation.

Ecofeminists have criticized deep ecology for failing to consider gender differences in the experience of the self and nature, for failing to examine the connection between the oppression of women and human treatment of nature, and for promoting a holism that, ecofeminists assert, disregards the reality and value of individuals and their relationships. Social ecologists have criticized deep ecology for failing to critique the relationship between environmental destruction on the one hand and social structure and political ideology on the other. In addition, a distrust of human interference with nature has led some deep ecology thinkers to present pristine wilderness, with no human presence, as the ideal. In rare and extreme cases, deep ecologists have implied a misanthropic attitude. In some instances, especially in early writings by deep ecologists, such criticisms have had considerable force. However, these problematic views are not essential to deep ecology, and some deep ecologists have developed a broadened view that overlaps with ecofeminism and social ecology and has expanded approaches to environmental ethics.

Further Reading

Bender, Frederic L. *The Culture of Extinction: Toward a Philosophy of Deep Ecology*. Amherst, N.Y.: Humanity Books, 2003.

Devall, Bill. *Simple in Means, Rich in Ends: Practicing Deep Ecology*. Salt Lake City, Utah: Peregrine Smith, 1988.

Diehm, Christian. "Darwin and Deep Ecology." *Ethics and the Environment* 19.1 (2014): 73–93.

Katz, Eric, Andrew Light, and David Rothenberg, eds. *Beneath the Surface: Critical Essays in the Philosophy of Deep Ecology*. Cambridge, Mass.: MIT Press, 2000.

Kober, Gal. "For They Do Not Agree in Nature: Spinoza and Deep Ecology." *Ethics and the Environment* 18.1 (2013): 43–63.

Snyder, Gary. *The Practice of the Wild*. Berkeley, Calif.: Counterpoint Press, 2010.

Deforestation

FIELDS OF STUDY

Agriculture; Animal Husbandry; Biology; Botany; Conservation Biology; Ecology; Ecosystems; Environment; Environmentalism; Farming; Forestry; Horticulture; Life Sciences; Preservation; Silviculture; Wilderness Conservation

SUMMARY

Deforestation, particularly in tropical regions, has given rise to concerns among environmentalists and scientists, in large part because of the role that tropical forests play in moderating global climate.

PRINCIPAL TERMS

- **biodiversity:** biological diversity in an environment as indicated by numbers of different species of plants and animals
- **clear-cutting:** a section of forest where all trees have been cut down for harvesting
- **deforestation:** loss of forestlands through encroachment by agriculture, industrial development, or non-sustainable commercial forestry
- **hectare:** a unit of surface, or land, measure equal to 100 acres, or 10,000 square meters: equivalent to 2.471 acres

- **slash-and-burn agriculture**: a method of agriculture used in the tropics, where forest vegetation is felled and burned, the land cropped for a few years, then the forest allowed to reinvade
- **United Nations Food and Agriculture Organization (FAO):** a specialized agency of the United Nations that leads international efforts to defeat hunger

Toward the end of the twentieth century, environmentalists became active in decrying the apparent accelerating pace of deforestation because of the loss of wildlife and plant habitats caused by the practice as well as its negative effects on biodiversity. By the 1990's research by mainstream scientists had confirmed that deforestation was indeed occurring on a global scale and that it posed a serious threat to global ecology. Although steps were taken to slow rates of deforestation in the early years of the twenty-first century, the loss of forestlands continued.

Deforestation as a result of expansion of agricultural lands or nonsustainable timber harvesting has occurred in many regions of the world at different periods in history. The Bible, for example, refers to the cedars of Lebanon. Lebanon, like many of the countries bordering the Mediterranean Sea, was thickly forested several thousand years ago. A growing population, overharvesting, and the introduction of grazing animals such as sheep and goats decimated the forests, which never recovered.

Similarly, the forests of Europe and North America have shifted in total area as human populations have changed over the centuries. When the European colonists arrived in the New World, they immediately began clearing the forests. Trees were harvested for building materials and export back to Europe or were simply felled and burned to clear space for farming. In North America, however, as agriculture became increasingly mechanized and farming shifted to the prairies, abandoned farms reverted to woodland. Environmental historians believe, in fact, that a greater percentage of land area in North America is now forested than was covered with trees prior to the arrival of European colonists. A similar phenomenon has taken place in many northern European countries as their populations have become increasingly urbanized.

As the European industrialized nations have gained forestland, however, the less developed countries in Latin America, Asia, and Africa have lost woodlands. While some of this deforestation has been caused by a demand for tropical hardwoods for lumber or pulp, the leading cause of deforestation in the twentieth and twenty-first centuries, as it was several hundred years ago, has been the expansion of agriculture. Growing demand by the industrialized world for agricultural products such as beef has led to millions of hectares of forestland being bulldozed or burned to create pastures for cattle. Researchers in Central America have watched with dismay as large beef-raising operations have expanded into fragile ecosystems in countries such as Costa Rica, Guatemala, and Mexico.

A tragic irony in this expansion of agriculture into tropical rain forests is that the soil underlying the trees is often unsuited for pastureland or raising other crops. Exposed to sunlight, the soil is quickly depleted of nutrients and often hardens. The once-verdant land becomes an arid desert prone to erosion that may never return to forest. As the soil becomes less fertile, thorny weeds begin to choke out the desirable forage plants, and the cattle ranchers move on to clear fresh tracts.

Slash-and-Burn Agriculture and Logging

Apologists for the beef industry often argue that their ranching practices are simply a form of slash-and-burn agriculture and do no permanent harm. It is true that many of the indigenous peoples in tropical regions have practiced slash-and-burn agriculture for millennia with only minimal impact on the environment. These farmers burn the understory, or low-growing shrubs and trees, to clear small plots of land. Any large trees that survive their fires are cut down with axes and then burned.

Anthropological studies have shown that the small plots these peasant farmers clear can usually be measured in square meters or feet, not in hectares or acres like cattle ranches, and are used for five to ten years. As fertility of the soil declines in one plot, the farmer clears a small plot next to the depleted one. The farmer's family or village gradually rotates through the forest, clearing small plots and using them for a few years, and then shifting to new ground, until they eventually come back to where they began one hundred or more years before. As long as the size of the plots cleared by peasant farmers remains small in proportion to the forest overall, slash-and-burn agriculture does not contribute significantly to deforestation. If the population of farmers grows, however, and more land must be cleared with each succeeding generation, as has been happening in many tropical countries, then even traditional slash-and-burn agriculture can be as ecologically devastating as the more mechanized cattle ranching operations.

Although logging is not the leading cause of deforestation, it remains a significant factor. Tropical forests are rarely clear-cut, as they typically contain hundreds of different species of trees, most of which may have no commercial value. Loggers may select only a few trees for harvesting from each stand. Selective harvesting is a standard practice in sustainable forestry, but just as loggers engaged in the disreputable practice of high-grading across North America in the nineteenth century, so loggers have high-graded in the late twentieth and early twenty-first centuries in Malaysia, Indonesia, and other tropical forests. High-grading is a practice in which loggers cut over a tract to remove the most valuable timber while ignoring the damage being done to the residual stand. The assumption is that, having logged over the tract once, the timber company will not be coming back. This practice stopped in North America, not because the timber companies voluntarily recognized their ecological damage, but because they ran out of easily accessible, old-growth timber to cut. Fear of a timber famine caused logging companies to create forest plantations and to undertake the practice of sustainable forestry. While global satellite photos indicate significant deforestation has occurred in tropical areas, enough easily harvested old-growth forest remains in some areas that there is no economic incentive for timber companies to switch to sustainable forestry.

Logging may also contribute to deforestation by making it easier for agriculture to encroach on forestlands. Logging companies build roads for their own use while harvesting trees, and farmers and ranchers later use these roads to move into the logged tracts, where they clear whatever trees the loggers have left.

Environmental Impacts
The extent of the problem of deforestation has long been a subject of debate. The United Nations Food and Agriculture Organization (FAO), which monitors deforestation worldwide, bases its statistics on measurements taken from satellite images. These data indicate that between 2000 and 2005 the net loss of forest area globally was 7.3 million hectares (18 million acres) per year, a reduction from 8.9 million hectares (22 million acres) per year in the decade between 1990 and 2000. In the period 2000-2005, South America and Africa had the greatest losses: South America lost 4.3 million hectares (10.6 million acres) of forest per year, and Africa lost 4 million hectares (9.9 million acres) per year. Environmental activists have been particularly concerned about forest losses in Indonesia and Malaysia, two countries where timber companies have been accused of abusing or exploiting native peoples in addition to engaging in environmentally damaging harvesting methods.

Researchers outside the United Nations have challenged FAO's data, with some scientists claiming the numbers are much too high and others providing convincing evidence that, if anything, FAO's numbers are too low. Few researchers, however, have tried to claim that deforestation on a global scale is not happening. In the 1990's the reforestation of the Northern Hemisphere, while providing an encouraging example that it is possible to reverse deforestation, was not enough to offset the depletion of forestland in tropical areas.

Deforestation affects the environment in a multitude of ways. The most obvious is in a loss of biodiversity. When an ecosystem is radically altered through deforestation, the trees are not the only thing to disappear. Wildlife decreases in number and in variety, and other plants also die. As forest habitat shrinks through deforestation, various plants and animals become vulnerable to extinction. Many biologists believe that numerous animals and plants native to tropical forests will become extinct as the result of deforestation before humans ever have a chance to become aware of their existence.

Other effects of deforestation may be less obvious. Deforestation can lead to increased flooding during rainy seasons. Rainwater that once would have been slowed or absorbed by trees instead runs off denuded hillsides, pushing rivers over their banks and causing devastating floods downstream. The role of forests in regulating water has long been recognized by both engineers and foresters. Flood control was, in fact, one of the motivations behind the creation of the federal forest reserves in the United States during the nineteenth century. More recently, disastrous floods in Bangladesh have been blamed on the logging of tropical hardwoods in the mountains of Nepal and India.

Conversely, trees can also help to mitigate drought. Like all plants, trees release water into the atmosphere through the process of transpiration. As the world's forests shrink, fewer greenhouse gases such as carbon dioxide will be removed from the atmosphere, less oxygen and water will be released into it, and the world will become a hotter, dryer place. Scientists and policy analysts alike are in agreement that deforestation is a major threat to the environment. The question is whether effective policies can be developed to reverse it.

Männikkö

Further Reading

Chew, Sing C. *World Ecological Degradation: Accumulation, Urbanization, and Deforestation, 3000 B.C.-A.D. 2000.* Walnut Creek, Calif.: AltaMira Press, 2001.

Dean, Warren. *With Broadax and Firebrand: The Destruction of the Brazilian Atlantic Forest.* Berkeley: University of California Press, 1997.

Geist, Helmut J., and Eric F. Lambin. "Proximate Causes and Underlying Driving Forces of Tropical Deforestation." *BioScience* 52, no. 2 (2002): 143-150.

Humphreys, David. *Logjam: Deforestation and the Crisis of Global Governance.* Sterling, Va.: Earthscan, 2006.

Palmer, Charles, and Stefanie Engel, eds. *Avoided Deforestation: Prospects for Mitigating Climate Change.* New York: Routledge, 2009.

Richards, John F., and Richard P. Tucker, eds. *World Deforestation in the Twentieth Century.* Durham, N.C.: Duke University Press, 1990.

Rudel, Thomas K., and Bruce Horowitz. *Tropical Deforestation: Small Farmers and Land Clearing in the Ecuadorian Amazon.* New York: Columbia University Press, 1994.

Sanchez, Ilya B., and Carl L. Alonso, eds. *Deforestation Research Progress.* New York: Nova Science, 2008.

Sponsel, Leslie E., Robert Converse Bailey, and Thomas N. Headland, eds. *Tropical Deforestation: The Human Dimension.* New York: Columbia University Press, 1996.

■ Desalination

FIELDS OF STUDY
Biology; Chemistry; Ecology; Environment; Environmentalism; Hydrology; Oceanography

SUMMARY
More than eighty countries around the world have problems obtaining sufficient potable water to serve their populations. Desalination offers a way to provide water to people living where sources of fresh water are scarce, but the process is associated with various negative environmental impacts.

PRINCIPAL TERMS
- **desalination:** process of removing minerals from salty water to make the water fit for humans to drink or for use in irrigation
- **distillation:** the process of purifying a liquid by successive evaporation and condensation'
- **ion exchange:** the process of reciprocal transfer of ions between a solution and a resin or other suitable solid
- **membrane filtration:** simple migration resulting from a concentration difference on the two sides of the membrane and in ultrafiltration is accelerated by pressure difference
- **osmosis:** the tendency of a fluid, usually water, to pass through a semipermeable membrane into a solution where the solvent concentration is higher, thus equalizing the concentration of materials on either side of the membrane
- **potable water:** water fit for drinking, free from contamination, without enough saline to be regarded as mineral water

Water is abundant in the world, but only about 3 percent of all water is potable—that is, fit for humans to drink. It is possible to remove salts from seawater or brackish water to make potable water, but the desalination process uses large amounts of energy. The costs of desalination depend on the type of feed water (seawater or brackish water) and its temperature, the method being used (membrane filtration, distillation, or ion exchange), the type of energy used (nuclear, petroleum, or solar), and the amount of water to be processed. Producing potable water through desalination is expensive compared with taking potable water out of the ground or from streams, ranging from about 50 cents to 70 cents per cubic meter of potable water produced (1 cubic meter is equal to about 35 cubic feet, or 264 gallons).

The nations that have established large desalination plants are bordered by oceans, have little potable water on their land, and have large amounts of cheap energy, such as petroleum, available for use. Middle Eastern countries produce the greatest amounts of potable water produced in the world through the desalination of seawater. Up to 49 billion liters (13 billion gallons) of potable water are produced each day by more than fifteen thousand desalination plants located in such places as North Africa, Saudi Arabia, the United Arab Emirates, Japan, Australia, and the United States.

The Jebel Ali desalination in Dubai, the United Arab Emirates, is expected eventually to provide up to 250 million cubic meters (8.8 billion cubic feet, or 66 billion gallons) of water per year. The plant uses the common multiflash distillation process, in which seawater is boiled at low pressure so that relatively low amounts of energy are needed. A new plant in Israel, called Sorek, is the largest and cheapest reverse-osmosis plant in the world and produces about 675,000 cubic meters of water daily. Desalinated seawater is now a mainstay of the Israeli water supply.

Among the large desalination plants in the United States are one in Tampa Bay, Florida, and one in El Paso, Texas. The El Paso plant uses the popular method of reverse osmosis to process undrinkable brackish waters, generating about 25 percent of the water used by the city. In the reverse osmosis process, membranes gradually purify the water as less salty water moves out of the membranes, leaving much saltier water behind.

Several negative environmental impacts are associated with the operation of desalination plants. For one thing, the plants use high amounts of energy, usually electricity generated by the burning of fossil fuels, a process that produces carbon dioxide, one of the gases associated with global warming. Desalination plants can also have more direct effects on the environment. The intake of seawater into a plant can kill organisms such as fish larvae and plankton, and the disposal into the ocean of warm residual waters very high in dissolved solids after processing may harm some animals. Plants can be designed to avoid the latter problem, however. The warm, concentrated brine waters can be mixed with cooler and less concentrated waters so that the water returned to the ocean is more similar in temperature and concentration to seawater, or the concentrated brine water can be dispersed over a large area in the ocean so that it changes the seawater composition and temperature very little.

Robert L. Cullers

Further Reading

Chiras, Daniel D. "Water Resources: Preserving Our Liquid Assets and Protecting Aquatic Ecosystems." In *Environmental Science*. 8th ed. Sudbury, Mass.: Jones and Bartlett, 2010.

Eltawil, Mohamed A., Zhao Zhengming, and Liqiang Yuan. "A Review of Renewable Energy Technologies Integrated with Desalination Systems." *Renewable and Sustainable Energy Review* 13 (2009): 2245-2262.

Escobar, Isabel, and Andrea Schäfer, eds. *Sustainable Water for the Future: Water Recycling Versus Desalination*. Oxford, England: Elsevier, 2010.

Karagiannis, Ioannis C., and Petros G. Soldatos. "Water Desalination Cost Literature: Review and Assessment." *Desalination* 223 (2008): 448-456.

National Research Council. *Desalination: A National Perspective*. Washington, D.C.: National Academics Press, 2008.

Websites

MIT Technology Review
https://www.technologyreview.com/s/534996/megascale-desalination/

Deserts and dunes

FIELDS OF STUDY

Ecology; Environment; Field Geography; Geology; Hydrogeology; Hydrology; Meteorology; Soil Science; Topology

SUMMARY

Deserts are arid or semiarid regions, defined as areas that receive an extremely small quantity of precipitation each year. Deserts and semideserts cover approximately one-third of the earth's total surface area and are found on every continent. Deserts and semideserts are characterized by unique geological and ecological features that are not present in more humid environments. These features include dunes, which are landforms of loose, shifting mounds or ridges made of sand or other sedimentation.

PRINCIPAL TERMS
- **aridisol:** a type of soil that contains little or no organic matter; typically, alkaline and composed mostly of salts such as chlorides, sulfates, and carbonates
- **deflation:** the removal of particles of clay, dust, sand, and rock from dry soil by strong winds
- **eolian:** of a deposit, arising from or carried by the action of the wind
- **fluvial:** of a deposit, arising from, carried by, or occurring in a river
- **playa:** a flat, dry lake-bed usually characterized by clay or salt deposits, where rainwater accumulates temporarily but eventually evaporates or moves underground; also called a pan or a dry lake
- **rain shadow:** a dry region that forms on the side of a mountainous area protected from the prevailing winds and, therefore, the moisture they carry
- **saltation:** a bouncing, jumping motion in which small, hard particles such as sand move over an uneven surface as a result of the flow of air or water
- **slip face:** the sloping side of a dune that faces away from the prevailing wind; also called lee slope
- **wash:** a narrow channel or streambed that is dry through most of the year but that can become partly or totally submerged during flash floods that occur during sudden heavy rains; also called wadi

Although the word "desert" immediately conjures an unmistakable image of a hot, sandy expanse, starkly empty of life, a great deal of variation makes up desert environments. While many are hot, some are freezing cold. Some border the ocean and others have annual rainy seasons. Three-fourths of desert surfaces are covered not by sand but rather by some other material, such as clay, rock, ice, or aridisol, a type of soil that contains little or no organic matter. (Aridisols are typically alkaline and composed mostly of salts such as chlorides, sulfates, and carbonates.)

Multiple definitions of deserts and systems for classifying them exist. One of the most widely used desert classification systems was developed by American scholar and field geographer Peveril Meigs, in the 1950's. Meigs sorted the earth's dry regions based on the amount of precipitation each receives in a year. Regions with twelve or more months in a row with no rainfall are known as extremely arid; regions that receive less than 250 millimeters (10 inches), of rainfall per year are known as arid; regions that receive between 250 and 500 millimeters (10–20 inches) of rainfall per year are known as semiarid. Only extremely arid and arid regions qualify as true deserts; semiarid regions, which are often unforested grasslands that border deserts, are sometimes known as steppes. Another way of defining a desert draws on a system developed by Russian-born German climatologist Wladimir Köppen, who proposed that a desert be identified as a region that loses more water each year through evaporation than it gains through precipitation.

Many deserts, despite being arid most of the year, do have some surface water. Ponds, lakes, and streams may form temporarily during brief heavy rainstorms, causing flash floods. Some deserts, however, are also crossed by permanent bodies of water, such as Egypt's Nile River or the Rio Grande in the Chihuahuan Desert; both rivers have their sources in high mountainous areas and traverse the desert on their way to the coast. In other deserts, subsurface water exists in the form of underground aquifers, or reservoirs of water, that sometimes rise to the surface and create oases, or isolated regions of fertile land.

Major Deserts of the World
Most the world's desert regions are located within two bands, or belts, each of which is within 25 degrees of the equator. In these belts, dry trade winds cause cloud cover to disperse, letting sunlight quickly heat the earth's surface. The Sahara Desert in North Africa is an example of a trade wind desert. The Sahara, which stretches from the Atlantic Ocean to the Red Sea, is the largest hot desert in the world. The central Sahara region, which occupies an area of about 9.1 million square kilometers (sq km), or about 3.5 million square miles, is the driest; its regions to the north and south receive more

precipitation and are more vegetated. The central Sahara is recorded as receiving less than 25 millimeters (1 inch) of rainfall per year, although precipitation patterns are extremely erratic there. There may be some years when this region receives no rainfall at all, but it is also possible that brief, intense thunderstorms may occur.

Much of the central Sahara is covered in ergs, or vast, flat, sandy areas. However, the region also is marked by a variety of surface landforms, including large dunes (such as Chech dune and Raoui dune); flat, high, stone plateaus; wide plains covered in gravel; dry riverbeds or wadis; low-lying salt flats; and the occasional oasis. A few mountain ranges, such as the Ahaggar (also known as the Hoggar) and the Tibesti mountains, also occur in this area. The Sahara is an extremely windy place, where temperatures of higher than 50 degrees Celsius (122 degrees Fahrenheit) have been recorded; winter and nighttime temperatures, though, may dip to below freezing.

The equatorial belt region also is home to a type of desert known as a rain shadow desert. These are dry areas that form between two mountain ranges that each block moist air currents from being blown toward the valley they enclose. The Gobi Desert is a rain shadow desert kept arid by the vast mountain ranges of the Himalayas; it spans northern China and southern Mongolia (*gobi* is the Mongolian word for "desert") and is bounded by, among other mountains, the Altai and Hangayn (or Khangai) to the north and the Yin and Qilian to the south.

Much of the Gobi, which occupies an area of about 1.3 million square kilometers (about 500,000 square miles), is not covered in sand, but dry sedimentary rock such as chalk and gypsum. These rock plains were formed 65 million or more years ago, and the Gobi has been the source of dinosaur and other fossils and of the remains of Stone Age societies. The climate in this region is marked by bitterly cold winters and warm summers. The total yearly rainfall in the Gobi ranges from about 200 millimeters (about 8 inches) in the northeast, which receives summer monsoons, to less than 50 millimeters (about 2 inches) in the extremely arid west.

Not all of the world's deserts, however, have formed near the equator. In cold environments like at the earth's poles, the air can no longer hold much moisture and there is little or no precipitation; any water that exists on the surface is not found in rivers or lakes but is locked in the form of ice. The inland plateau of the Antarctic receives little snow, and it is considered the world's largest desert at more than 14.2 million square kilometers (about 5.5 million square miles). Although the rest of the continent is marked by such varied surface features as mountains, glaciers, and coral reefs, the Antarctic desert is relatively featureless. The snow is packed into flat, firm plains marked only by small dune-like formations known as sastrugi (or zastrugi).

Dune Formation Processes
For a dune to form, a handful of key environmental factors must first be met. First, an abundant supply of dry sediment, usually sand, must exist. Second, there must be a way to transport that material, such as winds or moving water. These winds or water must be moving quickly enough to move the sediment, but not so powerfully as to destroy the mounds that are formed. Finally, there must be a place for the accumulated sediment to be deposited. The sand that makes up a dune may come from the eolian processes that erode mineral particles from the surface of rocks or may be composed of biological materials like the calcium-carbonate-rich crumbled shells or skeletons of marine organisms.

In a dune formed by eolian, or wind-driven, processes, sand particles may be transported in several different ways. Extremely tiny or light particles may be held in place vertically while they move long distances horizontally; their weight is fully supported by upward currents of air, while turbulent eddies of wind push them forward. This process is known as suspension. In saltation, particles move downwind in a continuous series of hops, skips, and jumps. Each time a particle jumps, it moves a short distance only, but as it moves it hits other particles and triggers them to hop forward as well. In the process of creep, saltating particles that hit large, heavy grains may push those grains slowly downwind.

All these processes cause sand particles to build up, though they do not necessarily create dunes. They may form sand sheets, or flat sandy plots of land, or ripples, which are low crests and troughs on the surface of a sandy surface (with the biggest particles at the crests). Ripples can grow into dunes under the continued action of the wind, which keeps moving particles up to the crest of the pile by saltation and creep, until the slip face becomes steep enough to collapse in a tiny avalanche. This collapsing sand becomes stabilized at an angle, known as the "angle of repose." As this cycle of buildup and collapse repeats, the dune inches forward in the direction of the wind.

The process by which underwater or riverbed dunes are formed through fluvial, or water-driven, processes, is similar. However, fluvial processes can carry larger particles in suspension than can be carried by eolian processes. In addition, saltation is less common underwater than it is in air, and suspension is more common. Finally, some particles are carried by rivers as dissolved sediments, which later recrystallize.

Types of Dunes
A minimum of five primary types of dunes—crescentic, linear, dome, parabolic, and star—have been identified by geologists, although many additional subtypes and variations exist. The basic category a dune falls into depends on its geometric shape, which is determined by factors such as wind direction and strength, sediment type and quantity, and the presence or absence of vegetation.

Crescentic dunes, as their name implies, roughly resemble a crescent moon in shape. Narrow curved tips form at the edges of the dune, pointing downwind; the dune's concave slip face (a slip face is a slope on the opposite side of the prevailing wind) is between these tips. Crescentic dunes are also known as barchans and are the most common dune shape on Earth. Typically, they form in areas where the land is flat and, the wind blows consistently from a single direction. These fast-moving dunes have been known to travel as quickly as 100 meters (about 328 feet) per year across a desert.

Sand ridges that are either long and straight or sinuous—curving back and forth—are called linear, or longitudinal, dunes. Particularly, sinuous dunes are sometimes known as seifs, from the Arabic word for "sword." (Ancient Eastern swords had curving blades.) Linear dunes often form when there are winds that blow from two directions; each wind forms ridges that are parallel to the direction in which it blows. Rare dome dunes are roughly oval or circular in shape and do not have a clear slip face; parabolic dunes are shaped like the letter U and often form in deserts that are near a coast. The long arms of such dunes point upwind and are stabilized by the presence of vegetation growing in the sand, while the convex curve of a parabolic dune is pushed forward by the wind. Finally, star dunes are composed of a tall central mound from which several long arms—a minimum of three—radiate outward. Each arm has its own slip face and is formed by a wind that blows in a different direction. Star dunes are common in certain parts of the Sahara Desert.

Any of these types of dunes may be further described as either simple, compound, or complex. Simple dunes have the minimum number of slip faces required to fall into a category; they usually form when the prevailing wind or winds remain relatively stable in strength and direction. Compound dunes, which tend to be large, comprise two or more smaller dunes with the same geometric shape, forming one atop the next. Complex dunes are like compound dunes, but they are formed of multiple smaller dunes with different geometric shapes. For complex dunes to form, the strength and direction of the wind or winds must change after the formation of the first dune.

Desertification
Desertification is defined as the degradation of formerly fertile land—that is, the decrease in the ability of what was once productive land to support the continued growth of vegetation. Desertification can be attributed to a variety of complex environmental factors, including the loss of nutrients and moisture in the soil, increased salinization of soil, and an increase in wind and water erosion. In the course of Earth's history, desert regions have formed, enlarged, and shrunk in size because of natural processes that had nothing to do with human activities. Desertification, however, is accelerating because of anthropogenic causes.

An estimated 70 percent of the world's dryland regions are experiencing either moderate or severe desertification. This phenomenon is caused by droughts and changes in temperature, cloud cover, and wind patterns from global warming, overgrazing, deforestation, irrigation, and other effects of unsustainable human land use. For example, livestock grazing on already semiarid land can remove vegetation that would otherwise prevent the erosion of soil by wind or water, and repeated cultivation of crops on the same land may deplete the supply of nutrients contained in the soil or make it harder for it to absorb water. Proximity to an existing desert is not necessary for desertification to occur.

M. Lee

Further Reading

Abraham, Athol D., and Anthony J. Parsons. *Geomorphology of Desert Environments*. New York: Springer, 2009.

Aleshire, Peter. *Deserts*. New York: Chelsea House, 2008.

Doran, Peter T., W. Berry Lyons, and Diane M. McKnight, eds. *Life in Antarctic Deserts and Other Cold, Dry Environments*. New York: Oxford University Press, 2010.

Laity, Julie. *Deserts and Desert Environments*. Hoboken, N.J.: Wiley-Blackwell, 2008.

Mather, Anne. "Arid Environments." In *Environmental Sedimentology*, edited by Chris Perry and Kevin Taylor. Malden, Mass.: Blackwell, 2007.

Maun, Anwar M. *The Biology of Coastal Sand Dunes*. New York: Oxford University Press, 2009.

Pye, Kenneth, and Haim Tsoar. *Aeolian Sand and Sand Dunes*. New York: Springer, 2008.

Verstraete, Michel M., Robert J. Scholes, and Mark Stafford Smith. "Climate Change and Desertification: Looking at an Old Problem Through New Lenses." *Frontiers in Ecology and the Environment* 7, no. 8 (October 2009): 421–428.

Dichloro-diphenyl-trichloroethane (DDT)

FIELDS OF STUDY
Agriculture; Biology; Chemistry; Ecology; Environmental Microbiology; Environmental Science; Public Health

SUMMARY
Dichloro-diphenyl-trichloroethane, better known as DDT, has been used extensively in agriculture and for control of insect-borne diseases worldwide. However, its persistence in the environment and ability to accumulate in the food chain have resulted in devastating consequences to wildlife. The harmful effects of DDT became a major focus for the emerging environmental movement during the 1960s.

PRINCIPAL TERMS

- **bioaccumulation:** process by which certain toxic substances, such as heavy metals, accumulate and continue accumulating in living organisms, posing threats to health life and the environment
- **biomagnification:** the process by which a compound, such as pollutants or pesticides, increases its concentration in the tissues of organisms as it travels up the food chain
- Dichloro-diphenyl-trichloroethane (DDT): **synthetic organochlorine insecticide**
- **Environmental Defense Fund (EDF):** a United States-based nonprofit, non-partisan environmental advocacy group, founded in 1967, known for its work on issues including global warming, ecosystem restoration, oceans, and human health, and advocates using sound science, economics and law toward successful environmental solutions
- **hydrocarbon:** an organic compound, such as acetylene or butane, containing only carbon and hydrogen, often occurring in petroleum, natural gas, coal, and bitumens
- **pyrethrum:** an insecticide made from the dried flower heads of several Old-World chrysanthemums

Cloud spraying for mosquitos at Norman Park, 1947 City Council workers are cloud spraying DDT (Dichloro-Diphenyl-Trichloroethane) to a mosquito breeding swamp at Norman Park.

Drought

FIELDS OF STUDY

Atmospheric Science; Climate Classification; Climate Modeling; Climate Zones; Climatology; Earth System Modeling; Ecology; Environmental Sciences; Environmental Studies; Hydroclimatology; Hydrology; Hydrometeorology; Meteorology; Physical Geography

SUMMARY

Drought is an unusually long period of below-normal precipitation. It is a relative rather than an absolute condition, but the result is a shortage of water for plant growth, affecting the people who live in that region and beyond. Drought is particularly disastrous for farmers and the practice of agriculture.

PRINCIPAL TERMS

- **adiabatic:** a change of temperature within the atmosphere that is caused by compression or expansion without transfer of heat into or out of the system
- **desertification:** the relatively slow, natural conversion of fertile land into arid land or desert
- **evapotranspiration:** the combined water loss to the atmosphere from both evaporation and plant transpiration
- **Palmer Drought Index:** a widely adopted quantitative measure of drought severity developed by W. C. Palmer in 1965
- **precipitation:** any form of liquid water or ice at falls from the atmosphere to the ground
- **Sahel:** the semiarid southern fringe of the Sahara in West Africa that extends from Mauritania on the Atlantic coast to Chad in the interior
- **soil moisture:** water that is held in the soil and that is therefore available to plant roots
- **subsidence:** in meteorology, the slow descent of air that becomes increasingly dry in the process, usually due to an area of high pressure

Droughts have had enormous impacts on human societies since ancient times. The most obvious effect is crop and livestock failure, which have caused famine and death through thousands of years of human history. Drought has resulted in the demise of some ancient civilizations and, in some instances, the forced mass migration of large numbers of people. Water is so critical to all forms of life that a pronounced shortage could, and has, decimated whole populations.

The effects of drought are profound. The dry conditions in the Great Plains of North America in the early 1930s in conjunction with extensive and improper farming activities resulted in the creation of the Dust Bowl, which at one point covered more than 200,000 square kilometers, or an area about the size of Nebraska. During the early 1960s, a severe drought affected the Mid-Atlantic states. Parts of New Jersey experienced sixty consecutive months of below-normal precipitation, so depleting local water supplies that plans were actively considered to bring rail cars of water into Newark and other cities in the northern part of the state, as the reservoirs that usually supplied the region were practically dry. The Sahel region south of the Sahara in West Africa had a severe drought beginning in the late 1960s and continuing into the early 1970s, creating an enormous negative impact on the local population, livestock, and vegetation. Hundreds of thousands of people starved, thousands of animals died, and many tribes were forced to migrate south to areas of more reliable precipitation. Large areas of eastern Africa in South Sudan, the Republic of Sudan and Ethiopia continue to experience an on-going state of drought. Drought and climate change in Syria are thought to have contributed to the brutal war ongoing since 2011.

Drought Characteristics

Almost all droughts occur when slow-moving air masses that are characterized by subsiding air movements dominate an area. Often, the air comes from continental interiors where the amount of moisture that is available for evaporation into the atmosphere is very limited. When these conditions occur, the potential for precipitation is low for several reasons. First, the humidity in the air is already low, as the continental air mass is distant from maritime (moist) influences. Second, air that subsides undergoes adiabatic heating at the rate of 10°C per 1,000 meters.

The term "adiabatic" refers to a change of temperature within a gas (such as the atmosphere) that occurs as a result of compression (descending air) or expansion (rising air), without any input or extraction of heat from external sources. For example, assume that air at a temperature of 0°C is passing over the Sierra Nevada in eastern California at an elevation of 3,500 meters. As the air descends and reaches Reno, Nevada, at an elevation of 1,200 meters, the higher atmospheric pressure found at lower elevations results in compression and heating at the dry adiabatic rate of 10°C per 1,000 meters, yielding a temperature in the Reno area of 23°C. Thus, adiabatic heating from subsiding air masses results in a decline in relative humidity and an increase in moisture-holding capacity. In addition, the movement of air under these conditions is usually unfavorable for vertical uplift and the beginning of the condensation process. The final factor that reduces precipitation potential is the decrease in cloudiness and corresponding increase in sunshine, which in turn leads to an increase in evapotranspiration demands that favor soil moisture loss.

Another characteristic associated with droughts is that once established, they appear to persist and even expand into nearby regions, resulting in desertification in extreme cases. This tendency is apparently related to positive feedback mechanisms. For example, the drying out of the soil influences air circulation and the amount of moisture that is then available for precipitation farther downwind. At the same time, the atmospheric interactions that lead to unusual wind systems associated with droughts can induce surface-temperature variations that, in turn, lead to further development of the unusual circulation pattern. Thus, the process builds on itself, causing the drought to both last longer and intensify. The situation persists until a major change occurs in the circulation pattern in the atmosphere.

Many climatologists concur with the concept that precipitation is not the only factor associated with drought. Other factors that demand consideration include moisture supply, the amount of water in storage, and the demand generated by evapotranspiration. Although the scientific literature is replete with information about the intensity, length, and environmental impacts of drought events, the role of individual climatological factors that can increase or decrease the severity of a drought is not fully understood.

Drought Identification

Research in drought identification has been changing over the years. Drought was once considered solely in terms of precipitation deficit. Although that lack of precipitation is still a key atmospheric component of drought, sophisticated techniques are now used to assess the deviation from normal levels of the total environmental moisture status. These techniques have enabled investigators to better understand the severity and length of drought events, as well as the extent of the affected area.

Drought has been defined in a number of ways. Some authorities consider it to be merely a period of below-normal precipitation, while others relate it to the likelihood of forest fires. Drought is also said to occur when the yield from a specific agricultural crop or pasture is significantly less than expected. It has also been defined as a period when soil moisture or groundwater decreases to a critical level.

Drought was identified early in the twentieth century by the U.S. Weather Bureau as any period of twenty-one or more days when precipitation was 30 percent or more below normal. Subsequent examination of drought events that were identified by this method revealed that soil moisture reserves were often elevated during these events to the extent that there was sufficient water to support vegetation. It was also determined that the amount of precipitation preceding the drought event was ample or even heavy. Thus, it became apparent that precipitation should not be used as the sole measure to identify drought. Subsequent research has shown that the moisture status of an area is affected by additional factors.

Further developments in drought identification during the middle decades of the twentieth century began to focus on the moisture demands that are associated with evapotranspiration in an area. Evaporation is primarily the process by which liquid water becomes water vapor at the surface and enters the atmosphere. To a lesser extent, this also includes the conversion of "solid water," as ice and snow, either directly by sublimation or through first melting into an intermediate liquid state. Transpiration refers to the loss of moisture by plants to the atmosphere. Although evaporation and transpiration can be studied and measured separately, it is convenient to consider them in applied climatological studies as the single process of evapotranspiration.

There are two ways to define evapotranspiration. The first is actual evapotranspiration, which is the actual or real rate of water-vapor return to the atmosphere from the earth and vegetation; this process could also be called "water use." The second is potential evapotranspiration, which is the theoretical rate of water loss to the atmosphere if one assumes continuous plant cover and an unlimited supply of water. This process could also be called "water need," as it indicates the amount of soil water needed if plant growth is to be maximized. Procedures have been developed that enable one to calculate the potential evapotranspiration for any area from monthly mean temperature and precipitation values.

Some drought-identification studies have focused on agricultural drought, looking at the adequacy of soil moisture in the root zone for plant growth. This procedure involved the evaluation of precipitation, evapotranspiration, available soil moisture, and the water needs of plants. The goal of this research was to determine drought probability based on the number of days when soil moisture storage is reduced to zero.

Evapotranspiration was also used by the Forest Service of the U.S. Department of Agriculture when it developed a drought index to be used by fire-control managers. The purpose of the index was to provide a measure of flammability that could create forest fires. This index has limited applicability to non-forestry users, as it is not effective for showing drought as an indication of total environmental stress.

Palmer Drought Index
One of the most widely adopted drought-identification techniques was developed by W. C. Palmer in 1965. The method, which became known as the Palmer Drought Index, defines drought as the period of time, usually measured in months or years, when the actual moisture supply at a given location is consistently less than the climatically anticipated or appropriate supply of moisture. The calculation of this index requires the determination of evapotranspiration, soil moisture loss, soil moisture recharge, surface runoff, and precipitation. The Palmer Drought Index values range from approximately +4.0 for an extremely wet moisture status class to -4.0 for extreme drought. Normal conditions have a value close to 0. Positive values indicate varying stages of abundant moisture, whereas negative values indicate varying stages of drought.

Although the Palmer Drought Index is recognized as an acceptable procedure for incorporating the role of potential evapotranspiration and soil moisture in magnifying or alleviating drought status, there have been some criticisms of its use. For example, the method produces a dimensionless parameter of drought status that cannot be directly compared with other environmental moisture variables, such as precipitation, which are measured in units (centimeters, millimeters) that are immediately recognizable. In addition, the index is not especially sensitive to short drought periods, which can affect agricultural productivity.

In order to address these shortcomings, other researchers use water-budget analysis to identify deviations in environmental moisture status. The procedure is similar to the Palmer method inasmuch as it incorporates the environmental parameters of precipitation, potential evapotranspiration, and soil moisture. However, the moisture status departure values are expressed in the same units as precipitation and are therefore dimensional. Drought classification using this index method ranges from approximately 25 millimeters for an above-normal moisture status class to -100 millimeters for extreme drought. The index would be close to 0 for normal conditions.

Significance
Drought is invariably associated with some form of water shortage, yet many regions of the world have regularly occurring periods of dryness. Three different forms of dryness have a temporal dimension described as perennial, seasonal, and intermittent. Perennially dry areas include the major deserts of the world, such as the Sahara, Arabian, and Kalahari. Precipitation in these areas is not only very low but also very erratic. Seasonal dryness is associated with regions where the bulk of the annual precipitation comes during a few months of the year, leaving the rest of the year without rain or other precipitation. Intermittent dryness is associated with those instances where the overall precipitation is reduced in humid regions or where the rainy season in seasonally dry areas does not occur or is shortened.

The absence of precipitation when it is normally expected creates variable problems. For example, the absence of precipitation for one week in an area where daily precipitation is the norm would be considered a drought. In contrast, it would take two or more years without any rain in parts of Libya in North Africa for a drought to occur. In those areas that have one rainy season, a 50 percent reduction in precipitation would be considered a drought. In regions that have two rainy seasons, the failure of one could lead to drought conditions. Thus, the word "drought" is a relative term, as it has different meanings in different climatic regions.

User demands also influence drought definition. Distinctions are often made among climatological, agricultural, hydrologic, and socioeconomic drought. Climatological, or meteorological, drought occurs at irregular periods of time, usually lasting months or years, when the water supply in a region falls far below the levels that are typical for that particular climatic regime. The degree of dryness and the length of the dry period are used as the definition of drought. For example, drought in the United States has been defined as occurring when there is less than 2.5 millimeters of rain in a forty-eight-hour period. In Great Britain, drought has been defined as occurring when there are fifteen consecutive days with less than 0.25 millimeter of rain for each day. In Bali, Indonesia, drought has been considered as occurring if there is no rain for six consecutive days.

Agricultural drought occurs when soil moisture becomes so low that plant growth is affected. Drought must be related to the water needs of the crops or animals in a particular place, since agricultural systems vary substantially. The degree of agricultural drought also depends on whether shallow-rooted or deep-rooted plants are affected. In addition, crops are more susceptible to the effects of drought at different stages of their development. For example, inadequate moisture in the subsoil in an early growth stage of a plant will have minimal impact on crop yield as long as there is adequate water available in the topsoil. However, if subsoil moisture deficits continue, then the yield loss could become substantial.

Hydrologic drought definitions are concerned with the effects of dry spells on surface flow and groundwater levels. The climatological factors associated with the drought are of lesser concern. Thus, a hydrological drought for a particular watershed is said to occur when the runoff falls below some arbitrary value. Hydrological droughts are often out of phase with climatological and agricultural droughts and are also basin-specific; that is, they pertain to the watersheds that they affect.

Socioeconomic drought includes features of climatological, agricultural, and hydrological drought and is generally associated with the supply and demand of some type of economic good. For example, the interaction between farming (demand) and naturally occurring events (supply) can result in inadequate water for both plant and animal needs. Human activities, such as poor land-use practices, can also create a drought or make an existing drought worse. The Dust Bowl in the Great Plains and the Sahelian drought in West Africa provide ready examples of the symbiotic relationship between drought and human activities.

In a sense, droughts differ from other major geophysical events such as volcanic eruptions, floods, and earthquakes because they are nonevents. They result from the absence of events (precipitation) that should normally occur. Droughts also differ from other geophysical events in that they often have no readily recognizable beginning and take some time to develop. In many instances, droughts are only recognized when plants start to wilt, wells and streams run dry, and reservoir shorelines recede.

There is wide variation in the duration and extent of droughts. The length of a drought cannot be predicted, as the irregular patterns of atmospheric circulation are not fully known and remain unpredictable. A drought ends when the area receives sufficient precipitation and water levels rise in the wells and streams. Because the severity and areal extent of a drought cannot be predicted, all that is really known is that they are an integral part of the overall natural system and that they will continue to occur.

Desertification

Desertification is defined as the degradation of formerly fertile land—that is, the decrease in the ability of what was once productive land to support the continued growth of vegetation. Desertification can be attributed to a variety of complex environmental factors, including the loss of nutrients and moisture in the soil, increased salinization of soil, and an increase in wind and water erosion. In the course of Earth's history, desert regions have formed, enlarged, and shrunk in size because of natural processes that had nothing to do with human activities. Desertification, however, is accelerating because of anthropogenic causes.

An estimated 70 percent of the world's dryland regions are experiencing either moderate or severe desertification. This phenomenon is caused by droughts and changes in temperature, cloud cover, and wind patterns from global warming, overgrazing, deforestation, irrigation, and other effects of unsustainable human land use. For example, livestock grazing on already semiarid land can remove vegetation that would otherwise prevent the erosion of soil by wind or water, and repeated cultivation of crops on the same land may deplete the supply of nutrients contained in the soil or make it harder for it to absorb water. Proximity to an existing desert is not necessary for desertification to occur.

Robert M. Hordon

Further Reading

Bryson, Reid A., and Thomas J. Murray. *Climates of Hunger.* Madison, Wisc.: University of Wisconsin Press, 1977.

Climate, Drought, and Desertification. United Nations World Meteorological Organization, 1997.

Fisher, R. J. *If Rain Doesn't Come: An Anthropological Study of Drought and Human Ecology in Western Rajasthan.* New Delhi, India: Manohar, 1997.

Frederiksen, Harald D. *Drought Planning and Water Resources Implications in Water Resources Management.* Washington, D.C.: World Bank, 1992.

Guffrey, Kelly. *Sowing the Seeds of Civil War: Regime Destabilization and the Adoption of Neoliberal Economic Policies in Syria by President Bashar al-Assad - Harm to Agriculture, Iraq Refugees, Severe Drought.* Washington, D.C.: United States Government, U.S. Department of Defense, 2018.

Mainguet, Monique. *Aridity: Drought and Human Development.* New York: Springer-Verlag, 2010.

Wilhite, Donald A., ed. *Drought and Water Crises: Science, Technology, and Management Issues.* Boca Raton, Fla.: CRC Press, 2005.

Wilhite, Donald A., and William E. Easterling, with Deborah A. Wood, eds. *Planning for Drought: Toward a Reduction of Societal Vulnerability.* Boulder, Colo.: Westview Press, 1987.

Workman, James G. *Heart of Dryness: How the Last Bushmen Can Help Us Endure the Coming Age of Permanent Drought.* London: Walker Publishing Company, 2009.

Dust Bowl disaster

FIELDS OF STUDY
Agriculture; Atmospheric Science; Climatology; Environmental Sciences; Environmental Studies; Hydroclimatology; Hydrology; Earth System Modeling; Physical Geography; Water Management

SUMMARY
The Dust Bowl ecological disaster of the 1930s revealed the damage that mechanized agriculture could cause if not accompanied by a program of soil management. Droughts periodically occur in the Great Plains of the United States. During such periods, winds pick up loose soil and create dust storms, especially during the spring months. During the twentieth century, new agricultural practices and overgrazing by cattle speeded soil erosion in the region.

PRINCIPAL TERMS
- **aquifer:** any geological formation containing or conducting ground water, especially one that supplies the water for wells or springs
- **drought:** a period of dry weather, especially a long one that is injurious to crops
- **Dust Bowl:** environmental disaster in the 1930s marked by huge dust storms in the southern region of the Great Plains of the United States
- **erosion:** the process by which the surface of the earth is worn away by the action of water, glaciers, winds, or waves
- **Great Plains Shelterbelt:** a project begun in 1934 to create windbreaks on the Great Plains of the United States in response to the severe dust storms of the Dust Bowl, which resulted in significant soil erosion and drought
- **soil conservation:** the prevention or reduction of soil erosion and soil depletion by protective measures against water and wind damage

Droughts periodically occur in the Great Plains of the United States. During such periods, winds pick up loose soil and create dust storms, especially during the spring months. Settlers reported numerous examples of this natural phenomenon during the nineteenth century. During the twentieth century, new agricultural practices and overgrazing by cattle speeded soil erosion in the region. Tractors and other machines allowed farmers to plow larger areas for planting wheat. In the process, they destroyed the natural grasses, the root systems of which had stabilized the soil. Because the wheat replaced the grasses, most farmers remained unaware that they were contributing to a coming catastrophe.

In 1931 a severe drought struck the Great Plains; it centered on the Texas and Oklahoma panhandles, northeastern New Mexico, eastern Colorado, and southwestern Kansas. The wheat crop withered in the fields, and its root systems were no longer able to support the soil. As the drought continued, soil particles that normally clustered together separated into a fine dust. When the winds blew in early 1932, they lifted the dust into the air, marking the beginning of the environmental disaster that a newspaper reporter later dubbed the Dust Bowl.

Although their number and severity increased, dust storms remained an issue of local and regional concern for the first two years. However, as the drought continued into 1934, the storms grew so immense that they caused damage in areas far from the plains. A storm that emanated from Montana and Wyoming in May, 1934, deposited an estimated twelve million tons of dust on Chicago, Illinois. Ships that were some 480 kilometers (300 miles) offshore in the Atlantic Ocean reported that dust from the same storm landed on their decks. Incidents such as these provoked national concern over the growing crisis on the plains.

Scientists identified two types of dust storms: those caused by winds from the southwest and those resulting from air masses moving from the north. While no less damaging, the more frequent southwest storms tended to be milder than the terrifying northern storms, which came to be known as "black blizzards." Huge walls of dust, sometimes more than 1.6 kilometers (1 mile) high, rolled across the plains at 100 kilometers per hour (62 miles per hour) or faster, driving frightened birds before them. The sun would disappear, it would become as dark as night, and frightened people would huddle in their homes, their windows often taped shut. On occasion, people stranded outside during these severe storms suffocated. Some black blizzards lasted less than one hour;

A farmer's son sits on a sand dune near his home in Liberal, Kansas, during the 1930's. (Library of Congress.)

others reportedly continued for longer than three days.

Most historians argue that the Dust Bowl was one of the worst ecological disasters in the United States, one that could have been mitigated had farmers practiced soil conservation in the years before drought struck. Instead, farms were ruined, causing some 3.5 million people to abandon the land. Many of them moved into small towns on the plains, while others journeyed to California in search of opportunity. Cattle and wildlife choked to death. Human respiratory illnesses increased markedly during the Dust Bowl era, and a number of people died from an ailment known as dust pneumonia. Anecdotal evidence indicates that many people grew depressed as the dust storms continued year after year.

Soil Conservation, Farming Practices and Ongoing Threats

The mid-1930s marked the peak of the Dust Bowl, with seventy-two storms that reduced visibility to less than 1.6 kilometers (1 mile) reported in 1937. The return of the rain in the late 1930s eased the crisis, and by 1941 the disaster was over. However, by that time ecologists and farmers had begun to undertake soil conservation measures in response to the crisis. The U.S. government provided expertise and financial support for many of these efforts. Farmers practiced listing, a plowing process that makes deep

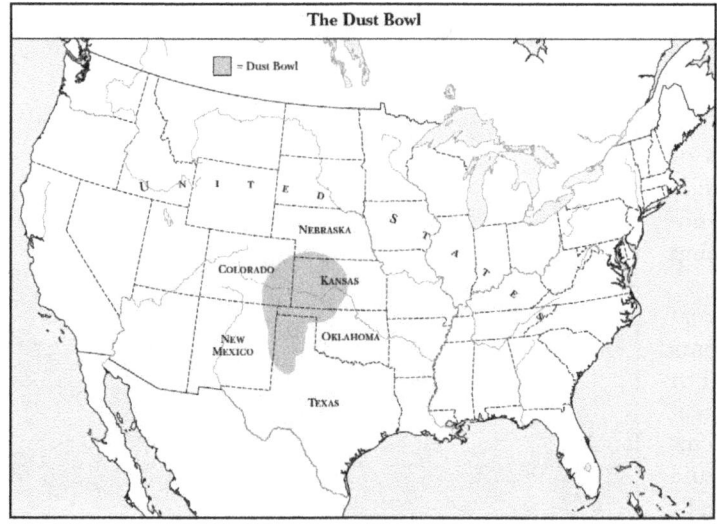

furrows to capture the soil and prevent it from blowing. Alternating strips of planted wheat with dense, drought-resistant feed crops such as sorghum slowed erosion by blocking wind and retaining moisture, which prevented the soil from separating into dust. On lands not farmed, natural grasses were planted to prevent erosion. The government also sponsored the Great Plains Shelterbelt, a program that used rows of trees to form windbreaks. Millions of trees were planted throughout the Great Plains, with more than 4,828 kilometers (3,000 miles) of shelterbelts created in Kansas alone.

Despite the experiences of the 1930s, once the drought ended many farmers returned to the farming practices that had damaged their fields. Soil conservation experts worried that the region would suffer a return of Dust Bowl conditions when the rains stopped. Their predictions came to pass in 1952, when another drought led to a series of dust storms, including several storms with wind gusts clocked at 129 kilometers (80 miles) per hour. That drought ended in 1957, but in accord with a twenty-year cycle, the region again faced a shortage of rainfall in the early 1970s. At that time some analysts confidently predicted that dust storms such as those seen in the 1930's were a thing of the past. They claimed that irrigation with aquifer water from deep wells would prevent soil erosion. However, shrewd observers pointed out that the fate of the region was now tied to a resource, aquifer water, that would become increasingly precious in the coming years. The possibility that the Great Plains could again witness a devastating ecological catastrophe like that of the 1930s remains.

Thomas Clarkin

Further Reading

Bonnifield, Paul. *The Dust Bowl: Men, Dirt, and Depression.* Albuquerque, N.M.: University of New Mexico Press, 1979.

Cunfer, Geoff. *On the Great Plains: Agriculture and Environment.* College Station, Texas: Texas A&M University Press, 2005.

Hurt, R. Douglas. *The Dust Bowl: An Agricultural and Social History.* Chicago: Burnham, 1981.

Lookingbill, Brad D. *Dust Bowl, USA: Depression America and the Ecological Imagination, 1929-1941.* Athens, Ohio: Ohio University Press, 2001.

Steinbeck, John. *The Grapes of Wrath.* 75th ed. New York: Viking, 2014

Worster, Donald. *Dust Bowl: The Southern Plains in the 1930s.* 25th anniversary ed. New York: Oxford University Press, 2004.

E

Earth Day

FIELDS OF STUDY
Commemorations; Ecology; Environmental Ethics; Environmentalism; Ethics; Public Policy

SUMMARY
Since its inception in 1970, participation in Earth Day activities has increased and helped to build awareness of a growing crisis.

PRINCIPAL TERMS
- **Arbor Day:** a day, varying in date but always in the spring, first observed in 1872, in certain states of the U.S. by the planting of trees
- **Civil Rights Movement:** the national effort by African Americans and their supporters in the 1950s and '60s to eliminate segregation and gain equal rights
- **Earth Day:** annual observance, inaugurated in 1970, intended to promote public concern for environmental problems
- **environmental racism:** patterns of racial prejudice regarding issues related to the environment, particularly environmental hazards
- **environmentalism:** advocacy for the preservation, restoration and/or improvement of the natural environment
- **National Environmental Policy Act (NEPA):** a national policy, enacted in 1970, to encourage harmony between humans and their environment, promote efforts to prevent or eliminate damage to the environment and biosphere, stimulate human health and welfare, and enrich understanding of the ecological systems and natural resources

In June 1969, Democratic U.S. senator Gaylord Nelson of Wisconsin, having observed how anti-Vietnam War demonstrations and "teach-ins" had influenced public opinion, conceived of the idea of a large teach-in to educate the general public about the importance of environmental issues. He suggested that such an event should be planned for April 22, a day when many states commemorated Arbor Day and a day that would not conflict with the timing of final exams on college campuses. Recognizing the potential of the idea, a small group of concerned citizens founded the organization Environmental Action to sponsor the event. They were able to raise the modest sum of $125,000, and a dynamic young law student, Denis Hayes, was put in charge of publicizing and coordinating activities. Senator Nelson and Republican congressman Paul N. McCloskey of California were named official co-chairs of the event.

Numerous historical factors contributed to the great success of the first Earth Day. By the late 1960s, an environmental movement was developing, and a growing number of organizations were helping to sensitize the public to environmental problems, and an unprecedented number of publications on environmental themes were being produced by prominent writers, such as Rachel Carson, Walter Udall, Lynn White, and Paul R. Ehrlich. Of even greater significance, Americans across the country were witnessing the harmful effects of environmental damage. In 1968 and 1969 members of Congress reflected public opinion as they considered nearly 140 bills related to environmental issues, and state and local jurisdictions created and voted on various other environmental laws. Almost four months before Earth Day, President Richard Nixon signed the National Environmental Policy Act of 1969, which required analysis and review of public projects. For many people, especially the young, the impressive achievements of the Civil Rights movement represented a model for reform based on a moral appeal. In addition, the spirit of youthful rebellion embodied in the antiwar movement was inspiring parallel movements throughout American society. In short, Senator Nelson could not have chosen a more auspicious context for the launching of his idea.

Conservative business and political leaders were not enthusiastic about the idea of Earth Day. Although President Nixon's press secretary announced the administration's support for the day, the president took no active role in any of the events. While some members of the administration suspected that the observance was a means for advancing the agenda of liberal Democrats, Nixon's secretary of the interior, Walter Hickel, urged Nixon to proclaim a national holiday and become an active participant. Hickel later wrote that he gave "marching orders" to Interior Department personnel to visit college campuses and that fifteen hundred employees of the department did so. The White House was embarrassed when the press reported that Controller General James Bentley had spent sixteen hundred dollars in public funds to send telegram warnings of a possible left-wing plot after he observed that Earth Day fell on the birthday of Soviet revolutionary leader Vladimir Ilich Lenin. Bentley was forced to apologize and pay for the telegrams with his own money.

April 22, 1970

In all measurable ways, the first Earth Day was a huge success. In New York City, Chicago, and Philadelphia large crowds gathered to hear speeches by politicians, poets, ecologists, and other concerned citizens. Some fifteen hundred college campuses, as well as ten thousand elementary and secondary schools, scheduled programs of one kind or another. The National Education Association estimated that about ten million schoolchildren participated in some kind of environmental activity for the day. Also, approximately two thousand communities planned environmental ceremonies of one kind or another.

An impressively diverse array of activities were conducted throughout the United States, and the atmosphere at many events was euphoric and theatrical. In Washington, DC, about ten thousand young people attended a rock concert in front of the Washington Monument. The University of Wisconsin held fifty-eight separate programs. To dramatize air-pollution problems caused by internal combustion engines, several universities held enthusiastic

NASA Earth America. By TheOriginalSoni (Own work)

automobile-wrecking events called "wreck-ins." Some localities also held "bike-ins." In New York City, Fifth Avenue was closed to motor vehicle traffic for two hours. Many idealistic people helped pro-environmental efforts. At the University of Washington, four hundred people planted trees and shrubs during a "plant-in" in an abandoned area near the campus. In Ohio, one thousand students from Cleveland State University gathered litter and loaded it into garbage trucks. In hundreds of communities, groups of Boy Scouts and Girl Scouts held cleanup campaigns and picked up litter.

The first Earth Day was an occasion for numerous speeches, including many by the best-known spokespersons for the environmental movement. Barry Commoner, Paul Ehrlich, Ralph Nader, and the aging René Dubos were among the speakers in greatest demand. Both houses of Congress were adjourned for the day, and many politicians were seen at various rallies. Senator Nelson spoke on nine university campuses in Wisconsin, California, and Colorado. Senator Thomas McIntyre, who delivered fourteen speeches in his home state of New Hampshire, set the record for the greatest number of speeches given by one person on that first Earth Day.

No major acts of violence marred the celebrations, but scattered incidents of militancy did occur. At Boston's Logan Airport, thirteen demonstrators were arrested for blocking traffic during a demonstration to protest a proposed expansion of the airport. In Washington, DC, about twenty-five hundred demonstrators assembled before the offices of the Department of the Interior to protest the approval of oil leases. Students at the University of California at Berkeley conducted a sit-in to register their disapproval of the presence on campus of job recruiters from Ford Motor Company, while at the University of Texas, twenty-six students were arrested for perching in trees to try to prevent the trees' destruction.

Earth Day 1990
In 1971 Earth Day was expanded into Earth Week, but the expanded observance was not successful. For a few years Earth Day attracted limited interest. By the mid-1980s, however, the celebration of Earth Day was regaining popularity, as environmentalists viewed the celebrations as a repudiation of President Ronald Reagan's conservative environmental policies. Leaders of the environmental movement wisely decided to concentrate their efforts on commemorating the twentieth anniversary of Earth Day, and Denis Hayes was chosen as the chairperson of the occasion.

On April 22, 1990, an estimated 200 million people in 140 countries participated in Earth Day. Organizers claimed that this was the largest grassroots demonstration in history. For this occasion, Hayes had the assistance of a large coalition of environmentalist and other socially conscious groups, and together they raised and spent about $4 million. The day was celebrated with marches, rallies, parades, concerts, and a large assortment of activities on all continents. Although the largest demonstrations were held in the developed industrial countries, scattered events also took place within the poorer and less developed regions of the world.

In Boston, a crowd of 200,000 people turned out; in New York City's Central Park, the various rallies attracted an estimated 750,000 participants; in Washington, DC, some 125,000 people participated in a demonstration on the National Mall; in St. Louis, an estimated 10,000 people planted 10,000 trees on the banks of the Mississippi River. Throughout the day, speeches were given by Hayes, Gaylord Nelson, Morris Udall, Barry Commoner, Bruce Babbitt, Senator Edward Muskie, and countless others. In a prime-time national television program called *The Earth Day Special*, aired by the American Broadcasting Company (ABC), Bette Midler played an abused Mother Earth who collapsed as a result of global warming, deforestation, and toxic poisoning. Although the events of the day were almost uniformly peaceful, ecoguerrilla groups attracted headlines by destroying oil-exploration gear and pouring sand in the fuel tanks of logging machinery.

At a time of controversy over issues such as protection of the northern spotted owl, President George H. W. Bush and his administration appeared distrustful of Earth Day 1990. Speakers at Earth Day events often criticized the Bush administration's policies However, President Bush, who had earlier referred to himself as "the environmental president," addressed several crowds via a telephone hookup. In comparison with those of twenty years before, the speeches of Earth Day 1990 were more somber and realistic, as it had become clear that environmental problems would not be quickly solved in a painless manner.

Earth Day 1990 was truly a global festival. In Brazil, a concert by Paul McCartney paid special attention to the environment. In West Germany, Green organizations sponsored the ceremonial planting of trees. Thailand's top rock band, the Carabano, held a concert with the theme "We Love the Forest." In Hong Kong a day-long educational entertainment featured singers, mimes, and exhibits of "green" consumer goods. Other activities included a roadway "lie-down" by five thousand protesters against car fumes in Italy, an 800-kilometer (500-mile) human chain across France, a trash-cleanup campaign on Mount Everest in Nepal, and a "flyby" of three thousand kites made by schoolchildren in Tours, France.

Impact of Earth Day
Yearly celebrations of Earth Day have become established. The day's silver anniversary in 1995 attracted considerable interest, but organizers decided to concentrate greater efforts on the thirtieth anniversary. With the help of Internet organization, some five thousand environmental groups working in 184 countries around the world joined together to celebrate Earth Day 2000.

The 1970 celebration of Earth Day tended to be a predominantly white, middle-class affair. Many African Americans were suspicious that the day would detract from issues of racial and economic justice.

Although such views did not completely disappear, they tended to decline with time. In 1969 polls indicated that only 33 percent of African Americans wanted the government to pay more attention to environmental issues; by 1976, approximately 58 percent of African Americans expressed the same viewpoint. By the time of Earth Day 1990, African American leaders had more evidence that pollution tends to be especially severe in areas where poor and marginalized people live, and thus they were able to use the day to publicize the issue of environmental racism.

Such celebrations as Earth Day are found to impact public attitudes and can solidify the commitment of environmental organizations. In a 1965 Gallup Poll, only 17 percent of the responding Americans said they considered the reduction of air and water pollution to be one of the most pressing problems demanding governmental action. Immediately after Earth Day 1970, the figure was 53 percent, but by 1980 it had fallen to 24 percent. Ironically, the environmental policies of President Ronald Reagan appeared to encourage the popularity of Earth Day during the 1980s, and the success of Earth Day 1990 is partially explained by a survey of the time in which 80 percent of Americans said they would support more strenuous environmental efforts, regardless of costs.

Without doubt, there are always faddish, trendy elements in the celebrations and speeches of Earth Day. Environmentalists are offended when some of the nation's worst polluters attempt to exploit the day, to use exposure at events as a form of commercial advertising. The important point, however, is that Earth Day promotes education and reflection about serious environmental problems that are experienced by people in their daily lives.

Thomas T. Lewis

Further Reading

Dunlap, Riley E., and Angela G. Mertig, eds. *American Environmentalism: The U.S. Environmental Movement, 1970–1990.* Philadelphia: Taylor & Francis, 1992.

Gottlieb, Robert. "The Sixties Rebellion: The Search for a New Politics." In *Forcing the Spring: The Transformation of the American Environmental Movement.* Rev. ed. Washington, D.C.: Island Press, 2005.

National Staff of Environmental Action. *Earth Day: The Beginning—A Guide for Survival.* New York: Bantam Books, 1970.

Nelson, Gaylord, with Susan Campbell and Paul Wozniak. *Beyond Earth Day: Fulfilling the Promise.* Madison, Wisc.: University of Wisconsin Press, 2002.

Scheffer, Victor. *The Shaping of Environmentalism in America.* Seattle, Wa.: University of Washington Press, 1991.

Shabecoff, Philip. *A Fierce Green Fire: The American Environmental Movement.* Rev. ed. Washington, D.C.: Island Press, 2003.

Switzer, Jacqueline Vaughn. *Green Backlash: The History and Politics of Environmental Opposition in the U.S.* Boulder, Colo.: Lynne Rienner, 1997.

■ Eat Local movement

FIELDS OF STUDY

Advocacy; Agencies; Agriculture; Commissions; Ecology; Environment; Environmentalism; Ethics; Food; Hunger; Organizations; Policy; Protest

SUMMARY

Proponents of the eat local movement assert that by buying and eating only locally grown or produced foods, people can enhance social and economic health while at the same time providing benefits to the environment, such as by supporting small farmers who employ organic and sustainable agricultural practices.

PRINCIPAL TERMS
- **carbon footprint:** the amount of greenhouse gases and specifically carbon dioxide emitted by something, such as a person's activities or a product's manufacture and transport, during a given period
- **community-sponsored agriculture (CSA):** a system that connects the producer and consumers within the food system more closely by allowing the consumer to subscribe to the harvest of a certain farm or group of farms
- **Eat Local Movement (locavore):** movement devoted to building locally based, self-reliant food economies in which sustainable food production, processing, distribution, and consumption are integrated

Filipino farm worker near Santa Maria, CA. (Library of Congress)

- **sustainable agriculture:** any of a variety of environmentally friendly farming methods that preserve an ecological balance by avoiding depletion of natural resources

One goal of the eat local movement is to reduce the carbon footprint of food by reducing the amount of energy it takes to get food from the field to the plate. According to proponents of the movement (often called locavores), the food in an average American meal travels at least 3,200 kilometers (2,000 miles) from its sources to the consumer. Those who endorse the practice of eating locally encourage people to eat only foods that are grown or produced near where they live. This, locavores assert, serves to keep people in better touch with the earth and the seasons and provides a connection between farm and table, reduces the carbon footprint of the food, sustains local economies, and encourages consumption of high-quality foods. Proponents of the movement generally seek out food that is grown in ways that are healthy, that respect the environment and those who work in the food-growing industry, and that are humane to animals. Although no single definition of "local" has been established, many proponents of the eat local movement suggest that people should buy and eat only those foods that are grown or produced within a radius of roughly 160 kilometers (100 miles) of where they live.

"Local food" includes food grown in one's own garden and food produced on local farms or by local community-sponsored agriculture (CSA) groups. In addition to reducing the expenditure of energy on shipping and other means of transporting food, the eat local movement is concerned with sustainable agriculture; buying and eating locally grown foods supports local small farms, many of which raise crops, maintain dairy cattle, and raise livestock for food according to the principles of organic farming, without using chemical fertilizers or pesticides and without the hormones or antibiotics often used by large food producers. Locavores point to fresher, better-tasting, lower-cost produce with higher amounts of vitamins and other healthy compounds that are lost when fruits and vegetables are picked green for shipping or are handled extensively before they reach the consumer. Locavores also note that cutting out the middlemen—the large food suppliers—means that local farmers receive a larger percentage of each dollar spent on food.

Often, eat local and CSA groups offer classes or other kinds of instruction on buying local foods in bulk during their peak season and preserving them for later use, such as by canning or drying; many also provide recipes that focus on the locally grown foods in their own areas. Eat local and CSA groups also generally encourage members and others to patronize restaurants that feature local foods that are in season.

Detractors of the eat local movement have noted that although eating locally may save energy by cutting down on the transportation of food, the overall environmental costs of growing food can remain quite high. Is it more environmentally friendly, for example, to grow fruits and vegetables outdoors in a sunny environment and then ship them to a store in another area than it is to use electricity and other energy sources to grow the same kinds of produce in greenhouses in less naturally sunny environments and then sell them locally? Debate on such issues is ongoing.

Marianne M. Madsen

Further Reading

Bendrick, Lou. *Eat Where You Live: How to Find and Enjoy Fantastic Local and Sustainable Food No Matter Where You Live.* Seattle, Wa.: Skipstone Press, 2008.

Cobb, Tanya Denckla. *Reclaiming Our Food: How the Grassroots Food Movement Is Changing the Way We Eat.* North Adams, Mass.: Storey Publishing, 2011.

Nabhan, Gary Paul. *Coming Home to Eat: The Pleasures and Politics of Local Food.* New York: W. W. Norton, 2009.

Pollan, Michael. *The Omnivore's Dilemma: A Natural History of Four Meals.* New York: Penguin Press, 2007.

Ecocentrism

FIELDS OF STUDY
Advocacy; Biology; Ecology; Ecosystems; Environment; Environmentalism; Ethics; Life Sciences; Philosophy and History of Science; Policy; Protest

SUMMARY
Ecocentrism is the view that the natural world is morally important and should be valued independent of present or future human interests; its importance should be reflected in human ethics and in human treatment of the earth.

PRINCIPAL TERMS
- **anthropocentrism:** philosophical viewpoint arguing that humans are the central or most significant entities in the world
- **authoritarianism:** principle of blind submission to authority, as opposed to individual freedom of thought and action
- **biocentrism:** an ethical point of view that extends inherent value to all living things
- **ecocentrism:** the view that the natural world is morally important and should be valued independent of present or future human interests
- **egalitarianism:** a belief in human equality especially with respect to social, political, and economic affairs
- **ethics:** the discipline dealing with what is good and bad and with moral duty and obligation
- **misanthropy:** hatred, dislike, or distrust of humankind

Sometimes called deep ecological ethics or dark green ethics, ecocentrism is the view that the natural world is morally important and should be valued independent of present or future human interests; its importance should be reflected in human ethics and in human treatment of the earth. The natural world includes the organisms of the earth as well as the earth itself and its elements and ecosystems. Ecocentrism thus differs from anthropocentrism, which claims that only human interests and values ultimately matter or that they matter more than any other interests and values, and biocentrism, which extends moral importance only to certain animals or living organisms in addition to humans. Formulation of the ecocentric ethic is usually credited to Aldo Leopold; it was later developed by Arne Naess and George Sessions.

Ecocentrism calls for a new ethical outlook based on the recognition that human life has emerged from and is dependent on the ecosystem. Humans are one constituent part of a larger whole; they are, along with other inhabitants, equal citizens of the earth, not the earth's masters. Such egalitarianism insists that human interests do not have automatic priority over the natural world. In some situations of conflict between human and nonhuman interests, priority should be granted to the nonhuman interests, as when commercial or economic development is restricted for the sake of habitat or species preservation.

Demands for equality between humans and nonhumans are often based on claims that the natural world has intrinsic value, but intrinsic value has been understood in different ways. One approach recognizes that organisms such as plants have identifiable interests, or ends, that should be respected alongside those of sentient beings. Another claims fundamental value for the existence and preservation of all species. The value of the biotic community is also identified with the robust, highly integrated functioning of the ecosystem and features such as integrity, stability, and beauty.

Some ecocentric thinkers worry, however, that an appeal to objective and abstract properties has too much in common with the kinds of reductive and scientific analyses of the world that have contributed to environmental degradation. An alternative strategy appeals for the rediscovery of a sense of reverence toward nature in its abundance and vitality. It points to the human experience of nature as an endless series of sensuous particulars, stories and emotional encounters with places, that imply that the value of nature can never be fully articulated.

Ecocentrism is also practical. It denies that environmental ethics should be based on enlightened human self-interest and efficient management. Human understanding of the earth's innumerable elements and systems, as well as humanity's effects on them, is limited. The appropriate response is skepticism toward adopting technological solutions to environmental problems (technocentrism). Ecocentrism favors familiar lower-tech and environmentally friendly means of addressing ecological problems. Further, some go beyond practical concern for the environment and seek reform of the social structures and economic systems, including capitalism, that have caused neglect of the ecosystem. Others emphasize that the goals of ecocentrism should be secured through piecemeal and local action and argument rather than by appeal to big facts.

Critics of ecocentrism accuse it of misanthropy and authoritarianism; they assert that it uses claims about the earth's intrinsic value dogmatically to secure priority for nature over individual people and communities. Some versions are also accused of mysticism and impracticality, insofar as they call for identification with a greater self or whole or an attitude of reverence toward something not fully understood.

Andrew Lambert

Further Reading

Boylan, Michael, ed. *Environmental Ethics.* Upper Saddle River, N.J.: Prentice Hall, 2001.

Curry, Patrick. *Ecological Ethics: An Introduction.* Malden, Mass.: Polity Press, 2006.

Schmidtz, David and Dan C. Shahar. *Environmental Ethics: What Really Matters, What Really Works.* 3rd ed. New York: Oxford University Press, 2018.

Sessions, George, ed. *Deep Ecology for the Twenty-first Century.* Boston: Shambhala, 1995.

Eco-fashion

FIELDS OF STUDY
Advocacy; Commercial Products; Design; Ecology; Environment; Environmentalism; Ethics; Fashion; Industries; Policy; Protest

SUMMARY
Eco-fashion designers seek to reduce the negative environmental impacts associated with traditional methods of textile and clothing production, which include air and water pollution.

PRINCIPAL TERMS
- **eco-fashion:** philosophy in the world of clothing design that is concerned with the use of sustainable materials produced in a socially responsible manner
- **Green Revolution:** increase in production of food grains, especially wheat and rice, in the mid-20th century, resulting from the introduction into developing countries of new, high-yielding varieties, and whose early dramatic successes were in Mexico and the Indian subcontinent
- **Global Organic Textile Standard (GOTS):** organization developed in 2006 with the aim of defining requirements recognized worldwide to ensure the organic status of textiles from harvesting of raw materials, environmentally and socially responsible manufacturing and labelling to provide credible assurance consumers
- **slow fashion:** the design and production of clothing made with sustainable, organic or recycled materials

The fashion industry's impacts on the environment began during the Industrial Revolution with the large-scale production of fabrics under conditions that included little regulation of pollution and little concern with production by-products. Since the advances in agricultural science of the Green Revolution in the mid-twentieth century, the growing of fibers used in textiles has relied increasingly on pesticides and synthetic fertilizers, a practice that has led to negative ecological impacts. Both revolutions made many kinds of textiles more widely available

Aveda Eco Fashion Week—Day 1—at Vancouver Salt Building in Olympic Village, Vancouver, Canada. February 23, 2011. By Jason Hargrove from Toronto, Canada

around the world. With the introduction of ready-to-wear clothing during the 1960s, consumer demand for product diversity and seasonal fashions rose dramatically. This demand led to practices that came to dominate the world market, causing widespread environmental and social harms.

Eco-fashion developed in the late twentieth century as an alternative design approach involving the use of organic, vintage, recycled, locally based, and natural materials to bring consumers ecologically and socially sustainable choices in clothing. Those who engage in eco-fashion practices focus on minimizing the use of hazardous chemicals and the production of waste by-products, maximizing efficiency in their use of energy and water, and establishing fair wages and production standards that are healthy for workers.

Eco-fashion design is evolving constantly to reduce the environmental and social impacts of textile production. Initially it targeted negative preexisting methods, such as the toxicity of pretreatment, dying, finishing, drying, and laundry processes in which emissions from formaldehyde, acids, and volatile organic compounds pollute the air and salts, surfactants, heavy metals, toxic chemicals, biocides, detergents, emulsifiers, and dispersants create aquatic toxicity. The manufacture and transportation of these products also require large amounts of electricity, principally generated from carbon dioxide-emitting fossil fuels. Scrap waste from garment assembly and unsold garments from retail are increasingly directed to landfills. Given that factory workers suffer from such methods owing to their exposure to chemicals, fiber dust, and polluted water, designers who support eco-fashion include healthy workplace standards and sufficient wages within their definition of eco-fashion.

The concept of sustainable fashion first appeared with the introduction of organic cotton as an alternative to conventionally grown crops in the United States. Brands such as Esprit and Vanity Fair began working to create large-scale sustainable clothing lines during the late 1980s, widening the market and developing environmentally sound practices. During the 1990s the Organic Trade Association (OTA) formed and worked with the Organic Fiber Council to adopt the Organic Fiber Certification Standard. In 1996 council member and environmental business leader Marci Zaroff coined and trademarked the term "ECOFashion" to brand and identify this market further. In the same year, the Patagonia apparel company emerged as a leader of eco-fashion, committing to the use of organic cotton in all its cotton clothing items. The Organic Exchange was created in 2002 to build a global community of farmers, manufacturers, brands, and retailers committed to producing organic fibers with sustainable practices. Organizations from four nations—the United Kingdom, the United States, Germany, and Japan—worked together to create the Global Organic Textile Standard (GOTS), a revised version of which was published in 2008; the GOTS provides a basis for the certification of fibers as organic.

The use of organic fabrics and the repurposing of existing clothing is called "slow fashion," a term coined in a 2007 article by Kate Fletcher, published in *The Ecologist*, where she compared the ethical fashion industry to the slow food movement, which defends biodiversity, awareness and responsibility in food consumption and opposes standardization of food – or fashion – tastes. Like the slow food movement, slow fashion seeks to improve the quality of life for humans, empty the landfills and help increase sustainable practices.

With growing market demand for environmentally conscious products of all kinds, greenwashing (deceptive claims that products are green or eco-friendly) on the part of companies that do not fully adhere to eco-fashion principles has made some observers skeptical of the industry. Additionally, the lack of transparency in clothing manufacture supply chains poses difficulties; disconnects often occur between designers and the agents who fill designer requests, and between designers and their manufacturing and packaging sites. Eco-fashion designers must research their supply chains carefully, and sometimes they must select more expensive and slower-paced alternative production sources to ensure that the principles of eco-fashion are not violated. For these reasons, eco-fashion companies may often be at a competitive disadvantage vis-à-vis conventional production lines. Eco-fashion companies, however, seek to reduce their long-term costs through augmentation of product longevity and versatility, closed-loop recycling, and participation in rental programs; all these strategies show promise in making eco-fashion increasingly competitive.

Elizabeth A. Barthelmes and Brian J. Gareau

Further Reading

Allwood, Julian M., et al. *Well Dressed? The Present and Future Sustainability of Clothing and Textiles in the United Kingdom.* Cambridge, England: University of Cambridge, Institute for Manufacturing, 2006.

Fletcher, Kate. *Sustainable Fashion and Textiles: Design Journeys.* Sterling, Va.: Earthscan, 2008.

Humphrey, Liz, and Nick Robins. *Sustaining the Rag Trade.* London: International Institute for Environment and Development, 2000.

Websites

Eco Fashion Brand is Upcycling Over 100,000 Sweaters Every Year - Slow Fashion
https://www.youtube.com/watch?v=0VRTilzmhg8&t=23s

The Ecologist: The Journal for the Post-Industrial Age:
https://theecologist.org/2007/jun/01/slow-fashion

Third Annual Eco-Friendly Fashion Show.
https://www.youtube.com/watch?v=Kd2YPnd7ins

■ Ecofeminism

FIELDS OF STUDY
Advocacy; Biology; Ecology; Ecosystems; Environment; Environmentalism; Ethics; Gender Studies; Life Sciences; Philosophy and History of Science; Policy; Protest

SUMMARY
Ecofeminism is an important movement within environmental philosophy, environmental activism, and environmental justice that addresses harms against nature and the patriarchal oppression of women, the feminization of nature and the naturalization of women, and the logic that connects all hierarchical relationships.

PRINCIPAL TERMS
- **deforestation:** clearing or thinning of forests by humans
- **ecofeminism:** philosophy that bridges the issues of feminism and environmentalism with the understanding that gender discrimination and environmental degradation are related manifestations of systematic oppression
- **environmental justice:** the fair treatment and meaningful involvement of all people regardless of race, color, national origin, or income in the developing, implementing and enforcing environmental laws, regulations and policies
- **essentialism:** any of various philosophies, most influential in continental Europe from about 1930 to the mid-20th century, that interpret human existence stressing its concreteness and its problematic character
- **nuclear proliferation:** the spread of nuclear weapons, nuclear weapons technology, or fissile material to countries that do not already possess them
- **pollution:** the introduction of contaminants into the natural environment that cause adverse change

A theoretical philosophy and an activist stance, ecofeminism understands that all oppression is linked by a shared logic and that historical, theoretical, and practical relationships exist between gender discrimination and environmental degradation. Ecofeminists assert that in order to address environmental harm, human beings need to attend to power-laden gender relationships; in order to address gender inequity, humans need to understand the logic that enacts hierarchical relationships. Ecofeminists argue that the feminine has long been associated with the natural, the body, and emotion, symbolized in metaphors such as the nurturing image of Mother Nature. Alternately, the masculine is tied to traits such as rationality and civilized (non-natural) progress. Shared logic perpetuates these dualisms—female/male, nature/culture, body/mind—which are overlaid with corresponding value judgments: Rationality, male traits, and culture are good; expressions of the body, female traits, and nature are bad. Ecofeminism seeks to understand and address these dualisms.

Ecofeminist artist Patricia Johnson collaborated with civil engineers to help design the Ellis Creek Water Treatment Facility in Petaluma, CA to create a major urban infrastructure within living nature. By Tim Williamsen (Own work).

Activist ecofeminism emerged during the 1970s with women-led groups who performed acts of civil disobedience to protest harms against nature such as nuclear proliferation, widespread pollution, and deforestation. These gatherings provoked reaction from academic feminists, who explored the theoretical relationships among multiple forms of degradation. Early ecofeminism was tied closely to the spirituality of nature- and female-centered religions, including paganism and Native American mythologies. Essentialist arguments for the "special" relationship between women and nature also permeated early scholarship.

One essentialist argument asserts that some female quality—the ability to bear children, to nurture life—imbues women with an innate sensitivity for and connection to environmental issues, locating female identity in biology. A second essentialist argument suggests that metaphysical gender-specific essences exist separate from biology and social constructions of gender. Another essentialism posits female identity ahistorically and thus assumes the oppressed "female" experience across time is a universal experience, regardless of class, race, ethnicity, or sexuality, which can serve to privilege the dominant female voice. Other essentialisms claim that particular ethnic and racial groups have an innate closeness with nature, often associated with cultural worldview, or imagine nature itself as fixed and unchanging.

While some ecofeminists argue that generalizations bring groups together in unity for a cause, and thus value essentialism's activist purpose, the more common rhetoric is anti-essentialist: Gender differences are shaped by culture and result from lived experience, not biology. Many scholars worry that collapsing the differences among women enacts the same dichotomies that ecofeminists strive to overcome, though with reversed value associations. Some believe that female identities are far more flexible and fluid than essentialist arguments describe; limited descriptions in turn limit the available prescriptions for social change. Scholars such as Donna Haraway believe that essentialist rhetoric wrongly focuses on mystical connections with nature and an oversimplified understanding of women rather than on the experiences of actual women. Ecofeminism has evolved to address these essentialist concerns, and discussion has come to center on material issues of justice related to the conditions that cause women to bear more severely the burdens of environmental harm, on the promise—based on women's social roles—of empowering women to address ecological

Vandana Shiva, international ecofeminist, author and antiglobalization activist. By Ajay Tallam.

disaster, on the logic that enables all forms of discrimination, and on the historical roots of hierarchical relationships.

Karen Warren has written about this "logic of domination" that founds hierarchical relationships. She argues for a critical ecofeminism that understands the impacts of anthropocentric rationality in the context of history, culture, and social structure. Other ecofeminists, including Carolyn Merchant, also trace the connected exploitations of women and nature to the impacts of reductionism—adopted by early scientific and religious institutions—on Western thought. In order to address the dichotomies perpetuated by this worldview, they explain, human beings need to address the logic that supports divisive relationships.

Ecofeminism in the early twenty-first century is more academic than activist, though it does drive small movements across the globe. Intellectually, it has splintered into several branches, including spiritual, essentialist, critical, transformative, radical, and materialist ecofeminism. Although ecofeminism has not launched the grand social change originally imagined by early thinkers, the shared concerns of ecological destruction and gender oppression continue to permeate environmental discourse and action.

Lissy Goralnik

Further Reading

Diamond, Irene and Gloria Feman Orenstein. *Reweaving the World: The Emergence of Ecofeminism.* San Francisco: Sierra Club Books, 1990.

Griffin, Susan. *Woman and Nature: The Roaring Inside Her.* New York: Harper and Row, 1980.

Kheel, Marti. *Nature Ethics: An Ecofeminist Perspective.* Lanham, Md.: Rowman & Littlefield, 2008.

Sturgeon, Noël. *Ecofeminist Natures: Race, Gender, Feminist Theory, and Political Action.* New York: Routledge, 1997.

Warren, Karen J. *Ecofeminist Philosophy: A Western Perspective on What It Is and Why It Matters.* Lanham, Md.: Rowman & Littlefield, 2000.

■ Ecological economics

FIELDS OF STUDY
Business; Commerce; Controversies; Debates; Ecology; Economics; Energy and Energy Resources; Environment; Environmentalism; Production; Public Policy; Renewable Resources

SUMMARY
Ecological economics seeks to replace the mainstream economics paradigm of economic welfare and continuous monetary growth with a model of intra- and intergenerational, holistic, sustainable well-being for both humankind and the environment.

PRINCIPAL TERMS
- **ecological economics:** interdisciplinary research field pursuing human and environmental well-being under the premise of recognizing the human economy as a subsystem that functions within the planet's ecological system
- **entropy:** the measurement of disorder, coined by German physicist Rudolph Clausius, such as ice melting in water, from to free, from ordered to disordered

- **holistic:** concerned with wholes or complete systems rather than with the analysis of, treatment of, or dissection into parts
- **recycling:** to treat, process or repurpose used or waste materials so as to reclaim for further usage
- **Stoic:** a member of a school founded by Greek philosopher (c. 300 BCE) holding that wise human beings should be free from passion
- **sustainable:** pertaining to a system that maintains its own viability by using techniques that allow for continual reuse
- **transdisciplinary:** relating to and crossing between more than one branch of knowledge

By utilizing transdisciplinary approaches, ecological economics investigates and tries to understand the complex interdependence of society, human economy, and ecological environment. Ecological economics acknowledges the fact that the interdependence of these systems is structured in such a way that the economic sector is part of the human society, which itself is embedded in the planet's ecosystem. The existence of both the economic sector and the society are impossible without the ecological system.

In the Western context, precursors of ecological economics can be traced to ancient Greece. The Greek philosopher, Plato (c.429–c.347 BCE) drafted a model of a more-or-less sustainable economy in his dialogue *Nomoi* (*Laws*); also, according to Stoic philosophy, men should live according to nature. Kenneth E. Boulding's seminal essay "The Economics of the Coming Spaceship Earth" (1966) cleared the ground for ecological economics, employing the concept of Spaceship Earth, which emphasizes that the planet—like a spaceship—is an (almost) closed system (insolation, the reception of solar energy by the earth, is an exception, along with other examples of radiation that reach the planet or are emitted by it). Herman E. Daly and Robert Costanza, among others, are considered the main initiators of both the International Society for Ecological Economics (founded in 1989) and the society's journal, *Ecological Economics*, which began publishing in 1989.

Transdisciplinary Approaches to Sustainable Well-Being

Although having its point of departure at the intersection of economics and natural (especially

Head shot of Jon D. Erickson, Professor of Ecological Economics at the University of Vermont. By Jdericks

ecological) sciences, ecological economics' transdisciplinary approach utilizes research findings from various other academic fields, such as philosophy—especially (environmental) ethics—history, political science, sociology, psychology, anthropology, biology, chemistry, geology, climatology, and physics. This approach is deemed essential, since mainstream economics has various epistemic shortcomings owing to its lack of multidisciplinary thought and its instrumental view of the ecosystem (in mainstream economics, ecosystemic goods and services are considered economic resources). In addition, such an approach is vital for addressing the well-being of individuals, the human society, and the planet's ecological systems in a sustainable way, which in turn means that well-being is important not only for the present time but also for future generations and future ecosystems.

Ecosystemic Dimensions and Policy Instruments
Ecological economics observes that continuing overpopulation beyond the earth's carrying capacity, unsustainable resource exploitation, overconsumption, and contamination beyond the sink function (the environment's assimilative ability to absorb pollutants) will increase entropy (ecosystemic disorganization) and reduce biodiversity as well as the aggregated quality of life on the planet. Particularly irreversible actions—such as the extinction of species and the overexploitation of nonrenewable resources—must be avoided. Therefore, the principle of uncertainty and the precautionary principle play important roles in the planning of economic actions and the implementation of policy instruments; it is, for example, extremely difficult to predict very accurately the long-term impacts of nuclear waste on the ecosystem. Hence, to facilitate intra- and intergenerational justice or equity and individual, social, and ecological sustainable well-being (instead of short-term economic welfare for a few), policy instruments are recommended.

Mainstream economics is associated with certain market failures—for example, inefficient allocation of services and goods such as clean water and air in some locations. In ecological economics, the prices of resources, products, and services have to speak the ecological truth. The price of an airplane ticket should reflect the monetarized impacts of, for example, air and noise pollution. The approach of calculating and charging ecologically correct prices is also known as the internalization of external effects. To limit emissions, governments, communities, unions of countries, and certain nonstate entities may issue pollution permits that producers can buy and trade. Ecological taxing and incentive systems are other measures that have been discussed.

Such measures are, however, not enough to prevent unsustainable external effects on the ecosystem. Concrete suggestions by ecological economists therefore include a decrease in the world population growth rate, strong limitations on the exploitation of nonrenewable resources, and sustainable use of renewable resources (that is, use of these resources in such a way that they can regenerate). Other recommendations include, but are not limited to, more extensive as well as stricter recycling and upcycling, a paradigm shift from conventional to organic farming, and a revaluation of economic mainstream values such as monetary wealth and economic growth.

Roman Meinhold

Further Reading
Boulding, Kenneth E. "The Economics of the Coming Spaceship Earth." 1966. In *Valuing the Earth: Economics, Ecology, Ethics*, edited by Herman E. Daly and Kenneth N. Townsend. Cambridge, Mass.: MIT Press, 1993.

Common, Michael S., and Sigrid Stagl. *Ecological Economics: An Introduction*. New York: Cambridge University Press, 2005.

Costanza, Robert, ed. *Ecological Economics: The Science and Management of Sustainability*. New York: Columbia University Press, 1991.

_____, et al. *An Introduction to Ecological Economics*. Boca Raton, Fla.: St. Lucie Press, 1997.

Daly, Herman E. *Ecological Economics and Sustainable Development*. Cheltenham, England: Edward Elgar, 2008.

Daly, Herman E., and Joshua C. Farley, eds. *Ecological Economics: Principles and Applications*. 2d ed. Washington, D.C.: Island Press, 2010.

Edwards-Jones, Gareth, Ben Davis, and Salman Hussain. *Ecological Economics: An Introduction*. Malden, Mass.: Blackwell, 2004.

Eriksson, Ralf, and Jan Otto Andersson. *Elements of Ecological Economics*. New York: Routledge, 2010.

Zografos, Christos, and Richard Howarth, eds. *Deliberative Ecological Economics: Ecological Economics and Human Well-Being*. New York: Oxford University Press, 2008.

Ecological footprint

FIELDS OF STUDY
Advocacy; Ecology; Energy and Energy Resources; Environment; Environmentalism; Ethics; Logic; Mapping; Mathematics; Measurement; Policy; Protest; Renewable Resources; Water Resources

SUMMARY
Nature provides for the needs of humans worldwide, but individuals are not always aware of how many natural resources they are using and how their consumption of those resources affects the environment

at large. The concept of the ecological footprint provides a valuable educational tool regarding environmental sustainability.

PRINCIPAL TERMS
- **carbon footprint:** the amount of greenhouse gases, specifically carbon dioxide emitted by a person's activities or a product's manufacture and transport during a given period
- **carrying capacity:** the maximum, equilibrium number of organisms of a species that can be supported indefinitely in a given environment
- **ecological footprint:** measure used to quantify and assess the impact of human activities on ecosystems
- **Global Footprint Network:** an international nonprofit organization founded inn 2003 whose stated mission is to "help end ecological overshoot by making ecological limits central to decision-making
- **sustainability:** the quality of not being harmful to the environment or depleting natural resources, thereby supporting long-term ecological balance

The concept of the ecological footprint (EF) emerged during the early 1990s as the favored measure of human beings' demands on nature. Ecological and carbon footprints are both matrices for measuring the impact of routine human activity on the environment, but the carbon footprint represents the total greenhouse gas (GHG) emission to the environment throughout a period by a person or an organization, whereas the ecological footprint measures human demand on the Earth's ecosystems.

Nature provides for the needs of humans worldwide, but individuals are not always aware of how many natural resources they are using and how their consumption of those resources affects the environment at large. The concept was first articulated by Mathis Wackernagel and William Rees of the University of British Columbia, who originally used the term "appropriated carrying capacity" but later adopted "ecological footprint" because it was more easily understood.

Measuring the EF of a person, group, or other entity essentially consists of comparing the entity's demands on nature with the earth's capacity to regenerate the resources used and to provide services. This measurement takes into consideration how much (biologically productive) land and water are (or would be) required to produce the resources the entity consumes and to absorb and render harmless its corresponding wastes, given current knowledge and

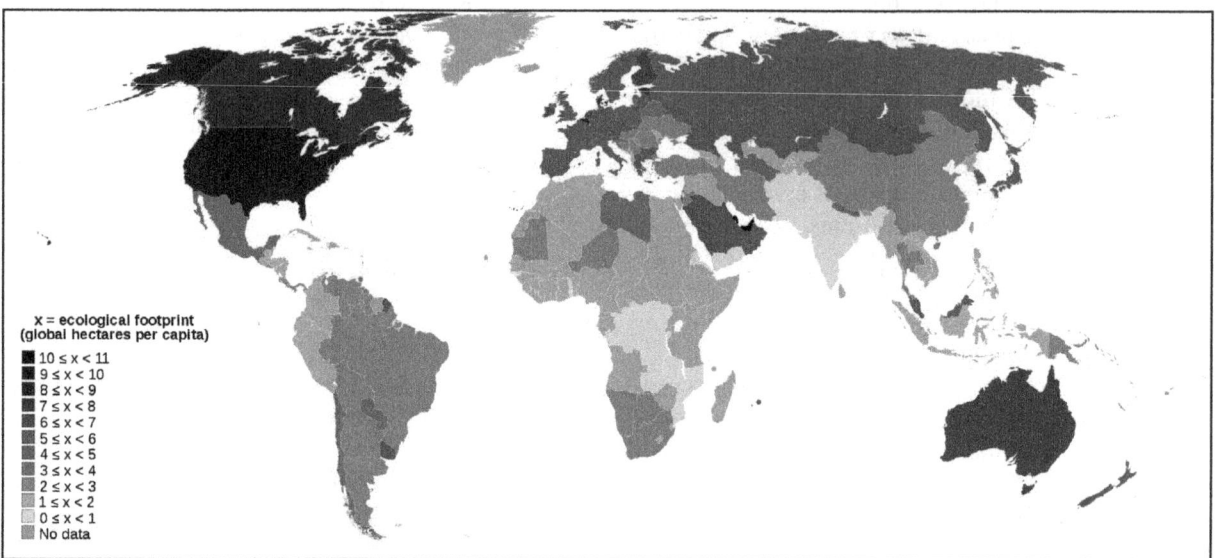

World map of countries shaded according to their ecological footprint in 2007 (published on 13 October 2010 by the Global Footprint Network). Lighter shades denote countries with a lower ecological footprint per capita and darker shaded for countries with a higher ecological footprint per capita. By Jolly Janner.

available technologies. The concept of EF can be used to estimate the use of resources by a population, a person, a region, a city, a country, a business or a sector of the economy, an organization or institution, or even a particular lifestyle. It makes possible comparisons of resource use among different countries and the calculation of per-capita EF measurements.

In 2006, the Global Footprint Network released the Ecological Footprint Standards, which was updated in 2009. These standards allow for consistent assessment of the ecological footprint across the public and private sectors. The data most commonly included in calculations concern carbon, food, housing, goods and services, waste, and recycling. Before the Global Footprint Network released its standards the many differences among the methodologies and accounting procedures used in calculating ecological footprints led some observers to question the reliability of EF measurements and even the concept's usefulness.

Although critics have increasingly focused on the limitations of the EF measure—particularly its accounting and calculation procedures—and have called for its continued improvement, the concept of EF is still widely considered to be useful. It can help people to understand the concept of environmental sustainability, educating them about carrying capacity and overconsumption and perhaps leading to the eventual alteration of world trends in consumption behaviors, as it becomes clear to increasing numbers of people that the lifestyles pursued by many in the developed world are unsustainable.

Nader N. Chokr

Further Reading

Chambers, Nicky, Craig Simmons, and Mathis Wackernagel. *Sharing Nature's Interest: Ecological Footprints as an Indicator of Sustainability*. 2000. Rpt. Sterling: Earthscan, 2007.

Fiala, Nathan. "Measuring Sustainability: Why the Ecological Footprint Is Bad Economics and Bad Environmental Science." *Ecological Economics* 67.4 (2008): 519–25.

Rees, William, and Mathis Wackernagel. *Our Ecological Footprint: Reducing Human Impact on the Earth*. Gabriola Island, Canada: New Society, 1996.

Vale, Brenda, and Robert James Dennis Vale. *Living within a Fair Share Ecological Footprint*. New York: Routledge, 2013.

Venetoulis, Jason. "Redefining the Footprint (Footprint 2.0)." *Sustainable Development: Principles, Frameworks, and Case Studies*. Ed. Okechukwu Ukaga, Chris Maser, and Mike Reichenbach. Boca Raton: CRC, 2010.

Venetoulis, Jason, and John Talberth. *Ecological Footprint of Nations: 2005 Update*. Oakland, Calif.: Redefining Progress, 2005.

■ Ecology as concept and in history

FIELDS OF STUDY

Biology; Classical Greek History; Ecology; Ecosystems; Environment; Environmentalism; History of Science; Humanities; Life Sciences; Philosophy

SUMMARY

The concept of ecology covers many broad areas and is the science that studies the relationships among organisms and their biotic and abiotic environments. It emerged as a discipline in the late nineteenth century, gaining prominence in the latter half of the twentieth century as general interest in and awareness of environmental issues increased.

PRINCIPAL TERMS

- **biosphere:** the part of the earth's crust, waters, and atmosphere that supports life
- **British Ecological Society:** a leading organization since 1913, whose stated goal is "to make the best scientific evidence accessible to decision-makers"
- **Ecological Society of America:** a nonprofit organization founded in 1915 whose stated goals are to to promote ecological science by improving communication among ecologists, raise public awareness; increase resources for ecological science, and enhance communication between the ecological community and policy-makers"
- **ecology:** from the Greek *oikos*, meaning house, dwelling or habitation; the interdisciplinary scientific study of interactions among organisms and their environments
- **ecosystems:** committed to making the best scientific evidence accessible to decision-makers
- **Gaia hypothesis:** a model of the earth as a self-regulating organism, an alternative to a mechanistic model

Charles Darwin. (Library of Congress).

The global sum of ecosystems is known as the biosphere, and this is where ecological theory has been used to explain regulatory phenomena at the planetary scale. One of the best-known holistic theories to explain this is the Gaia hypothesis, as developed by James Lovelock. Lovelock's thesis, like Aldo Leopold's seminal "land ethic," affirms that the biosphere together with its atmospheric environment forms a single entity or natural system.

Ecology crosses between many disciplines, including ecophysiology, the study of how physiological functions of organisms influence the way they interact with the environment; eco-mechanics, which uses physics and engineering principles to examine the interactions of organisms with their environment and with other species; behavioral ecology, which examines the role of behavior in enabling a species to adapt to its environment; community ecology, or synecology, which explores the interrelationships among species within ecological communities; and ecosystem ecology, which is concerned with the flows of energy and matter through the components of the ecosystem.

The study of ecological topics arose in ancient Greece, but these studies were part of a catchall science called natural history. The earliest attempt to organize an ecological science separate from natural history was made by Carolus Linnaeus in his essay *Oeconomia Naturae* (1749; *The Economy of Nature*, 1749), which focused on the balance of nature and the environments in which various natural communities exist. Although the essay was well known, the eighteenth century was dominated by biological exploration of the world, and Linnaeus's science did not develop.

Early Ecological Studies

The study of fossils led some naturalists to conclude that many species known only as fossils must have become extinct. However, Jean-Baptiste Lamarck argued in his *Philosophie zoologique* (1809; *Zoological Philosophy*, 1914) that fossils represent the early stages of species that evolved into different, still-living species. In order to refute this claim, geologist Charles Lyell mastered the science of biogeography and used it to argue that species do become extinct and that competition from other species seems to be the main cause. In his book *On the Origin of Species by Means of Natural Selection: Or, The Preservation of Favoured Races in the Struggle for Life* (1859) English naturalist Charles Darwin blends his own research with the influence of Linnaeus and Lyell in order to argue that some species do become extinct, but existing species have evolved from earlier ones. Lamarck had underrated and Lyell had overrated the importance of competition in nature.

Although Darwin's book was an important step toward ecological science, Darwin and his colleagues mainly studied evolution rather than ecology. However, German evolutionist Ernst Haeckel realized the need for an ecological science and coined the name *oecologie* in 1866. It would be another three decades before steps were actually taken to organize this science. Virtually all of the early ecologists were specialists in the study of particular groups of organisms, and it was not until the late 1930's that some efforts were made to write textbooks covering all aspects of ecology. Since the 1890's most individual ecologists have viewed themselves as plant ecologists, animal ecologists, marine biologists, or limnologists.

Nevertheless, general ecological societies were established. The first was the British Ecological Society, which was founded in 1913 and began publishing the *Journal of Ecology* in the same year. Two years later, ecologists in the United States and Canada founded the Ecological Society of America, which launched the journal *Ecology* in 1920. The British Ecological Society and the Ecological Society of America have been the leading organizations in ecology ever since, though other national and regional societies have also been established. More specialized societies and journals also began appearing; for example, the Limnological Society of America was established in 1936 and expanded in 1948 into the American Society of Limnology and Oceanography. It publishes the journal *Limnology and Oceanography*.

Although Great Britain and Western Europe were active in establishing ecological sciences, it was difficult for their trained ecologists to obtain full-time employment that utilized their expertise. European universities were mostly venerable institutions with fixed budgets; they already had as many faculty positions as they could afford, and these were all allocated to the older arts and sciences. Governments employed few, if any, ecologists. The situation was more favorable in the United States, Canada, and Australia, where universities were still growing. In the United States, the universities that became important for ecological research and the training of new ecologists were mostly in the Midwest. The reason was that most of the country's eastern universities were similar to European ones in being well established with scientists in traditional fields.

Ecology After 1950
Ecological research in the United States was not well funded until after World War II. With the advent of the Cold War, science was suddenly considered important for national welfare. In 1950 the U.S. Congress established the National Science Foundation, and ecologists were able to make the case for their research along with the other sciences. The Atomic Energy Commission had already begun to fund ecological research efforts by 1947, and under its patronage the Oak Ridge National Laboratory and the University of Georgia gradually became important centers for radiation ecology research. (In 1966, five years after its formation, the Institute of Radiation Ecology at the University of Georgia would become simply the Institute of Ecology. In 2007, it would be renamed the Eugene P. Odum School of Ecology in memory of the University of Georgia professor widely regarded as the father of modern ecology.)

Another important source of research funding was the International Biological Program (IBP), which, though international in scope, depended on national research funds. Officially established in 1964, it began operations in 1967 after an extended planning phase. Even though no new funding sources were created after the IBP ended in 1974, its existence meant that more research money flowed to ecologists than in previous years.

Ecologists learned to think big. Computers became available for ecological research shortly before the IBP got under way, and so computers and the IBP became linked in ecologists' imaginations. The first Earth Day in 1970 helped awaken Americans to the environmental crisis, and they expected ecologists to advise on environmental policy. The IBP encouraged a variety of studies, but in the United States, studies of biomes (large-scale environments) and ecosystems were most prominent. The IBP-funded biome studies were grouped under the headings of desert, eastern deciduous forest, western coniferous forest, grassland, and tundra (a proposed tropical forest program was never funded). Even after the IBP ended, a number of the biome studies continued at reduced levels.

Ecosystem studies were also large in scale, at least in comparison with many previous ecological studies, though smaller in scope than biome studies. The goal of ecosystem studies was to gain a total understanding of how individual ecosystems—such as a lake, a river valley, or a forest—work. IBP funds enabled research students to collect data and to use computers to process the data. However, ecologists could not agree on what data to collect, how to compute outcomes, and how to interpret the results. Therefore, thinking big did not always produce impressive results.

Plant and Animal Ecologies
Because ecology is enormous in scope, the discipline was bound to experience growing pains. It arose at the same time as the science of genetics, but because genetics is a cohesive science, it reached maturity much sooner than ecology. Ecology can be subdivided in a wide variety of ways, and any collection of ecology textbooks shows how diversely it is organized by different ecologists. Nevertheless, self-identified

professional subgroups tend to produce their own coherent findings.

Plant ecology progressed more rapidly than other subgroups and has retained its prominence. In the early nineteenth century, German naturalist Alexander von Humboldt's many publications on plant geography in relation to climate and topography were a powerful stimulus to other botanists. By the early twentieth century, however, the idea of plant communities was the main focus for plant ecologists. Henry C. Cowles began his studies at the University of Chicago in geology but switched to botany and studied plant communities on the Indiana dunes of Lake Michigan. He received his doctorate in 1898 and stayed at that university as a plant ecologist. He trained others in the study of community succession.

Frederic E. Clements received his doctorate in botany in the same year from the University of Nebraska. He carried the concept of plant community succession to an extreme by taking literally the analogy between the growth and maturation of an organism and that of a plant community. His numerous studies were funded by the Carnegie Institute in Washington, D.C., and even ecologists who disagreed with his theoretical extremes found his data useful. Henry A. Gleason was skeptical; his studies indicated that plant species that have similar environmental needs compete with each other and do not form cohesive communities. Although Gleason first expressed his views in 1917, Clements and his disciples held the day until 1947, when Gleason's individualistic concept received the support of three leading ecologists. Debates over plant succession and the reality of communities helped increase the sophistication of plant ecologists and prepared them for later studies on biomes, ecosystems, and the degradation of vegetation by pollution, logging, and agriculture.

Animal ecology emerged from zoology. A good illustration of the transition is the career of Stephen A. Forbes, professor of zoology and entomology at the University of Illinois and head of the State Laboratory of Natural History. His responsibilities focused his attention on the practical uses of zoology for agriculture and for fish and wildlife management; he also had a theoretical interest in both evolution and ecology. He brought together these various interests in his 1887 essay "The Lake as a Microcosm."

One important aspect of the early history of animal ecology was the attempt to understand and describe the growth or decline of animal populations mathematically. Mathematics is a universal language, and the fluctuation of animal populations is a universal problem. Therefore, this aspect of ecology developed globally rather than regionally. It was also possible to use the same mathematical methods to study population changes from the standpoints of ecology, evolution, and genetics. This situation promoted a lively exchange and rapid progress in the development of population ecology in the United States, Great Britain, Australia, Italy, and the Soviet Union. The great challenge was to develop equations that could help predict the pattern of population fluctuations. It turned out to be easier to develop mathematical models than to understand or predict the fluctuations of real populations. Nevertheless, these efforts eventually paid off in the ability of fish and wildlife biologists to gauge the level of harvesting that could maintain stable populations versus the level that would cause a population to decline.

Limnology and Marine Ecology
Limnology, the scientific study of bodies of fresh water, is important for managing freshwater fisheries and water quality. The Swiss zoologist François A. Forel coined the term and also published the first textbook on the subject in 1901. He taught zoology at the Académie de Lausanne and devoted his life's research to understanding Lake Geneva's characteristics and its plants and animals. In the United States in the early twentieth century, the University of Wisconsin became the leading center for limnological research and the training of limnologists. There, zoologist Edward Birge and fellow faculty member Chancey Juday pioneered North American limnology with their extensive field studies. The university, which has retained its preeminence in the field, established its Center for Limnology in 1982.

Marine ecology is viewed as a branch of either ecology or oceanography. Early studies were made either from shore or close to shore because of the great expense of committing oceangoing vessels to research. The first important research institute was the Statione Zoologica at Naples, Italy, founded in 1874. Its successes soon inspired the founding of others in Europe, the United States, and other countries. Karl Möbius, a German zoologist who studied oyster beds, was an important pioneer of the community concept in ecology. Great Britain dominated the seas during the nineteenth century and made the first substantial commitment to deep-sea research by equipping the HMS *Challenger* as an oceangoing laboratory that sailed the world's seas from 1872 to 1876. Its scientists collected so many specimens and such quantities of data that they called upon marine scientists in other countries to help them write the fifty large volumes of reports (1885-1895). The development of new technologies and the funding of new institutions and ships in the nineteenth century enabled marine ecologists to monitor the world's marine fisheries and other resources and provide advice on harvesting marine species.

The twentieth century brought advances in deep-sea exploration technology that allowed marine ecologists to gain a much more profound understanding of marine ecosystems. Pioneering work in oceanic acoustic research during World War I was followed by the development of acoustic sounding devices and other deep-marine electronic oceanographic instruments. Naturalist William Beebe and engineer Otis Barton ushered in the era of manned deep-sea exploration in the 1930's with their bathysphere dives. The next decade saw the advent of deep-ocean camera systems. Further advances included the first successful missions of untethered research submersibles in the 1950's and the unmanned remotely operated oceanographic systems that emerged in the 1960's. A landmark discovery came in 1977 when a manned submersible expedition explored the distinctive hydrothermal vent ecosystems along the mid-ocean ridge on the ocean floor near the Galápagos Islands.

Space-Based Ecological Observation

Since the late twentieth century satellite-based monitoring capabilities have provided researchers with invaluable tools for assessing and monitoring ecosystems while enabling them to increase their understanding of the earth as a collection of integrated systems. Remote-sensing instrumentation aboard orbital platforms provides scientists with both localized details and an overall global view of terrestrial and marine ecosystems. Regular, repeated collection of data from an area affords a means for monitoring how that area is changing over time. Because the data are collected in digital form, they are comparatively easy to integrate with other data. Satellite coverage includes remote areas that would be difficult or impossible to monitor by other means.

Since the United States launched the first satellite in the Landsat series in 1972, a host of environmental satellites equipped with a variety of remote-sensing instruments have been deployed by a number of nations. Notable among these is the Earth Observing System (EOS) series of satellites. A project of the U.S. National Aeronautics and Space Administration (NASA), EOS comprises several orbital missions to gather data on various factors influencing the earth's climate system and their interactions. The first EOS satellite, Terra (originally called EOS AM-1), was launched in 1999.

Optical instruments that detect different wavelengths within the ultraviolet, visible, and infrared spectral regions, along with radars, light detection and ranging instruments (lidars), and more, provide researchers with insights into a wide range of earth systems. Remote sensing via satellite can be used to monitor and study surface temperatures, snowfield conditions, soil moisture, the contours of the surfaces and floors of the oceans, ocean temperatures, air chemistry, air quality, water quality, sedimentation and organic matter in bodies of water, thermal pollution, weather patterns, urban influences on local climate, global climate change, alterations in land cover and land use, human activity and infrastructure, terrestrial and marine animal population size and distribution, habitat loss, plant health and response to stressors, crop yields, deforestation, coral reef conditions, algal blooms, droughts, floods, fires, major oil spills, disease outbreaks and pest infestations, and encroachment of invasive species.

In combination with ground-based ecological studies, satellite-based research facilitates investigations exploring how individual ecosystems contribute to larger ecologies and how pervasive factors such as an increase in global temperature would affect those individual ecosystems. At a time when scientists are striving to understand the global impacts of human activity, the ability to view ecosystems and their functions within a worldwide context is critical.

Light Versus Deep Ecology

Ecology that tends to take a "managerial" approach to the environment, with the status quo almost implicitly accepted, is sometimes characterized as light ecology, in contrast with deep ecology, which is considered to be more progressive if somewhat abstract and idealistic. The founding father of deep ecology, Arne Naess, considered it as having fundamental ethical implications and going beyond the transformation of technology and politics to a transformation of humanity.

Contradictory positions between light and dark ecology can initially be reconciled through a focus on the seminal writings of the environmental guru Aldo Leopold. He wrote of how the land ethic rests on a single unifying premise: "that the individual is a member of a community of interdependent parts." Leopold's vision served to enlarge the boundaries of community to include soils, water, plants, and animals. Especially since Leopold's writings were rediscovered during the 1960's, his land ethic thesis has become a central tenet of environmental thinking, and the symbiotic relationship he proposes between human beings and nature has remained the dominant orthodoxy across much ecological thinking.

Philosophical theories of the value of nature and ecology that cut across these deep and shallow divisions can be grouped into three broad categories: anthropocentrism, inherentism, and ecocentrism. Anthropocentrism recognizes nature and ecology in all its manifestations primarily as a resource that simply contributes to human value and can be used in whatever way is beneficial to human beings, as masters of their ecological environment. Inherentism recognizes that the very concept of value is innately human; thus this philosophical notion remains at odds with the tenets of deep ecology. Ecocentrism, in contrast, aims to challenge all ideas that suggest that human beings are somehow exempt from the natural processes and natural laws within which all other animals have to live.

It may be stating the obvious to observe that modern societies are based on anthropocentric principles, but unless this view is challenged, ecocentrists assert, ecological and environmental problems will not be solved. Ecocentrists argue that societies have to move toward a more earth-centered approach, which puts nature first. They point out the hidden social processes underlying developed societies' comfortable lifestyles and the price paid in damage to the natural environment as a result. Some go so far as to suggest, as Philip W. Sutton has, that the Western world is suffering from the "disease of over-consumption." Furthermore, ecocentrists use the well-known theories of German sociologist Ulrich Beck regarding risk management, together with the Gaia hypothesis and other holistic models, to help frame current ecological thinking about the pressing environmental dangers facing the planet, most notably global warming.

Pat Brereton and Frank N. Egerton; updated by Karen N. Kähler

Further Reading

Golly, Frank B. *A History of the Ecosystem Concept in Ecology.* New Haven, Conn.: Yale University Press, 1993.

Leopold, Aldo. *A Sand County Almanac, and Sketches Here and There.* 1949. Reprint. New York: Oxford University Press, 1987.

Molles, Manuel Carl. *Ecology: Concepts and Applications.* 5th ed. Boston: McGraw-Hill, 2009.

Morris, Christopher. "Milestones in Ecology." In *The Princeton Guide to Ecology*, edited by Simon A. Levin. Princeton, N.J.: Princeton University Press, 2009.

National Research Council Space Studies Board. *Earth Science and Applications from Space: National Imperatives for the Next Decade and Beyond.* Washington, D.C.: National Academies Press, 2007.

Odum, Eugene P., and Gary W. Barrett. *Fundamentals of Ecology.* 5th ed. Belmont, Calif.: Thomson Brooks/Cole, 2005.

Porritt, Jonathon. *Seeing Green: The Politics of Ecology Explained.* New York: Blackwell, 1984.

Real, Leslie A., and James H. Brown, eds. *Foundations of Ecology: Classic Papers with Commentaries.* Chicago: University of Chicago Press, 1991.

Taylor, Paul. *Respect for Nature: A Theory of Environmental Ethics.* Princeton, N.J.: Princeton University Press, 1986.

■ Ecotourism

FIELDS OF STUDY

Advocacy; Business; Commerce; Ecology; Economics; Environment; Environmentalism; Ethics; Policy; Production; Protest; Tourism

SUMMARY

Supporters claim that ecotourism dollars help save endangered wilderness areas that might otherwise be subject to indiscriminate exploitation of natural resources. Detractors note that even the most conscientious ecotourism can contribute to the destruction of fragile ecosystems and cultures, and poor ecotourism practices can wreak even more havoc.

PRINCIPAL TERMS

- **conservation:** careful preservation and protection of natural resources to prevent exploitation, destruction, or neglect
- **ecosystems:** a group of interconnected elements formed by the interaction of a community of organisms with their environment
- **ecotourism:** environmentally, socially, and culturally responsible recreational travel intended to preserve ecosystems and improve the well-being of local populations
- **Global Sustainable Tourism Council (GSTC):** an agency within the United Nations that establishes and manages global sustainable standards for public policy-makers, destination managers, hotels, and tour operators.
- **Rainforest Alliance:** a nonprofit network of farmers, foresters, communities, scientists, governments, environmentalists, and businesses dedicated to conserving biodiversity and ensuring sustainable livelihoods
- **tourism:** the practice of travelling for recreation
- **UNESCO World Heritage Center:** the coordinator within the United Nations for all matters related to cultural and natural legacies from the past, what exists in the present and what will be passed on to future generations worldwide considered irreplaceable sources of life and inspiration
- **United Nations World Tourism Organization (UNWTO):** the leading international tourism organization, which promotes tourism as a driver of economic growth, inclusive development and environmental sustainability

Improvements in travel after World War II, especially the development of jet aircraft, dramatically increased the numbers of tourists in all areas of the globe. With this trend came increased interest in visiting exotic locations to enjoy unspoiled landscapes, view unusual wildlife, and participate in recreational adventures. According to the United Nations World Tourism Organization (UNWTO), by the late twentieth century tourism had become a main income source and the top export category for many developing countries.

The rise in tourism as a leisure activity brought economic benefits such as development and employment opportunities to many pristine areas, but it was sometimes accompanied by negative social, cultural, and environmental impacts. Local communities and lifestyles were sometimes displaced, and ecosystems were altered with the building of hotels, roads, and other amenities for guests. The growing numbers of tourists threatened the very vistas and animals that lured visitors in the first place.

Despite these problems, environmentalists recognize tourism as a means to benefit preservation efforts. While low-income countries or regional governments have the option of exploiting their natural resources to provide revenue, preserving those resources can provide an ongoing alternative source of income—tourist dollars—that gives governments an incentive to protect wilderness areas. Coupled with a "no-impact" ethic, ecotourism is seen as a method of saving ecosystems that are quickly disappearing. This can also foster a conservation mindset among the local population and visitors.

Ideally, ecotourism operations employ practices that have minimal negative impacts on the environment and local cultures. Tours focus on natural destinations and rotate the routes they travel and the sites they visit. Participants gain an understanding of their surroundings and how human activity—including their own—affects the ecosystem. Local communities and indigenous populations are involved in managing ecotourism and reap economic benefits from it. The revenues produced by ecotourism are used to help preserve the natural environment.

Ecotourism proponents point to the regions that have successfully used ecotourism to preserve environments and support local communities. Ecotourism in the Ecuadoran rain forest helped staved off oil exploration and provided income to native peoples in the area. A former director of a mountain gorilla project in Africa credits ecotourism with the survival of mountain gorillas and their habitats; gorilla ecotourism has also provided significant revenue for local communities. In Costa Rica, the market demand for pristine wilderness has led to the establishment of national parks and protected areas over more than 25 percent of the nation's territory by 2014. In Kenya, hundreds of millions of annual tourist dollars provide a powerful incentive to ensure the survival of the country's elephant and rhinoceros populations.

Negative Impacts

Ecotourism is not without its drawbacks. Observers in Costa Rica, for example, have noted that although some national parks are large, most visitors want to see specific sites, which leads to overcrowding, trail erosion, and pollution at those sites. Also, scientists have noted changes in the behavioral patterns of local wildlife that appear to be linked to human activity. In Africa, the proximity of ecotourist groups to mountain gorillas puts the great apes at risk from human infectious diseases such as measles, polio, influenza, and tuberculosis.

Growth in ecotourism also promotes development outside protected areas, with attendant environmental degradation. In addition, not all of the people who participate in ecotourism activities have a deep understanding of the no-impact philosophy and a full appreciation of its importance; some of these people contribute to negative impacts through their actions in sensitive areas.

To complicate matters, some purported ecotourism is little more than greenwashed tourism. The burgeoning popularity of ecotourism has led to a proliferation of companies offering purported ecotours that actually fail to employ sustainable practices. In the absence of regulation or even consensus on what constitutes ecotourism, some operators sell their products as ecotours although they do not meet the standards of the term as it is usually understood. One Costa Rican tourism project touted as an ecodevelopment included environmentally unfriendly amenities such as a shopping center and a golf course.

Studies indicate that local communities often do not benefit from activities in their surrounding areas touted as ecotourism. In many countries, foreign interests own tourist facilities and recreational sites, thus ensuring that profits flow out of the local area. Environmentally insensitive tourism can displace native populations into marginal lands or drive them from a subsistence lifestyle into poverty-wage service jobs. In Nepal, local families earn little money while serving as porters for tourists. In areas where locals do profit, problems can still arise. Some communities in Costa Rica, for example, have moved from a subsistence to a market economy, a transition that belies the ethic of maintaining the integrity of local cultures.

Indigenous tribal communities in South America, Asia, and southern and eastern Africa, have been forcibly displaced from their homelands, which were conserved for ecotourism. Conservation measures such as protections against poaching can result in indigenous peoples being barred from traditional hunting or fishing grounds.

Critics maintain that the concept of ecotourism is inherently flawed. They argue that ecotourists merely pave the way for mass tourists, people who demand the comforts of home, such as hot showers, electricity, and plastic shopping bags, while they visit remote areas. Moreover, the developing nations that offer ecotourist attractions are often the least able to invest the funds necessary to counter the negative impacts of tourism. Only a small percentage of tourist dollars may go toward the management of natural resources.

Opponents of ecotourism assert that it is merely a variant of tourism that will inevitably despoil the very areas it is intended to protect. The deluge of tourists visiting the Galápagos Islands, for example, has overwhelmed the Ecuadoran government's ability to manage them. The annual number of visitors to the islands surpassed the government's target limit of 25,000 people decades ago; by 2015, the number of visitors per year had swelled to more than 200,000.

Economic development to accommodate the tourist traffic caused appreciable damage to the fragile island environment, while tourists and the vessels that transport them have brought exotic, potentially invasive species with them inadvertently. In 2007, the UNESCO World Heritage Centre added the Galápagos Islands to its list of endangered sites in part because of the damage done by tourism, though the islands were removed from the list in 2010. Environmental advocates recommend that potential ecotourists carefully review the literature of any organization that offers ecotours to be sure that its practices and philosophy are in keeping with the goals of environmental and cultural preservation.

Emerging Standards

Interest in ecotourism's role in sustainable development and concerns regarding the detrimental effects of ecotourism's mismanagement led to the first World Ecotourism Summit, held in Quebec, Canada. A joint initiative of the UNWTO and the United Nations Environment Programme (UNEP), the summit was held in 2002, designated by the United Nations as the International Year of Ecotourism. The summit laid the groundwork for the Global Sustainable Tourism Criteria (GSTC), introduced in 2008 by the United Nations Foundation, the UNWTO, UNEP, and the Rainforest Alliance. The first international criteria for sustainable tourism practices, these voluntary standards are based on four key elements of sustainable tourism: effective sustainability planning, maximum social and economic benefits for local communities, minimum negative impacts on cultural heritage, and minimum negative impacts on the environment. The criteria are meant not only for ecotourism but also to guide the tourism industry in general toward sustainable practices.

In 2010 the Global Sustainable Tourism Council (GSTC), a global membership body sponsored in part by the UNWTO, began developing an accreditation program for the world's existing ecotourism certification bodies to bring ecotourism businesses into compliance with universal standards. The GTSC's accreditation criteria, the second version of which was published in 2013, use measurable indicators of environmental and socioeconomic impacts, cultural affects, and sustainability—such as electricity and energy consumption per serviced area, freshwater consumption and waste production per guest per night, and the quality of water discharged from on-site wastewater treatment facilities—to distinguish true ecotourism businesses from greenwashed enterprises. The GSTC has also devised sets of criteria for hotels, guides, and tourism destinations themselves.

Thomas Clarkin; updated by Karen N. Kähler

Further Reading

Fennell, David A. *Ecotourism.* New York: Routledge, 2008.

France, Lesley. *The Earthscan Reader in Sustainable Tourism.* 1997. Sterling, U.K.: Earthscan, 2002.

Ghazali, Suriati, and Morshidi Sirat. *Global Ecotourism and Local Communities in Rural Areas.* Pulau Pinang, Malaysia: University Sains Malaysia Press, 2011.

Honey, Martha. *Ecotourism and Sustainable Development: Who Owns Paradise?* Washington, DC: Island Press, 2008.

McLaren, Deborah. *Rethinking Tourism and Ecotravel.* Boulder, Colo.: Kumarian Press, 2003.

Miller, Andrew P. *Ecotourism Development in Costa Rica: The Search for Oro Verde.* Lanham, Md.: Lexington Books, 2012.

Patterson, Carol. *The Business of Ecotourism: The Complete Guide for Nature and Culture-Based Tourism Operators.* Bloomington, Ind.: Trafford Publishing, 2007.

Schellhorn, Matthias. "Development for Whom? Social Justice and the Business of Ecotourism." *Journal of Sustainable Tourism* 48.1 (2010): 115–36.

Weaver, David B. *Ecotourism.* Hoboken, N.J.: Wiley Press, 2008.

_____, ed. *The Encyclopedia of Ecotourism.* Wallingford, England: CAB International, 2001.

Websites

PRI.
www.pri.org/stories/2017-10-19/tourism-harming-gal-pagos-islands.

The Guardian.
www.theguardian.com/global-development/2016/aug/28exiles-human-cost-of-conservation-indigenous-peoples-eco-tourism.

Electronic waste

FIELDS OF STUDY
Commercial Products; Ecology; Emissions; Environment; Environmentalism; Hazardous Materials; Industries; Pollution; Technology and Applied Science; Toxic Waste

SUMMARY
As consumers replace outdated or broken electronic equipment or appliances, the old items are discarded, and many end up in landfills. These discarded electronic devices often contain hazardous materials that can leach into the environment.

PRINCIPAL TERMS
- **arsenic:** a grayish-white element having a metallic luster, vaporizing when heated, and forming poisonous compounds
- **cadmium:** a highly toxic silver-white, malleable, ductile, metallic chemical element occurring as a sulfide or carbonate in zinc ores
- **electronic waste:** electronic equipment or parts of equipment discarded when broken or obsolete
- **industrial waste:** the waste produced by industrial activity including any material rendered useless during a manufacturing process such as that of factories, industries, mills, and mining operations
- **landfill:** a low area of land built up from deposits of solid refuse in layers covered by soil or the solid refuse itself.
- **planned obsolescence:** a product deliberately designed to have a specific life span

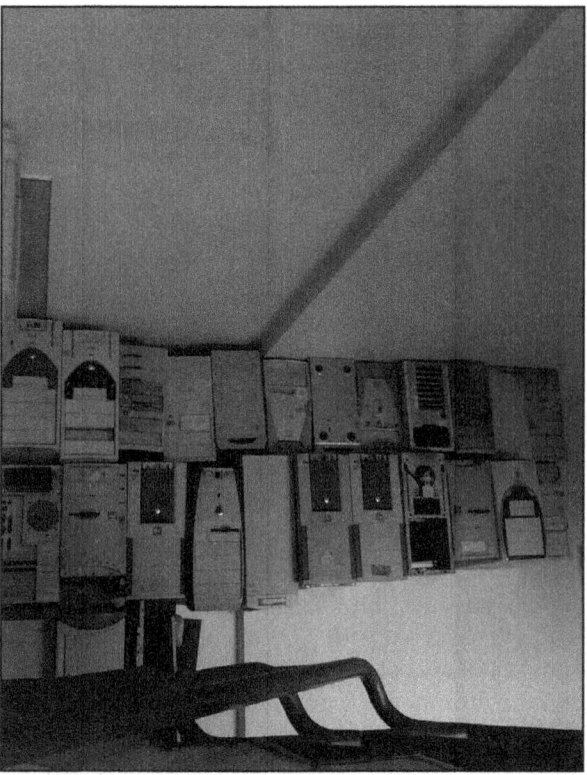

Waste disposal from a computer organization. By Shyamlal T. Pushpan.

Electronic waste, or e-waste, is generated when consumers discard broken or obsolete electronic equipment or parts of such equipment. Industrial waste that is produced during the manufacture of electronic equipment is also sometimes referred to as e-waste. Electronic waste raises environmental concerns because many electronic devices contain numerous heavy elements, such as lead and mercury, and other toxic elements, such as cadmium and arsenic. Additionally, some devices, such as smoke detectors, can include radioactive elements. Some electronic devices also contain plastics that have been treated with fire-retardant chemicals, many of which can be toxic when released into the environment.

Unlike the disposal of many other consumer products containing hazardous materials, the disposal of electronic devices did not receive regulatory attention until the early twenty-first century. When these devices were first introduced, they existed in limited quantity and were expensive. If such a device began to malfunction or stopped working, it would be repaired. By the late twentieth century, however, electronic devices such as computers, microwave ovens, televisions, and cell phones had become common and comparatively inexpensive. The cost of repairing a broken device was often comparable to the cost of simply replacing it.

The pace of technology, too, was such that newer devices often had more desirable features than older ones, and many consumers elected to replace broken electronics rather than repair them. Some manufacturers began to construct electronic devices that were non-serviceable—that is, they could not be repaired—so consumers would have to buy new ones when the old ones no longer worked. Manufacturers began to make some devices in which batteries could not be changed when they died; some sold devices to consumers with the knowledge that the devices would not function with later generations of the technology. Such planned obsolescence exacerbated the problem of electronic waste.

At first, most electronic waste simply ended up in landfills. However, by the 1990s, public awareness of the environmental hazards posed by electronic waste began to rise. Grassroots movements put pressure on governments and industry to curb the landfilling of electronic waste. According to a 2016 Environmental Protection Agency (EPA) report on the recycling of electronic products, twenty-five states and the District of Columbia had passed laws designed to reduce electronic waste, generally by establishing recycling centers devoted to electronics or by requiring retailers or manufacturers to implement take-back programs for used electronics. A number of states have also banned electronic waste from entering landfills altogether. However, according to a 2009 report by the Environmental Protection Agency, only 25 percent of electronics considered to be ready for disposal are recycled. In a 2016 report, the EPA estimated that as compared to the total US consumer electronics generated in 2014, 41.7 percent was recycled.

Electronic waste is difficult and expensive to recycle. Many electronic devices have materials in them that are valuable if recovered—such as lead, copper, and small amounts of gold, platinum, and silver—but recovering these materials is a labor-intensive process. For this reason, much of the electronic waste collected for recycling in the United States is shipped to developing countries, where unskilled and poorly paid laborers break apart the devices to get to the commercially useful materials. Often this work is done without the kind of oversight and regulation required by the environmental laws of industrialized nations, creating environmental and health hazards in those developing countries.

International efforts to limit the environmental damage done by electronic waste have led some developing countries to limit imports of e-waste or to regulate its disposal, leaving developed nations with fewer options for disposing of their e-waste. Manufacturers and governments have sought ways to raise revenues to address the expensive process of recycling electronic waste. To handle the cost of electronic waste disposal, some companies charge fees to accept old electronics. California has legislated disposal fees that consumers must pay when they purchase new electronic devices.

Raymond D. Benge Jr.

Further Reading

Consumer Reports. "Where to Recycle Electronics, Free." June 2009, 11.

Hester, Ronald E., and Roy M. Harrison, eds. *Electronic Waste Management.* Cambridge, England: RSC, 2009.

Jozefowicz, Chris. "Waste Woes." *Current Health*, January 2, 2010, 24–27.

Sthiannopkao, Suthipong, and Ming Hung Wong. "Handling E-Waste in Developed and Developing Countries: Initiatives, Practices, and Consequences." *Science of the Total Environment* 463–464 (2013): 1147–153.

United Nations Environment Programme. *Recycling: From E-Waste to Resources.* Berlin: Oktoberdruck, 2009.

Wood, Molly. "Recycling Electronic Waste Responsibly: Excuses Dwindle." *New York Times.* New York Times, 31 Dec. 2014. Web. 30 Jan. 2015.

Websites

Atlantic.
http://www.theatlantic.com/technology/archive/2016/09/the-global-cost-of-electronic-waste/502019/.

United States Environmental Protection Agency
www.epa.gov/sites/production/files/2016-12/documents/electronic_products_generation_and_recycling_2013_2014_11282016_508.pdf. Accessed 22 Mar. 2018.

Endangered species

FIELDS OF STUDY
Biology; Botany; Conservation; Controversies; Ecology; Ecosystems; Environment; Environmentalism; Horticulture; Life Sciences; Marine Biology; Public Policy; Zoology

SUMMARY
Natural causes as well as pollution, habitat fragmentation and destruction, and other environmental stresses imposed by human activity can drive species toward extinction. Once a population's size declines past a certain point, various factors will eventually wipe out the population entirely. Implementing protective policies can save a declining species from extinction and, ideally, enable it to recover and thrive.

PRINCIPAL TERMS
- **Dichloro-diphenyl-trichloroethane (DDT):** a colorless, tasteless, almost odorless crystalline chemical compound, originally developed as an insecticide, ultimately becoming infamous for its environmental impacts.
- **ecosystem:** all the living things in an area and how they affect each other and the environment
- **Endangered Species Act:** one of dozens of US environmental laws passed in the 1970s, designed to protect critically imperiled species from extinction as a "consequence of economic growth and development untempered by adequate concern and conservation"
- **endangered species:** plants and animals whose numbers are so reduced they are at risk of becoming extinct
- **extinction:** the complete local, regional, or global die-off of a species
- **habitat:** the place or environment where a plant or animal naturally or normally lives and grows
- **pollination:** the act of transferring grains from the pollen-bearing (male) part of a flower's stamen to its pistil (female)
- **World Wide Fund for Nature**: an international non-governmental organization founded in 1961, working in the field of the wilderness preservation, and the reduction of human impact on the environment.

Extinction of a species does not occur in a vacuum. Causes, typically environmental, are many and often complex. Likewise, because of the many intricate, interconnected relationships existing within ecosystems, the loss of any member may have a ripple effect, eventually having profound negative results. For example, the extinction of a single insect, bird, or bat species may result in the extinction of one or more plant species dependent on the animal species for pollination. If the plant is a critical item in the diet of certain animals, those too may be adversely affected.

Paul R. Ehrlich and Anne H. Ehrlich introduce their book *Extinction: The Causes and Consequences of the Disappearance of Species* (1981) by referring to fictitious "rivet poppers"—workers whose job it is to remove rivets from the wings of airplanes. The expectation is that many rivets could be removed without the wings falling off. By analogy, the Ehrlichs consider many world leaders—politicians, bureaucrats, industrialists, engineers, religious leaders, and even some scientists—to be rivet poppers. Through their policies and practices, these leaders espouse programs that will, by design or neglect, result in the loss of endangered species. Ecosystems, by their nature, are somewhat redundant: They are likely to continue to function even after the loss of several species. Ecologists refer to this capacity as "resistance." Ecosystems also possess resilience, or the ability to recover after disturbances, including those in which species are lost. However, just as one would not wish to fly in an airplane from which even a few rivets have been removed, it seems only prudent to take reasonable steps to prevent endangered species from becoming extinct.

Extinction is the conclusion of a long, gradual process typically involving a considerable span of time. When a species undergoes a drastic reduction in the extent of its range, accompanied by a reduction in the number of individuals, it may be designated as a rare species. As this trend continues, the species is likely to be considered threatened prior to being recognized as endangered.

Factors Contributing to Species Loss

The issue of species loss is complex and affected by many factors yet does not threaten all species equally. Whether because of their intrinsic nature or environmental conditions, some species are naturally more

Endangered and Threatened Species, 2008

	Mammals	Birds	Reptiles	Amphibians	Fishes	Snails	Clams	Crustaceans	Insects	Arachnids	Plants
Total listings	357	275	119	32	151	76	72	22	61	12	747
Endangered species	325	254	79	21	85	65	64	19	51	12	599
United States	69	75	13	13	74	64	62	19	47	12	598
Other countries	256	179	66	8	11	1	2	—	4	—	1
Threatened species	32	21	40	11	66	11	8	3	10	—	148
United States	12	15	24	10	65	11	8	3	10	—	146
Other countries	20	6	16	1	1	—	—	—	—	—	2

Source: Data from U.S. Department of Commerce, *Statistical Abstract of the United States, 2009*, 2009.
Note: Numbers reflect species listed by U.S. government as "threatened" or "endangered"; actual worldwide totals of species that could be considered threatened or endangered are unknown but are believed to be higher.

predisposed to becoming endangered or extinct than others. As one would expect, species with smaller numbers of individuals are more vulnerable than those with more—though even huge populations can rapidly collapse due to human influence, as in the infamous case of the passenger pigeon. Each species has a critical population size. Once the numbers fall below that size, the species is especially subject to extinction. Natural populations undergo year-to-year fluctuations in numbers; therefore, a small population will "crash" more readily than a large one.

Several categories of animal and plant species are at high risk of becoming endangered or extinct. Among these are species restricted to special habitats. Most such animal or plant species, by becoming tolerant of an unusual situation, lose their ability to compete in a more general one. One example is island species: If threatened by humans, predators, competing exotic species, or diseases, native island species cannot easily escape. A disproportionate number of animals native to islands have become extinct. Large species with low reproductive rates are also at risk. Large species require more space than do smaller species; therefore, the number of large specimens occupying a given area is lower than the number of smaller ones. Also, most large species, whether whales or trees, are likely to reproduce less often than smaller ones. Even when large species are protected, it is difficult for them to increase their numbers.

Neotropical migratory birds such as warblers, orioles, and tanagers winter in tropical Central or South America or the Caribbean and breed in eastern North America. Their migratory pattern is advantageous in that they can take advantage of the availability of summer food in the north while escaping harsh conditions in winter. Migration, however, is a process that is fraught with danger. As the tropical forests in which they spend the winter are destroyed and the temperate forests in which they breed are fragmented, neotropical migrants may be threatened; thus, they are subject to double jeopardy.

Among the other at-risk species are those at the end of long food chains. Animals such as hawks, owls, and various cat species suffer when any of the links in their food chain are affected. Also, they may be more subject to damage by toxic substances such as environmentally persistent pesticides because of chemical amplification along the food chain. Finally, species of economic value are also in a precarious situation. Many animals have been hunted to extinction; an often-cited example is the passenger pigeon. Plants used medicinally, such as ginseng, have been subjected to over-collecting. Regulatory protections are in place to control the harvesting of American ginseng, which has been dug in eastern North America for centuries.

Conservation and Management

In order to preserve biodiversity and not lose species that are important to the health and existence of an ecosystem, wildlife conservation and management practices must be put into effect. There are three basic approaches to wildlife conservation and management: the species approach, the ecosystem approach, and the wildlife management approach.

The species approach involves giving endangered species legal protection, protecting and managing their habitats, propagating species in captivity, and reintroducing species into safe habitats. In 1903 President Theodore Roosevelt established the first wildlife refuge in the United States. The refuge, located on Pelican Island on the east coast of Florida, was developed to protect the brown pelican, which was in decline. (Only in 2009 was the species officially declared to be out of danger.) Since then the National Wildlife Refuge System has grown to more than 550 refuges and other units. Habitats in the United States are also protected through the national park and forest systems and the National Wilderness Preservation System. In addition to the government, private conservation organizations such as the National Audubon Society, the Sierra Club, and the Nature Conservancy have been of tremendous value in acquiring and protecting sensitive landscapes.

According to statistics reported by the International Union for Conservation of Nature and Natural Resources (IUCN), by 2018 over 200,000 protected areas had been established around the world, an increase from approximately 30,000 in 2000. These areas, which include strict nature reserves and wilderness areas, national parks, natural monuments, habitat and species management areas, protected landscapes and seascapes, and managed resource protected areas, represented almost 15 percent of the planet's land surface and 7 percent of total ocean area. The IUCN also maintains a Red List of Threatened Species, ranking the conservation status of individual species on a scale from "least concern" to "extinct"; "endangered" and "critically endangered" are the stages preceding "extinct in the wild."

Other forms of the species approach to saving diversity include gene banks, seed banks, botanical gardens, and zoos. The seeds of many endangered plant species are preserved in climatically controlled environments. The organization Botanic Gardens Conservation International estimates that more than eighty thousand plant species are in cultivation in the world's botanic gardens. Many botanical gardens, such as Kew Gardens in England, are repositories for plant species that are endangered or have even ceased to exist in the wild. Some of these plants are reintroduced into native habitats after being cultivated for decades in these gardens or seed banks.

Egg pulling and captive breeding are two methods that zoos and animal research centers use for preserving endangered animal species. Egg pulling involves collecting eggs from endangered species in the wild and hatching the eggs in zoos or research centers, as was done with California condors beginning in 1983. Endangered species still in the wild are sometimes captured and put into research centers to breed in a controlled environment. When the captive populations become large enough, some of the individuals are reintroduced into protected habitats. The Arabian oryx, a large antelope species that originated in the Middle East, was hunted to extinction in the wild; however, the species survived thanks to captive breeding programs that began in San Diego, Los Angeles, and Phoenix zoos. The oryx has since been reintroduced into its native habitats, although the captive population outnumbers the population in the wild.

The second approach to saving biodiversity is the ecosystem approach, which emphasizes preserving balanced populations of species within their native habitats. It involves establishing legally protected wilderness areas and wildlife reserves. An important part of making sure that the habitat is safe is to eliminate all alien or invasive species from the area. The Minnesota Zoo has formed a partnership with other organizations to help protect certain animals in their native habitats, notably the desert black rhino and Hartmann's mountain zebra. Instead of moving these animals to Minnesota, the zoo supports conservation efforts to study and protect these animals in their native habitats in Namibia.

The third approach to preserving biodiversity is the wildlife management approach. When it is decided which species or group of species will be managed in a given area, a management plan is put into effect. Steps in the plan include investigating and determining the kinds of cover, food, water, and space the targeted species requires. Action is then taken to grow the plants that provide the needed cover and food for the species.

Hunting and International Cooperation
Legal and illegal commercial hunting has led to the extinction or near extinction of many animal species. Despite policies to regulate hunting, poaching

remains a lucrative business, particularly in underdeveloped countries. Some threatened and endangered species are killed for their hides, horns, or other ornamental or medicinal parts, while others are captured and smuggled alive, as there is a market for exotic pets and decorative plants.

Species, primarily game species, are managed through the establishment of laws that regulate hunting and hunting quotas. Hunters are required to have licenses and to use only certain types of hunting equipment and are permitted to hunt only during certain months of the year. Limits are set on the size, number, and sex of animals that can be hunted any game refuge.

Management plans and international treaties have been developed to protect migrating game species, such as waterfowl. In North America, waterfowl such as ducks, geese, and swans nest in Canada during the summer and migrate to the United States and Central America in the fall and winter. The United States, Canada, and Mexico have signed agreements to protect these waterfowl from overhunting and habitat destruction.

Some wildlife refuges in the United States include human-built nesting sites, ponds, and nesting islands for waterfowl. The US National Wildlife Refuge System includes thirty-eight wetland management districts administering more than twenty-six thousand waterfowl production areas, which contribute to the protection of migratory birds. In 1986, amid concerns regarding record lows in waterfowl populations, the United States and Canada entered into an agreement (later joined by Mexico) to attempt to restore the continental waterfowl population and associated habitats.

In July 1975, a wide-reaching treaty to protect endangered species, the Convention on International Trade in Endangered Species of Wild Fauna and Flora (CITES) went into force. This agreement, which by 2018 had been agreed to by 183 parties, extends varying protections to more than thirty-five thousand plant and animal species. Under CITES, some endangered and threatened species cannot legally be commercially traded, either alive or as products. Others can be traded, but only by persons who obtain the proper export licenses. One of the best-known results of the CITES agreement is the 1989 ban (subsequently weakened) on the international trade in ivory. The ban was enacted to halt the decline of the African elephant, which had dwindled in population from 2.5 million animals in 1950 to approximately 350,000 at the treaty's inception. The end of the international ivory trade has been seen as a major factor in improving elephant population levels, although illegal trade and poaching remain problems.

In 1980 the World Wildlife Fund (now the World Wide Fund for Nature), the United Nations Environment Programme, and IUCN developed a world conservation strategy. The plan, which was expanded in 1991, seeks to preserve biological diversity, combine wildlife conservation and sustainable development, encourage rehabilitation of degraded ecosystems, and monitor sustainability of ecosystems. Fifty countries established national conservation programs in response to this plan.

US Laws

Important US laws that control imports and exports of endangered wildlife and wildlife products began with the Lacey Act of 1900, passed in response to an egret population decline resulting from the commercial value of their feathers as decoration. The Lacey Act prohibited transporting live or dead wild animals or their parts across state borders without a federal permit. Later came the Endangered Species Preservation Act of 1966, then the Endangered Species Act (ESA) of 1973, which has been amended several times. The ESA was unique in that, where previous wildlife regulations had focused primarily on game animals, the ESA program focused on identification of all endangered species and populations in order to save biodiversity, regardless of the species' usefulness to humans. The act classifies endangered species as those that are in immediate danger of extinction and threatened species as those that are likely to become endangered in any habitat in the future. Some species are classified as being locally threatened even though they can be found in large numbers in some parts of their former habitats. The US Fish and Wildlife Service is required to prepare a recovery plan for each species that the ESA lists as officially endangered. In 2018 the ESA listings for endangered and threatened US species included over 1,450 animals and 940 plants.

The ESA provides that a listed species cannot be harassed, harmed, pursued, hunted, shot, trapped, killed, captured, or collected, either on purpose or by accident. It further prohibits importing or exporting endangered species, as well as possessing, selling, transporting, or offering to sell any endangered species. Violators face fines and imprisonment. In 1995 the US Supreme Court ruled to extend further protection for endangered species by ruling that habitat essential for species survival must be protected, whether on public or private land.

Critics view the ESA as a major stumbling block to economic progress. A classic example is the delay that occurred in the late 1970s in the construction of Tellico Dam in the mountains of eastern Tennessee because of the presence of the snail darter, a small endangered fish. The dam was ultimately completed, and other populations of the fish were unexpectedly found elsewhere, resulting in the snail darter's being removed from the endangered list. Some people who criticize the ESA assert that business and public interests should take precedence over the protection of wildlife, especially plants and animals of no obvious value to human beings.

Some of the ESA's recovery plans have proved successful. Bald eagles, which numbered only 800 in 1970, were able to rebound to a population of at least 5,000 nesting pairs in the contiguous United States and approximately 70,000 total individuals in North America by 2018, largely because of the US ban on the pesticide Dichloro-diphenyl-trichloroethane (DDT). The American alligator was listed as an endangered species in 1967 after its population declined because of habitat destruction and high demand for alligator meat and products made from alligator hides. Because of ESA protection, the alligator was reestablished in its southern range; it was removed from the endangered list in 1987. Most of the ESA success stories have involved such "charismatic megafauna." Less glamorous endangered species, such as fungi, wildflowers, liverworts, mosses, and insects, receive less attention even though their roles in ecosystems may be more important. In spite of ESA protection, a number of species remain critically endangered, largely because their standing was so precarious by the time they were listed as protected. ESA listed status has not saved some species from going extinct.

Other Protective Measures

The loss of aquatic species has generally attracted less attention than the extinction of land species. With the realization of the importance of healthy freshwater and marine species, however, governments have begun establishing marine preserves. Fishing, construction, tourism, pollution, and other human disturbances are closely regulated and restricted in these areas. The National Marine Sanctuary Program, which was developed in 1972 in the United States, has established thirteen sanctuaries and a number of marine national monuments. The most extensive of these, the Papah naumoku kea Marine National Monument in the northwest Hawaiian Islands, is the single largest conservation area in the United States and the world's largest fully protected marine area. It includes both marine and terrestrial habitats.

International measures to protect species from destruction or exploitation include the 1946 International Convention for the Regulation of Whaling.

Overwhaling worldwide caused a huge decline in whales, from an estimated 4.4 million in 1900 to approximately 1 million by the end of the twentieth century. Overharvesting of whales affected almost every whale species of commercial value. In 1946 the International Whaling Commission (IWC) was established to set annual whaling quotas to prevent commercial overharvesting and the extinction of whales. However, many whaling countries ignored the suggested quotas. In 1971 the United States stopped all commercial whaling and banned imports of all whale products. In 1974 the IWC began to regulate whaling according to the principle of maximum sustainable yield. When a given species of whale—such as the right whale, bowhead whale, or blue whale—fell below the optimal population for such a yield, the IWC issued a ban on hunting that species.

Other international agreements to protect endangered species include the Convention on the Conservation of Migratory Species of Wild Animals (Bonn Convention), which entered into force in 1983, and the Convention on Biological Diversity (CBD), which entered into force in 1993. The Bonn Convention is the only international treaty that focuses on the conservation of terrestrial, marine, and avian migratory species, their habitats, and their migration routes. The CBD, which opened for signature at the 1992 Earth Summit in Rio de Janeiro, Brazil, concerns the significance of biodiversity for future generations, the sovereignty of each nation over its resources, and each nation's need and right to conserve and protect its own biodiversity. The treaty details a plan that directs industrialized countries to help fund projects for the protection of biodiversity within developing countries; it also stresses the right of national governments to decide who may have access to their resources. The treaty provides for a sharing of technologies, particularly biotechnologies that have been developed from plants originating in developing countries, thereby giving developing countries substantial benefits from any technologies that are developed based on theses countries' genetic resources. Agenda 21, a comprehensive international plan of action adopted at the Earth Summit, addresses the need for nations to "promote the rehabilitation and restoration of damaged ecosystems and the recovery of threatened and endangered species."

Thomas E. Hemmerly, Toby Stewart, and Dion Stewart; updated by Karen N. Kähler

Further Reading

Bräutigam, Amie, and Martin Jenkins. *The Red Book: The Extinction Crisis Face to Face.* Gland, Switzerland: IUCN, 2001.

Chiras, Daniel D. "Preserving Biological Diversity." In *Environmental Science.* 8th ed. Sudbury, Mass.: Jones and Bartlett, 2010.

Groom, Martha J., Gary K. Meffe, and Carl Ronald Carroll. *Principles of Conservation Biology.* 3d ed. Sunderland, Mass.: Sinauer, 2006.

Mackay, Richard. *The Atlas of Endangered Species.* Rev. Sterling, Va.: Earthscan, 2009.

McNeely, Jeffrey A., et al. *Conserving the World's Biological Diversity.* Washington, D.C.: Island Press, 1990.

Vié, Jean-Christophe, Craig Hilton-Taylor, and Simon N. Stuart, eds. *Wildlife in a Changing World: An Analysis of the 2008 IUCN Red List of Threatened Species.* Gland, Switzerland: IUCN, 2009.

Wagner, Viqi, ed. *Endangered Species: Opposing Viewpoints.* Detroit: Greenhaven Press, 2008.

Websites

United States Environmental Protection Agency
www.epa.gov/endangered-species/about-endangered-species-protection-program.

The IUCN Red List of Threatened Species, International Union for Conservation of Nature and Natural Resources. www.iucnredlist.org/.

Convention on International Trade in Endangered Species of Wild Fauna and Flora. www.cites.org/eng/disc/what.php.

■ Environmental determinism

FIELDS OF STUDY

Advocacy; Biology; Climatology; Ecogeography; Ecology; Ecosystems; Environment; Environmentalism; Ethics; Life Sciences; Philosophy and History of Science; Policy; Protest

SUMMARY

Environmental determinists hold that human activity, culture, and character are ultimately determined by environmental factors. From this point of view, the environment, rather than human activity, is the dominant party in the relationship between humans and environment.

PRINCIPAL TERMS

- **climate:** changes in the earth's weather patterns
- **environmental determinism:** the theory that features of the environment ultimately determine human culture, character, and societal development
- **free will:** voluntary choice or decision
- **geography:** a science that deals with the description, distribution, and interaction of the diverse physical, biological, and cultural features of the earth's surface

The notion of environmental determinism achieved considerable prominence during the late nineteenth and early twentieth centuries, especially through the work of geographer Ellen Churchill Semple. The "environment" in this context is understood to be the natural physical features of a setting—the geographical features, such as climate and landform, as opposed to factors directly or indirectly the result of human activity (such as social and economic factors). According to environmental determinists, these environmental factors determine the course of human character, action, and cultural development. Semple, for example, wrote that the climate of northern Europe causes the development of an "energetic, provident, serious, thoughtful rather than emotional, cautious rather than impulsive" character in the peoples resident there; in contrast, tropical climates lead to the development of laziness because the warm weather and ready availability of food make survival easy and do not encourage the development of a keen work ethic.

Because of this tendency toward categorization of peoples according to environmental ancestry, environmental determinism became associated with racist and imperialist attitudes. It is important to note, however, that environmental determinism is not a normative moral theory (that is, it is not a theory about how human beings "ought" to behave). It simply attempts to explain differences in human development through differences in environmental ancestry. It passes no judgment on the moral merits or otherwise of any particular character trait or culture and does not in itself justify discrimination or imperialism.

Environmental determinism has been criticized for being overly simplistic and for not providing sufficient support for its central claim. It seems implausible that one set of factors should have an ultimate determinative influence on human development, and environmental determinists have been criticized for providing very little supporting evidence for such a strong claim. Such criticism has led some environmental determinists to concede that other factors play some role, but they still believe that environmental factors are dominant.

Environmental determinism is often thought to deny human free will. The natural environment into which a person is born is not a matter of that person's choosing, and so if environment determines individuals' outlooks, values, and thereby their decisions, then humans are not free agents and not morally responsible for their behavior—they are just victims of their circumstances.

Some of these concerns are misplaced, however. Environmental determinism does not deny that human beings are rational deliberators who weigh and assess options and then make decisions based on those deliberations. Further, environmental determinism need not deny that an individual's decisions play a crucial role in determining the course of that person's character development. The most that an environmental determinist is committed to is that the course of an individual's deliberations and thus the role of those deliberations in the development of the individual's character will have been ultimately determined by environmental factors. That is quite different from the claim that human beings' deliberations are pointless and inert. They are one part of the mechanism by which the environment determines the course of human development. Whether this still challenges the status of humans as fully free, morally responsible agents turns on the finer details of what free will is taken to involve.

Gerald K. Harrison

Further Reading

Sauer, Carl. *Carl Sauer on Culture and Landscape: Readings and Commentaries.* Edited by William M. Denevan and Kent Mathewson. Baton Rouge: Louisiana State University Press, 2009.

Semple, Ellen Churchill. *Influences of Geographic Environment, on the Basis of Ratzel's System of Anthropo-Geography.* 1911. Ann Arbor, Mich.: University Microfilms International, 1993.

Sutton, Mark Q., and E. N. Anderson. *Introduction to Cultural Ecology.* Lanham, Md.: Altamira Press, 2010.

■ Environmental ethics

FIELDS OF STUDY
Advocacy; Ecology; Environment; Environmentalism; Ethics; Philosophy and History of Science; Policy; Protest

SUMMARY
The issue of human beings' relationship to the natural world—including humans' responsibility for stewardship of the land and natural resources, and for protection or preservation of plant and animal life—has long been a part of philosophical inquiry. Points of view in the field of environmental ethics continue to evolve.

PRINCIPAL TERMS
- **animal liberationist:** one who campaigns for animal rights, often using direct action
- **anthropocentrism:** the belief that human beings are the most important entities in the universe

- **ecocentrism:** a philosophy or perspective that places intrinsic value on all living organisms and their natural environment, regardless of their perceived usefulness or importance to human beings
- **ecofeminism:** a movement or theory that applies feminist principles and ideas to ecological issues
- **environmental ethics:** field of inquiry concerned with evaluating the ethical responsibilities humans have for the natural world
- **stewardship:** the careful, responsible management of something entrusted to one's care, in this case, the earth

Clearcutting in Southern Finland.

The ethical responsibilities that humans have for the natural world, including natural resources, have been examined from many, often conflicting, perspectives, including anthropocentrism, individualism, ecocentrism, and ecofeminism. Each perspective has strengths and weaknesses.

Anthropocentrism is a human-centered philosophy that holds that moral values should be limited to humans and should not be extended to other creatures or to nature. A justification for this perspective is that moral relationships are sets of reciprocal rules followed by humans in their mutual relationships. Nonhumans are excluded from moral relationships because they lack comprehension of these rules. Some anthropocentrists argue that, from an evolutionary perspective, successful species should not work for the net good of other species; species that have done so in the past have become extinct.

Some anthropocentrists oppose restrictions on the use of natural resources because such restrictions may have negative impacts—for example, the loss of jobs or the loss of products beneficial to humans. Others stress that the natural world is a critical life-support system for humans and advocate effective environmental controls so that it will maintain its full value for present and future generations. This anthropocentric regard for the environment is based on the practical value of the natural world for meeting human needs rather than on a belief that the natural world has intrinsic value.

Those who hold an individualist philosophy believe that humans should extend moral concern to individual animals of certain species. Individualists include advocates of animal liberation and animal rights. Individualists accept that all humans have intrinsic value; they also argue that because some animals share morally relevant qualities valued in humans, these animals should be extended moral concern. Animal liberationists define the capacity for pleasure and pain (sentience) as the morally relevant feature to be considered. Animal rightists value more complex qualities, including desires, consciousness, a sense of the future, intentionality, and memories; they commonly associate these qualities with most mammals. Individualists generally are not concerned with the use of natural resources unless that use involves a direct threat to individuals of a species deserving moral concern, as through hunting or trapping.

Ecocentrism is based on the belief that the natural world has intrinsic value; it includes both the land ethic and deep ecology perspectives. Land ethic advocates believe that moral concern should be extended to the natural world, including natural units such as ecosystems, watersheds, and bioregions. Land ethic advocates emphasize respect (rather than rights) for the natural world. Ecocentrists may justify a land ethic by noting that all living creatures have a common origin and history on the planet and are ecologically connected and interdependent. The notions of common origin and history, as well as interdependence, are viewed as analogous to the human concept of family. Ecocentrists view humans as members of a large family comprising all of nature. Because family relationships entail not only privileges but also responsibilities for the well-being of the other family members and their environment, it follows that humans have responsibility for the natural world.

Impact on land health is an important criterion by which natural resource use is assessed in a land ethic. Characteristics of land health include the occurrence of natural ecological functioning, good soil fertility, absence of erosion, and having all the original species properly represented at a site (biodiversity). From a land ethic perspective, natural resource use should minimize long-term impacts on land health or should even enhance land health.

Deep ecology is often viewed as an ecosophy—an ecological wisdom that calls for a deep questioning of lifestyles and attitudes. Some guidelines regularly cited by deep ecologists include living lives that are simple in means but rich in ends, honoring and empathizing with all life-forms, and maximizing the diversity of human and nonhuman life.

Ecofeminists believe that many environmental problems are tied to human beings' desire to dominate nature, and this desire is closely linked with the problem of the domination of women and other groups in society. Ecofeminists believe that these problems would decline with a transformation in societal attitudes from dualistic, hierarchical, and patriarchal thinking to an emphasis on enrichment of underlying relationships and greater focus on egalitarian, empathetic, and nonviolent attitudes. Ecofeminism emphasizes less intrusive and more gentle use of natural resources.

Many Westerners have reexamined established cultural and religious perspectives for inspiration and insights in developing an environmental ethic. Native American cultures are often viewed as a source of moral insights on the human relationship to the environment. While it is difficult to generalize, given the many diverse Native American cultures, several perspectives appear common to many Native American groups: a strong sense of identity with a specific geographic feature, such as a river or a mountain; the notion that all of the world is inspirited and has being, life, and self-consciousness; and a strong sense of kinship with the natural world. Such Native American views are commonly associated with reduced environmental impacts and harmonious relationships with the natural world.

Judaism, Christianity, and Islam share common traditions; each contains elements upon which scholars have drawn for insights into environmental responsibility. Some scholars emphasize portions of the biblical book of Genesis, where the world is seen as God's creation for the free use and enjoyment of humans. Subjugation and use of nature are acceptable, but the land also must be appreciated and protected as belonging to God. Others emphasize the special role of humans as caretakers or advocate close relationships to the natural world, as exemplified by Saint Francis of Assisi. Attitudes toward the natural world and the use of natural resources vary widely among different groups of Jews, Christians, and Muslims. Some Eastern philosophies, such as Daoism and Buddhism, contain insights for environmental ethics. Both encourage a caring behavior toward nature.

Richard G. Botzler

Further Reading

Armstrong, Susan J., and Richard G. Botzler, eds. *Environmental Ethics: Divergence and Convergence.* New York: McGraw, 2003.

Barnhill, David Landis and Robert Gottlieb, eds. *Deep Ecology and World Religions: New Essays on Sacred Ground.* Albany, N.Y.: SUNY Press, 2001.

DesJardins, Joseph R. *Environmental Ethics: An Introduction to Environmental Philosophy.* Stamford, Conn.: Thomson, 2006.

Keller, David R., ed. *Environmental Ethics: The Big Questions.* Hoboken, N.J.: Blackwell, 2010.

Pojman, Louis P., and Paul Pojman, eds. *Environmental Ethics: Readings in Theory and Application.* Stamford, Conn.: Thomson Wadsworth, 2008.

Sterba, James P., ed. *Earth Ethics: Introductory Readings on Animal Rights and Environmental Ethics.* Upper Saddle River, N.J.: Prentice Hall, 2000.

Environmental justice and racism

FIELDS OF STUDY
Advocacy; Climatology; Climate Change; Commerce; Controversies; Ecology; Environment; Economics; Ethics; Industry; Labor; Public Policy; Racial Justice

SUMMARY
The Intergovernmental Panel on Climate Change (IPCC) particularly emphasizes a climate justice approach to global warming solutions, as the risks associated with climate change are generally greater for disadvantaged people and communities at all levels of development around the world.

PRINCIPAL TERMS
- **climate justice:** a human-centered approach to addressing climate change that prioritizes the rights of the most vulnerable people and shares the burdens and benefits of climate change equitably
- **environmental racism:** patterns of racial prejudice regarding issues related to the environment, particularly environmental hazards
- **lead poisoning:** chronic intoxication produced by the absorption of lead into the system, characterized by fatigue, abdominal pain, nausea, diarrhea, loss of appetite, anemia, a dark line along the gums, and muscular paralysis or weakness of limbs
- **sustainable development**: development that meets the needs of the present without compromising the ability of future generations to meet their own needs and aims to achieve social equality and protect the environment

The issues of environmental justice and environmental racism have become increasingly important elements in debates regarding industrial and government environmental practices that have impacts on low-income and other disadvantaged communities. Although the goals of a clean environment for all and an end to racist practices are attractive to many people, achieving these goals can be difficult, because human beings tend to desire both justice and manufactured goods that are linked to environmental degradation.

During the 1960s, most Americans involved in environmental activities were white and members of the upper or middle classes. Issues such as conservation and preservation of natural resources were of little interest to members of low-income and minority groups, who were often more concerned with civil rights and the improvement of economic conditions. In the 1970s and 1980s, however, as concerns mounted about lead poisoning, the dangers of hazardous waste dumps, and the effects of soil and water pollution, minority leaders began to take notice. Researchers found that garbage dumps and other contaminated sites were disproportionately located near communities with higher-than-average percentages of minority group residents. A 1983 study of landfill and incinerator sites in Houston, Texas, for example, showed that these facilities were usually found near African American neighborhoods. This and other studies led to a grassroots movement during the 1980s to address the problem known as environmental racism.

Activists contend that racism continues to prompt governments to issue permits for waste facilities in low-income and minority areas. In some cases, communities have welcomed such facilities as sources of employment. In addition, members of these communities are less likely to possess the knowledge and the resources to oppose regulatory decisions. Through public protests and political pressure, activists seek to bring national attention to the problem, with the hope of influencing the decisions of local and national policy makers.

Grassroots organizations can point to some successes. Residents of one California community successfully pressured their town council to implement a program to screen for environmental lead poisoning. When citizens in Halifax County, Virginia, learned that the federal government was considering their community as the location for a nuclear waste depository, they formed a group to fight the proposal. More than fourteen hundred residents, both African American and white, voiced their opposition at a public meeting, and shortly thereafter, government officials dropped the county as a potential depository site. In 1986, the residents of Revelator, Louisiana, received cash settlements and relocation assistance from a nearby chemical manufacturer after they sued the company for damages to health and property caused by emissions from the chemical plant. These and other victories indicate the potential power of grassroots movements that seek environmental justice.

Such successes do not mean that these movements are without their critics, however. Some argue that the concept of environmental justice is so broad and vague that it cannot serve as a guide for policy makers. Moreover, they maintain that the available evidence regarding environmental racism is flawed. They contend that studies have failed to determine whether harmful facilities have been located in already existing minority communities or the communities coalesced around the facilities.

Proponents of environmental justice reject these arguments as further evidence of injustice and racism. They claim that major corporations that hope to maximize profits have a vested interest in attacking the movement, which, if successful, would significantly raise their costs of production and waste disposal. Activists also complain that the national news media consistently ignore environmental racism, favoring instead sensational stories that do not examine the deeper institutional causes of the environmental disasters featured in the headlines. Finally, proponents also fault the mainstream environmental movement for its fixation on preservation issues.

Despite continuing debates over the meaning of environmental justice and the reality of environmental racism, some politicians have perceived the issues involved as deserving of legislation. In 1992, members of the US Congress sponsored the Environmental Justice Act, which would have required the Environmental Protection Agency (EPA) to identify and monitor areas with high levels of toxic chemicals. The measure failed, but the legislators' effort brought heightened attention to the issue. Two years later, President Bill Clinton issued Executive Order 12898, which requires federal agencies to pursue environmental justice and acknowledges the existence of environmental racism. The order has had limited impact, but it has drawn more attention to the issue.

One major problem regarding accusations of environmental racism centers on the matter of proof. Courts require evidence that the alleged racism is intentional, and in most cases, this is impossible to prove. However, in 1997 a federal judge ruled that suits can be filed based on "disparate impact," which means that the effect of racial discrimination, regardless of the intent, can be used to assess responsibility. This decision was a victory for activists, as it gave them increased opportunities to pursue remedies in the courts. However, some observers noted that the ruling would prompt industrial interests, many of which would not be guilty of polluting, to avoid siting their facilities in minority communities out of fear of expensive lawsuits.

The issue of environmental justice has garnered support because the stated goals of advocates—a clean environment for all and an end to racist industrial and government practices—are attractive to many people. Achieving those goals has proven difficult, however. People tend to desire both justice and the manufactured goods that cause environmental degradation. While most people oppose racism, they understandably have no desire to relocate polluting industries into their own neighborhoods. In addition, as the debate about disparate impact indicates, conflicts over environmental racism may have the unintended consequence of denying low-income people jobs that they desire.

Hurricane Katrina and Hurricane Harvey
In 2005, Hurricane Katrina, an extremely destructive and deadly Category 5 hurricane made landfall on Florida and Louisiana and virtually decimated the city of New Orleans and its surrounding areas.

Katrina refocused America's attention on the legacy of racial segregation and poverty in the Gulf South. Statistics show that the storm's impacts weighed more heavily upon racial minorities and the poor and that the recovery of socially and economically vulnerable storm victims has lagged mainstream society. Over the course of many years, racial segregation in this region established patterns of settlement for many African Americans in less desirable flood-prone areas. Industries attracted by cheap land and weak resistance clustered around minority communities, and segregation and poverty forced African Americans into those areas. While the civil rights movement of the 1960s began to dismantle racial segregation in public accommodations and for individual rights, in the 1980s, the environmental justice movement began to address inequities in community health and resource allocation. Despite these efforts, however, Hurricane Katrina encountered a Gulf

South still heavily burdened with social and economic disparities.

In the aftermath of Hurricane Harvey in 2017, which caused a particularly severe amount of damage in the city of Houston, Texas, some commentators focused on the disparate impact of the natural disaster due to social and economic inequality in the city. Those in wealthier neighborhoods had more resources to rely on in surviving and recovering from the hurricane, while those located in low-income neighborhoods (many with communities of color) faced greater struggles; for instance, because these neighborhoods often contain more unstable and even dangerous buildings and facilities, they are more deleteriously impacted by the pollutants spread through excessive flooding. In 2018, the EPA released a report showing evidence that people living in neighborhoods of color are at greater risk for being exposed to polluted air.

Thomas Clarkin

Further Reading

Arnold, Craig Anthony. *Fair and Healthy Land Use: Environmental Justice and Planning*. American Planning Association, 2007.

Bullard, Robert D. and Beverly Wright. *Race, Place and Environmental Justice After Hurricane Katrina: Struggles to Reclaim, Rebuild, and Revitalize New Orleans and the Gulf Coast*. Boulder, Colo.: Westview Press, 2009.

Checker, Melissa. *Polluted Promises: Environmental Racism and the Search for Justice in a Southern Town*. New York: New York University Press, 2005.

Pellow, David Naguib, and Robert J. Brulle, editors. *Power, Justice, and the Environment: A Critical Appraisal of the Environmental Justice Movement*. Cambridge, Mass.: The MIT Press, 2005.

Rhodes, Edwardo Lao. *Environmental Justice in America: A New Paradigm*. Bloomington, Indiana: Indiana University Press, 2003.

Sandler, Ronald, and Phaedra C. Pezzullo, editors. *Environmental Justice and Environmentalism: The Social Justice Challenge to the Environmental Movement*. Cambridge, Mass.: The MIT Press, 2007.

Shallcross, Tony, and John Robinson, editors. *Global Citizenship and Environmental Justice*. Amsterdam, Netherlands: Rodopi, 2006.

Websites

The Atlantic
www.theatlantic.com/politics/archive/2018/02/the-trump-administration-finds-that-environmental-racism-is-real/554315/.

Time
time.com/4923381/hurricane-harvey-human-disaster/.

■ Erosion

FIELDS OF STUDY

Agriculture; Animal Husbandry; Biology; Botany; Conservation Biology; Conservation; Ecology; Ecosystems; Environment; Environmental Engineering; Environmentalism; Horticulture; Life Sciences; Renewable Resources

SUMMARY

The loss of soil fertility through erosion is incalculable, as are the secondary effects of pollution of surrounding waters and increase of sedimentation in rivers and streams.

The control of erosion is vital because soil loss is one of the most serious threats to world food security.

PRINCIPAL TERMS

- **erosion**: the loss of topsoil through the actions of wind and water, and the efforts undertaken to mitigate such loss
- **eutrophication:** a body of water overly enriched with minerals and nutrients which induce excessive growth of plants and algae
- **herbicides:** a substance or preparation for killing plants, especially weeds.
- **nitrates:** a chemical compound that includes nitrogen and oxygen and used as fertilizers in agriculture
- **pesticides:** a chemical preparation for destroying plant, fungal, or animal pests.
- **phosphates:** a fertilizing material containing compounds of phosphorus
- **topography:** the natural features of land, especially. the shape of its surface, or the science of mapping those features
- **topsoil:** the fertile, upper part of the soil

In the United States alone, some two billion tons of soil erode from cropland on an annual basis. About 60 percent, or 1.2 billion tons, is lost through water erosion, and the remainder is lost through wind erosion. This is equivalent to losing 0.3 meter (1 foot) of topsoil from 810,000 hectares (2 million acres) of cropland each year. Although soil is a renewable resource, soil formation occurs at rates of just a few inches per hundred years, which is much too slow to keep up with erosive forces. The loss of soil fertility through erosion is incalculable, as are the secondary effects of pollution of surrounding waters and increase of sedimentation in rivers and streams.

Erosion removes the topsoil, the most productive soil zone for crop production and the plant nutrients it contains. Erosion thins the soil profile, which decreases a plant's rooting zone in shallow soils, and can disturb the topography of cropland sufficiently to impede the operation of farm equipment. Through erosion, nitrates, phosphates, herbicides, pesticides, and other agricultural chemicals are carried into surrounding waters, where they contribute to cultural eutrophication. Erosion also causes the deposit of increased sedimentation in lakes, reservoirs, and streams, which eventually require dredging.

There are several types of wind and water erosion. The common steps in water erosion are detachment, transport, and deposition. Soil particles become detached from soil aggregates, and the particles are carried, or transported, away; in the process, the particles scour new soil particles from aggregates. Finally, the soil particles are deposited when the water flow slows. In splash erosion, raindrops impacting the soil can detach soil particles and hurl them considerable distances. In sheet erosion, a thin layer of soil is removed by tiny streams of water moving down gentle slopes. This is one of the most insidious forms of erosion because the effects of soil loss are imperceptible in the short term. Rill erosion is much more obvious because small channels form on a slope. These small channels can be filled in by tillage. In contrast, ephemeral gullies are larger rills that cannot be filled by tillage. Gully erosion is the most dramatic type of water erosion. It leaves channels so deep that even equipment operation is prevented. Gully erosion typically begins at the bottoms of slopes, where the water flow is fastest, and works its way with time to the top of a slope as more erosion occurs.

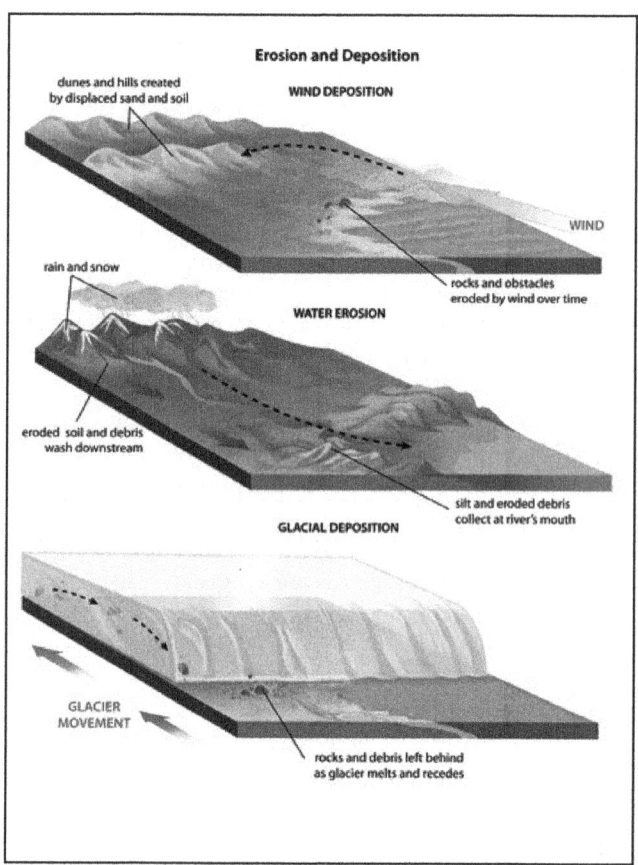

Erosion & Deposition: Erosion and deposition are earth processes which break down and transport substances or structures. Soil is an example of a substance whose compounds may be dissolved by outside forces. An example for a structure is a mountain range whose surface breaks away. Many forces initiate the physical and chemical changes and cause particles and structures to move. They include the wind, water, glaciers, precipitation, extreme weather, earth movements, and more. © EBSCO

Wind erosion generally accounts for less soil loss than does water erosion, but in states such as Arizona, Colorado, Nevada, New Mexico, and Wyoming, it is actually the dominant type of erosion. Wind speeds 0.3 meter (1 foot) above the soil that exceed 16 to 21 kilometers (10 to 13 miles) per hour can detach soil particles. These particles, typically fine to medium-size sand grains fewer than 0.5 millimeters (0.02 inches) in diameter, begin rolling and then bouncing along the soil, progressively detaching more and more soil particles by impact. The process, called saltation, is responsible for 50 to 70 percent of all wind erosion. Larger soil particles are too big to become suspended and continue to roll along the soil. Their movement is called surface creep.

The most obvious display of wind erosion is called suspension, when very fine silt and clay particles detached by saltation are knocked into the air and carried for enormous distances. The Dust Bowl of the 1930's was caused by suspended silt and clay in the Great Plains of the United States. It is also possible to see the effects of wind erosion on the downward sides of fences and similar obstacles. Wind passing over these obstacles deposits the soil particles it carries. Other effects of wind erosion are tattering of leaves, filling of road and drainage ditches, wearing of paint, and increasing incidence of respiratory ailments.

Control Measures

The four most important factors affecting erosion are soil texture and structure, roughness of the soil surface, slope steepness and length, and soil cover. Several passive and active methods of erosion control involve these four factors. Wind erosion, for example, is controlled through the creation of windbreaks, rows of trees or shrubs that shorten a field and reduce the wind velocity by about 50 percent. Tilling the land perpendicular to the wind direction is also a beneficial practice, as is keeping the soil covered by plant residue as much as possible.

Water erosion is controlled through several practices. In the United States, farmers can get help from the federal government-sponsored Conservation Reserve Program to protect highly erosive, steeply sloped land. Tilling land along the contours of slopes aids in preventing erosion, as does the shortening of long slopes by terracing, which also reduces the slope steepness. Permanent grass waterways can be planted in areas of cropland that are prone to water flow. Likewise, grass filter strips can be planted between cropland and adjacent waterways to impede the velocity of surface runoff and cause suspended soil particles to sediment and infiltrate before they can become contaminants.

Conservation tillage practices such as minimal tillage and no-till farming, or zero tillage, have been widely adopted by farmers as a simple means of erosion control. As the names imply, these are tillage practices in which as little disruption of the soil as possible occurs and in which any crop residue remaining after harvest is left on the soil surface to protect the soil from the impacts of rain and wind. The surface residue also effectively impedes water flow, which results in less suspension of soil particles. Because the soil is not disturbed, practices such as no-till farming also promote rapid water infiltration, which reduces surface runoff. Zero tillage is rapidly becoming the predominant practice in southeastern states such as Kentucky and Tennessee, where rainfall levels are high and erodible soils occur.

Mark Coyne

Further Reading

Blanco, Humberto, and Rattan Lal. *Principles of Soil Conservation and Management.* New York: Springer, 2008.

Faulkner, Edward. *Plowman's Folly.* Norman: University of Oklahoma Press, 1943.

Field, Harry L., and John B. Solie. "Erosion and Erosion Control." In *Introduction to Agricultural Engineering Technology: A Problem-Solving Approach.* New York: Springer, 2007.

Plaster, Edward. *Soil Science and Management.* Clifton Park, N.Y.: Delmar Cengage Learning, 2008.

Schwab, Glen, et al. *Soil and Water Conservation Engineering.* Clifton Park, N.Y.: Delmar Cengage Learning, 2005.

■ Estuaries

FIELDS OF STUDY

Climate; Coastal Engineering; Ecology; Environment; Erosion Control; Fisheries Science; Hydrology; Land-Use Management; Oceanography; Sustainable Development

SUMMARY

Estuaries are ecosystems that support a rich variety of plant and animal species, as well as provide ideal conditions for human communities. However, the health of estuaries is delicate, and stress from pollution or human manipulation can resonate throughout the estuary ecosystem with negative impacts on the plants, animals, and humans who live there.

PRINCIPAL TERMS

- **ecosystem:** a biological community of interacting organisms and their physical environment

- **estuary:** the tidal mouth of a river
- **pollution:** the introduction of waste such as garbage, sewage, or pesticides into an ecosystem
- **salinity:** the amount of salt in a body of water
- **sustainable development:** economic development that is conducted without depletion of natural resources

An estuary is a coastal body of water where a river meets a tidal body of water. There are four different types of estuaries: coastal-plain estuaries, which are formed when rising sea levels fill in an existing river valley; tectonic estuaries, which result from the shifting of the Earth's crust; bar-built estuaries, which are defined by the presence of a barrier island or sandbar that protects the estuary from the ocean; and fjord estuaries, which are created when glaciers recede and the ocean fills the carved-out valley left behind. Estuaries are frequently characterized by their brackish water, a mix of fresh and salt water, and their role in maintaining ocean ecosystems by filtering out sediments from rivers.

Estuaries are also among the most productive ecosystems on earth, and they are home to many species of plants and animals that are specifically adapted for life in a coastal, brackish environment. Some species, such as horseshoe crabs, salmon, or striped bass, only spend part of their lifecycles in estuaries, where they feed or reproduce, but others, like oysters, live there permanently. Estuaries have also historically provided ideal sites for human settlement, as they provide ports that facilitate trade on the ocean as well as access to freshwater. Many major cities have developed around estuaries, such as New York City and Tokyo.

However, human activity can also create problems for estuary ecosystems through pollution and overfishing. Changes in the balance between freshwater and saltwater in an estuary can also cause problems for the species who live there. These changes can be due to drought, flood, or water extraction for human communities or agriculture, and they lead to vegetation transition or reductions in the populations of oysters or crabs.

The Chesapeake Bay, which spans the Atlantic Coast of the United States from Maryland to southeast Virginia, demonstrates how these forces can threaten estuarine environments. The Chesapeake Bay is the largest estuary in the United States, and several large rivers flow into it, including the Susquehanna, the Chester, the Potomac, the Rappahannock, and the James. More than 3,600 species live within the Chesapeake Bay and its surrounding watershed, including about 350 species of fish, 173 species of shellfish, and numerous species of birds, mammals, reptiles and amphibians, bay grasses, and lower-food-web species, including both bottom-dwelling and free-floating plant and animal communities. The Chesapeake Bay is also widely known for its thriving seafood industry; however, since the 1970s, overfishing and deteriorating environmental conditions in the bay have caused decreases in the populations of fish, other wildlife, and plants in the watershed.

The Chesapeake Bay has experienced environmental pressures related to population growth, land-use policies, air and water pollution, overfishing, invasive species, and climate change. An estimated 18.1 million people lived within the Chesapeake Bay watershed as of 2016, and this population is predicted to reach 20 million by 2030. This large population contributes to environmental stress through the development of homes, businesses, and infrastructure, which destroys habitats and increases the pollutants entering the bay. Excess nutrients and sediment from agricultural and industrial runoff have contributed to marine dead zones in the bay, areas where oxygen has been depleted and vital sunlight cannot reach bottom-dwelling organisms. Efforts have been undertaken to improve the health of the Chesapeake Bay by restoring water quality through more careful management of land use, updating wastewater treatment plants, reducing harmful pollutants in agriculture and development, restoring bay grass and wetland habitats, improving fishery management, establishing stewardship and education programs, and enacting protective legislation.

The Chesapeake Bay is also vulnerable to the impacts of climate change. The water temperature in the bay has already increased 1.2 °F between 1960 and 2014, and the water level is expected to rise between 1.3 and 5.2 feet before the end of the twenty-first century. Rising sea levels lead to the inundation of coastal ecosystems, which also make human communities more vulnerable to storms. Throughout the Chesapeake Bay watershed, species like the brook trout are threatened by increases in water temperature. To mitigate some of these threats, officials from the six watershed states, the District of Columbia, and the federal government signed the *Chesapeake Bay Watershed Agreement* in 2014, which established shared goals for reducing contamination in the bay and improving its water quality.

Julia Kendrick and Courtney A. Smith

Further Reading

Blazer, Vicki, et al. *Technical Report December 2012: Toxic Contaminants in the Chesapeake Bay and Its Watershed—Extent and Severity of Occurrence and Potential Biological Effects.* US Environmental Protection Agency, US Geological Survey, and US Fish and Wildlife Service, 2012.

Crane, Julian R. and Ashton E. Solomon, editors. *Estuaries: Types, Movement Patterns and Climatical Impacts.* New York: Nova Science Publishers, Inc., 2010.

Kennish, Michael J. *Encyclopedia of Estuaries.* New York: Springer, 2016.

F

■ Fisheries

FIELDS OF STUDY
Biology; Business; Commerce; Ecology; Economics; Ecosystems; Environment; Environmentalism; Food; Hunger; Life Sciences; Limnology; Marine Biology; Oceanography; Zoology

SUMMARY
The status of the world's fisheries is a chief concern of many environmental nonprofit organizations, governmental agencies, and international entities such as the United Nations and multinational trade organizations because of fisheries' crucial importance in protecting marine environments. The collapse of fisheries poses serious threats to biodiversity, local and global economies, food security, and the ecological health of the oceans and of small fresh- and saltwater bodies.

PRINCIPAL TERMS
- **aquaculture:** underwater agriculture that cultivates aquatic animals and plants in natural or controlled marine or freshwater
- **coral reefs:** stony formations created by the depositing of exoskeletons by colonies of certain marine polyps
- **environmental justice:** the fair treatment and meaningful involvement of all people regardless of race, color, national origin, or income in the development, implementation, and enforcement of environmental laws, regulations, and policies
- **fisheries:** locations where one or more species of fish are caught by human fishers
- **mangrove:** a shrub or small tree that grows in coastal saline or brackish water and that occur worldwide in tropic or subtropic areas
- **marine ecosystem:** the interaction of plants, animals, and the sea or oceans environments
- **seagrass:** any of various grass-like marine plants that grow underwater in salt water
- **UN Food and Agriculture Organization (FAO):** a specialized agency of the United Nations that leads international efforts to defeat hunger

Fisheries, in the most basic way, signify the relationship between humans and fish. Since the dawn of civilization, humans have taken fish for consumption from all accessible bodies of water, traditionally without much concern about depleting resident populations. In modern times, however, overfishing has led to the collapse of major consumed fish species from stocks that once seemed ever-replenishing. Perhaps the most poignant example is the Northern Atlantic cod fishery collapse during the early 1990s. Although measures were taken to limit catch size some thirty years previously, and most of the fishers in the region obeyed the new rules, the fishery still collapsed. This event was seen as a warning signal to the world; it caused policy makers to take greater notice of the epidemic of overfishing and sparked worldwide efforts to take more effective measures to protect fisheries as important resources.

Overfishing is a serious threat to biodiversity, local and global economies, food security, and the ecological health of the oceans and of small fresh- and saltwater bodies. According to the United Nations Food and Agriculture Organization (FAO), at least 30 percent of the world's fisheries had been completely exploited—that is, overfished—by 2018, and approximately 90 percent were fully fished. It has been predicted that stocks of all fished species will collapse near the middle of the twenty-first century if trends continue. The proper management of fisheries is essential to the maintenance of sustainable

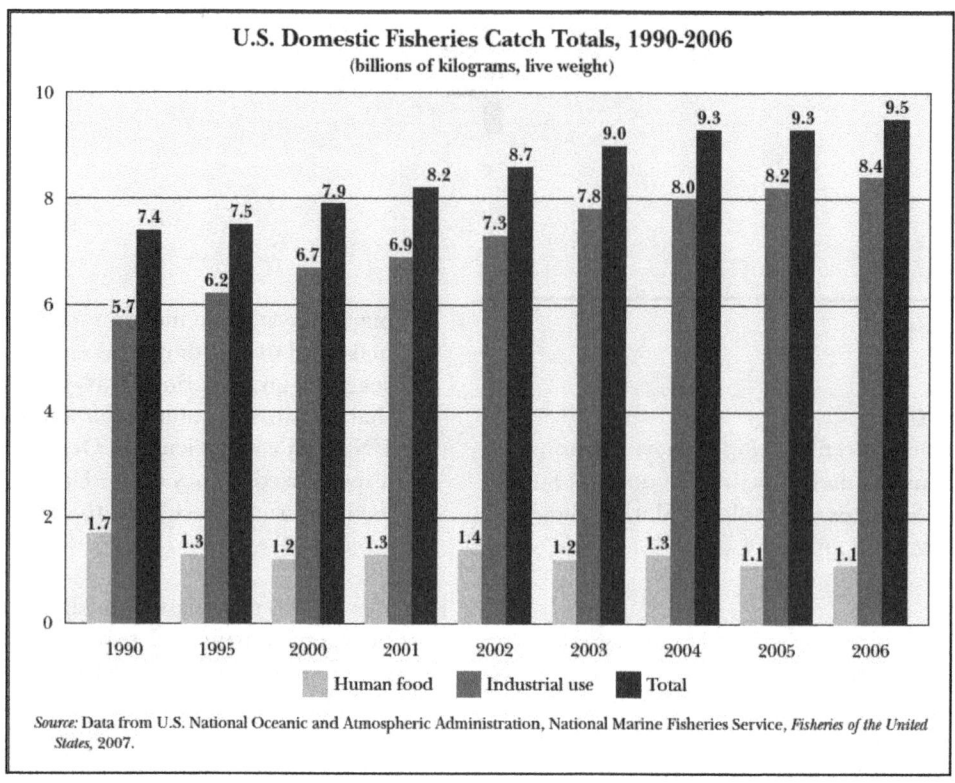

populations of commercial species of fish for both ecosystem integrity and the use of future generations.

Sustaining edible fish populations has benefits to both the environment and to those who rely upon fish-derived nutrients for sustenance. According to the FAO, fish and shellfish provide essential nutrition for 4.3 billion people and at least 50 percent of total animal protein and minerals for people in some of the world's poorest countries, including Bangladesh and Cambodia. Aside from the economic returns of the fishing industry, many rely on fish as a major food source, and further fisheries collapses could result in food crises within reliant communities. This means that responsible management of fisheries is also an environmental justice issue, as mismanagement caused by desire for higher monetary returns in the short term could deprive the world's hungry of an important source of nutrition.

In addition, many of the most overfished species are high on the food chain, meaning they prey upon smaller species of fish and play crucial roles in maintaining the equilibria of other fish populations and the larger marine ecosystem. The collapse of further fisheries would therefore not only spell disaster for communities that depend on a continuous supply of fish but also cause the degradation of marine ecosystems through a trickle-down effect in the food web.

Ecological Importance

Traditionally, the relationship between humans and the fish they catch and consume has been basically like any predator-prey relationship. In this way is dissimilar to the relationship between humans and domesticated food animals, which are bred in captivity, kept in managed numbers, and are in no threat of extinction. The latter relationship most closely resembles a form of mutualism, in which both populations are maintained, although one partner in this case is exploited by the other. Regardless of any ethical issues raised by the practice, domesticated animals live in populations sustained by humans. Fish, in contrast, have historically been taken wild from nature, a practice that in itself does nothing to sustain hunted populations.

Issues of regulation also plague the capture fisheries industry. Many targeted species are located in remote marine areas without national ties, where excessive fishing occurs frequently and is mostly undocumented. Species under the threat of extinction can be taken without repercussion, and many are sold under the monikers of more common fish. Also, unlike the case of domesticated animals, the harvesting of fish compromises many different species. While a certain species may be targeted, various others may be taken as bycatch. Even if bycatch is released according to laws and regulations, many die in the process. Meanwhile, once one hunted species has collapsed, commercial fishing can shift to another species, usually without major recognition from global consumers. Commercial fishing therefore poses a threat to multiple fish species at once.

A good example of the multilevel ecosystem effects of irresponsible fishing is the tuna, a carnivorous fish; a number of related species of tuna are found around the world. Tuna have a propensity to swim close to dolphin pods, presumably because the dolphins afford protection from sharks and other predators. Fishers have traditionally relied on dolphin pods to lead them to schools of tuna, and dolphins have often been caught and strangled in the fishers' nets. When the public at large became aware of this practice, the resulting outcry led to the adoption of technology designed to reduce dolphin catch. Also important ecologically, a widespread decline of tuna species could potentially cause unnatural explosions in tuna prey populations. If dolphin deaths were to approach a large number, dolphin populations would also be in danger, and species that are connected with them in the food web would likewise be subject to change. Often, the decline of many commercial fish species could result in similar top-down trophic (nutrition-related) effects.

The growth in demand for seafood, along with the disruption to fish populations caused by traditional capture fisheries, has led to a major rise in fish farming and other forms of aquaculture. By 2018, aquaculture provided more than 50 percent of all seafood taken for human consumption, and that figure was expected to continue to increase. While such practices do represent a means of sustaining fish populations, they also bring their own set of environmental concerns. Fish farms can be considerable sources of pollution due to high concentrations of nutrients and waste in the water, and also can use high levels of energy. Additionally, scientists have noted that large populations of farmed fish could pose a threat to genetic diversity, especially if individuals escape into the wild.

Socioeconomic Importance

The negative effects of overfishing are felt most severely in developing countries where the populations depend on fishing for sustenance and nutrition. The populations of many traditionally fished species have declined because of unrestricted extraction, and aquatic ecosystems on which communities rely for good catches year after year have become increasingly degraded. Millions of people in developing countries depend on fish for sustenance, and these numbers are only expected to increase as populations increase.

Features of coastal lands such as estuaries, coral reefs, mangrove forests, and sea grass beds are essential for many fish to spawn and provide the fish with shelter and habitat as well. The health of such areas is an important element of productive fisheries and maintaining healthy fisheries in turn ensures the perpetuity and value of these areas. Mangrove forests are natural barriers to destructive waves caused by storms, and they reduce coastal erosion by holding sediments within their roots. Coral reefs and sea grass beds harbor high levels of marine biodiversity and host a variety of ecosystem functions. Estuaries help with nutrient cycling and provide spawning grounds for fish and habitat for migratory birds. Environmentalists seek to protect such ecosystems as an integral part of protecting fish stocks, and therefore fisheries and all their interrelated economic activity.

However, efforts to preserve fish populations and sustain their vital socioeconomic role face complex challenges. It can be difficult to balance the long-term needs of the entire ecosystem against the immediate demands of the human population, especially when those demands relate to basic sustenance. A paradox arises in which economic factors make fishing vitally important, and yet fishing threatens the environmental basis and economic viability of the entire system. Similarly, the need to protect ecosystem components such as estuaries and reefs is often pressured by demand for development either directly related to the fishing industry or indirectly linked through the growing human population sustained to some degree by seafood. Further, all these issues face not only direct degradation due to human activities, but also broader complications brought about by climate change.

Solutions to Overfishing

Fish farming, or aquaculture, is one option for alleviating the damage done by commercial fishing. In contrast to the capture of wild fish, in aquaculture commercial aquatic animals are raised under controlled conditions for maximum yield. Aquaculture is akin in many ways to agriculture in theory and practice. As some commercial species can be bred in captivity, aquaculture offers the possibility of relieving wild populations from overfishing, potentially allowing their numbers to recover.

Fish farming operations are generally designed with yield in mind, however, and the environmental impacts of aquaculture can often be similar to those for which industrial agriculture operations are criticized. The use of antibiotics and poor waste handling by aquaculture operations lead to the pollution of waterways and, in the case of marine operations, the oceans as well. Pharmaceutical pollution is known to cause a plethora of negative effects in marine organisms, and high waste loads lead to algal blooms and other devastating downstream problems.

The feed for carnivorous farmed fish must come from other fish, and although some breeds are being developed that tolerate grain feed, a large percentage will still be fish-based, thus making the whole process arguably energy-inefficient and failing to seal a more sustainable circle of inputs and outputs. Further, fish that escape from aquaculture operations pose threats to the native aquatic ecosystems in which they are raised, as they are more often than not invasive. Regardless of the potential hazards to the environment, aquaculture was one of the world's fastest-growing food-production systems in the early twenty-first century.

Responsible consumerism in seafood is a growing movement, spearheaded in the United States by the Monterey Bay Aquarium's Seafood Watch Program, which offers consumers information—through print pocket guides and the aquarium's website—on the various kinds of fish available, noting which are sustainable choices and which are better avoided. The reasoning is that if consumers purchase mostly seafood that originates from well-managed fisheries, seafood from overfished areas will be subject to negative market pressures.

Some observers have argued that the most important change that can be made to protect the world's fisheries is a shift to sustainable development in poor countries and increased food security for people who must rely on overfished resources to survive. Only when people are guaranteed adequate nutrition can they make choices about their diets. Even if continuing a destructive practice will ultimately mean an end to resource production altogether, communities that depend on seafood protein will fish where they must to survive.

Jamie Michael Kass

Further Reading

Black, Kenneth D., ed. *Environmental Impacts of Aquaculture.* Boca Raton, Fla.: CRC Press, 2000.

Field, John G., Gotthilf Hempel, and Colin P. Summerhayes, eds. *Oceans 2020: Science, Trends, and the Challenge of Sustainability.* Washington, D.C.: Island Press, 2002.

Hart, Paul J. B., and John D. Reynolds, eds. *Handbook of Fish Biology and Fisheries.* Malden, Mass.: Wiley-Blackwell, 2002.

Miller, G. Tyler, Jr., and Scott Spoolman. "Sustaining Aquatic Biodiversity." In *Living in the Environment: Principles, Connections, and Solutions.* Belmont, Calif.: Brooks/Cole, 2009.

Reddy, M. P. M. *Ocean Environment and Fisheries.* Enfield, N.H.: Science Publishers, 2007.

Walters, Carl J., and Steven J. D. Martell. *Fisheries Ecology and Management.* Princeton, N.J.: Princeton University Press, 2004.

Websites

CNBC.
www.cnbc.com/2018/02/22/commercial-fishing-covers-half-of-global-oceans-study.html.

NOAA Fisheries, National Oceanic and Atmospheric Administration
www.fisheries.noaa.gov/topic/commercial-fishing.

■ Floodplains

FIELDS OF STUDY

Biology; Ecology; Ecosystems; Environment; Environmentalism; Geography; Hydrology; Islands; Life Sciences; Limnology

SUMMARY

Floodplains occupy an important part of landforms around rivers covering large areas, especially within humid and tropical climatic settings. They also house riparian wetlands, acting as buffers to flooding. Floodplains provide habitat for many land and aquatic life-forms.

PRINCIPAL TERMS

- **algae:** any of numerous groups of chlorophyll-containing, mainly aquatic eukaryotic organisms ranging from microscopic single-celled forms to multicellular forms 100 feet (30 meters) or more, distinguished from plants by the absence of true roots, stems, and leaves
- **delta:** the deposit by running water of clay, silt, sand, gravel, or similar detrital material at the mouth of a river
- **floodplains:** low-lying areas adjacent to river channels that become partially or completely covered with water when the rivers overflow their banks
- **geomorphology:** the study of the characteristics, origin, and development of landforms
- **riffles:** a shallow landform in a flowing channel
- **sinuosity:** a bending or curving shape or movement
- **watershed:** the ridge or crest line dividing two drainage areas

Marshes and flood-meadows on the Narew's floodplains, on road between villages Siekierki and Góra. By Athantor (Own work).

Floodplains filter water and provide silt and nutrients that make them fertile places. Perhaps the most famous examples are the fertile floodplains of Egypt's Nile Valley region, which have supported civilizations for several millennia. Floodplains also provide fresh water and backwaters to wetlands, and they also dilute salts, thereby improving the health of the habitat for fish, bird, and plant populations that inhabit the floodplains.

Floodplains are good for food production such as rice cultivation. Farmers graze their livestock on the grasslands in floodplains, and fresh fruits and cash crops are grown in floodplains, which are often very fertile and easy to cultivate. In tropical settings, timber is harvested on floodplains, and nontimber forest resources such as animals and plants are used for foods and medicines as well as construction materials.

Types and Ecology

Riverine flooding can cover vast areas, many of which are among the most diverse biologically productive ecosystems on earth. Three types of floodplains are identified based on temperature: temperate stochastic, temperate seasonal, and tropical seasonal. Within floodplains, algae appear to provide the most important source of primary production within the grazer web. The flow regime is very important in determining the physical habitat for biotic composition. The shape, size, and the formation of features such as deltas, riffles, runs, pools, and backwaters that tend to shift are linked to the flow regimes of rivers. Certain aquatic life-forms have their early life stages in floodplains, and the types of fauna and flora within floodplains can be as diverse as in any other ecosystems. Owing to the highly dynamic nature of floodplain terrains, varied species may be seen on the same floodplains over the course of years.

Floodplains contain several kinds of geomorphologic features, including oxbow lakes, point bars, areas of dead water, and braided channels. Swamps, among other types of riparian wetlands, can also be found in floodplains. Floodplains can be classified into different types depending on their morphology. Several methods of classification are used, but the simplest and most common is based on the fluvial styles: gravel-dominated, sand-dominated with high sinuosity, and sand-dominated with low sinuosity.

Environmental Threats

As global warming increases, some floodplains may see more flooding, which will greatly affect local populations because of land subsidence and increases in water level. For example, Bangkok, the capital of Thailand, is sinking at a rate of 2 to 5 centimeters (0.79 to 1.97 inches) per year because of sediment compression and compaction owing to increased human activities. The elevation of New Orleans, Louisiana, is also dropping at a rate of about 5 centimeters per year.

Urbanization affects the hydrology of floodplains, either by reducing water through withdrawal or by adding to it through importation. It also alters the water chemistry by introducing chemicals, sediments, and other form of pollutants, including increases in temperature, all of which affect the biotic richness of floodplains. The nutrients brought into play through flood activities can also be altered through changes in land-use patterns. Changes within a drainage basin (watershed) affect the production and supply of organic materials in floodplains. High levels of biodiversity can provide stability to floodplain ecosystems and help protect them from human-caused impairments.

The expansion of urban areas into floodplains and wetlands alters the onset, duration, distribution, speed, quantity, and quality of floodwaters. Among the human activities that lead to increases in flooding are deforestation and the removal of stabilizing vegetation along riverbanks. Human-built structures along or near rivers affect the flow direction, resulting in deflection of the water, or reduce storage. Storm drains, housing developments, and pavements increase the rate of rainfall runoff to rivers, thereby increasing the rate of flooding. The straightening of river channels increases the rate at which water is transported. Another human activity that affects floodplains is the dumping of sediment loads from farms or construction sites into rivers, which decreases channel depth and increases the area covered by floodwaters. With increasing changes in land use, a watershed approach to floodplain management becomes imperative.

Impairment of floodplain waters can have adverse effects on coastal ecosystems, as these waters end up in lakes or oceans. The quality of the water in rivers has a great impact on the quality of the water in nearby coastal areas; the waters of the Amazon, for example, can be traced several miles into the Atlantic Ocean. Contaminants carried in such waters ultimately affect large biological populations.

Solomon A. Isiorho

Further Reading

Bridge, John S. *Rivers and Floodplains: Forms, Processes, and Sedimentary Record.* Malden, Mass.: Blackwell, 2003.

Millius, Susan. "Losing Life's Variety." *Science News*, March 13, 2010, 20-25.

Richards, Keith, James Brasington, and Francine Hughes. "Geomorphic Dynamics of Floodplains: Ecological Implications and a Potential Modelling Strategy." *Freshwater Biology* 47, no. 4 (2002): 559-579.

Tockner, Klement, and Jack A. Stanford. "Riverine Flood Plains: Present State and Future Trends." *Environmental Conservation* 29, no. 3 (2002): 308-330.

Fossil fuels

FIELDS OF STUDY

Business; Chemistry; Climatology; Commerce; Ecology; Economics; Emissions; Energy and Energy Resources; Environment; Environmentalism; Geology; Geosciences; Hazardous Materials; Pollution; Toxic Waste

SUMMARY

Fossil fuels—coal, oil, and natural gas—are found in the Earth's crust. They are the result of the decomposition of the remains of dead plants and animals under heat and pressure as they were covered with sediment, becoming part of the earth's crust, as either landmass or seabed. These fuels are nonrenewable resources; they have required millions of years to form. The use of such fuels is the cause of significant environmental degradation.

PRINCIPAL TERMS
- **fossil fuels:** fuels formed over long spans of time from buried dead organisms carbon
- **hydrocarbon:** a simple class of organic chemicals obtained from the earth's two large reservoirs of organic material: petroleum and coal
- **global warming:** increase in the average surface and ocean temperature of the Earth since 1850 and the projected persistence of the trend
- **carbon monoxide:** colorless, odorless, poisonous gas that burns with a pale-blue flame, produced when carbon burns with insufficient air
- **acid rain:** environmentally harmful precipitation – rain, snow or sleet – containing high concentrations of pollutants from coal smoke, chemical manufacturing and smelting, released into the atmosphere
- **Kyoto Protocol:** an international treaty extending the 1992 United Nations Framework Convention on Climate Change that commits states to reduce greenhouse gas emissions, based on the scientific consensus that global warming is occurring an is extremely likely that human-made

Fossil fuels are the most widely used sources of energy production throughout the world and are essential to human activity in modern society. The use of such fuels, however, is the cause of significant environmental degradation through air and water pollution and habitat destruction. The carbon dioxide emitted by the burning of fossil fuels has been linked to global warming.

Fossil fuels are composed of high percentages of carbon and hydrocarbons, and the ratio of carbon to hydrogen varies considerably from one type of to another. A gas such as methane has a low ratio and burns quickly, whereas a substance such as coal, composed almost entirely of carbon, has a lower ratio and burns more slowly. When burned, all fossil fuels produce large amounts of energy, and this characteristic led them to play a significant role in the industrialization and modernization of the world. According to the Energy Information Association, in 2007 fossil fuels accounted for the production of 86.4 percent of the energy consumed worldwide. Although fossil fuels can meet the energy production needs of the world, they are a resource that is diminishing, and their extraction and use both cause considerable environmental problems.

The major concern regarding the use of fossil fuels is their release of carbon dioxide (CO_2), one of the greenhouse gases that has been linked to global warming. According to the U.S. Department of Energy, the burning of fossil fuels produces almost twice as much CO_2 as natural processes can absorb each year.

Environmental Impacts of Coal

The three types of coal—anthracite, bituminous, and lignite—are retrieved either by deep-shaft underground mining or by opencast (surface) mining. Both types of mining cause considerable damage to the area mined, as they destroy land and pollute air and rivers. The pollution from lignite mining is particularly harmful to forests.

When coal is burned it emits several harmful substances, including sulfur dioxide, nitrogen oxide, mercury, particulates, and carbon dioxide. Sulfur dioxide, nitrogen oxide, and particulates contribute to the formation of acid rain, which can cause respiratory illness. Mercury that enters rivers, streams, or lakes combines into the chemical methylmercury. It is highly toxic to water plants, to fish, and to animals and people who consume the fish. It is, however, the CO_2 produced by the burning of that is of the greatest environmental concern. CO_2 is the major that causes global warming, and coal-fired power plants are responsible for the greatest amount of CO_2 released into the air. The transport of coal further contributes to the of pollutants into the air. Although some coal is transported as through pipelines, most coal is transported by train using diesel-fueled locomotives, which in turn emit more CO_2 and other pollutants.

Environmental Impacts of Oil

Crude oil or is composed of hydrogen and carbon compounds. It is a liquid form of fossilized derived from the decomposition of dead plants and animals found in underground reservoirs in sedimentary basins on land areas and in seabeds. Oil is extracted from the earth by pumping, using wells and oil rigs, and it is transported long distances through pipelines or in ships. The major danger to the environment from these activities is the occurrence of oil spills, which have serious impacts on wildlife, especially marine life, seabirds, and sea mammals.

Gasoline, diesel fuel, jet fuel, and liquid propane gas are all derived from oil. It is also the feedstock or raw material from which plastics, polyurethane, and many other products are made. Oil plays a vital role in the everyday life of the modern world. Gasoline, diesel, and jet fuel are the most commonly used sources of energy in transportation. Gasoline is the primary fuel used in private cars; diesel fuel is used in freight trucks, in train engines, and in such heavy equipment as construction and farm vehicles. Diesel is also the fuel of choice for other kinds of machinery used in agriculture and construction. The generators used to provide electricity to such facilities as hospitals and nursing homes in times of emergency are usually powered by diesel fuel. Propane, which is a cleaner-burning fuel than either gasoline or diesel, is used in indoor equipment.

Oil's greatest impacts on the environment come from its use as fuel. In 2008, consumption of oil worldwide amounted to 85.4 million barrels per day. When oil is burned in any of its forms, it emits various harmful substances into the atmosphere. These include CO_2, carbon monoxide (CO), sulfur, and lead. Lead in is classified as a carcinogen. Sulfur dioxide contributes to the formation of acid rain, which is harmful to animals, plants, and human beings. The nitrogen oxide and volatile compounds found in emissions from the burning of oil are among the causes of ground-level ozone. Many of these pollutants contribute to lung irritation, asthma, bronchitis, and lung disease.

Environmental Impacts of Natural Gas

Natural gas is composed of hydrogen and carbon; it is primarily methane. Like coal and oil, natural gas is formed over millions of years from the decomposing remains of plants and animals covered by sand and subjected to heat and pressure. Deposits of natural gas are found in both landmasses and seabeds. Extraction of the gas, achieved through drilling and the establishment of wells, has impacts on the wildlife in the area of the drilling through the disruption of habitat. Natural gas is transported by pipelines to refineries. At both drilling sites and along the pipelines, leaks can occur that can result in serious explosions.

Natural gas has many uses. It is used to generate electricity, as a fuel in industry, and for heating homes and powering home appliances. Natural gas also serves as a raw material for producing a wide variety of products ranging from fertilizers to medicines. When burned, natural gas emits fewer pollutants, especially sulfur and nitrogen, than either oil or coal. It does, however, produce CO_2.

Efforts to Reduce Negative Impacts

Fossil fuels continue to play an important role in all major areas of the world's economy. The generation of electricity, transportation, construction, and food production are all highly dependent on the use of fossil fuels. Governments around the world and the fossil-fuel industries are involved in ongoing efforts to combat the adverse effects of fossil fuels on the environment. Both by passing legislation within their own borders, such as the Clean Air Act in the United States, and by entering into international agreements such as the Kyoto Protocol, governments have set limits on the amounts of pollutants that may be emitted from fossil fuels and have set targets for reducing emissions. Through research and new technology, as well as programs, the fossil-fuel industries have worked to reduce the negative impacts of fossil fuels on the environment.

Because coal-fired power plants are the greatest emitters of CO_2, the primary greenhouse gas, citizens' groups in several countries, especially the United Kingdom, have argued for the elimination of coal mining and of the use of coal as a fuel. Methane has replaced coal as the fuel of choice in a number of power plants. The coal industry has responded with strong efforts to develop technologies aimed at reducing CO_2 emissions from the burning of coal and improving mining techniques to reduce adverse effects on land and communities in the areas where coal is mined. The reclamation of land at strip mines is one of these efforts.

In the United Kingdom, the technique of extracting coal through underground coal has been investigated. The procedure involves injecting a mixture of steam and oxygen down a borehole through which gas is extracted from the coal and brought to the surface. This technique, however, may cause contamination of underground water supplies in onshore locations and the collapse of the burned-out coal seams both on- and offshore.

The term "clean coal" was popularized in 2008 by coal industry groups, at a time when Congress was contemplating climate change legislation. While the term is deliberately vague, it is often understood to mean coal plants that capture the carbon dioxide emitted from smokestacks and bury it underground as a way of limiting global warming.

Coal-fired power plants using a technology known as integrated gasification combined cycle (IGCC) have also been introduced; IGCC plants convert coal into a synthetic gas before it is used to generate electricity. These plants have the capability of reducing pollutant emissions significantly, including the capture and separation of up to 95 percent of CO_2, which is then stored underground. The technology, known as "carbon capture and storage," is still new and is costly and complicated. Only one coal plant in the United States, the Petra Nova project in Texas, captures CO_2 in this fashion.

Other technologies and procedures implemented by the oil and gas industries to reduce the negative effects of their products on the environment include horizontal drilling techniques that can increase the area from which oil or gas can be extracted from one well. Reducing the number of wells drilled reduces the impacts on habitat and wildlife. In addition, the oil and gas industries use double-hulled tankers and double-lined pipelines to help reduce oil spills. In response to governmental mandates, oil companies have funded research that has resulted in the reformulation of gasoline and diesel fuels to reduce the emissions they produce. Regardless, mining for coal remains a highly polluting practice, often damaging streams and waterways.

Shawncey Webb

Further Reading

Archer, David. *Global Warming: Understanding the Forecast.* Malden, Mass.: Wiley-Blackwell, 2006.

Higman, Christopher, and Maarten van der Burgt. *Gasification.* Boston: Elsevier, 2008.

Kelley, Ingrid. *Energy in America: A Tour of Our Fossil Fuel Culture.* Lebanon, N.H.: University Press of New England, 2008.

Martin, Raymond S., and William L. Leffler. *Oil and Gas Production in Nontechnical Language.* Tulsa, Okla.: Pennwell, 2005.

Pirani, Simon. *Burning Up: A Global History of Fossil Fuel Consumption.* London, U.K.: Pluto Press, 2018.

Shogren, Jason F. *The Benefits and Costs of the Kyoto Protocol.* Washington, D.C.: AEI Press, 1999.

Williams, A., et al. *Combustion and Gasification of Coal.* New York: Taylor & Francis, 1998.

Websites

Department of Energy.
https://www.energy.gov/fe/science-innovation/office-clean-coal-and-carbon-management

New York Times.
https://www.nytimes.com/2017/08/23/climate/what-clean-coal-is-and-isnt.html

G

■ Genetic modification

FIELDS OF STUDY
Agriculture; Biology; Biochemistry; Ecology; Ecosystems; Environmental Biotechnology; Environmental Microbiology; Environmentalism; Health; Inventions; Life Sciences; Medicine; Technology and Applied Science

SUMMARY
The genetic modification of bacteria has made possible the manufacture of medically important human proteins such as insulin and growth hormone. Genetically altered bacteria have also been used as a means to introduce into plants genetic material that increases the plants' resistance to disease, pests, or freezing.

PRINCIPAL TERMS
- **biotechnology:** the use of living organisms or other biological systems in the manufacture of drugs or other products or for environmental management, including the use of genetically engineered bacteria to produce human hormones
- **deoxyribonucleic acid (DNA):** an extremely long macromolecule that is the main component of chromosomes and the material that transfers genetic characteristics in all life forms
- **enzymes:** any of various proteins originating from living cells capable of producing chemical changes in organic substances by catalytic action, as in digestion
- **etiological:** of or relating to causes or origins
- **genetically modified organisms:** bacteria that humans have manipulated at the genetic level to possess specific properties or carry out certain functions
- **plasmids:** small, circular, double-stranded DNA molecules distinct from a cell's chromosomal DNA and existing naturally in bacterial cells

The ability to alter bacteria genetically is the outcome of several independent discoveries. In 1944 Oswald Avery and his coworkers demonstrated gene transfer among bacteria using purified deoxyribonucleic acid (DNA), a process called known as transformation. In the 1960's the discovery of restriction enzymes permitted the creation of hybrid molecules of DNA. Such enzymes cut DNA molecules at specific sites, allowing fragments from different sources to be joined within the same piece of genetic machinery. Restriction enzymes are not species-specific in choosing their targets. Therefore, DNA from any source, when treated with the same restriction enzyme, will generate identical cuts. The treated DNA molecules are allowed to bind with each other, while a second set of enzymes called ligases are used to fuse the hybrids. The recombinant molecules may then be introduced into bacteria cells through transformation. In this manner, the cell acquires whatever genetic information is found in the DNA. Descendants of the transformed cells will be genetically identical, forming clones of the original.

The most common forms of genetically altered DNA are bacterial plasmids, small circular molecules separate from the cell chromosome. Plasmids may be altered to serve as appropriate vectors for genetic engineering, usually containing an antibiotic resistance gene for selection of only those cells that have incorporated the DNA. Once the cell has incorporated the plasmid, it acquires the ability to produce any gene product encoded on the molecule. The resulting artificially produced DNA is called recombinant DNA. The first bacterium to be genetically altered through recombinant DNA technology for medical purposes,

Escherichia coli, contained the gene for the production of human insulin. Prior to creation of the insulin-producing bacterium in the 1970's, diabetics were dependent on insulin purified from animals. In addition to being relatively expensive, insulin obtained from animals produced allergic reactions among some diabetes patients. Insulin obtained from genetically altered bacteria, by contrast, is identical to human insulin. Subsequent recombinant DNA research has led to the manufacture of a variety of human proteins, including human growth hormone, parathyroid hormone, several kinds of interferons, many monoclonal antibodies, hepatitis B surface antigen, clotting factors, and granulocyte colony-stimulating factor.

Genetically altered bacteria may also serve as vectors for the introduction of genes into plants. The bacterium *Agrobacterium tumefaciens*, the etiological agent for a plant disease called crown gall, contains a plasmid known as Ti. Following infection of the plant cell by the bacterium, the plasmid is integrated into the host chromosome, becoming part of the plant's genetic material. Any genes that were part of the plasmid are integrated as well. Desired genes can be introduced into the plasmid, promoting pest or disease resistance within plants infected by the bacterium.

In April 1987, scientists in California sprayed strawberry plants with genetically altered bacteria to improve the plants' freeze resistance; this event marked the first deliberate release of genetically altered organisms in the United States to be sanctioned by the Environmental Protection Agency (EPA). The release of the bacteria represented the climax of more than a decade of public debate over what would happen when the first products of biotechnology became commercially available. Fears centered on the creation of bacteria that might radically alter the environment through elaboration of gene products not normally found in such cells. Other concerns included the creation of super bacteria with unusual resistance to conventional medical treatment.

Despite these fears, approval for further releases of genetically altered bacteria soon followed, and the restrictions on release were greatly relaxed. By 1991 permits for field tests of more than 180 genetically altered plants and microorganisms had been granted. Between 1987 and 2004 more than 10,000 trials were conducted at more than 39,000 sites, and more than sixty biotechnology products entered the market. Among future planned tests are clinical trials of the use of a modified *Streptococcus mutans* in fighting dental cavities. Scientists have modified this bacterium, responsible for tooth decay in its unaltered form, so that it does not produce the lactic acid that ordinarily erodes tooth enamel. In animal tests, it has been found that the modified bacterium eventually replaces the *S. mutans* naturally occurring in the mouth.

In general, "red biotechnology" (the application of biotechnology in medicine) tends to generate less controversy than "green biotechnology" (use of biotechnology in food production). In the United States, anyone intending to produce or import genetically altered microorganisms for commercial purposes must submit a notice to the EPA, which assesses whether the organism constitutes an unreasonable risk to human health or the environment.

Richard Adler; updated by Karen N. Kähler

Further Reading

Drlica, Karl. *Understanding DNA and Gene Cloning: A Guide for the Curious*. 4th ed. Hoboken, N.J.: John Wiley & Sons, 2004.

Food and Agriculture Organization and World Health Organization. *Safety Assessment of Foods Derived from Genetically Modified Microorganisms*. Geneva, Switzerland: World Health Organization, 2001.

Han, Lei. "Genetically Modified Microorganisms: Development and Applications." In *The GMO Handbook: Genetically Modified Animals, Microbes, and Plants in Biotechnology*, edited by Sarad R. Parekh. Totowa, N.J.: Humana Press, 2004.

Lynas, Mark. *Seeds of Science: Why We Got it So Wrong on GMOs*. London: Bloomsbury Sigma, 2018.

Shetterly, Caitlin. Modified: GMOs and the Threat to Our Food, Our Land, Our Future.

New York: G.P. Putnam's Sons, 2016.

Stemke, Douglas J. "Genetically Modified Organisms: Biosafety and Ethical Issues." In *The GMO Handbook: Genetically Modified Animals, Microbes, and Plants in Biotechnology*, edited by Sarad R. Parekh. Totowa, N.J.: Humana Press, 2004.

Watson, James D., et al. *Recombinant DNA: Genes and Genomes—A Short Course*. 3d ed. New York: W. H. Freeman, 2007.

Glacial melting

FIELDS OF STUDY
Atmospheric Sciences; Climate and Climate Change; Climatology; Ecology; Environment; Environmentalism; Meteorology

SUMMARY
The most widespread indication that the earth has been warming over time has been the steady erosion of ice in the Arctic and Antarctic and on mountain glaciers. Although a few exceptions do exist, the worldwide erosion of ice leaves little doubt that that the earth has experienced steadily increasing warming for at least a century.

PRINCIPAL TERMS
- **climate change:** change of climate attributed directly or indirectly to human activity altering the composition of the global atmosphere and to natural climate variability observed over comparable time periods
- **glacial melting:** gradual erosion of polar ice caps and mountain glaciers
- **glaciation:** the forming, existence, or movement of glaciers over the surface of the earth
- **global warming:** increase in the average surface and ocean temperature of the Earth since 1850 and the projected persistence of the trend
- **Intergovernment Panel on Climate Change (IPCC):** created in 1988 by the World Meteorological Organization and the United Nations Environment Programme, the objective of the IPCC is to provide governments at all levels with scientific information that they can use to develop climate policies
- **United States Geological Survey (USGS):** a scientific agency of the United States government that studies the U.S. landscape, its natural resources, and the natural hazards that threaten it

According to a report by Hamish D. Pritchard and his colleagues that appeared in the journal *Nature* late in 2009, accelerated ice flow (dynamic thinning), measured by high-resolution laser altimetry, reaches all latitudes in Greenland and has intensified on key Antarctic grounding lines, penetrating far into the interior of each ice sheet, spreading as ice shelves thin by ocean-driven melt. Mountain glaciers are in rapid retreat around the earth, with very few exceptions. The Intergovernmental Panel on Climate Change's models project that between one-third and one-half of existing mountain glacial mass could disappear over the next hundred years. Sometime during the twenty-first century, the last glacier may melt in Montana's Glacier National Park.

Mass loss from the mountain glaciers, as well as massive ice shelves and sheets along the coasts of Greenland and Antarctica, is accelerating faster than expected, contributing to sea-level rises. As seas gradually rise because of glacial melting and thermal expansion along some coasts, isostatic rebound, the gradual rising of land masses that have been compressed by glaciers in the past, has provoked the rise

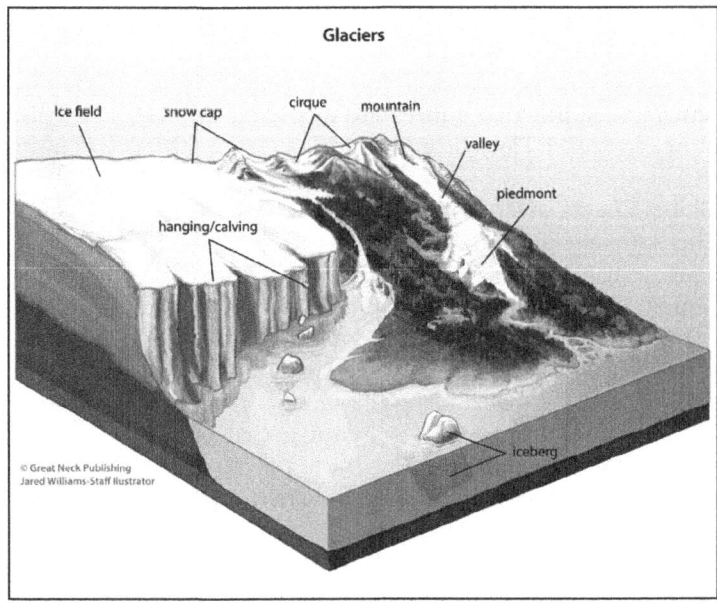

Glaciers: Glaciers have many features including ice fields, snow caps, cirques, mountains, valleys, piedmonts, icebergs, and hanging or calving portions. © EBSCO

of Scotland's coast following the last ice age. The Great Lakes, remnants of melted glaciers, also experience some isostatic rebound.

Mountain Glaciers in Rapid Retreat
Climbers have been plucked from the Matterhorn in the Swiss Alps as thawing mountainsides crumbled under them. During the summer of 2003, Mont Blanc, Europe's tallest mountain, was closed to hikers and climbers because its deteriorating snow and ice were too unstable to allow safe passage. The mountain was crumbling as ice that once held it together melted during a record-warm summer in Europe. Scientists have estimated that by 2025 glaciers in the Swiss Alps will have lost 90 percent of the volume they contained a century earlier.

The U.S. Geological Survey (USGS) has collected a digital library that describes the state of more than 67,000 glaciers around the world. Using historical photographs, images from space satellites, precision laser measurements, and other tools, the USGS has created an archive that tells a story of glacial retreat around the world. The rate of ice loss doubled from 1988 to 2002. Generally, the only glaciers gaining mass have been in wet maritime areas of the world, such as parts of Norway and Sweden, where melting has been offset by increased snowfall, another facet of climate change. Alaska's Hubbard Glacier is advancing so swiftly that it threatens to seal off the entrance to Russell Fjord near Yakutat.

Global Effects of Ice Melt
The snow-and-ice crown of Mount Kilimanjaro in equatorial Africa may vanish by the mid-twenty-first century. Kilimanjaro will no longer live up to its name, which in Swahili means "mountain that glitters." Nearby Mount Kenya's ice fields lost three-quarters of their mass during the twentieth century. By the end of the twenty-first century, Glacier National Park in Montana may lose the last of its permanent glaciers. The original 150 glaciers within Glacier National Park had been reduced to 37 by 2002, and most of these were small remnants of once-mighty ice masses.

Millions of people around the world may face severe water shortages as glaciers around the world melt. Ecuador, Peru, and Bolivia, where major cities rely on glaciers as their main source of water during dry seasons, could be among the most intensely affected. Areas of the Himalayas face grave danger of flooding initially, as glaciers melt, followed by water shortages from the ebbing of glacier-fed rivers in the region that supply water to one-third of the world's population, mainly in India and China.

Evidence from Antarctica suggests that melting ice may flow into the sea much more easily than earlier believed, perhaps leading to an accelerating rise in sea levels. A study published in the journal *Science* in 2003 suggested that seas might rise as much as several meters during the next several centuries, given projected global warming based on continuing usage of fossil fuels at the levels of the early twenty-first century. Additional studies have provided further evidence of this trend, as the important Pine Island Glacier of the West Antarctic Ice Sheet has been receding into the ocean at a faster pace. This mass of ice could influence the entire region in years to come.

Bruce E. Johansen

Further Reading
Alley, Richard B., et al. "Ice-Sheet and Sea-Level Changes." *Science* 310 (October 21, 2005): 456-460.

Dutch, Steven I., ed. *Encyclopedia of Global Warming*. 3 vols. Pasadena, Calif.: Salem Press, 2010.

Hansen, James. "Defusing the Global Warming Time Bomb." *Scientific American*, March 2004, 68-77.

Jamail, Dahr. *The End of Ice: Bearing Witness and Finding Meaning in the Path of Climate Disruption*. New York: The New Press, 2019.

Knight, Peter G., ed. *Glacier Science and Environmental Change*. Malden, Mass.: Blackwell, 2006.

Lynas, Mark. *High Tide: The Truth About Our Climate Crisis*. New York: Picador, 2004.

_____. *Six Degrees: Our Future on a Hotter Planet*. New York: HarperCollins, 2008.

Pritchard, Hamish D., et al. "Extensive Dynamic Thinning on the Margins of the Greenland and Antarctic Ice Sheets." *Nature* 461 (October 15, 2009): 971-975.

Global warming

FIELDS OF STUDY

Atmospheric Sciences; Climate and Climate Change; Climatology; Atmospheric Structure and Dynamics; Ecology; Environment; Environmentalism; Land-Use Ecology; Meteorology

SUMMARY

The findings of scientists concerning the causes of global warming are extremely important in that they provide guidance for policy makers. Harmful consequences may result if the anthropogenic, catastrophic theory of global warming is correct and policy makers do not take the political and economic actions necessary to address the problem; conversely, harmful consequences may result if the theory is wrong but major political and economic decisions are made in the belief that it is correct.

PRINCIPAL TERMS

- **anthropogenic:** caused or produced by humans
- **Atlantic Multi-Decadal Oscillation:** a coherent mode of natural variability occurring in the North Atlantic Ocean with an estimated period of 60-80 years, based upon the average anomalies of sea surface temperatures (SST) in the North Atlantic basin
- **Celsius/Fahrenheit:** Celsius, predominate in Europe, and Fahrenheit, predominate in the US and its territories, are different scales to measure temperature
- **climate change:** change of climate attributed directly or indirectly to human activity altering the composition of the global atmosphere and to natural climate variability observed over comparable time periods
- **climatology:** the science that deals with the phenomena of climates or climatic conditions.
- **El Niño/La Niña:** a natural part of the global climate system, occurring when the Pacific Ocean and the atmosphere above it change from their neutral ('normal') state for several seasons. El Niño events are associated with a warming of the central and eastern tropical Pacific, while La Niña events are the reverse, with a sustained cooling of these same areas
- **global warming:** increase in the average surface and ocean temperature of the Earth since 1850 and the projected persistence of the trend
- **Pacific Decadal Oscillation:** a robust, recurring pattern of ocean-atmosphere climate variability centered over the mid-latitude Pacific basin

According to the Intergovernmental Panel on Climate Change (IPCC), the overall global temperature during the twentieth century increased by a little less than 1 degree Celsius (1.8 degrees Fahrenheit). This involved an increase of about 0.5 degree Celsius (0.9-degree Fahrenheit) from 1910 to 1945 and a similar increase from about 1975 to 2000 or so (peaking in 1998), with a slight decrease in the intervening years. (The figures are approximate because of uncertain data and yearly fluctuations, occasionally as large as 0.25 degree Celsius up or down, and the complexity of adjusting the raw temperature data.) Explaining these increases and projecting future trends and their consequences are the key issues addressed by scientists who examine global warming.

In 2007, the IPCC summarized the following findings of that year's Assessment of Climate Change

- Warming of the climate system is unequivocal.
- Most of (50% of) the observed increase in globally averaged temperatures since the mid-20th century is very likely (confidence level 90%) due to the observed increase in anthropogenic (human) greenhouse gas concentrations.
- Hotter temperatures and rises in sea level "would continue for centuries" even if greenhouse gas levels are stabilized, although the likely amount of temperature and sea level rise varies greatly depending on the fossil intensity of human activity during the next century.
- The probability that this is caused by natural climatic processes alone is less than 5 percent.
- World temperatures could rise by between 1.1 and 6.4° Celsius (2.0 and 11.5° Fahrenheit) during the twenty-first century....
- Sea levels will probably rise by 18 to 59 centimeters (7.08 to 23.22 inches).
- There is a confidence level 90% that there will be more frequent warm spells, heat waves and heavy rainfall.

- There is a confidence level 66% that there will be an increase in droughts, tropical cyclones and extreme high tides.
- Both past and future anthropogenic carbon dioxide emissions will continue to contribute to warming and sea level rise for more than a millennium.
- Global atmospheric concentrations of carbon dioxide, methane, and nitrous oxide have increased markedly as a result of human activities since 1750 and now far exceed pre-industrial values over the last 650,000 years.

Climate Cycles and Human Causes

Two basic theories have been posited regarding the source of the warming, both of which could easily be partially correct. Some see the warming as basically natural (as many scientists agree is probably the case for the pre-1945 warming, which predates the large increase in atmospheric carbon dioxide). In fact, short-term natural causes have been documented, such as volcanic eruptions (the Mount Pinatubo eruption in 1991 was followed by a strong temperature down-spike in 1992) and the El Niño/La Niña weather cycle (the very strong El Niño of 1998 resulted in a large temperature up-spike). In addition, some long-term fluctuations, such as the Pacific Decadal Oscillation and the Atlantic Multidecadal Oscillation, affect global as well as local temperatures. In addition, solar energy is not constant; there are slight cyclical variations that correlate with sunspot activity. These do not seem to be sufficient to explain the post-1975 warming; Patrick J. Michaels and Robert C. Balling, Jr., in their 2009 book *Climate of Extremes: Global Warming Science They Don't Want You to Know*, estimate that natural causes explain only 25 percent of the post-1975 warming (compared to 75 percent of the earlier warming). Some scientists, however, think natural answers can be found for the rest.

These scientists think the current warming is natural and cyclic, a Modern Warm Period to follow the Medieval Warm Period and Little Ice Age (which ended about 1850). A wide array of historical temperature proxies illustrates that there is a roughly 1,500-year cycle, with shorter heating and warming sub-cycles. The proxies include ice cores dating back hundreds of thousands of years, six thousand boreholes (from all continents), seabed and lake-bed sediment cores, tree rings and tree lines, cave stalagmite cores, peat bogs, and records. These do show occasional remarkable shifts, global temperature increasing nearly 1 degree Celsius for about a decade at the end of the Younger Dryas (11,500 years ago) for reasons still unknown (there was an increase in greenhouse gases after the rise).

Environmental scientist S. Fred Singer has estimated that the Medieval Warm Period exceeded the current warming (so far). Others dispute this. Climatologist Michael E. Mann has argued that global temperatures changed only slightly during these periods, far less than in the twentieth century.

Among the many possible anthropogenic, or human-caused, influences on climate change is land use. The effects of land use are important, but they are for the most part local; cropland is warmer than forest, and urban areas are much warmer than cropland (the urban effect). Overgrazing can lead to desertification, which makes the land warmer.

The strongest effect comes from the production of greenhouse gases such as carbon dioxide. Water vapor is an extremely important greenhouse gas, and methane (much of it from rice paddies and livestock raising) is far more powerful than carbon dioxide, but a trend toward increasing atmospheric methane halted during the mid-1990's. Carbon dioxide is the greenhouse gas that is the cause of greatest concern. Since the beginning of the twentieth century, the amount of carbon dioxide in the atmosphere has increased from about 290 parts per million to almost 390. The increase in the greenhouse effect is far smaller than the increase in greenhouse gases, however, particularly in areas of high humidity, owing to the atmospheric equivalent of the law of diminishing returns. Greenhouse gas warming is strongest at night and therefore in winter, and in upper latitudes; in the atmosphere it leads to a warmer and cooler stratosphere.

Climate Models

Computer climate models can be used for historical research as well as future projections. Extremely complex general circulation models provide detailed information on how natural warming or greenhouse gas warming is likely to affect climate all over the planet and into the atmosphere, and the models' projections can be tested against observational data. Such testing is as necessary for the findings produced by computer models as for the findings produced by any other scientific experiments; results must be shown to be replicable by others, and the data must be freely available for examination by others. One problem with the testing of data from climate models is that observational data are often too recent (satellite tracking of hurricanes began in 1970, for example, and satellite measurement of Arctic sea ice in 1979) to allow scientists to determine reliably whether changes represent coincidental long-term oscillations or result from the current warming trend.

In the early twenty-first century, most climate models project a linear global surface temperature increase from 2 to 3 degrees Celsius (3.6 to 5.4 degrees Fahrenheit) per century (occasionally much more owing to positive feedback effects, such as increased evaporation leading to increased humidity). Early models exaggerated the warming and could not match the previous history. Later models that added in sulfate aerosols were more accurate, but they failed to predict the absence of net warming since 1998 or the relative lack of warming in the Southern Hemisphere; the problem was that the models used one unknown to check another. Some of the early error may have resulted from negative feedbacks, such as clouds (low-level cumulus and stratus clouds reflect solar light, cooling the planet), which the models often ignored. Climate models predict different specific results from natural and greenhouse gas warming, and many observations (most notably the overall cooling of Antarctica) tend to support the latter, though not entirely.

Michael E. Mann and Lee R. Kump have praised three projections made by James E. Hansen while he director of the National Aeronautics and Space Administration's Goddard Institute for Space Studies and presented testimony to the U.S. Congress on climate change in 1988. The most severe scenario (A) predicted an increase of just over 1 degree Celsius in the following thirty years, comparable to the high-end projection of 3 degrees per century but started to diverge from reality within a few years as too high. The middle scenario (B) projected an increase of less than 1 degree, roughly comparable to an increase of 2 degrees per century, and the low scenario (C) projected an increase of about 0.25 degree in thirty years, probably less than 1 degree per century. Scenarios B and C tracked closely with each other, and with the actual observed data, up to 2005. Reports since then (including the Goddard Institute's own December 2009, estimate) show that scenario C has been the most accurate.

Consequences and Solutions

Global warming may have many possible effects. The Medieval Warm Period, though beneficial to European and Arctic agriculture, often led to drought elsewhere (including the drought suspected of having caused the collapse of the Native American Anasazi culture). Similar effects can be seen in the twenty-first century; the decline in the snowpack on Africa's Mount Kilimanjaro is apparently more a result of increased local aridity than of global warming.

Warming also leads to a sea-level rise of 1 to 2 centimeters (0.4 to 0.8 inch) per decade, which could increase if the vast Greenland and Antarctic ice packs melt significantly (most models predict more snow in the interiors and more meltwater on the edges of these ice packs, and observations confirm this), which could also seriously alter key ocean currents. Also with warming, warm-weather crops can be grown further north and warm-weather habitats invade cold-weather habitats. Some scientists fear that global warming will lead to more frequent or more severe extreme weather events (particularly tropical cyclones), but there has been no observational evidence of such a trend (for example, North Atlantic hurricanes declined after the severe 2005 season).

Suggested approaches to addressing global warming include both adapting to the heat (and the effects of the heat) when it occurs (and meanwhile devoting resources to solving other problems) and trying to reduce the increase in warming. The latter can have no effect on the natural component of temperature rise and will be unnecessary if the increase is small.

Some proposed solutions aimed at reducing the greenhouse gas emissions linked with global warming are questionable. In particular, the substitution of ethanol for fossil fuels has drawbacks: Growing the crops needed to produce can in some cases decrease food production or increase cropland at the expense of forestland. Given that forests help to remove carbon dioxide from the atmosphere, the net result may be an increase in greenhouse gases.

Among the most useful and affordable ways to reduce greenhouse gases may be to increase the numbers of hybrid and electric vehicles in relation to gasoline-fueled vehicles, to improve energy conservation by individuals and industries, and to reduce reliance on the burning of fossil fuels for electricity by developing alternative sources of power. Because stronger proposed changes would involve serious economic dislocations, calls for such changes generally include long time lines for achievement. Per-capita carbon dioxide emissions have declined slightly from a 1979 peak; if the trend continues, they will level off with global around 2050.

Increasing Urgency
In 2018, the IPCC released an urgent Special Report on the impacts of a global increase in temperature by 1.5 °C compared to levels before the Industrial Revolution. The 2015 Paris Agreement within the United Nations Framework Convention on Climate Change (UNFCCC) aimed to keep the increase in average global temperate to less than 2 °C above pre-industrial levels; however, the IPCC Special Report outlines the difference that even .5 °C could make to conditions on earth. Compared to an increase in 1.5 °C, a 2 °C increase is projected to escalate risks from extreme temperatures and droughts. It is likely to further destabilize marine ice sheets, leading to amplified vulnerability of islands and coastal areas to sea level rise. The odds of transforming terrestrial ecosystems increases by 50% at 2 °C compared to 1.5 °C, which would then lead to greater loss of biodiversity. Limiting global warming to 1.5 °C decreases the risks of global warming for humans in terms of likely food and water shortages, although the IPCC Special Report makes it clear that any increase in global warming is expected to negatively affect human health.

In order to limit global warming to 1.5 °C, the IPCC Special Report cites models that involve drastic reductions in carbon and methane emissions, amounting to a decline by about 45% from 2010 levels by 2030, and reaching net zero emissions by 2050. Suggested pathways for achieving this include investing in low-carbon energy technologies, changes in urban planning to support low-emission transportation, and supporting sustainable development, which balances economic growth, environmental protection, and social prosperity for poor or disadvantaged populations. The IPCC particularly emphasizes a climate justice approach to global warming solutions, as the risks associated with climate change are generally greater for disadvantaged people and communities at all levels of development around the world.

Timothy Lane

Further Reading
Alley, Richard B. *The Two-Mile Time Machine: Ice Cores, Abrupt Climate Change, and Our Future.* Princeton, N.J.: Princeton University Press, 2000.
Dutch, Steven I., ed. *Encyclopedia of Global Warming.* 3 vols. Pasadena, Calif.: Salem Press, 2010.
Fagan, Brian. *The Great Warming: Climate Change and the Rise and Fall of Civilizations.* New York: Bloomsbury, 2008.
Houghton, John Theodore. *Global Warming: The Complete Briefing.* 4th ed. New York: Cambridge University Press, 2010.
Mann, Michael E., and Lee R. Kump. *Dire Predictions: Understanding Global Warming.* New York: DK, 2008.
Michaels, Patrick J., and Robert C. Balling, Jr. *Climate of Extremes: Global Warming Science They Don't Want You to Know.* Washington, D.C.: Cato Institute, 2009.
Room, Joseph. *Climate Change: What Everyone Needs to Know.* New York: Oxford University Press, 2018.
Singer, S. Fred, and Dennis T. Avery. *Unstoppable Global Warming: Every 1,500 Years.* Updated ed. Lanham, Md.: Rowman & Littlefield, 2008.
Weart, Spencer W. *The Discovery of Global Warming.* Rev. ed. Cambridge, Mass.: Harvard University Press, 2008.

■ Globalization

FIELDS OF STUDY
Business; Commerce; Ecology; Economics; Environment; Environmental Economics; Environmentalism; Government; International Relations; Politics; Production; Treaties

SUMMARY
Globalization in the early twenty-first century rests on a free trade or neoliberal economic model that favors open markets and global competition among states and nonstate actors in the world economy. Intense competition among developing nations to secure investment and jobs from huge transnational corporations pushes ecological interests in those countries to the background of their political agendas. Corporate interests in the developed world

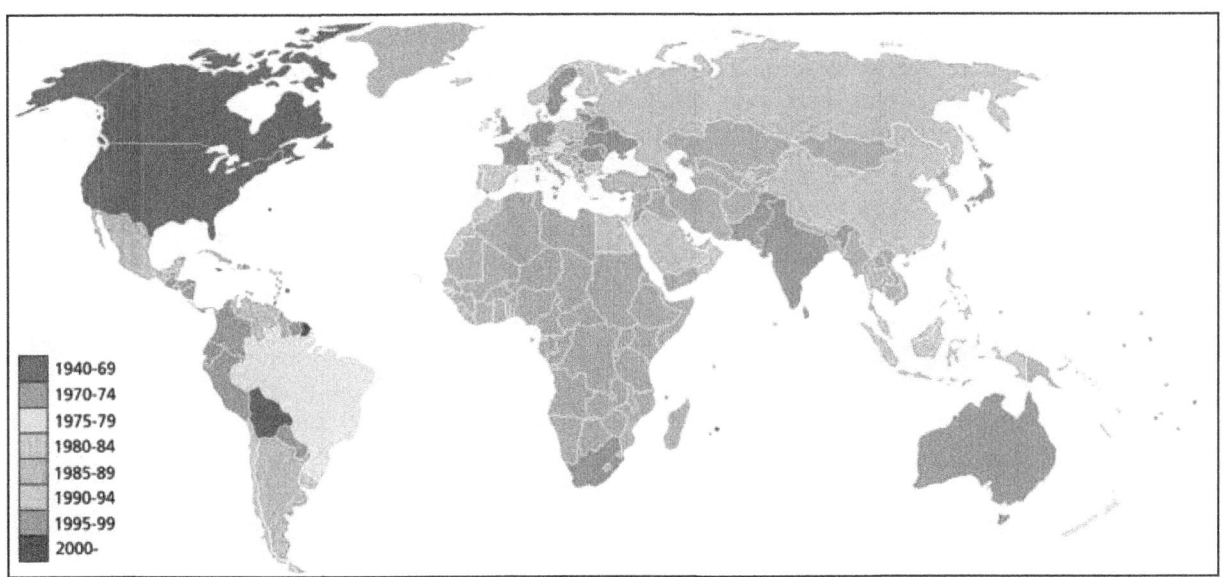

McDonald's World locations map. By BG00 (Mctoto).

tend to suppress movements for ecological reform that would cut into corporate profits.

PRINCIPLE TERMS
- **globalization:** intercontinental integration of regional economies, cultures, and political and financial systems, driven by the transnational exchange and circulation of labor, ideas, technologies, products, services, languages, and popular culture
- **League of Nations:** an organization, predating the United Nations, established in 1920 for international cooperation, initiated by the victorious Allied powers at the end of World War I
- **neoliberalism:** ideas associated with *laissez-faire* economic liberalism, and free market capitalism, including privatization, austerity, deregulation, free trade, and reductions in government spending in order to increase the role of the private sector in economy and society
- **transnational corporations**: among the world's biggest economic institutions, these for-profit enterprises are marked by engagements in business activities in two or more countries outside the nation of origin and by management decisions based on regional or global alternatives
- **World Bank:** an international financial institution, established in 1944, that provides loans to countries for capital projects and whose most recent stated goal is the reduction of poverty
- **World Trade Organization:** a global international organization established in 1995 and run by its member nations who deal with the rules of trade between nations

The modern state emerged in Western Europe in 1648, with the Peace of Westphalia, which ended the Thirty Years' War among Europe's various princes and ended the political struggle between the Roman Catholic Church and the European state, with the secular state arising as the sovereign and independent actor on the world stage. In this state-centric model, each state was recognized by other states as having the exclusive right to determine domestic policy, and each was expected to address common issues and to resolve conflicted interests through negotiations with other states in a process referred to as international relations.

After World War II, the United Nations was established (in October 1945) to usher in a new era of post-international or global relations. The United Nations followed the model of the League of Nations, created in 1919 after World War I under the Treaty of Versailles, "to promote international cooperation and to achieve peace and security." The extent and appalling nature of the World War II crimes of the Nazis against Jews and other victim groups forged a consensus among members of the global community regarding shared norms of behavior, while structural reconstruction of Europe and elsewhere after the war welded global economic networks; these factors together ultimately undermined the import of state sovereignty in favor of global cooperation.

Cooperation

New global realities demanded that states cooperate with each other to deal with common threats, develop markets, exchange technologies, manage conflict, and share power with rising nonstate actors, nongovernmental organizations (NGOs). Soon, NGOs took the form of economic organizations such as transnational corporations, advocacy organizations such as Amnesty International and Greenpeace, and service organizations such as the Red Cross and Doctors Without Borders. However, the post-World War II era also saw the emergence of less benevolent nonstate actors networking globally. Terrorist organizations expanded globally as did transnational criminal syndicates.

The twenty-first century model of globalization has many varied and interwoven aspects, and it has been evolving for a very long time, congealing most intensely since the mid-twentieth century. Globalization is primarily an economic system and is characterized by globalized financial markets, trade networks, foreign investment, and capital flows, but it has many aspects, including political globalization (effected through the proliferation of international and regional coalitions of states and nonstate actors), military globalization (networks of military force and alliances), environmental globalization (global efforts to address environmental degradation and global warming at the supranational level), and cultural globalization (through an "acculturation" process whereby people's everyday lives are fundamentally altered by the exchange of foods, people, ideas, technologies, and other products).

Regulating global trade

The General Assembly of the United Nations is the forum where members of the global community meet to negotiate political interests and address common problems and challenges. Trade practices in the global economy are regulated by three international institutions: the World Trade Organization, which regulates global trade and rules on disputes in the global marketplace; the World Bank, which makes short-term, high-interest loans to economically struggling nations; and the International Monetary Fund, which intervenes in debt-bearing nations to reorient their trade practices and financial policies. Global cooperation has fostered trade and economic development across the planet, but economic prosperity—in the developing world as well as in the developed world—is often purchased at the cost of human rights and ecological devastation. Globalization also fosters international debate on these problems, however, and has sparked global movements for "fair trade" practices and Green movements that have culminated in ongoing cooperation and international agreements to restrict polluting practices.

Controversies

Globalization is the subject of heated debate around the world, among politicians and economists as well as among scientists and environmental activists. From the standpoint of environmental ethics, globalizing trade practices have had devastating effects on the earth's natural environment. Regional neglect and the pollution of air, land, and ocean waters are driven by the "race to the bottom" phenomenon that pits developing countries against each other in efforts to lure global investors. Critics argue that the existing system is simply a broader-reaching, more profitable model of colonialism, a neocolonialism, whereby governments act as mere salespersons, promoting the profits of their corporations in a global marketplace.

Critics charge also that developing countries have no fighting chance in the global trade game, and so the rich get richer through the growing exploitation of the global poor and the devastation of the environment, in both the developed and the developing nations. Globalists, in contrast, assert that "free trade" promotes freedom and democracy, and that even as global inequality rises, poverty can be reduced through free trade. They argue that problems such as environmental degradation and global warming should be viewed as opportunities for entrepreneurial innovation and new economic ventures, and not as problems to be addressed through political intervention and legal restrictions.

Wendy C. Hamblet

Further Reading

Baylis, John, et al., editors. *The Globalization of World Politics: An Introduction to International Relations.* Oxford UP, 2013.

Bhagwati, Jagdish. *In Defense of Globalization.* Oxford UP, 2004.

Braun, Joachim von, and Eugenio Diaz-Bonilla, editors. *Globalization of Food and Agriculture and the Poor.* Oxford UP, 2007.

Friedman, Thomas L. *The World Is Flat: A Brief History of the Twenty-first Century.* 2nd rev. ed., Farrar, Straus and Giroux, 2008.

Glyn, Andrew. *Capitalism Unleashed: Finance, Globalization, and Welfare.* Oxford UP, 2006.

Kaplinsky, Raphael. *Globalization, Poverty, and Inequality: Between a Rock and Hard Place.* John Wiley, 2013.

Klein, Naomi. *This Changes Everything: Capitalism vs. The Climate.* New York: Simon & Schuster, 2015.

Mander, Jerry, and Edward Goldsmith. *The Case Against the Global Economy: And for a Turn Toward the Local.* Sierra Club Books, 1996.

McCulloch, Jock, and Geoffrey Tweedale. *Defending the Indefensible: The Global Asbestos Industry and Its Fight for Survival.* Oxford UP, 2008.

McGrew, Anthony, and Paul Lewis, editors. *Global Politics: Globalization and the Nation-State.* John Wiley, 2013.

Grasslands

FIELDS OF STUDY
Agriculture; Conservation Biology; Ecology; Environment; Environmentalism; Land Management; Landscape Ecology; Land-Use Management; Livestock Management

SUMMARY
Grasslands are a major feature of Earth's ecology, as they comprise 37 percent of land area. While grazing is of mutual benefit to plants and animals, overgrazing is ultimately detrimental to both the plant and animal populations, as well as to grassland ecosystems. Maintaining a balance between grazing animals and the plants on which they feed prevents deleterious consequences.

PRINCIPAL TERMS
- **Bureau of Land Management:** an agency within the United States Department of the Interior that administers more than 247.3 million acres (1,001,000 kilometers) of public lands, constituting one-eighth of the country's landmass
- **grasslands:** an area, as a prairie, in which the natural vegetation consists largely of perennial grasses
- **habitat:** the specific part of the environment occupied by the individuals of a species
- **industrialization:** is the process by which an economy is transformed from primarily agricultural to one based on the manufacturing of goods
- **overgrazing:** when plants are exposed to intensive grazing for extended periods of time, or without the necessary recovery periods population: a group of all the individuals of one species
- **United States Forest Service:** an agency of the U.S. Department of Agriculture that administers the nation's 154 national forests and 20 national grasslands
- **urbanization:** An increase in a population in cities and towns versus rural areas

Grazing is the consumption of any plant species by any animal species; grasslands are ecosystems where grasses and other nonwoody vegetation predominate. While grazing is of mutual benefit to plants and animals, overgrazing is ultimately detrimental to both the plant and animal populations, as well as to grassland ecosystems. Maintaining a balance between grazing animals and the plants on which they feed prevents deleterious consequences.

Grasslands are a major feature of Earth's ecology, as they comprise 37 percent of land area. Grasslands characterized by the presence of low plants, mostly grasses, and are distinguished from woodlands, tundra, and deserts. Grasslands experience sparse to moderate rainfall and are found in both temperate and tropical zones. Grassland plants coevolved over millions of years with the grazing animals that depended on them. Wild ancestors of cattle and horses, as well as antelope and deer, were found in Eurasian grasslands. On the North American prairie, bison and antelope prospered. Wildebeest, gazelle, zebra, and buffalo dominated African savannas, whereas the kangaroo was the preponderant grazer in Australia. Grasslands occupied vast areas of the world more than ten thousand years ago, before the development of agriculture and industrialization, and the subsequent explosive growth of the human population.

Grazing is a symbiotic relationship whereby animals gain their nourishment from plants, which in turn benefit from the activity. Grazing is also an important contributor to human food security, as it provides most of the nutrition for ruminants that in turn, supply humans with meat and milk. Grazing removes the vegetative matter required for grasses to grow, facilitates seed dispersal, and disrupts mature plants, permitting young plants to take hold. Urine and feces from grazing animals recycle nutrients to the plants. The grassland ecosystem also attracts other animals, including invertebrates, birds, rodents, and predators. The grasses, grazing animals, and grassland carnivores, such as wolves or cat species, constitute a food chain.

Grasses are generally well suited to periods of low rainfall because of their extensive root systems and can go dormant during periods of drought. Humans have been an increasing presence in grassland areas, where more than 90 percent of modern crop production occurs, and much urbanization and industrialization have taken place. The remaining grasslands, unsuitable for crops because of inadequate rainfall or difficult terrain, are used for grazing by domesticated or wild herbivores. In addition, many woodland areas around the world have been cleared and converted to grasslands where animals can graze.

Impacts of Overgrazing

Continued heavy grazing of a given area leads to deleterious environmental consequences. Even repeated removal of leaf tips will not adversely affect the regeneration of grasses, provided that the basal zone of the plant remains intact. Whereas animals can generally safely eat the upper half of the grass shoot, if they ingest the lower half, which sustains the roots and fuels regrowth, they will eventually kill the plants. Overgrazing leads to denuding of the land, invasion by less nutritious plant species, erosion caused by decreased absorption of rainwater by soil, and starvation of animal species. Because the loss of plant cover changes the reflectance of the land, climate changes can follow that make it virtually impossible for plants to return, with desertification an ultimate consequence.

The number of animals is not the only factor in overgrazing; the timing of the grazing can also be detrimental. Grasses require time to regenerate, and continuous grazing will inevitably kill them. Consumption too early in the spring can stunt their development. Semiarid regions are particularly prone to overgrazing because of low and often unpredictable rainfall; regrettably, these are the areas of the world to which much livestock grazing has been relegated, because the moister grassland areas have been converted to cropland.

Overgrazing by wildlife, as well as livestock, can also be deleterious. The Kaibab Plateau deer disaster in Arizona is one such example, where removal of natural predators and livestock that competed with the deer for food led first to a deer population explosion, then to overgrazing by the deer, followed by starvation and large die-offs within the deer population. Protection of elk and bison in Yellowstone National Park has similarly led to high populations, excessive grazing, and changes to the environment. Only the provision of winter feed has prevented the die-offs that would otherwise naturally ensue. Ironically, winter feeding has perpetuated the problem by maintaining these populations at levels higher than grazing can sustain. Feeding has also encouraged the animals to congregate in unusually large numbers, which has contributed to the spread of disease. In 2010 a coalition of conservation groups lost a lawsuit to stop the supplemental feeding of elk and bison on the National Elk Refuge in Wyoming.

Grasslands Management

Grassland areas need not deteriorate if they are properly managed, whether for livestock, wild animals, or both. The land's carrying capacity, or the number of healthy animals that can be grazed indefinitely in an area, must not be exceeded. Because of year-to-year changes in weather conditions and hence food availability, determining carrying capacity is not simple; worst-case estimates typically have been used as guidelines to minimize the risk of exceeding carrying capacity. The goal should be a grassland rendered and kept healthy by optimizing, not maximizing, the number of animals. For private land, optimizing livestock numbers is in the long-term self-interest of the landowner. For publicly held land, managed in common or with unclear or disputed ownership, restricting animals to the optimum level is particularly difficult to achieve. Personal short-term benefit often leads to long-term disaster, in a phenomenon known as the tragedy of the commons.

Managing grasslands involves controlling the numbers of animals and enhancing the habitat. Cattle and sheep can be physically restricted through the use of herding and fencing, although requiring such restrictions can be difficult to achieve through political means. Much more problematic is controlling wildlife when natural predators have been eliminated and hunting is severely restricted. As for habitat improvement, the prudent use of chemical, fire, mechanical, and biological approaches can increase carrying capacity for domesticated and wild herbivores. Removing woody vegetation by burning or mechanical means can increase grass cover, fertilizing can stimulate grass growth, and reseeding with desirable species (plants native to the particular region) can enhance the habitat. Effective grassland management also requires matching animals with the grasses on which they graze.

An approach to grazing known as holistic management may have the potential not only to stave off ecosystem damage of grasslands but also to reverse desertification. This approach operates on the essential principle that, because herbivores and perennial grasses evolved together, the grasses will thrive only in combination with herbivores grazing and roaming naturally. Contrary to common wisdom regarding best management practices for grazing, holistic management involves grazing livestock in ultradense, constantly moving herds that mimic big-game grazing patterns. The livestock till the soil with their hooves and fertilize it with their excrement. By grazing the grasses, they allow sunlight to reach the grasses' growth buds; by contrast, when grazing is so restricted that the vegetation can die upright, the growth buds are shielded from the sun and the entire plant dies the following year. The common management practice of allowing grazed land an extended period to rest and recover, then, may not promote a resurgence of vegetative cover; rather, this practice may cause the land to remain barren and dry.

In 1992 holistic management pioneer Allan Savory began a program in Zimbabwe, Africa, in which livestock herds were increased by 400 percent on 2,630 hectares (6,500 acres) of land that had been barren for hundreds of years. By 2010, after years of holistic planned grazing, this area had become healthy grassland with open water. Other holistic management practitioners around the globe have enjoyed similar successes.

Grazing in the United States

There are roughly 312 million hectares (770 million acres) of rangelands (grasslands, forests, wetlands, and other ecosystems that are suitable for grazing) in the United States, more than half of which are privately owned. The federal government manages 43 percent, and the remainder is under state and local government control.

Laws pertaining directly to grazing in the United States include the Taylor Grazing Act of 1934, the Federal Land Policy and Management Act of 1976, and the Public Rangelands Improvement Act of 1978. The Taylor Grazing Act introduced measures to control the unregulated grazing practices of homesteaders that had led to overgrazing, enhanced erosion, damage to streams and springs, and the land's reduced productivity; however, rancher needs still tended to take precedence over range condition. Four decades later, heightened environmental awareness led to passage of the Federal Land Policy and Management Act, which established a multiple-use mandate for land management agencies to serve present and future generations in their practices. Not long after came passage of the Public Rangelands Improvement Act, which sought to improve the condition of public rangelands so that they might meet their potential for grazing and other uses. U.S. laws pertaining to environmental quality and endangered species also have impacts on rangeland management.

Both the U.S. Forest Service and the Bureau of Land Management implement a regulatory system of permits, rental fees, herd size limits, and grazing seasons. They must maintain a balance among several often-conflicting objectives: providing forage for grazing and browsing animals, ensuring the land's long-term health and productivity, protecting watersheds, managing wildlife habitat, administering permitted mineral and energy resource exploration and extraction, offering recreational opportunities, and preserving the land's distinctive character and aesthetic appeal. In order to meet the array of resource needs, rangeland management agencies inventory, classify, and monitor rangeland conditions. Where rangeland health needs improvement, they implement measures to restore ecosystem functions. Public land decision makers must take into account a variety of factors that affect rangelands, including severe and extensive wildfires, invasive plant species, rural residential development driven by population increases, and global climate change.

James L. Robinson, updated by Karen N. Kähler

Further Reading

Chiras, Daniel D., and John P. Reganold. "Rangeland Management." In *Natural Resource Conservation: Management for a Sustainable Future*. 10th ed. Upper Saddle River, NJ: Prentice Hall, 2010.

Du Toit, Johan, Richard Kock, and James Deutsch. *Wild Rangelands: Conserving Wildlife While Maintaining Livestock in Semi-Arid Ecosystems*. Chicester, England: Wiley, 2012.

Gibson, David J. *Grasses and Grassland Ecology*. New York: Oxford University Press, 2009.

Gordon, Iain J., and Herbert H. T. Prins, eds. *The Ecology of Browsing and Grazing*. New York: Springer, 2010.

Lemaire, G., et al., eds. *Grassland Ecophysiology and Grazing Ecology*. Wallingford, UK: CABI Publishing, 2007.

Lye, Lin Heng. *Sustainability Matters: Environmental Management in Asia*. Singapore: World Scientific, 2010.

Manske, Llewellyn L. *Perpetually Sustainable Grazingland Ecosystems*. Dickinson, North Dakota: North Dakota State University, 2013.

Manske, Llewellyn, and Sheri A, Schneider. *Biologically Effective Management of Grazinglands*. Dickinson, ND: North Dakota State University, 2014.

Vallentine, John F. *Grazing Management*. Amsterdam/New York: Elsevier Science, 2016.

Woodward, Susan L. *Grassland Biomes*. Westport, CT: Greenwood Press, 2008.

■ Green buildings

FIELDS OF STUDY

Architecture; Construction; Ecology; Energy and Energy Resources; Environment; Environmental Engineering; Environmentalism; Land-Use Management; Landscape Architecture; Renewable Resources; Technology and Applied Science

SUMMARY

Residential and commercial buildings generate more than 30 percent of the world's emissions of carbon dioxide, a greenhouse gas that has been linked to global warming. Green buildings reduce carbon emissions substantially and provide significant environmental benefits by reducing solid waste, efficiently using energy and other resources, reducing air and water pollution, and conserving natural resources.

PRINCIPAL TERMS

- **Earth Day:** annual event, founded in 1970 and celebrated worldwide on April 22 to demonstrate support for environmental protection
- **Environmental Defense Fund:** a US-based nonprofit advocacy group, working to find environmental solutions with scientists and policy specialists worldwide
- **Gaia hypothesis:** theory developed in 1970 proposing that living organisms interact with their inorganic surroundings on Earth to form a complex system that helps maintain and perpetuate life on the planet
- **green buildings:** Structures designed and constructed to increase resource efficiency and reduce negative impacts on human health and the environment
- **photovoltaic:** of, relating to, or utilizing the generation of a voltage when radiant energy falls on the boundary between dissimilar substances
- **U.S. Green Building Council (USGBC):** a private, nonprofit organization founded in 1993 to promote sustainability in building design, construction, and operation

A picture of a house fitted with both photovoltaic (PV) and thermodynamic panels. The PV-panels are the blue ones, located along the right edge of the roof, and the thermodynamic panels are the black boxes with horizontal lines (pipes) directly above the right-most skylight (window). By KVDP (Own work).

Although energy efficiency and sustainability were not major concerns at the time, early green buildings originated during the mid-nineteenth century. The Galleria Vittorio Emanuele II in Milan, Italy, designed in 1861, and the Crystal Palace in London, England, built in 1851, both used underground air cooling and roof ventilators to control the interior temperature.

From the 1930s to the 1960s technological advances such as the inventions of reflective glass, structural steel, and air-conditioning resulted in the proliferation of high-rise buildings that consumed huge amounts of cheap fossil fuels. During the 1960s, however, environmental consciousness grew, and visionaries began defining green building. During this period scientist James Lovelock formulated the Gaia hypothesis, a holistic concept of the earth as a single, complex organism. In 1969 landscape architect Ian L. McHarg published *Design with Nature*, which helped define green architecture.

Beginnings of the Movement

On the first Earth Day, in April 1970, millions of Americans showed their concern about the environment. The 1973 and 1979 oil crises demonstrated the need for the nation to seek energy from diversified sources and become less dependent on fossil fuels. The US government and many corporations began investing in research into methods of energy conservation and alternative energy sources.

During the 1980s architect Malcolm Wells designed green underground and earth-sheltered buildings. In 1982, physicist Amory Lovins and his wife, environmentalist Hunter Lovins, emphasized the basic green principle of using regional resources in founding their Rocky Mountain Institute, a nonprofit resource policy center that promotes resource efficiency and global security. Beginning during the mid-1980s, popular environmental organizations—such as the Sierra Club, Greenpeace, the Nature Conservancy, and Friends of the Earth—became increasingly active. Growing awareness of the problem of sick building syndrome raised concerns regarding the indoor environments of some workplaces.

In 1984 architect William McDonough designed a headquarters building for the Environmental Defense Fund in New York City using a high-performance building approach (the building was completed in 1985). During the late 1980s Pliny Fisk III designed Blueprint Farm—a green agricultural community—in Laredo, Texas, using recycled materials, wind power, and photovoltaic panels.

Milestones During the 1990s

In 1992, the first local green building program began in Austin, Texas, and the US Environmental Protection Agency (EPA) launched the Energy Star program, a voluntary energy-efficiency labeling program for consumer products. By 2012, Energy Star labels were appearing on sixty-five product categories and Energy Star ratings had become the standard for major appliances, homes, commercial buildings, and heating systems. By 2012, 1.4 million Energy Star–qualified homes had been built throughout the United States. Many other countries adopted the Energy Star idea, including Japan, Taiwan, China, New Zealand, South Africa, and the nations of the European Union.

In 1993 Bill Clinton's presidential administration began the successful "Greening of the White House" initiative, and the nonprofit US Green Building

Council (USGBC) was created to promote the construction of environmentally responsible, healthy, and profitable buildings. USGBC is a national, voluntary consensus coalition with members from all sectors of the building industry. In 1995, USGBC began developing its green building certification program, known as LEED (for Leadership in Energy and Environmental Design), which became available for public use in 2000. This voluntary system provides third-party certification that certain standards have been met in the construction of high-performance, sustainable buildings, with an emphasis on reducing carbon dioxide emissions and increasing energy efficiency. LEED certification covers a wide range of existing and new commercial and residential buildings, including offices, schools, medical facilities, private homes, and stores.

Environmental Benefits

The key areas measured in the LEED certification process reflect the environmental benefits of green building. Sustainable site development involves preserving natural resources for future generations and can include reusing existing buildings, planting around buildings, roof gardens, and underground or earth shelters. Building for water savings and efficiency involves monitoring water supplies and usage, recycling gray or previously used water, and constructing rainwater catchment systems. To improve energy and atmosphere efficiency, buildings can use geographically and climatically appropriate energy resources, including renewable energy. Efforts to conserve materials and resources include using renewable, recycled, local, chemical-free, nonpolluting, and durable materials. Indoor environmental quality can be improved with nontoxic materials, adequate ventilation and insulation, energy-efficient temperature controls, and materials that emit few or no volatile organic compounds.

In the twenty-first century, as environmental knowledge and building technologies continue to improve, the green building movement is gaining worldwide momentum. Given that commercial and residential buildings generate more than 38 percent of carbon dioxide emissions and represent 68 percent of total electricity consumption in the United States, the benefits of green building have become increasingly obvious. By 2015 more than forty countries had developed their own LEED initiatives, including Australia, Brazil, Canada, France, India, Israel, Mexico, the United Arab Emirates, and the United Kingdom. Green buildings offer a number of economic benefits in the long term, including reduced costs for heating, cooling, and electricity.

Alice Myers

Further Reading

Fisanick, Christina, ed. *Eco-rchitecture*. Detroit: Greenhaven Press, 2008.

GreenSource. *Emerald Architecture: Case Studies in Green Building*. New York: McGraw-Hill, 2008.

Johnston, David, and Scott Gibson. *Green from the Ground Up: A Builder's Guide—Sustainable, Healthy, and Energy-Efficient Home Construction*. Newtown, Conn.: Taunton Press, 2008.

Kruger, Abe, and Carl Seville. *Green Building: Principles and Practices in Residential Construction*. Clifton Park, N.Y.: Delmar, 2013.

McHarg, Ian L. *Design with Nature*. 25th Anniversary Edition. Hobokien, N.J.: Wiley, 1995

United States Environmental Protection Agency. "Sources of Greenhouse Gas Emissions: Commercial and Residential Sector Emissions." *EPA.gov*. Environmental Protection Agency, 17 Apr. 2014. Web. 30 Jan. 2015.

Yudelson, Jerry. *The Green Building Revolution*. Washington, DC: Island, 2008.

■ Groundwater pollution

FIELDS OF STUDY

Agriculture; Animal Husbandry; Biology; Ecology; Ecosystems; Emissions; Environment; Environmentalism; Hazardous Materials; Life Sciences; Pollution; Toxic Waste; Water Resources

SUMMARY

Many public and private water supplies rely on wells that tap important groundwater reserves. Pollution of groundwater leads to changes in water quality that can affect groundwater use for a given purpose.

PRINCIPAL TERMS

- **groundwater pollution:** degradation, by chemicals and other substances, of the water found below the surface of the earth

- **hydrostatic:** of or relating to fluids at rest or to the pressures they exert or transmit
- **nitrogen:** a colorless, tasteless, odorless element that, as a diatomic gas, is relatively inert and constitutes 78 percent of the atmosphere and is a constituent of organic compounds found in all living tissues
- **pollutants:** any substance, as certain chemicals or waste products, that renders the air, soil, water, or other natural resource harmful or unsuitable for a specific purpose
- **topography:** the detailed mapping or charting of the features of a relatively small area, district, or locality
- **transmissivity:** a measure of the ability of a material or medium to transmit electromagnetic energy, as light

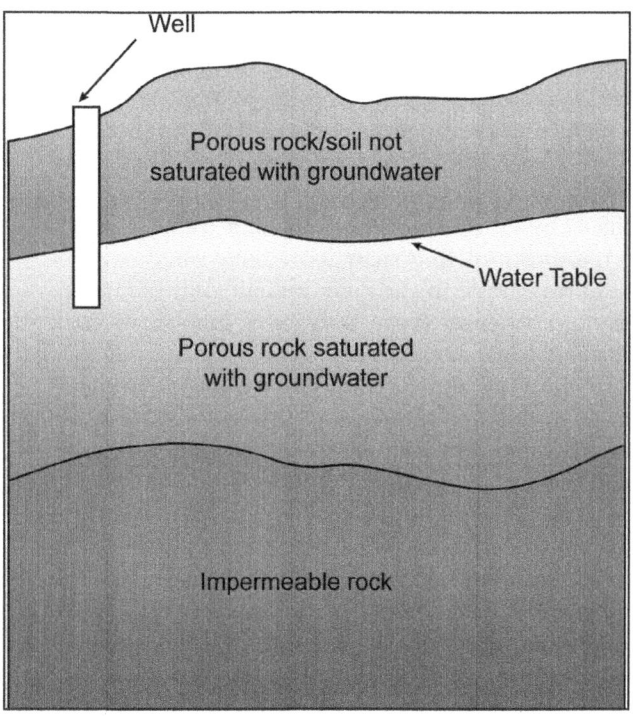

Groundwater - A visualization of the Hubbard Model from a simulation conducted under the INCITE program. By US Government.

Humans require vast amounts of fresh water for use in homes, livestock operations, agriculture, and industrial processes. Groundwater is an important source of fresh water. The pollution of groundwater by human activity can contaminate water-supply wells, making the water they provide unacceptable for drinking and other purposes. This can lead to a need for new water supplies that may not be readily available or easily accessible. In some instances, polluted groundwater interacts with surface water, thus contaminating the surface-water environment as well.

Groundwater constitutes a small but significant portion of the world's overall water supply. Much of the earth's surface is covered by water, but an estimated 97.2 percent of it exists as salt water. Since fresh surface water may account for as little as 0.009 percent of the earth's water, groundwater is a significant source of readily available fresh water. Groundwater occurs in the saturated zone of the earth, which is the area below the surface where pores between particles—void spaces in the soil or rock—are filled with water. In some places, groundwater may be encountered near the surface, but in other areas, such as arid regions, it can be quite deep below the surface. Groundwater flows from areas of high hydrostatic head to areas of low hydrostatic head. Shallow groundwater often mimics topography, flowing downhill toward streams and lakes.

The soil and rock through which groundwater flows consist of particles of varying size, which help determine the classification of the soil or rock and how well water will move through the material. Sand-sized particles are seen in unconsolidated sandy soils or sandstones. Smaller particles may form silty or clayey soils or their bedrock equivalents of siltstones and shales. In the saturated zone, groundwater saturates the pores and voids between the particles. The size of the pores and the degree to which they are interconnected affect hydraulic conductivity—a measure of the ability of water to move through the rock or soil.

Transmissivity is the measure of the ability of an aquifer to transmit water and is a measure of the hydraulic conductivity multiplied by the saturated thickness of the aquifer. Therefore, a thick aquifer with relatively poor hydraulic conductivity might be able to transmit as much water as a thinner aquifer composed of materials with greater hydraulic conductivity. Groundwater is recharged by rainwater percolating through the soil, snowmelt, and rivers and streams.

Threats to Groundwater

Humans produce a wide array of pollutants and combinations of pollutants. The degree and extent to which individual pollutants can affect groundwater quality is dependent on a large number of variables, which can include the amount of contaminant introduced into the environment, the time frame in which it is introduced, its toxicity, its mobility, whether it will readily degrade in the environment, and the chemical and physical characteristics of the soil or rock through which it will pass.

Even something as common as nitrogen can lead to pollution in groundwater. Nitrogen can be mobile in the environment in the form of dissolved nitrates and nitrites. Sources for pollution include septic tanks, leaks from sewage treatment plants and lagoons, and animal wastes. Nitrogen is also an important component of many fertilizers used in agriculture, and such fertilizers may become dissolved by rainwater and percolate down into groundwater. In high enough concentrations, nitrates can make water unacceptable for human consumption. At even higher concentrations, the water can become unacceptable for livestock and other animals.

Gasoline spills and leaks from underground storage tanks are relatively common sources of groundwater pollution. Some of the dissolved-phase components of gasoline are quite mobile in the environment; however, many are also susceptible to biological degradation. Gasoline and other substances less dense than water can float on the surface of groundwater, but seasonal fluctuations in the water table can smear such contaminants in the soil, potentially making them more difficult to remove. Other contaminants, such as chlorinated solvents, can be denser than water and have the capacity to sink into aquifers.

Although metals as a group are generally not considered very mobile and tend to be adsorbed onto soils, some are quite mobile, and contamination by heavy metals can be a relatively common form of groundwater contamination. Although less common, radiological contamination of groundwater can be a concern. Groundwater often moves slowly, but radioactive half-lives can be quite long.

Raymond U. Roberts

Further Reading

Appelo, C. A. J., and D. Postma. *Geochemistry, Groundwater, and Pollution.* New York: Balkema, 2005.

Chiras, Daniel D. "Water Pollution: Sustainably Managing a Renewable Resource." In *Environmental Science.* Sudbury, Mass.: Jones and Bartlett, 2010.

Heath, Ralph C. *Basic Groundwater Hydrology.* Reston, Va.: U.S. Geological Survey, 2004.

Sampat, Payal. *Deep Trouble: The Hidden Threat of Groundwater Pollution.* Washington, D.C.: Worldwatch Institute, 2000.

Todd, David Keith, and Larry W. Mays. *Groundwater Hydrology.* Hoboken, N.J.: John Wiley & Sons, 2005.

Younger, Paul L. *Groundwater in the Environment: An Introduction.* Malden, Mass.: Blackwell, 2007.

H

Habitat destruction

FIELDS OF STUDY
Agriculture; Animal Husbandry; Biology; Botany; Conservation Biology; Ecology; Ecosystems; Environment; Environmental Technology; Environmentalism; Forestry; Horticulture; Life Sciences; Marine Biology; Oceanography; Zoology

SUMMARY
The destruction of habitat represents a pressing threat to global biodiversity, as habitat loss leads to species extinctions. In the twenty-first century, most habitat destruction is occurring in developing nations, where overpopulation and poverty contribute to the need to convert forestland to agriculture.

PRINCIPAL TERMS
- **deforestation:** clearing or thinning of forests by humans to make the land available for other uses
- **ecosystem:** a relatively self-sufficient group of communities and their abiotic environment
- **environment:** the habitat created by the interaction of the abiotic and biotic parts of an ecosystem
- **habitat destruction:** degradation of a natural landscape so that it becomes functionally incapable of supporting its native species
- **habitat:** the specific part of the environment occupied by the individuals of a species
- **monoculture:** the use of land for growing only one type of crop
- **population:** a group of all the individuals of one species
- **species:** a group of similar organisms that are capable of interbreeding and producing fertile offspring

Habitat destruction occurs when human beings remove or significantly alter the land or aquatic communities where animal species dwell. Human civilization was built on the practice of altering land for human purposes, such as the burning of forest for pasture, logging for construction materials, the draining of marshland for development, and mining for resource extraction. In the twenty-first century, however, the density of human population on the earth, combined with modern industrial technology, means that humans' alterations by humans of the natural landscape greatly exceed the ability to render land productive for human needs. Further, the most biodiverse ecoregions of the world, tropical rain forests and coral reefs, have seen rapid increases in

Red Panda - Colchester Zoo, Colchester, Essex, England - September 2008. By Keven Law from Los Angeles, USA.

habitat destruction. Tropical forests are down to about 1 billion hectares (2.5 billion acres) from the nearly 1.6 billion hectares (4 billion acres) they occupied approximately two hundred years ago. One-fifth of coral reefs have been destroyed, and another one-fifth have been severely degraded.

Habitat destruction is most often caused by expansion of agriculture. Planting crops or raising livestock requires wide expanses of bare soil or grassy plane. When suitable land is unavailable naturally, many land types can be converted to agricultural uses: Wetlands can be drained, forests can be logged, deserts can be irrigated. Although the productivity of such converted land may be relatively high, biodiversity and ecosystem functionality drop harshly. Modern agriculture is often monocultural (devoted to a single type of crop), without topographical complexity, and devoid of any plants not being grown for either sale or consumption. Therefore, the available niches for native species are nearly all removed.

Other causes of habitat destruction include mining, ocean trawling, oil prospecting, urban sprawl, and infrastructure development. These practices directly destroy ecosystems, but other practices indirectly degrade surrounding ecosystems. Desertification is caused by overgrazing of livestock and excessive extraction of groundwater, which render the land unusable, usually affecting communities that already live in resource-impoverished landscapes. Deforestation on a small scale can cause ecosystem collapse by dividing a forest into fragments, rendering the land unfit to support animals with large ranges and plants with wide dispersal needs. Coral degradation is rarely caused by direct destruction; rather, coral is negatively affected by increases in water temperature and changes in water chemistry resulting from climate change and industrial pollution.

Depending on terrain characteristics and climate patterns, landscapes naturally acquire ecosystem types that are functional to their locations. When humans alter these land types for unnatural purposes, the consequences can be deleterious. Modern examples include the levy system in New Orleans, which replaced an extensive natural wetland buffer; when Hurricane Katrina hit the city in 2005, the levy system failed, and the city was flooded. The devastation that resulted when Haiti was struck by a massive earthquake in 2010 was magnified by the high rate of deforestation, and therefore high rate of soil erosion, in that nation. Maintaining functional natural landscapes is increasingly seen as a priority, and restoration efforts are being developed to restore native habitat types even in urban areas.

Developing Countries

Since the mid-twentieth century, most of the world's habitat destruction has taken place in developing countries, as developed countries have already exhausted their most accessible resources and altered much of their land for development. For example, nearly 50 percent of wetlands in the United States and 60–70 percent of European wetlands have been destroyed.

The main factors contributing to land alteration in the developing world are poverty, overpopulation, lack of sustainable technology, and adherence to cultural practices. For example, many communities in the developing world frequently cook with charcoal, as electricity and natural gas are in short supply. Charcoal is acquired through the burning of forestland, and the results are mass deforestation and air pollution. New technologies that do not require large monetary investments, such as solar ovens and permaculture techniques, are increasingly being offered as solutions to poor communities that rely on habitat destruction to survive.

Many farmers in the developing world are wary of foreign technology and are unwilling to risk the failure of a crop to adopt a new technology, even if it could result in increased production. Because of this wariness, education is needed to help farmers ease into the use of new technologies that can benefit them, and the environment, in the long term. For ecological conservation to be sustainable, it must be beneficial for both the farmers and the environment.

Future Solutions

The ability of humans to alter the land for their own gain is one of the main reasons humans have been able to expand over the planet and live in nearly every climate. Since the dawn of large civilizations, humans have increasingly acquired the means to reap more and more from the environment, sometimes altering it so severely that they have eradicated whole communities of species. Mass extinctions occurred in prehistoric times during the early years of humanity, but the species that died off at that time were mostly those targeted for food. In contrast, in modern times entire ecosystems are destroyed for resources and agriculture, and this can only have increasingly negative effects on both the natural world and human habitations.

In order to measure the direct importance for humans of many landscapes, scientists have defined the ecosystem services provided by the landscapes. These services, such as erosion prevention, storm buffering, soil productivity, and wildfire prevention, are often given monetary equivalents to introduce them into a system of economics.

Other approaches to reducing habitat destruction include the use of modern techniques to increase agricultural productivity that can reduce the need to clear more land for farming. Planting native species, creating tree-shaded spaces, and increasing topographical heterogeneity on agricultural land can increase biodiversity without having a large impact on production. In addition, the incorporation of natural areas around cities to buffer weather, prevent erosion, and safeguard watersheds may save much money in repairs and provide protection against disasters.

Jamie Michael Kass

Further Reading

Barbault, R., and S. D. Sastrapradja. "Generation, Maintenance, and Loss of Biodiversity." *Global Biodiversity Assessment.* Ed. V. H. Heywood. New York: Cambridge UP, 1995.

Cincotta, Richard P., and Robert Engelman. *Nature's Place: Human Population Density and the Future of Biological Diversity.* Washington, DC: Population Action Intl, 2000.

Pullin, Andrew S. "Effects of Habitat Destruction." *Conservation Biology.* New York: Cambridge UP, 2002

Tibbetts, John. "Louisiana's Wetlands: A Lesson in Nature Appreciation." *Environmental Health Perspectives* 114 (2006): A40–A43.

■ Hazardous wastes

FIELDS OF STUDY
Biology; Chemistry; Conservation Biology; Ecology; Ecosystems; Emissions; Environment; Environmental Chemistry; Environmentalism; Hazardous Materials; Life Sciences; Pollution; Public Health; Toxic Waste Management

SUMMARY
Although many national governments have taken steps to regulate the disposal of hazardous wastes, some have not, and many such wastes continue to be produced all over the world. Improper disposal of hazardous wastes creates serious problems for the environment.

PRINCIPAL TERMS

- **dioxin:** any of several persistent toxic heterocyclic hydrocarbons that occur especially as by-products of various industrial processes and waste incineration
- **Environmental Protection Agency (EPA):** an independent federal agency, created in 1970, that sets and enforces rules and standards to protect the environment and control pollution
- **incineration waste:** waste products of industrial society that pose dangers to human heath and the environment
- **hazardous waste:** waste products of industrial society that pose dangers to human health and the environment
- **polychlorinated biphenyls**: group of man-made organic chemicals consisting of carbon, hydrogen and chlorine atoms
- **Resource Conservation and Recovery Act (RCRA):** the public law, enacted in 1976, that creates the framework for the proper management of hazardous and non-hazardous solid waste
- **Superfund:** a United States federal government program, established as the Comprehensive Environmental Response, Compensation, and Liability Act in 1980, designed to fund the cleanup of sites contaminated with hazardous substances and pollutants

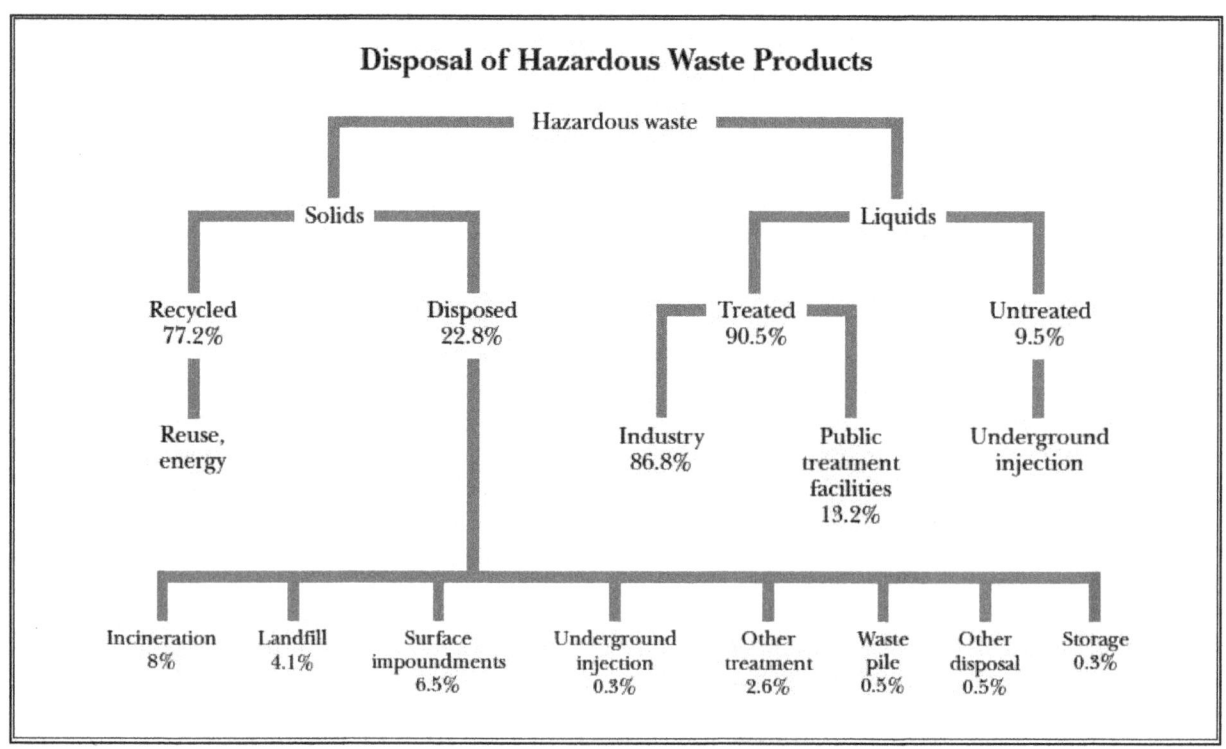

In the United States, hazardous wastes are legally defined as materials that have ignitable, corrosive, reactive, or toxic properties. In the early 1990s, approximately 97 percent of all hazardous waste in the United States was produced by 2 percent of the waste generators. Remediation and cleanup of these wastes involve substantial economic cost. Since the 1970s the United States and other Western democracies have tried to regulate hazardous waste disposal. Hazardous wastes are also a serious problem in the former Soviet Union and other Eastern European nations.

Environmental Problems

Improper disposal of hazardous waste can lead to the release of chemicals into the air, surface water, groundwater, and soil. High-risk wastes are those known to contain significant concentrations of constituents that are highly toxic, persistent, highly mobile, or bioaccumulative. Examples include dioxin-based wastes, polychlorinated biphenyls (PCBs), and cyanide wastes. Intermediate-risk wastes may include metal hydroxide sludges, while low-risk wastes are generally high-volume, low-hazard materials.

Radioactive waste is a special category of hazardous waste, often presenting extremely high risks, as do biomedical and mining wastes.

Hazardous waste presents varying degrees of health and environmental hazards. When combined, two relatively low-risk materials may pose a high risk. Factors that affect the health risk of hazardous waste include dosage received; age, gender, and body weight of those exposed; and weather conditions. The health effects posed by hazardous waste include cancer, genetic defects, reproductive abnormalities, and central nervous system disorders.

Environmental degradation resulting from hazardous waste can render various natural resources, such as croplands and forests, useless and can harm animal life. For example, chemicals can leach out of improperly stored waste and into groundwater. Hazardous wastes may also generate long-lasting air pollution, water pollution, or soil contamination. In the past, before standards were in place for managing hazardous wastes, such materials were often buried or stored in unattended drums or other containers. This situation created threats to the environment and human health when the original containers began to leak, and the materials leached into the soil and the water supply.

Love Canal

The events that took place at Love Canal demonstrate the environmental damage and dangers to human health posed by the improper disposal of toxic wastes. The discovery and identification of dangerous chemical wastes in the Love Canal neighborhood of Niagara Falls, New York, in 1976, transformed a community where the residents' livelihoods depended on the chemical companies long established in the area.

Beneath a 2-foot clay cap over the canal lay 43.6 million pounds of eighty-two different chemical residues. Included were benzene, a chemical known to cause anemia and leukemia; lindane, exposure to which results in convulsions and excess production of white blood cells; chloroform, a carcinogen that attacks respiratory, nervous, and gastrointestinal systems; trichloroethylene, a carcinogen that attacks genes, livers, and nervous systems; and methylene chloride, which can cause recurring respiratory distress and death. The most dangerous chemical in the waste, however, was dioxin, a component of the 200 tons of trichlorophenol dumped in the canal. Dioxin is one of the most powerful known carcinogens.

Breaks in the clay cap and walls of the canal created openings through which the toxic chemicals eventually flowed, when during the winter of 1975-1976, heavy snowfall caused the groundwater level to rise, filling the uncovered canal. Portions of the landfill subsided, and waste storage drums surfaced in several locations. Surface water, heavily contaminated with chemicals, was found in the backyards of houses bordering the canal.

Houses were boarded up and abandoned, a school was left empty and falling, warning signs were posted, and the entire area was fenced off. The completeness of the human and ecological devastation at Love Canal made it the standard against which all subsequent chemical waste disasters have been compared. The toxic terror generated by Love Canal was caused by chemical waste buried at a time when the term "pollution" was not yet part of the American vocabulary.

Methods for Handling Wastes

The technologies and methods used in dealing with hazardous solid and liquid wastes continue to evolve. Several approaches have had positive impacts on the environment and the consumption of natural resources. One is the reduction of the volume of waste material through efforts to generate less of it. Another approach is to recycle hazardous materials as much as possible. A third means of dealing with hazardous waste is to treat it to render it less harmful; often, such treatment also reduces its volume. Least desirable among methods of addressing the problem is the storage of hazardous wastes in landfills. The Environmental Protection Agency (EPA) has established standards for the responsibility and tracking of hazardous wastes, based on the principle that waste generators are responsible for their waste "from cradle to grave." This principle requires that waste generators and disposal sites keep extensive records.

The costs for the cleanup and remediation of hazardous waste are substantial and are likely to continue to grow. This situation is particularly true in Eastern Europe and the nations of the former Soviet Union, where the magnitude of past dumping of hazardous materials is slowly becoming apparent. Meanwhile, less industrialized nations generally are ignoring the issue of hazardous waste, thereby setting themselves up for future difficulties.

US Legislation

The Love Canal case shows how informed and active citizens can influence legislators and policy makers to address environmental problems. Lois Gibbs, president of the Love Canal Homeowners Association, united the community and became an effective and persuasive advocate for families seeking government aid. Gibbs involved herself in December 1977, when her son, Michael, began to experience asthma and seizures just four months after he entered kindergarten at the 99th Street School. She went door-to-door and questioned other residents about their health to discover the full extent of contamination. In 1979 Gibbs traveled to Washington, D.C., where she testified before Congress on behalf of the Love Canal home owners.

The chemical disaster at Love Canal left behind a legacy of lawsuits and bitterness. In October 1983, a tentative settlement of the billions of dollars in lawsuits was reached by lawyers of the Hooker Electrochemical Corporation, the city of Niagara Falls, the Niagara Falls Board of Education, Niagara County, and former residents of the Love Canal area.

In November 1980, Congress passed legislation to deal with the cleanup of toxic wastes. The Comprehensive Environmental Response, Compensation, and Liability Act (CERCLA), commonly referred to as Superfund, established a $1.6 billion fund for the cleanup of hazardous substances to be administered by the EPA. The money was to be used when "no responsible party could be identified or when the responsible party refuses to or is unable to pay for such a cleanup."

The 1984 amendments to the federal Resource Conservation and Recovery Act (RCRA) included a thorough overhaul of hazardous waste legislation. Previously exempt sources that generated between 100 and 1,000 kilograms (220 and 2,200 pounds) of hazardous waste per month were brought under RCRA provisions. Congress further tried to force the EPA to adopt a bias against the landfilling of hazardous waste with a "no land disposal unless proven safe" provision. The amendments also added underground storage tanks for gasoline, petroleum, pesticides, and solvents to the list of sources to be regulated and remediated. In addition to RCRA, the Superfund provides for the cleanup of all categories of abandoned hazardous waste sites except for radioactive waste sites. Several other statutes (and ensuing EPA regulations) have dealt with these aspects of the hazardous waste problem. The cleanup of existing sites will continue to be a troubling problem, while the cleanup and disposal of radioactive waste will be a major issue for the future. Household waste, which is not regulated by RCRA, often includes small quantities of hazardous materials such as pesticides. Many of these materials are still being landfilled in the early twenty-first century, as individual consumers remain ignorant of the proper ways to dispose of such wastes.

Waste Incineration

The waste-minimization philosophy expressed in RCRA is a sound long-range strategy for dealing with hazardous waste. However, some materials will continue to be deposited in landfills. Incineration offers one solution to the problem of volume of material but poses issues of air quality and disposal of the highly toxic ash remaining. As some firms have found, minimizing their waste stream affords them economic benefits while conserving natural resources.

The incinerators used in burning waste products vary in type depending on the kinds and amounts of wastes to be processed. Solid household waste usually generates a lot of ash; if large amounts of such waste are incinerated, ash must be continuously extracted. This may be done with a moving grate incinerator, in which the grate is a conveyor belt. Waste is dumped onto the grate's front end, and ash and clinkers (unburned solids) are removed at the back end. Air is forced up through the grate to cool the grate and to aid combustion. If necessary, the grate can also be water-cooled. Air is also injected above the grate to ensure complete combustion of the gases. European law concerning waste incineration requires that the gases reach at least 850 degrees Celsius (1,560 degrees Fahrenheit) for at least 2 seconds to guarantee the breakdown of toxic organic material. If the gases are not hot enough, an oil burner is used, so wastes with relatively low fuel value can be treated in a moving grate incinerator.

Incineration has several benefits as a method of waste disposal: It reduces the volume of waste by about 95 percent while producing useful amounts of heat, and it can be used to sterilize medical waste and to neutralize dangerous chemicals. Waste incineration can also produce environmentally harmful by-products if it is not conducted carefully, and for this reason the practice has not achieved widespread acceptance in the United States.

Peter Neushul and John M. Theilmann

Further Reading

Fletcher, Thomas H. *From Love Canal to Environmental Justice: The Politics of Hazardous Waste on the Canada-U.S. Border.* Orchard Park, N.Y.: Broadview Press, 2003.

Gibbs, Lois. *Love Canal: My Story.* Albany: State University of New York Press, 1982.

Grisham, Joe. *Health Aspects of the Disposal of Waste Chemicals.* New York: Pergamon, 1986.

Hill, Marquita K. "Hazardous Waste." In *Understanding Environmental Pollution.* New York: Cambridge University Press, 2010.

LaGrega, Michael D., Philip L. Buckingham, and Jeffrey C. Evans. *Hazardous Waste Management.* New York: McGraw-Hill, 2001.

McKinney, Michael L., Robert M. Schoch, and Logan Yonavjak. "Municipal Solid Waste and Hazardous Waste." In *Environmental Science: Systems and Solutions.* Sudbury, Mass.: Jones and Bartlett, 2007.

Royte, Elizabeth. *Garbage Land: On the Secret Trail of Trash.* New York: Little, Brown, 2005.

Hydraulic fracturing (fracking)

FIELDS OF STUDY

Alternative Energy Sources; Climate Engineering; Deforestation; Ecology; Ecosystems; Environment; Environmentalism; Fossil Fuels; Geology; Hazardous Waste Management; Hydraulic Engineering; Seismology; Water Pollution Control

SUMMARY

Fracking, also called hydraulic fracturing, is a drilling process whereby fluid is pushed into the ground at high pressure to fracture shale and other poorly porous rocks. The fracturing of the shale creates fissures in the rock, which then allows oil and gas to flow out of the rock formation and into a wellbore, where oil and natural gas is then extracted.

PRINCIPAL TERMS
- **Environmental Protection Agency (EPA):** a governmental organization in the United States, established in 1970, whose overall goal is to protect human health and the environment; issues regarding air quality, water quality, and land use.
- **methane:** a colorless odorless flammable gaseous hydrocarbon, CH_4, a product of decomposed organic matter and he carbonization of coal
- **natural gas:** a flammable gas, consisting mostly of methane that occurs naturally underground, and can be used as fuel.
- **shale plays:** shale formations that contain significant amounts of natural shale gas
- **shale:** a fine grained, thin layered sedimentary rock that forms from compaction of silt and mineral particles; black shale contains organic material that breaks down to form natural gas and oil

- **wellbore:** the drilled hole, also called a borehole, which the process of fracking creates

Modern day hydraulic fracturing—commonly known as fracking—began in the late 1940s, when Floyd Farris of Stanolind Oil and Gas began to study the relationship between gas and oil output in the United States. On March 17, 1949, petroleum production experts in Oklahoma performed the first commercial application of hydraulic fracking. On the same day, the Halliburton and Stanolind Oil Companies successfully fractured another oil well in Holliday, Texas.

By the 1980s technological advances in shale formation identification led to increased fracking in shale plays within the United States. Fracking also increased in the 1990s after George Mitchell created new fracking technology combining hydraulic fracturing with a horizontal drilling procedure. Fracking expanded in the early 2000s with an EPA report stating that hydraulic fracking created no threat to underground drinking water supplies. That report had numerous detractors, leading to the current EPA stance of trying to gain a deeper understanding of the potential impact of fracking on drinking water resources, as well as "factors that may influence those impacts." Presently, there is no conclusive EPA statement regarding the effect of fracking on water, communities, or the greater environment, nor is there research on the long-term impact of fracking on personal health, the environment, or water quality.

Fracking

Fracking is a way to increase natural oil and gas production. Basically, fracking is a drilling process whereby a well is drilled vertically at a surface depth of 1 to 2 miles. The vertical well is then encased in steel or cement to ensure that well particles and oil don't leak into the groundwater. Once the vertical well reaches a deep layer of rock, horizontal drilling may begin. Horizontal drilling occurs along oil bearing rock layers, as far as 1 mile (1.6 km) from the vertical wellbore. After the well has been drilled vertically and horizontally, fracking fluid is pumped down the well at extremely high pressure (approximately 9,000 pounds per square inch or 62 megapascals). The pressure fractures the surrounding rock, creating fissures and cracks through which oil and gas can flow. A pipe at the center of the horizontal portion allows oil and gas to come to the surface and be collected. Seismologists have determined that the fracking process poses risks of microearthquakes (with magnitudes below 2), although out of 100,000 wells, the largest induced earthquake thus far recorded was magnitude 3.6. Fracking produces large amounts of wastewater, disposed of by injection into deep wells, which pose a much higher risk.

Statistics, Pros, and Cons
Fracking as a process has increased dramatically over the past decade. In the year 2000, there were over 275,000 natural gas wells drilled in the United States. In 2010, that number doubled to over 500,000; and every year, approximately 12,000 new wells are drilled within shale plays across North America. Among the major shale plays in the United States are the Barnett Shale Play in Texas, the Marcellus Shale Play in the Northeastern United States, the Anadarko-Woodford Play in Oklahoma, the Granite Wash Play in Texas and Oklahoma, the Niobrara Shale Play in the Rocky Mountains, the Bakken Shale Play in Montana, North Dakota, and Canada, and the Eagle Ford Shale Play in South Texas.

Although the process of fracking leads to high levels of domestic oil production, lower gas prices, and an increase in jobs; areas where fracking is common do pay a steep price.

Well drilling usually occurs in low income, rural areas. Land owners in these areas see the economic benefits of fracking, as they are usually paid over $2,000 an acre for permission to use land for well drilling/fracking. However, after trees are cleared and the ground is leveled, trucks and drilling equipment take over the landscape, exposing residents to increased traffic, dust, and ground tremors from artificial and natural/seismic activity.

Environmental activists also claim that fracking affects drinking water and creates a route for potentially carcinogenic chemicals to escape into the water and air. Although proponents of fracking say fracking is a safe and economical source of clean energy, critics respond with specific examples of its dangers.

For example, in 2011, a fracking well in Pennsylvania malfunctioned, spewing thousands of gallons of contaminated fracking fluid onto the ground for half a day. In 2011, researchers from Duke University tested drinking water at sixty fracking sites throughout Pennsylvania and New York and found that drinking water near fracking wells had dangerously high levels of methane. Fracking wells may also release carcinogenic chemical compounds such as benzene, ethylbenzene, toluene, and hexane into the air. Long-term exposure to these chemicals may cause birth defects, neurological issues, blood disorders, and certain cancers.

As fracking increases in the United States, the responsibility to ensure the safety of the land and the population around fracking sites increases as well. The EPA has called for complete transparency regarding both the benefits and long-term effects of fracking.

Although the idea of fracking is slow to spread abroad because of environmental concerns, some countries, such as Canada, India, and China, are actively pursuing research on advanced fracking techniques, so that their countries can also reap the benefits of local natural gas and oil production. In the United States, fracking continues to be a hotly debated environmental and political issue, and the EPA continues to work with individual states and stakeholders to ensure that natural gas extraction does not come at the expense of public health and quality of life.

Gina Riley

Further Reading
Ladd, Anthony E., ed. *Fractured Communities: Risk, Impacts, and Protest Against Hydraulic Fracking in U.S. Shale Regions.* New Brunswick, N.J.: Rutgers University Press, 2018.

Raimi, Daniel. *The Fracking Debate: The Risks, Benefits and Uncertainties of the Shale Revolution.* New York: Columbia University Press, 2017.

Websites
Environmental Protection Agency
https://www.epa.gov/hydraulicfracturing.

Indicator species

FIELDS OF STUDY
Biology; Botany; Climate Change; Conservation Biology; Ecology; Ecosystems; Environment; Environmentalism; Horticulture; Lichenology; Life Sciences; Marine Biology; Zoology

SUMMARY
Indicator species are useful for monitoring the impacts of human activities on the environment, particularly in assessing cumulative effects for which more direct measures are not available. Sometimes the investigation of population fluctuations in species not known to be indicators leads to recognition of previously unrecognized environmental problems. Indicator species are also used to monitor the progress of environmental remediation efforts.

PRINCIPAL TERMS
- **bioassay:** determination of the biological activity or potency of a substance by testing its effect on the growth of an organism
- **cosmopolitan species:** a species whose geological distribution is exhibited in all regions if not most regions of the globe, but generally does not include areas of extreme weather, such as the Antarctic and Arctic
- **cyanobacteria:** any of a major group of photosynthetic bacteria that are single-celled but often form colonies in the form of filaments, sheets, or spheres and are found in diverse environments
- **endemic:** natural to or characteristic of a specific people or place
- **indicator species:** animal and plant species whose presence, relative abundance, or conditions are diagnostic for some factor in the environment
- **lichen:** a composite organism that arises from algae or cyanobacteria living among filaments of multiple fungi species in a mutualistic relationship

Certain species of plants and animals exhibit strong responses to environmental factors, which are not necessarily human-made or deleterious. Observing these species in the field provides a convenient method for initial detection of factors of interest. Usually direct measurements are necessary if the information is to be used for determining environmental policy. Rather than using single species, many environmental surveys employ groups of species, defined either by taxonomic categories or by form and function, that respond similarly to a given environmental stressor.

Characteristics

The most useful indicator species in environmental monitoring are those that are cosmopolitan (that is, occur in a variety of habitat types over a wide area), are common, and easily recognized. Rare and endangered species, and those that are narrowly endemic, make poor indicators. One of the strengths of using indicator species is that it allows a person without extensive training or specialized equipment to survey many sites rapidly and identify those that merit more detailed monitoring. This advantage is lost if a species is rare enough to be absent from sites where no pollution or other degradation is present, or if the species is difficult to recognize in the field.

The more specific the response, the better the indicator. A combination of pollution, physical disturbance, and climate change may be causing a general decline of plants and animals. Under those conditions, an epiphytic lichen that concentrates pollutants from the air would be a good indicator of atmospheric pollution, while an introduced weedy species of herb might be a better indicator of disturbance, and a common native insect would be a better indicator of the overall effect of environmental degradation on food webs. If the mechanism of a pollutant's action is known, scientists may look for specific metabolic changes in a variety of species.

If a dominant or keystone species also has sensitivities making it a useful indicator species, its value in survey work is strengthened. A dominant species (in terrestrial ecosystems, a plant) is the one with the largest biomass. A keystone species is one whose removal would profoundly affect other members of the community—for example, a predator that keeps the most common herbivore in check. Changes in the health or relative abundance of a dominant or keystone species have disproportionate effects on the functioning of an ecosystem.

*Black Bryony (*Tamus communis*), Warren House Gill Black Bryony is Britain's only member of the Dioscoreaceae or Yam family, which has a number of tropical climbing species some with large edible tubers. By Andrew Curtis*

Examples in Environmental Monitoring

One of the earliest biological responses to environmental degradation to be recognized was that of epiphytic lichens to industrial air pollution. Lichens, a symbiotic association of a fungus, an alga, and in some cases a cyanobacterium, absorb water and nutrients directly from the air or rainwater. Acid rain, high levels of nitrogen, and heavy metals all affect lichen growth in ways that are quite species-specific. In forested areas such as Central Europe and the northwestern United States, where unpolluted mature forests support a diverse lichen flora, total lichen cover, relative abundance of certain species, and the chemical makeup of lichen thalli all provide a cumulative picture of air quality over a number of years. The cumulative effect is helpful to investigators because continual monitoring of air quality can be prohibitively expensive, and pollution is often episodic in nature.

Lichenologists recognized in the nineteenth century that members of the lichen family Stictaceae, which contain cyanobacteria, were very sensitive to air pollution. Only in the early twenty-first century did forest management biologists discover that these lichens are important sources of nitrogen in coniferous forests and that their conservation is a matter of concern in its own right.

Planktonic organisms, both plants and animals, are useful for monitoring pollution and temperature changes in the oceans. Some species concentrate particular pollutants. Relative and absolute abundance can be determined from dragnet samples. If a pollutant interferes with a particular metabolic function, analysis of enzyme levels and metabolic by-products in a mass sample of many different species can provide a direct measure of that pollutant's impact on the biosphere, including animals much farther up the food chain. Such bioassays detect levels of toxic compounds high enough to be of biological concern but too low to detect from direct analysis of seawater samples.

In many areas of the world attempts are under way to remediate environmental damage, restoring, as much as possible, original natural environments. The presence of indicator species is one measure of whether such efforts, which are never complete, are considered successful. In wetlands restoration projects in California, biologists have used the presence of thriving breeding populations of clapper rail as an indication that a healthy wetland ecosystem has been reestablished. This native bird is fairly common and tolerates moderate disturbance but had disappeared from large areas because of draining and severe pollution.

Mapping extensions and contractions in the ranges of individual species of plants and of vegetation types has been useful in reconstructing past climate fluctuations and in rounding out the picture of progressive global warming since the mid-twentieth century. In the early twenty-first century, broad-leaved evergreens, which are characteristic of regions with mild winters and overall drier climate, began extending their ranges both in Europe and in western North America. Both the alpine and Arctic tree lines are slowly advancing. These effects are reminders that change is not necessarily negative: People living near the Arctic tree line welcome the milder winters and more vigorous forest growth.

Martha A. Sherwood

Further Reading

Conti, M. E., ed. *Biological Monitoring: Theory and Applications—Bioindicators and Biomarkers for Environmental Quality and Human Exposure Assessment.* Billerica, Mass.: WIT Press, 2008.

Dunne, Niall. "Global Warming. Tracking the Effects of Climate Change on Plants." *Plant and Garden News* 18, no. 3 (2003): 1–4.

Jovan, Sarah. *Lichen Bioindication of Biodiversity, Air Quality, and Climate: Baseline Results from Monitoring in Washington, Oregon, and California.* Portland, Oreg.: Pacific Northwest Forest Experiment Station, 2008.

Spellerberg, Ian F. *Monitoring Ecological Change.* New York: Cambridge University Press, 2005.

Intergenerational Environmental Justice

FIELDS OF STUDY

Advocacy; Controversies; Debates; Ecology; Environment; Ethics; Philosophy and History of Science; Protest; Public Policy

PRINCIPAL TERMS

- **climate justice:** a human-centered approach to addressing climate change that prioritizes the rights of the most vulnerable people and shares the burdens and benefits of climate change equitably
- **deforestation:** clearing or thinning of forests by humans
- **distributive justice:** justice concerned with the apportionment of privileges, duties, and goods based on individual merits and in the best interests of society
- **fossil fuels:** a fuel (such as coal, oil, or natural gas) formed in the earth from plant or animal remains
- **Industrial Revolution:** in modern history, the process of change—beginning in Britain in the 18th century—from an agrarian and handicraft economy to one dominated by industry and machine manufacturing
- **intergenerational justice:** the sense of obligation or fair play that one generation of humanity holds toward the generations that follow and precede
- **Intergovernmental Panel on Climate Change (IPCC):** created by the United Nations and the World Meteorological Association in 1988 to provide policymakers with regular scientific assessments on climate change, its implications and potential future risks, as well as to put forward adaptation and mitigation options
- **sustainable development:** development that meets the needs of the present without compromising the ability of future generations to meet their own needs and aims to achieve social equality and protect the environment

SUMMARY

In addition to societal issues such as how the young should treat the elderly or whether one generation should pay for the education of the next, the concept of intergenerational justice encompasses numerous questions regarding the environment. Is it fair or just for the current generation to exploit natural resources to the point where those resources may become exhausted? Is it fair or just for today's society to fill landfills with garbage or the atmosphere with pollutants that tomorrow's citizens will have to clean up? Although the answers to such questions regarding an implicit social contract reaching across generations may seem self-evident, not everyone agrees that the members of each generation have a moral obligation to leave the world a better place than they found it.

The laws and policies that are created in relation to environmental issues are in part influenced by legislators' and policy makers' views about the necessity of pursuing intergenerational justice.

The Intergovernmental Panel on Climate Change (IPCC), established by the United Nations and the World Meteorological Association in 1988, particularly emphasizes a climate justice approach to global warming solutions, as the risks associated with climate change are generally greater for disadvantaged people and communities at all levels of development around the world.

In addition to societal issues such as how the young should treat the elderly or whether one generation should pay for the education of the next, the concept of intergenerational justice encompasses numerous questions regarding the environment. Is it fair or just for the current generation to exploit natural resources to the point where those resources may become exhausted? Is it fair or just for today's society to fill landfills with garbage or the atmosphere with pollutants that tomorrow's citizens will have to clean up? Although the answers to such questions regarding an implicit social contract reaching across generations may seem self-evident, not everyone agrees that the members of each generation have a moral obligation to leave the world a better place than they found it.

Some economists and policy analysts have argued in favor of what might appear to be shortsighted selfishness on the part of the current generation of humanity. They point to past ecologically unsound practices, such as overreliance on fossil fuels, and assert that the technological progress that humans have made can be attributed to their need to respond to problems created by the selfish behavior of past generations. Using this line of reasoning, they claim that it is unnecessary for current generations to preserve natural resources, curb population growth, or reduce industrial pollution. Frequently coupled to this argument is the statement that past generations showed no restraint or consideration for intergenerational justice, and current generations should be equally free to engage in selfish behavior. This latter argument is sometimes referred to as "mutual unconcern" between generations.

The flaw in pursuing a policy of mutual unconcern is that it is based on an assumption of continual technological and scientific progress. While it may be historically true that technological advances allowed past generations to substitute new resources for depleted ones, such as the substitution of coal for fuel when deforestation rendered charcoal scarce in Great Britain during the Industrial Revolution, humans cannot presume that science will always provide technical solutions to environmental problems. The historical record is rife with examples of technical solutions that, in the long run, generated more problems than they solved.

Further, engaging in unsound or damaging practices while arguing that the next generation will find a way to clean up the resulting mess fails on moral grounds. People should recognize that current actions do have significant impacts on the future. The fact that current generations may live to see the consequences of their actions should not release them from the moral obligations implicit in the social contract. The idea of distributive justice within a generation suggests, for example, that it is immoral for the wealthy to exploit the poor; that same concept of distributive justice suggests that rather than pursuing a policy of mutual unconcern, intergenerational justice mandates mutual concern, particularly regarding the environment.

Nancy Farm Männikkö

Further Reading

Gosseries, Axel, and Lukas H. Meyer, eds. *Intergenerational Justice.* New York: Oxford University Press, 2009.

Hiskes, Richard P. *The Human Right to a Green Future: Environmental Rights and Intergenerational Justice.* New York: Cambridge University Press, 2009.

Mitchell, Sherry and Larry Dossey. *Sacred Instruction: Indigenous Wisdom for Living Spirit-Based Change.* Berkeley, Calif: North Atlantic Books, 2018

Intergovernmental Panel on Climate Change (IPCC)

FIELDS OF STUDY
Climate Change; Global Warming; Government; International Policy; Meteorology and Atmospheric Sciences; Organizations and Agencies; Sustainable Development

PRINCIPAL TERMS
- **adaptation:** process of human adjustment to actual or expected changes in climate in order to mitigate harm or take advantage of beneficial opportunities
- **climate system:** comprised of the atmosphere, the hydrosphere, the cryosphere, the lithosphere, and the biosphere, the climate system evolves over time in response to interactions between the spheres and external influences such as solar variations or anthropogenic forces
- **global warming:** estimated average increase in global surface temperature, expressed relative to pre-industrial levels, over a 30-year period centered on a single year or decade
- **greenhouse gas (GHG):** naturally-occurring and anthropogenic gases that make up the atmosphere and absorb and emit radiation
- **Industrial Revolution:** a period of rapid industrial growth fueled by the combustion of fossil fuels beginning in the second half of the eighteenth century
- **net zero CO_2 emissions:** *balance of anthropogenic CO_2 emissions with anthropogenic removals over a time period*
- **Paris Agreement:** agreement under the United Nations Framework Convention on Climate Change (UNFCCC), adopted in Paris in 2015 at the 21st session of the Conference of Parties (COP). A primary goal of the agreement is to keep global warming below 2 °C
- **sustainable development:** development that meets the needs of the present without compromising the ability of future generations to meet their own needs and aims to achieve social equality and protect the environment

SUMMARY
Since its creation in 1988, the IPCC has proven to be one of the most credible sources of information regarding the state of scientific knowledge about climate change, its impacts, and possible policy responses. Its input has been instrumental in the global effort to reduce greenhouse gas emissions and develop strategies to meet the challenges of climate change.

The first call for an international effort to study the effects of anthropogenic climate change on the global community was issued by the World Meteorological Organization (WMO) at its 1979 World Climate Conference. The statement urged world governments to use scientifically generated knowledge to direct policy initiatives designed to slow the progression of global warming. In 1985, the Advisory Group on Greenhouse Gases (AGGG) was established by the International Council of Scientific Unions (ICSU; later the International Council for Science), the WMO, and the United Nations Environment Programme (UNEP). The group was to periodically evaluate scientific data relating to climate change, following a joint conference. These events created the impetus for the establishment of the Intergovernmental Panel on Climate Change (IPCC).

In 1987, the WMO and UNEP agreed that an organization should be created to coordinate an ongoing international effort to evaluate the results of scientific research on the climatic and socioeconomic effects of greenhouse gas (GHG) emissions. This organization would both evaluate research findings and suggest appropriate and effective policy responses based on those findings. The WMO established the IPCC in 1988 and gave it its mission. The panel was to develop a strategy to increase scientific information on global warming, use that information to assess possible policy initiatives for addressing climate change, evaluate policies already in place or proposed, and report its findings to governments and international organizations. UNEP and the United Nations General Assembly endorsed the IPCC.

The IPCC convenes annually in meetings that include hundreds of government officials and researchers from government agencies and nongovernmental organizations (NGOs) from countries that are members of the WMO and UNEP. At these meetings, the IPCC's objectives and activities are determined, the election of its chair is held, and members of its bureau, Task Force Bureau, secretariat, working groups, and Task Force on National Greenhouse Gas Inventories are selected. The thirty-member bureau, with representatives from all regions of the world, oversees the three working groups, and the Task Force Bureau directs the work of the task force. Working Group I (WG I) analyzes the scientific evidence regarding the causes of climate change, Working Group II (WG II) deals with the effects of climate change, and Working Group III (WG III) examines possible ways to reduce the negative effects. The task force is charged with developing better ways of measuring and reporting countries' GHG emissions. In addition, temporary special topic groups may be formed as necessary. The secretariat oversees and organizes all IPCC functions.

Activities of the IPCC

The main activities of the IPCC involve producing assessment reports (ARs) and methodology reports. The First Assessment Report (FAR) was requested by the U.N. General Assembly in 1989 to form the basis for the creation of the United Nations Framework Convention on Climate Change (UNFCCC). The FAR was completed in 1990. WG I concluded that anthropogenic GHG emissions, principally those of carbon dioxide (CO_2), would increase and would be responsible for global warming and sea-level rise during the twenty-first century. WG II asserted that this would have negative impacts on land and water ecosystems, coastal areas and cities, forestry and agriculture, and weather. WG III suggested short- and long-term policy responses. An IPCC supplementary report, prepared in 1992 to provide updated information for the newly created UNFCCC, supported the conclusions of the FAR.

Further IPCC ARs were published in 1995, 2001, 2007, and 2014. Each contained the three working group sections, with summaries for policy makers, as well as a synthesis report that summarizes the overall findings. These ARs reiterated and expanded the findings of the FAR and expressed greater confidence in the accuracy of the simulation models used to project future climate change, including the ability to better distinguish between natural and anthropogenic GHG emissions. The findings of these improved models indicated that human activity is the primary cause of past and future increases in global warming.

The Fifth Assessment Report, *Climate Change, 2014* (AR5), was the most strongly stated, asserting that the evidence for anthropogenic global warming had grown since the AR4. As a result, there are significantly more impacts in recent decades that can be attributed to climate change, such as coastal erosion and rising sea levels, transforming terrestrial and marine ecosystems, and changes in food production and human health. The AR5 makes it clear that continued warming will result in significant changes in all aspects of the global climate system, increasing the likelihood of irreversible and widespread impacts for both humans and the environment. It further stated that these effects will continue for hundreds of years under all simulated scenarios, even if GHG emissions could be limited to their current levels. All ARs undergo a rigorous review and revision process by the working groups, government officials, expert scientists, and the panel before receiving final approval.

In addition to the ARs produced by the working groups, the task force produced *Revised 1996 IPCC Guidelines for National Greenhouse Gas Inventories* to inform governments about the available methods for measuring GHG emissions. In 2006, an updated version was prepared that detailed improvements in software and methods for measuring emissions. The IPCC sponsors expert meetings and workshops to find more effective ways to do its work. It also creates special topic and function groups, such as the Task Group on Data and scenario Support for Impacts and Climate Analysis, which provides the latest data from climate change studies that use different models for use by those who evaluate their impacts and develop response strategies.

In 2018, the IPCC released a Special Report on the impacts of a global increase in temperature by 1.5 °C compared to levels before the Industrial Revolution. The 2015 Paris Agreement within the UNFCCC aimed to keep the increase in average global temperate to less than 2 °C above pre-industrial levels; however, the IPCC Special Report outlines the difference that even .5 °C could make to conditions on earth. Compared to an increase in 1.5 °C, a 2 °C increase is projected to escalate risks from extreme temperatures and droughts. It is likely to further destabilize marine ice sheets, leading to amplified vulnerability of islands and coastal areas to sea level rise. The odds of transforming terrestrial ecosystems increases by 50% at 2 °C compared to 1.5 °C, which would then lead to greater loss of biodiversity. Limiting global warming to 1.5 °C decreases the risks of global warming for humans in terms of likely food and water shortages, although the IPCC Special Report makes it clear that any increase in global warming is expected to negatively affect human health.

In order to limit global warming to 1.5 °C, the IPCC Special Report cites models that involve drastic reductions in carbon and methane emissions, amounting to a decline by about 45% from 2010 levels by 2030, and reaching net zero emissions by 2050. Suggested pathways for achieving this include investing in low-carbon energy technologies, changes in urban planning to support low-emission transportation, and supporting sustainable development, which balances economic growth, environmental protection, and social prosperity for poor or disadvantaged populations. The IPCC particularly emphasizes a climate justice approach to global warming solutions, as the risks associated with climate change are generally greater for disadvantaged people and communities at all levels of development around the world.

Controversies Involving the IPCC

The IPCC's activities and processes have not been without their critics. Some have said that the ARs are inaccurate, alarmist, and politically driven. For example, Christopher W. Landsea, science and operations officer at the National Hurricane Center, resigned from participation in AR4 in protest when a lead author of both the 1995 AR and AR4, Kevin E. Trenberth, director of the Climate Analysis Section at the National Center for Atmospheric Research, asserted that increased hurricane activity was caused by global warming, which Landsea strongly disputed. Landsea claimed that AR4's conclusions were a product of political pressure and scientific community consensus, and that they stated the research questions as though they had empirical support, which, in his opinion, they did not. In 2005, the British House of Lords expressed concerns that the IPCC was exaggerating the future magnitude and impacts of global warming. It commissioned a report to perform a more objective cost/benefit analysis regarding the IPCC's suggested responses and the expense of possible damage. Ironically, the resulting Stern Review concluded that IPCC ARs and reports from other sources had underestimated the future risks of global warming.

In fact, another general criticism of the IPCC has been that it is too conservative in its analytical approach and downplays the need for immediate aggressive action. Critics cite other analyses that forecast higher future increases in GHG emissions, global temperatures, and sea-level rise than do the IPCC. Some claim that political pressure from the United States resulted in Robert T. Watson being replaced by Rajendra K. Pachauri as IPCC chair in 2002, because the latter was considered to be more desirable by conservative politicians and large oil companies. At the heart of these concerns are the IPCC review, revision, and acceptance processes, which are viewed by some as being unnecessarily cumbersome, leading to more conservative, bureaucratically acceptable findings.

Critics have objected to line-by-line reviews and revisions of the summary sections of the ARs by the panel, with some saying that they downplayed and distorted findings that had been validated through the peer-review process. The heads of major atmospheric research and meteorological organizations have responded by saying that this was a media campaign attacking rigorously reviewed data compilations and summarizations because of the biases of the critics. The IPCC has stood by its position that the revisions make the ARs more understandable, rather than being politically motivated, and that the strength of the AR processes lies in its objective, cautious, analytical approach, following the ideals of the scientific method.

Critics have also asserted that the submission deadlines for AR material are too far in advance of their publication. As a result, in order to allow for the lengthy review processes, the latest research findings are excluded. On the eve of the release of AR4, Pachauri responded by acknowledging that important research that projected more pronounced future global warming had been completed since the submission deadlines, indicating that aggressive policy initiatives should be initiated sooner than recommended in the report.

Context

The goal of the IPCC is to foster international cooperation between scientists working on climate modeling, those assessing climate impacts, and policy advisers in order to determine the causes and dangers of climate change and develop sensible response strategies. This has proven to be a challenge because of the disagreements and debates among scientists and the frequent tension between political dynamics and research findings and conclusions. Nevertheless, the IPCC's work has had significant impacts.

Since 1988, multitudes of scientists have volunteered to participate in the panel's work without compensation, resulting in a dramatic increase in global warming research, and the IPCC has ensured that participants are from both developing and industrialized countries. Over 3,750 contributing and lead authors and expert reviewers from more than 130 countries contributed to AR4. The FAR brought about the creation of the UNFCCC and the Kyoto Protocol, which has been ratified by 183 nations.

Criticisms notwithstanding, a strong consensus of support for the IPCC has emerged within the global scientific establishment. Most scientific organizations from all parts of the world, including the United States' National Research Council, the Network of African Science Academies, the European Geosciences Union, and the Royal Meteorological Society have voiced strong support for the IPCC, calling it the foremost authority on the state of scientific knowledge relating to climate change. IPCC's ARs have become a primary source of information in climate change policy debates worldwide, and many conservative political factions that had denied the existence of anthropogenic global warming for decades have now conceded that it is real, in the face of the growing body of scientific evidence.

In 2007, the IPCC was awarded the Nobel Peace Prize for its efforts to compile, analyze, and disseminate scientific data about climate change as the basis for formulating strategies for mitigating against its negative global impacts. The co-winner of this award was former United States vice president Al Gore, who was also recognized for his work on global warming. The IPCC will undoubtedly continue to be an important force in the ongoing international efforts to understand and counteract the current and future effects of climate change. The Sixth Assessment Report (AR6) is scheduled to be finalized in 2022, in time for the first stock to take place under the Paris Agreement of 2015.

Jack Carter; updated by Julia Kendrick

Further Reading

Brolin, Bert. *A History of the Science and Politics of Climate Change: The Role of the Intergovernmental Panel on Climate Change.* New York: Cambridge University Press, 2008.

DiMento, Joseph F. C., and Pamela M. Doughman, eds. *Climate Change: What It Means for Us, Our Children, and Our Grandchildren.* Cambridge, Mass.: The MIT Press, 2007.

IPCC, 2014: *Climate Change 2014: Synthesis Report. Contribution of Working Groups I, II and III to the Fifth Assessment Report of the Intergovernmental Panel on Climate Change* [Core Writing Team, R.K. Pachauri and L.A. Meyer (eds.)] IPCC, Geneva, Switzerland. IPCC, 2018: Global Warming of 1.5 °C.

Skodvin, Tora. *Structure and Agent in the Scientific Diplomacy of Climate Change: An Empirical Case Study of Science-Policy Interaction in the Intergovernmental Panel on Climate Change.* Norwell, Mass.: Kluwer Academic, 2000.

Websites

Intergovernmental Panel on Climate Change.
http://www.ipcc.ch

International Nature Preservation Policies

FIELDS OF STUDY

Conservation; Controversies; Ecology; Ecosystems; Environment; Environmental Law; Environmentalism; Government; International Relations; Landscape Ecology; Ornithology; Politics; Preservation; Public Policy; Renewable Resources; Zoology

SUMMARY

Governments generally promote their policies concerning nature preservation through the passage of legislation. In the United States, changes in nature preservation policy over time have reflected the changes that have taken place in the attitudes of the public toward the need for government protection for the natural environment.

PRINCIPAL TERMS
- **conservation:** planned management of a natural resource to prevent exploitation, destruction, or neglect
- **Convention on International Trade in Endangered Species:** an international agreement between governments. Its aim is to ensure that international trade in specimens of wild animals and plants does not threaten their survival
- **Endangered Species Conservation Act:** signed into law in 1973, designed to protect critically imperiled species from extinction
- **Forest Reserve Act:** law, passed in 1891, that allowed the President of the United States to set aside forest reserves from the land in the public domain
- **preservation policies:** decisions made, and regulations put in place by any level of government to undertake the protection of natural resources
- **preservation:** careful protection of something and planned management of a natural resource to prevent exploitation

The Industrial Revolution, which diminished traditional agriculture while encouraging urbanization and technology, began straining the relationship between humanity and natural resources during the early nineteenth century. As the technological advances being made in Europe rapidly spread west, environmental damage and natural resource depletion escalated in the United States, inspiring the American conservation movement. The conservation and environmental movements have continued to exert tremendous influence on policy making.

American artist George Catlin first proposed setting aside land for wildlife and American Indians during the nineteenth century, and in 1864 geographer George Perkins Marsh published *Man and Nature: Or, Physical Geography as Modified by Human Action*, the first influential book to address the human impact on nature. The Homestead Act of 1862 greatly encouraged expansion in the western United States by giving more than 405 million hectares (1 billion acres) of land to settlers, a policy that often resulted in barren landscapes. Destructive logging methods were employed, land was rapidly cleared for agriculture, and large-scale fires raged. Some western grasslands experienced such excessive grazing that many regions had not recovered their full productivity by the end of the twentieth century.

As farmers' journals described "wearing out" several homesteads during westward journeys, naturalist Henry David Thoreau and essayist Ralph Waldo Emerson countered with writings that fueled increasing support for nature conservation. The public expressed considerable outrage regarding the near elimination of several wildlife species that previously had existed in massive numbers, such as bison, deer, elk, and beaver. This led to legislation that created the world's first national park in 1872 at Yellowstone, Wyoming, followed by an 1873 petition to Congress by the American Association for the Advancement of Science to curtail the inefficient use of natural resources such as water, soil, forests, and minerals.

The 1891 Forest Reserve Act began the establishment of natural forests, and the 1900 Lacey Act initiated wildlife protection by regulating commercial hunting. The 1894 Buffalo Protection Act provided recognition that a previously abundant natural resource could rapidly become an endangered species. As naturalist John Muir championed numerous wilderness preservation projects and became the first president of the Sierra Club in 1892, federal legislation in the United States began classifying natural resources as renewable or nonrenewable. Renewable resources can be regenerated and even improve under proper management but can be depleted or eliminated if misused. Nonrenewable resources are present only in fixed amounts and will not regenerate regardless of human efforts. Examples of renewable resources include plants, animals, soils, and inland waters; nonrenewable resources include minerals and fossil fuels. The founding of private conservation organizations such as the American Forestry Association in 1875, the American Ornithologists' Union in 1883, the Boone and Crockett Club in 1887, and the New York Zoological Society in 1895 increased public influence on conservation legislation at all levels of government.

Theodore Roosevelt and Franklin D. Roosevelt

President Theodore Roosevelt initiated habitat protection for wildlife in 1903 when he set aside Pelican

Island in Florida's Indian River as a federal bird sanctuary. Through such initiatives as the 1908 White House Governors' Conference on Conservation, Roosevelt's politics and personality helped establish more than fifty wildlife refuges, five national parks, and eighteen national monuments, and increased the area of national forests by more than 69.7 million hectares (150 million acres). Roosevelt's policies required that certain public lands be held in trust for the "good of the country" and separated many public domain regions from commercial interests.

During and immediately following Theodore Roosevelt's presidency, Forest Service chief Gifford Pinchot and Interior Secretary James Garfield implemented more unified policies governing natural resource planning that relied on scientific principles, leading to development of the discipline of conservation biology. Many nonrenewable resources were then protected from exploitation by private industry by the 1920 Mineral Leasing Act.

Political debates and administration changes during the Great Depression of the 1930s shelved more environmental legislation until President Franklin D. Roosevelt signed the Taylor Grazing Act in 1934, whereby all public domain lands would be managed as part of the public trust. The dry Dust Bowl years of the Great Plains states during the early 1930s severely depleted migratory bird populations, motivating a renewed surge of public conservation activity and passage of the 1934 Duck Stamp Act, which tacked a conservation fee for the acquisition of wetlands onto waterfowl hunting licenses.

In 1933 the Soil Erosion Service (later called the Soil Conservation Service), the Civilian Conservation Corps, and the Tennessee Valley Authority were established to provide water and soil conservation assistance to landowners as farmland in the Midwest continued to deteriorate under improper agricultural practices. Franklin Roosevelt's Civilian Conservation Corps provided unemployed Americans with more than two million jobs planting trees and building irrigation systems and dams. More federal involvement was initiated after dust clouds from the dry soil of midwestern farmland blew east all the way to Washington, DC. In 1940, the US Congress enacted the Bald Eagle Protection Act to protect the national bird.

Post-World War II Conservation

Advances in technology and economic development in the turbulent era following World War II, combined with the postwar baby boom, put additional stressors on environmental resources. President Harry Truman began a national program for water-pollution control, with later legislation requiring states to set and enforce standards for natural rivers. In attempts to reduce insect-borne disease and increase food production, Dichloro-diphenyl-trichloroethane (DDT) and other synthetic pesticides were developed and had considerable initial success, causing the near-complete disappearance of malaria and the production of bumper crops.

The 1947 Forest Pest Control Act provided for the detection and chemical destruction of insects that carried diseases harmful to humans, but the numerous new and powerful experimental substances being invented caused other severe environmental problems, which in many cases caused more damage than those that the pesticides were created to prevent. Grassroots public outcry stimulated several federal restrictions on chemicals such as DDT, with many citizens alerted to these dangers by former US Fish and Wildlife Service biologist Rachel Carson's 1962 book *Silent Spring*. Carson is credited with warning mainstream America about the health and environmental hazards posed by pesticides and other toxic chemicals. Her work stimulated further writings that described human threats to the environment, including *The Population Bomb* (1968), by Paul R. Ehrlich, and *The Limits to Growth* (1972), by Donella H. Meadows, Dennis L. Meadows, Jørgen Randers, and William W. Behrens III.

All forms of environmental pollution greatly increased during the 1950s and 1960s. Television beamed graphic examples of environmental problems into public view, notably the mercury poisoning at Minamata Bay, Japan; killer smog episodes in London, England, and Los Angeles, California; and the 1967 *Torrey Canyon* oil spill in the English Channel. As the prices of land and water rights skyrocketed, the Land and Water Conservation Fund, which was set up by federal legislation during the 1960s to increase outdoor recreation space in the United States, generated revenues from offshore drilling leases. Several catastrophic environmental events occurred in 1969, including toxic waste fires on the Cuyahoga River in Cleveland, Ohio, and a coastal oil spill near Santa Barbara, California. Public pressure regarding these and other concerns led to passage of the National Environmental Policy Act of 1969 (NEPA), which became law on January 1, 1970. During the development of this precedent-setting act, Congress discovered that more than eighty governmental units had activities directly affecting the environment, but no government policies were in place to coordinate and review such activities.

Private individuals and organizations such as the Sierra Club, the Nature Conservancy, the Wilderness Society, the National Wildlife Federation, and the National Audubon Society began lobbying for more laws to establish nature preservation areas for both renewable and nonrenewable natural resources. Two highly visible social programs that influenced public opinion were conducted by the Nature Conservancy: Oklahoma's Tallgrass Prairie National Preserve "Adopt a Bison" program and Montana's Pine Butte Swamp Preserve, where dinosaur fossils were discovered in 1978. The Endangered Species Preservation Act of 1966 and the Endangered Species Conservation Act of 1969 did not directly protect any species, but they led to later legislation that did.

The 1970s and 1980s

Following unanimous passage of NEPA by Congress over President Richard Nixon's objection, the 1970s saw the passage and often complicated enforcement of several laws regulating nature preservation. The Environmental Protection Agency (EPA) was established in 1970, followed later that year by passage of significant amendments to the 1963 Clean Air Act. Important pollution-control measures were then implemented by the 1972 Water Pollution Control Act, the 1973 Endangered Species Act, the 1976 Toxic Substances Control Act, the 1976 National Forest Management Act, and the 1977 Clean Air Act amendments. The Endangered Species Act is considered the most effective and wide-reaching act ever passed by Congress to protect natural ecosystems.

Key legislation supporting nature preservation that was passed during the 1980s included the 1980 National Acid Precipitation Act, the 1980 Alaska National Interest Lands Conservation Act (which enabled the size of the refuge and park systems to double), and the 1987 amendments to the 1972 Clean Water Act. Surveys conducted during the late 1980s revealed that more than 60 percent of wildlife areas in the United States were permitting activities that were harmful to wildlife, with the most destructive practices, such as military activities and drilling, not falling under legal jurisdiction of the US Fish and Wildlife Service. This era also saw increased public interest in nature preservation following events such as the 1986 Chernobyl nuclear power plant catastrophe and the 1989 *Exxon Valdez* oil spill, as well as controversies concerning acid rain, tropical deforestation, the harvesting of old-growth timber, and the discovery of a trend toward global warming.

Private corporations that had previously sacrificed important wildlife habitat began to realize that the environment was an important issue to American consumers. In response to pressure from consumers, employees, and stockholders, many businesses implemented stewardship programs designed to protect natural resources and allow public enjoyment of their underdeveloped lands.

The 1990s and Beyond

Many nature preservation goals first proposed during the Industrial Revolution finally began to be realized with the systematic creation and maintenance of healthy forests; the prevention of timber depletion and siltation of streams; the provision of food, cover, and protection for wildlife; and the establishment of places where human beings could escape growing urbanization by taking part in outdoor recreation. Water reservoirs could now control flooding, provide clean water for humans and livestock, keep the soil fertile for agriculture, provide irrigation, and generate power. Water treatment plants were now effective in keeping rivers clean by processing wastes from urban sewage, while fish hatcheries provided supplemental stocks to natural and human-made reservoirs, streams, and lakes. Scenic easements along riverbanks aided antipollution efforts and reduced erosion, while green spaces required by city zoning regulations held soil and became available for community use while maintaining the environment's natural beauty. Mass interurban transportation systems moved people efficiently, and footpaths and bicycle trails offered outdoor recreation and cultural opportunities. Continual management of resources was more successful in keeping delicate ecosystems in balance, with ongoing environmental efforts including the seeding of wildlife foods, controlled burning to destroy unwanted vegetation, and the closing of wildlife habitats during mating and birthing seasons.

Conservationists-turned-environmentalists greatly influenced nature preservation policies during the 1990s as President George H. W. Bush passed legislation in 1990 that amended the 1970 Clean Air Act to focus more on reducing acid rain and emissions from fossil fuels and nitrogen oxide. However, an activist citizens' commission formed in 1992 by the Defenders of Wildlife found that the United States was "falling far short" of meeting the urgent needs of nature preservation. The public response to this information helped lead to passage of the 1997 National Wildlife Refuge System Improvement Act, which shifted the priorities of nature preservation systems toward the formation of multiple-use environments. This key legislation redefined the mission statement regarding conservation of habitats for fish, wildlife, and plants; designated priority public uses such as hunting, fishing, wildlife observation and photography, and environmental education and interpretation; and required that "environmental health" be maintained on public lands. The principle of multiple use, however, continues to allow mining, drilling, grazing, logging, and motorized recreation, as well as military training such as bombing, tank, and troop exercises, on lands designated for nature preservation.

By 2016, however, despite surveys finding that the American public values preserving nature through efforts such as maintaining national parks, experts argued that the National Park Service (NPS), responsible for overseeing this task, has been consistently underfunded over several years. While the number of people visiting the parks has steadily increased, the congressional budget granted to the NPS has proven inadequate to allow the organization to properly staff the parks and conduct necessary maintenance projects.

International Efforts

Cooperative international nature preservation efforts began with the 1918 Migratory Bird Treaty signed by the United States and Great Britain (for Canada), and later Mexico. The International Union for Conservation of Nature and Natural Resources, founded in 1948, represented the interests of 116 countries toward protecting endangered and threatened "living resources." The United Nations Conference on the Human Environment, hosted by Sweden in 1972, was instrumental in establishing nature preservation as an international concern. Utilizing concepts from this conference, the US Congress passed the 1983 International Environmental Protection Act, which included landmark legislation incorporating wildlife and plant conservation and biological diversity as objectives when the United States provides assistance to developing countries.

The 1973, Convention on International Trade in Endangered Species (CITES) involved the cooperation of more than one hundred nations to regulate the import and export of natural resources. Earth Day, first held on April 22, 1970, as a campus-based event encompassing an estimated twenty million people across the United States, began to be combined with other concerns, and annual observances of Earth Day have continued to demonstrate massive support for conservation issues. The international Greenpeace Foundation was formed in 1971 with the aim of applying pressure to governments and organizations to stop such practices as the testing of nuclear weapons and the dumping of radioactive and toxic wastes.

Sentiments favoring the preservation of nature began taking political form in Europe during the early 1980s, notably with the formation of the Green Party in Germany. International collaboration on environmental preservation issues that influenced later legislation included the 1987 Montreal Protocol to protect the ozone layer, the 1992 United Nations Conference on Environment and Development (also called the Earth Summit) in Brazil, and the 1994 United Nations Population Conference in Egypt. The 1992 Earth Summit in Rio de Janeiro, Brazil, which was the largest international meeting ever held (representatives of 178 nations attended), emphasized an approach to nature conservation that focused on sustainable growth and utilitarian solutions. A paper that resulted from the Earth Summit titled "World Scientists' Warning to Humanity" warned that if current consumption rates continued, the earth's resources would be reduced to the point where the world would be "unable to sustain life in the manner we now know."

This monumental event was followed by another UN summit in 1997 in New York City, attended by representatives of more than fifty nations, during which the progress made in the intervening years was reviewed. Many analysts noted that although progress had been made, the local and global issues surrounding nature preservation and environmental problems were continuing to grow in complexity, and many kinds of environmental damage may be irreversible. Although the attendees did agree to take further action on issues of nature preservation, they made few concrete commitments. The five-year review process continued with the 2002 World Summit on Sustainable Development, which was held in Johannesburg, South Africa; the United States did not participate in this conference, which also produced few concrete commitments.

Daniel G. Graetzer

Further Reading

Chiras, Daniel D., and John P. Reganold. *Natural Resource Conservation: Management for a Sustainable Future.* 10th ed., San Francisco: Cummings/Pearson, 2010.

De Steiguer, J. E. *The Origins of Modern Environmental Thought.* Tucson, Ariz.: University of Arizona Press, 2006.

Dowie, Mark. *Losing Ground: American Environmentalism at the Close of the Twentieth Century.* Cambridge, Mass.: MIT Press, 1995.

Kline, Benjamin. *First Along the River: A Brief History of the United States Environmental Movement.* 3rd ed., Lanham, Md.: Rowman & Littlefield, 2007.

O'Neill, John. *Ecology, Policy, and Politics: Human Well-Being and the Natural World.* New York: Routledge, 1993.

Websites

U.S. News and World Report.
www.usnews.com/news/articles/2016-08-19/americans-value-national-parks-but-funding-is-lacking.

L

■ Land-use policy

FIELDS OF STUDY
Business; Commerce; Controversies; Ecology; Economics; Environment; Environmental Law; Environmental Engineering; Environmentalism; Land-Use Management; Political Science; Public Policy; Standards and Practices; Urban Planning

SUMMARY
Policy makers around the world have become increasingly aware of the potential environmental impacts of government and community decisions regarding land use. Since the late twentieth century, approaches to land-use planning have trended toward ecological conservation and preservation.

PRINCIPAL TERMS
- **conservation:** protection of plants, animals, and natural areas, especially from damaging effects of human activity and a plan for avoiding the unnecessary use of natural materials such as wood, water, or fuel
- **land-use:** the way in which a society organizes, plans, and manages social and physical activities on the landscape
- **planned unit development (PUD):** a designed grouping of both varied and compatible land uses, such as housing, recreation, commercial centers, and industrial parks, all within one contained development or subdivision
- **preservation:** to keep alive or in existence, safe from injury or harm and make lasting
- **transferable development rights (TDR):** a tool for controlling urban sprawl by concentrating development
- **zoning:** the act of setting rules for the use of land and the types of structures that can be built on it

Human societies are organized to survive in specific environments. The adaptive nature of culture allows a society to respond to changes in the environment and to cause even more changes. Once human beings began to establish cities, where their lifestyles became relatively sedentary, formal land-use policies arose, and decisions about how to manage land became critical to survival. Even during the earliest periods of the Mayan, Egyptian, and other city-state civilizations, land-use control was implemented for religious, agricultural, hunting, and residential purposes. Religious proscriptions dictated the appropriate appearances of structures. Plagues, warfare, and resource distribution demonstrated the need to isolate structures as well as groups of people. Crowding and cultural conflicts made it necessary for groups to create processes to make government decisions about who gets to use particular resources. Land-use policies are even more necessary in the contemporary world of competing interests, growing populations, and diminishing natural resources. Increased scientific knowledge about environmental impacts has fueled the need to ensure that appropriate land-use decisions are made in the interests of survival through long-term resource management.

Roots of Land-Use Policy
The earliest land-use policies took the form of religious prohibitions and mandates as the priest class interpreted the needs of the gods for sacred spaces. Kings exercised a divine mandate in interpreting just what could be allowed in sacred areas, while market forces dictated the use of less important secular space. Later, after kings began losing their divine authority, they could still regulate the use of secular space in the name of promoting social order. Social class structure continued to support the allegiance to class-based authority.

After the Renaissance, European societies believed that while God no longer directly intervened in day-to-day activities, his presence was felt in the need to maintain social order through hierarchies. It was believed that this social order must also be reflected in physical space, or in the way in which a landscape was arranged. Accordingly, it was only proper that buildings, towns, and landscapes reflect particular patterns in form and ownership. In the American colonies that became the United States, examples of kingly intervention in land-use decisions can be found in the marking of potential mast pine trees in New England with the king's broad arrow, the designation of village squares, and the reservation of lots for the king's agents. However, the rise of a democratic society provided a shift in decision-making authority to elected representatives.

By the mid-nineteenth century, early planning laws in the United States began to regulate urban tenement housing and prohibit "obnoxious uses." In 1893 the World's Columbian Exposition in Chicago promoted the exchange of planning and design concepts in exhibits by landscape architect Frederick Law Olmsted, artist Saint Gaudens, and others. By 1895 Los Angeles had an ordinance prohibiting the locating of steam shoddying plants (plants that reprocessed scrap wool) within 30.5 meters (100 feet) of a church. In addition to ordinances and the designation of parks, planned communities were initiated. The publication of Ebenezer Howard's *Tomorrow: A Peaceful Path to Real Reform* (1898) launched the garden city movement. In an 1899 decision, the Massachusetts Supreme Court upheld a building height limitation, and by 1909 zones of limits for buildings were upheld in the U.S. Supreme Court.

By the early twentieth century, organized planning efforts were under way in some large cities. Hartford, Connecticut, established a planning board in 1907. By 1909 Wisconsin had passed the first state enabling act for planning, and Los Angeles provided the first American use of zoning to direct future development in a series of multiple zoning ordinances. In 1913 Massachusetts became the first state to make planning mandatory for local governments. Newark hired the first full-time city planner in 1914, and the first comprehensive zoning code in the United States was enacted in New York City in 1916. By 1925 Cincinnati, Ohio, had become the first major U.S. city to adopt a comprehensive plan, and Burlington, Vermont, had authorized a municipal planning commission.

Policies After World War II

In the 1940s planning for war and postwar public housing brought the U.S. government to review the land-use planning process. The mustering of resources for World War II demonstrated the value of planning, and many towns implemented town plans as postwar prosperity began. In 1962 the Chicago Area Transit Study showed the applicability of cost-benefit studies to planning for suitable development.

In the 1970s Americans became increasingly concerned with the need to control development. Performance standards were upheld in the courts as one mechanism to manage growth through land-use policy. In 1971 the concept of transferable development rights (TDRs) was introduced to help preserve urban landmarks. Creative land-use tools such as conservation easements, controlled access rights, planned unit development (PUD) density credits, and overlay districts were increasingly used in the 1970's as communities expanded their regulatory schemes. In their zeal to manage growth, some communities enacted land-use policies that were discriminatory. In an early decision on discriminatory land-use regulation, in 1975 the New Jersey Supreme Court struck down a restrictive Mount Laurel zoning ordinance on the basis that it did not allow a regional "fair share" of low- and middle-income housing.

By the late 1980s a enough court cases existed to reduce the likelihood that communities would enact discriminatory ordinances, but a slump in the U.S. economy fueled the challenge of some land-use ordinances and policies as unauthorized "takings"—the erosion or loss of landowner rights through excessive regulation without due compensation. Yet most land-use regulations are not generally found to be true takings because at least some economic use of land is allowed. Still, the issue of a taking is frequently raised when regulations deprive an owner of one or more desired uses. The economic climate has a strong influence on the reaction of a population to issues such as takings and to land-use regulation in general.

Land-use regulation is generally viewed as the control of two categories: subdivision (physical size or boundary change) and development (physical use or alteration within the boundary). Changing landownership or subdivision is a land use because it affects the management of resources on the land and can fragment habitat. Most land that has been subdivided stays that way; it is unusual for such land to revert to an original larger tract. Land that is used or developed through construction, clearing, or other alteration, including various forms of land management such as agriculture, also undergoes a change to the natural path of succession. In fact, enough past alteration through introduction of new species or direct physical action has so altered some landscapes that it is almost impossible to determine their true "natural" condition. It is this perspective that, along with concern regarding environmental impacts from yet-to-occur changes, is most commonly used to justify land-use policies.

Conventional and Conservation Planning
Conventional land-use planning assumes the desirability of economic growth through new development and therefore tends to favor revenue generators. Tools such as zoning are the main techniques for conventional planning. A government land-use plan is implemented through a series of development regulations that differ according to zones. A series of base maps are prepared after the completion of natural, social, and infrastructural resource inventories. The maps can be viewed as both opportunities for and constraints on growth. Individual base maps, when combined with the community's or region's goals and objectives for growth, result in a land-use map containing zones. Each zone reflects a category: commercial, industrial, residential, recreational, historical, governmental, agricultural, special, and others that are suggested by the inventory and goal processes. Each category can contain a variety of subcategories based on lot size and range of allowable uses.

The size, configuration, and pattern of the arrangement of lots can all affect growth. For example, large lots reduce the number of houses, while small lots increase the number of houses, reduce the amount of open space, and increase the fragmentation of landownership. The community uses the planning process to agree upon the development capacity of given areas. By using various tools and approaches, a community can seek a balance between achieving appropriate densities and maintaining natural and aesthetic resources; the objective is to achieve sustainability.

Conservation planning is one approach to improving land use within conventional planning. In conservation planning, structures and uses such as septic systems are located away from valued or critical natural resources on a tract. Conventional planning, even if conservation-oriented, can still lead to checkerboard or highway strip patterns of development that are harmful to open space. Clustering is a technique that goes one step further by attempting to preserve or conserve open space by treating it as a natural resource. Other cited benefits of clustering are the fostering of a sense of community, reduction of urban sprawl, and the presentation of traditional village appearances.

In clustering, dwelling units (or commercial structures) are grouped together to allow a larger uninterrupted area of open space (generally 25 to 30 percent), which is often maintained in a residential or commercial subdivision as common land under shared ownership. Some communities encourage clustering through policies that allow a greater density of units or square feet of construction. Thus a 20-hectare (50-acre) tract that might be approved for construction of five houses in a traditional checkerboard subdivision of ten lots might be allowed to contain up to fifteen units of housing if the houses (or condominiums) are clustered and if a specified percentage of the parcel is preserved as open space. The open space might be a separate 8-hectare (20-acre) lot that is deeded to the prospective unit owners as shared or common land to be managed in a certain way.

Conservation planning can be employed on a city- or statewide level through the administration of tiered levels of permits and other development review processes coupled with a system of land-use planning in which designated areas are conserved. An example is Oregon's establishment of Greenline boundaries, which limit growth at the edges of cities. Trails and greenbelt corridors also reflect conservation planning, but these require considerable coordination and cooperation when more than one community or state is involved.

Ecological Planning
Ecological land-use planning expands on conventional land-use planning by taking a more integrative perspective of ecosystem dynamics and applying it over a greater period of time than the normal five-year period. To take this comprehensive approach, planners require significant amounts of data about a wide variety of resources and a fairly stable set of goals and objectives in a relatively constant sociopolitical setting. Ecological planning provides the

benefit of long-range dynamic planning while attempting to prevent, rather than remedy, problems. However, it is more costly than conventional planning in initial expenditures. The Netherlands and some other northern European countries have a long history of ecological land-use planning, particularly in response to increased population pressures in areas of finite land resources.

The decrease in numbers of rural inhabitants and the growing numbers of suburbanites in the United States in the latter half of the twentieth century caused an increase in the consideration of various planning techniques and tools. Vermont adopted a statewide land-use policy law in which individual development decisions are made by a regional volunteer citizen panel at quasi-judicial hearings in which dispute resolution techniques and consensus building are encouraged. Vermont has found this case-by-case process to work quite well despite the lack of a comprehensive statewide plan and despite an expanding population. Oregon uses a similar process. Florida's Environmental Land and Water Management Act of 1972 allows the state to designate areas of critical interest that local governments must consider when enacting local policies. It also requires state review of development projects that are large or have regional impacts.

The cumulative adverse environmental effects of many small subdivision and construction projects can significantly outweigh the effects of larger projects that are more intensely regulated. By the end of the twentieth century most U.S. states had begun to recognize this and had increased local control of land use. Most had also implemented statewide natural resource programs that reflected ecological understanding of the interactions between changing land use and the need to manage wildlife and other natural resources.

Forms of Land-Use Regulation

Land use can be controlled through incentives or restrictive processes. Incentive-based land-use control approaches include direct funding, grants, tax abatements, trade-offs such as credits for clustering, and other forms of positive feedback. Restrictive land-use control is the form more commonly recognized and employed, notably through the use of permits, licenses, environmental assessments, taxes, and direct prohibitions. Restrictions may be imposed by government or by landowners via covenants or easements. Government restrictive regulations have two forms: prescriptive, in which the objectives and specifications are precisely articulated; and subscriptive, in which the outcomes are specified but the individual means of achieving them are left to the discretion of the owner or community. Critical land uses and issues, such as matters of public health, are more likely to be prescriptively regulated. Less critical matters might be handled by subscriptive means, also called performance-based planning. Even tax structuring can be performance-based when land is taxed based on actual use rather than potential use. This can reduce the pressure for commercial development of high-value or expensive properties.

Most communities and governments use a combination of the two forms of land-use control and the two forms of regulations. Much of land-use policy concerns how the combination is achieved. Issues of land-use control can become highly politicized, as in the wise-use movement or the controversy involving the northern spotted owl and logging policy in the northwestern United States. In such cases, reaching enough community consensus to support consistent regulation can be difficult to achieve because of the differing cultural and social values held by the parties involved.

Although the United States and Canada have contributed greatly to the literature on land-use planning and the development of innovative techniques, North America has relatively weak land-use controls, as does Australia. Northern Europe and countries such as Japan are known for their comprehensive land-use controls. In the United States, as in many countries, the redistribution of population and wealth, together with the restructuring of public lands, necessitates constant reexamination of land-use policies and regulations. For developing nations, land-use policies become particularly critical in the evaluation of trade-offs between natural and land-based resources on one hand and economic well-being on the other. Pressures exist for these countries to exploit their natural resources while striving to achieve the prosperity they see in more developed countries such as those in North America and Northern Europe. As global markets have expanded, the need for international dialogue on land-use issues has also grown.

Robert M. Sanford and Hubert B. Stroud

Further Reading

Arendt, Randall. *Rural by Design: Maintaining Small Town Character.* Chicago: American Planning Association, 1994.

Arnold, Craig Anthony. *Fair and Healthy Land Use: Environmental Justice and Planning.* Chicago: American Planning Association, 2007.

Goetz, Stephan J., James S. Shortle, and John C. Bergstrom, eds. *Land Use Problems and Conflicts: Causes, Consequences, and Solutions.* New York: Routledge, 2005.

Johnston, Robert J., and Stephen K. Swallow, eds. *Economics and Contemporary Land Use Policy: Development and Conservation at the Rural-Urban Fringe.* Washington, D.C.: Resources for the Future, 2006.

Marsh, William M. *Landscape Planning: Environmental Applications.* 5th ed. Hoboken, N.J.: John Wiley & Sons, 2010.

Porter, Douglas R. *Managing Growth in America's Communities.* 2d ed. Washington, D.C.: Island Press, 2008.

Sargent, Frederic O., et al. *Rural Environmental Planning for Sustainable Communities.* Washington, D.C.: Island Press, 1991.

Light pollution

FIELDS OF STUDY
Astronomy; Bioenergy Technology; Ecology; Electrochemistry; Energy Efficiency; Environment; Environmentalism

SUMMARY
In addition to its adverse effects on the work of astronomers and the wasted energy it represents, light pollution can have deleterious impact on animals, including humans.

PRINCIPAL TERMS
- **astronomy:** the study of the sun, moon, stars, planets, comets, gas, galaxies, gas, dust and other non-Earthly bodies and phenomena
- **circadian rhythm:** the cyclical 24-hour period of human biological activity
- **diurnal:** of, relating to, or occurring in the daylight
- **light pollution:** human-caused illumination of areas beyond where light is intended and/or wanted
- **nocturnal:** of, relating to, or occurring in the night

In addition to its adverse effects on the work of astronomers and the wasted energy it represents, light pollution can have deleterious impact on animals, including humans.

Like most pollution, the most obvious effect of light pollution is aesthetic. Stray light from human-made lighting sources makes the sky bright at night, which makes stars and constellations less easily visible. The field of astronomy is adversely affected by light pollution for this reason; in fact, in many highly populated areas, astronomers are unable to see most galaxies and nebulae, even with the use of telescopes. Because of the impact of this situation on their field of study, astronomers have been among the most vocal critics of light pollution, but environmentalists have increasingly joined in efforts to address the problem, noting that it has negative impacts on the environment as well.

Light pollution is often an unintended consequence of intentional lighting. The light from human-made sources, if not properly directed, often goes beyond the areas that are meant to be illuminated. Many people who install outdoor lighting fixtures do not even think about how the light they generate affects others. In this way, light pollution is much like noise pollution. Light pollution is also like noise pollution in that for many years it was thought of solely as a nuisance issue, a problem without any real consequences for the environment.

Environmental Effects
One of the biggest environmental impacts associated with light pollution is the waste of energy represented by light that goes beyond where it is intended to go. When light shines beyond the area that needs to be illuminated, the light source is using more energy than is necessary to do the intended job. It has been estimated that the energy used for outdoor lighting could be cut in half if all lighting fixtures were appropriately shielded to direct their light more precisely. Aside from the economic benefits, such a reduction in energy consumption would have a significant

Los Angeles skyline at night. By Aaron Logan

impact on the need for electricity generation, which often has negative environmental effects.

Light pollution has also been found to be detrimental to wildlife. Nocturnal animals are adapted to life in the dark, and the lack of full darkness in areas with high levels of light pollution can influence the activities of such animals. For example, it can affect the relationship between predator and prey, as the prey cannot hide in darkness as they normally would. Some studies have shown that migratory animals, particularly birds, are also affected by light pollution, and other research has found that increased light affects the breeding practices of many animals. Excess nighttime lighting can affect plant growth as well.

Humans also appear to be physically affected by light pollution. An increasing body of evidence indicates that the human circadian rhythm developed to include a certain number of hours of darkness per day. Disruptions of a person's circadian rhythm can result in serious health problems, ranging from sleeplessness to irritability and depression.

Mitigation
Outdoor lighting is essential for safety and security, but not all outdoor lighting generates the same level of light pollution. Reducing light pollution can be as easy as replacing conventional outdoor lighting fixtures with fixtures that are shielded so that they direct light where it is wanted and limit the amount of light that goes in other directions. Such fixtures can achieve the same level of illumination in intended areas with lower-wattage bulbs than are needed in conventional fixtures, resulting in energy savings in addition to reductions in stray light. Shielded light fixtures tend to cost more than unshielded fixtures, but the savings in electricity costs may offset the greater initial expense over time. Retrofitting existing lights with shielding on a large scale, however, can be prohibitively expensive.

The biggest impediment to the reduction of light pollution is a general lack of awareness and understanding of the effects of light pollution among the public. The deleterious effects of air pollution and water pollution are generally well understood, but the effects of light pollution—other than on astronomy—are often not immediately recognized. Because of this lack of awareness, outdoor lighting is often installed by nonexperts. Municipalities, businesses, and home owners could reduce light pollution by consulting with lighting engineers who can design lighting plans to ensure enough illumination of desired areas with minimal waste of light and energy.

Raymond D. Benge, Jr.

Further Reading
Bakich, Michael E. "Can We Win the War Against Light Pollution?" *Astronomy*, February 2009, 56–59.

Gallaway, Terrel, Reed N. Olsen, and David M. Mitchell. "The Economics of Global Light Pollution." *Ecological Economics* 69, no. 3 (2010): 658–65.

Klinkenborg, Verlyn. "Our Vanishing Night." *National Geographic*, November 2008, 102–23.

Luginbuhl, Christian B., Constance E. Walker, and Richard J. Wainscoat. "Lighting and Astronomy." *Physics Today*, December 2009, 32–37.

Mizon, Bob. *Light Pollution: Responses and Remedies.* New York: Springer, 2002.

Rich, Catherine, and Travis Longcore, eds. *Ecological Consequences of Artificial Night Lighting.* Washington, D.C.: Island Press, 2006.

M

Marine life

FIELDS OF STUDY
Atmospheric Sciences; Biology; Climate and Climate Change; Climatology; Ecology; Ecosystems; Ecosystems; Environment; Life Sciences; Marine Biology; Meteorology; Oceanography; Zoology

SUMMARY
Rising sea levels, the results of melting ice and thermal expansion of seawater, are likely to constitute the most notable challenge related to global warming for many millions of people around the world.

PRINCIPAL TERMS
- **acidification:** The act or process of making something sour (acidifying), or changing into an acid
- **calcification:** the deposition of calcium salts in body tissues
- **carbon dioxide:** a colorless, odorless, incombustible gas resulting from the oxidation of carbon
- **climate change:** change of climate attributed directly or indirectly to human activity altering the composition of the global atmosphere and to natural climate variability observed over comparable time periods
- **global warming:** increase in the average surface and ocean temperature of the Earth since 1850 and the projected persistence of the trend
- **Pliocene:** pertaining to an epoch of the Tertiary Period, occurring from 10 to 2 million years ago, and characterized by increased size and numbers of mammals, by the growth of mountains, and by global climatic cooling
- **salinity:** of, containing, or resembling common table salt; salty or salt-like

Rising temperatures affect sea levels and the salinity and acidity of the oceans, which in turn have impacts on aquatic life and the lives of people and animals that live near shorelines. Rising sea levels, the results of melting ice and thermal expansion of seawater, are likely to constitute the most notable challenge related to global warming for many millions of people around the world.

Many scientists have projected that the earth's temperature will rise at least 3 degrees Celsius (5.4 degrees Fahrenheit) during the twenty-first century, bringing it to a level near that of the middle Pliocene, three million years ago, when the seas were 15 to 35 meters (50 to 115 feet) higher than at the beginning of the century. The carbon dioxide (CO_2) level at that time reached 425 parts per million (ppm); by 2010 the CO_2 level was at 390 ppm, increasing at a rate of about 2 to 3 ppm per year. In 2013 the CO_2 level surpassed 400 ppm, and in 2017 it surpassed 410 ppm for the first time in millions of years.

In an article published in the March 2004, issue of *Scientific American*, James E. Hansen, then director of the National Aeronautics and Space Administration's Goddard Institute for Space Studies, warned that if recent growth rates of CO_2 emissions and other greenhouse gases continue during the next fifty years, temperature increases could provoke large rises in sea level, with potentially catastrophic effects. According to Hansen, a temperature increase of 2 to 3 degrees Celsius (3.6 to 5.4 degrees Fahrenheit) could raise sea levels by about 25 meters (82 feet) within a few centuries. In late March, 2006, a report in the journal *Science* stated that melting ice, principally from Greenland and the West Antarctic ice sheet, could contribute to a rise in sea levels of several meters within a century—an upward revision of previous estimates. By the 2010s, some areas had already seen noticeable shifts in sea level. According to the US Environmental Protection Agency (EPA), from 1993 to 2016 global average sea level rose by 0.11 to 0.14 inches each year, about two times faster than long-term historical trends.

Climate Change and Marine Life in the Geologic Record

Scientists recognize at least five major global mass extinction events, of which the Permian extinction, which took place 251 million years ago at the Permian-Triassic boundary, was the most devastating. At that time, approximately 95 percent of marine species and 70 percent of land species became extinct in three distinct pulses over a period of about eighty thousand years.

While several theories have been proposed to account for the Permian extinction, among the best supported of these theories is one that attributes it to drastic climate change triggered by massive volcanic eruptions in Siberia. According to this theory, each eruption caused brief cooling episodes due to volcanic dust and sulfuric acid aerosols in the atmosphere; once the dust and aerosols dispersed, this cooling was followed by a period of warming caused by the huge amounts of carbon dioxide, sulfur dioxide, and other greenhouse gases (GHGs) that were released along with the lava. Over the course of a million years, repeated eruptions, dwarfing anything humans have experienced in their brief tenure on Earth, eventually overwhelmed the planet's capacity to self-correct.

Another theory postulates that the ocean depths became increasingly oxygen depleted, favoring the growth of bacteria that produce hydrogen sulfide. High pressures and cold temperatures in the abyss allowed the hydrogen sulfide to build up, only to be released in a gigantic "burp" of highly toxic fumes. Still another theory points to the storage of large quantities of methane in the form of clathrates in deep-sea sediments, suggesting that this methane was abruptly released when warming raised the temperature of the deeper regions of the ocean by 5 degrees Celsius. In addition to being toxic and a powerful GHG, methane is explosive at concentrations as low as 5 percent. Both of these events—deoxygenation of the oceans and the release of trapped methane—could have similarly been triggered by the Siberian eruptions and the significant global warming that resulted.

Whatever the cause, the Permian extinction, which devastated every group of plants and animals, was extremely abrupt by geological standards. Sedimentary rocks dating from after the cataclysm are nearly bare of fossils for the first ten million years of the Triassic period.

Rising Acidity in the Oceans

In 2003, scientists Ken Caldeira and Michael E. Wickett noted in the journal *Nature* that CO_2 levels were rising in the oceans more rapidly than at any time since the age of the dinosaurs. As the oceans absorb more CO_2 and become more acidic, their capacity to hold more CO_2 in the future is strongly reduced.

Concerns about rising ocean acidification were reinforced in January 2009, when 155 scientists from twenty-six nations, organized by several international groups under the aegis of the United Nations, issued the Monaco Declaration, warning of severe damage to the oceans by rising acidity. In June 2009, seventy science academies around the world called for the inclusion of ocean acidity on the agenda of international climate change studies, noting that the oceans already had become more acidic than at any time during the past 800,000 years.

Carbon dioxide is being injected into the oceans much more quickly than nature can neutralize it. Seawater is usually alkaline, about 8.2 pH. The pH scale is logarithmic, so a 0.1 decrease in pH, the change since the beginning of the Industrial Revolution, indicates a 30 percent increase in the concentration of hydrogen ions.

Scientists have investigated what continued ocean acidification might do to animals with calcium shells. One study investigated 328 colonies of massive *Porites* corals on the Great Barrier Reef off Australia; these corals in the past have grown to more than 6 meters (20 feet) tall over decades to centuries. Results from sixty-nine sections of the reef found that calcification had declined 14.2 percent between 1990 and 2005, impeding the corals' growth by 13.3 percent. Scientists continued to monitor this situation through the 2010s, noting that carbon dioxide levels continued to increase.

The Ocean Food Web

Warming sea surface temperatures may interfere with phytoplankton production, with impacts rippling through the food web. Cooler, upwelling ocean water breaks through warm surface waters less frequently, reducing the nutrients available for plants and animals living in the oceans. By the 1990s, such decreases in productivity were detected near the California coast, where scientists have documented a measurable decrease in the abundance of zooplankton, the second level in the food web. By the 1990s, the abundance of zooplankton was 70 percent lower than it had been during the 1950s. A 2014 suggested that climate change would decrease phytoplankton and zooplankton biomass by approximately 6 and 11 percent, respectively. However, scientists noted that much of plankton ecology was still not understood, making accurate predictions about climate change's effects on plankton populations difficult. In 2015 the results of the first global plankton survey were reported, giving researchers a baseline to work with and confirming that plankton are sensitive to ocean temperature and other issues related to climate change.

For decades, scientists also have documented the effects of climate change on marine animals. For example, studies have noted decreased reproduction and increased mortality in seabirds and marine mammal populations in warming water. A 1999 report published by the World-Wide Fund for Nature and the Marine Conservation Biology Institute noted that the population of the sooty shearwater, a seabird, off the California coast declined 90 percent during the late 1980s and early 1990s, and the population of the Cassin's auklet declined 50 percent. Zooplankton populations declined markedly at the same time. In Alaska, a severe decline in shearwaters from 1997 to 1998 "was clearly due to starvation," according to the report. Other studies have found marked increases in disease incidence among species such as turtles, echinoderms, mollusks, and various marine mammals. Marine plants can also suffer from ocean warming, with declines in species such as eelgrass posing drastic changes to habitats and entire ecosystems.

The threats posed by climate change to ocean ecology and biodiversity have the potential to drastically impact human populations. Seafood is a vital nutrition source for billions of people worldwide, and the warming ocean puts the vitality of global fisheries at risk. In addition to directly harming many marine species relied upon by humans for food, climate change also compounds problems caused by overfishing.

Main Threats

The main threats to the abundance and diversity of marine life derive from Earth's human population explosion and its concomitant overexploitation and pollution of coastal waters. The exploding population is also a major factor in global warming. In terms of direct threats posed to marine life from rising temperatures, dwindling sea ice in the Arctic and especially the Antarctic is probably the most clear-cut. While anthropogenic climate change is undoubtedly a factor in coral-reef destruction and the decline of fisheries, it is likely not the only one. As scientists learn more about long-term cycles involving ENSO and analogous oscillating pressure and current systems in other oceans, a better understanding of the relationship of current extreme events to long-term trends should emerge.

Bruce E. Johansen and Martha A. Sherwood

Further Reading

Alley, Richard B., et al. "Ice-Sheet and Sea-Level Changes." *Science* 310 (October 21, 2005): 456-460.

Caldeira, Ken, and Michael E. Wickett. "Oceanography: Anthropogenic Carbon and Ocean pH." *Nature* 425 (September 25, 2003): 365.

Hansen, James. "Defusing the Global Warming Time Bomb." *Scientific American*, March 2004, 68-77.

Holland, Jennifer S. "Acid Threat." *National Geographic*, November, 2001, 110-111.

Lynas, Mark. *Six Degrees: Our Future on a Hotter Planet.* New York: HarperCollins, 2008.

Mathews-Amos, Amy, and Ewann A. Berntson. *Turning Up the Heat: How Global Warming Threatens Life in the Sea.* Gland, Switzerland: World Wide Fund for Nature/Marine Conservation Biology Institute, 1999.

Websites

Ocean & Climate Platform.
ocean-climate.org/?page_id=3830&lang=en.

United States Environmental Protection Agency.
www.epa.gov/climate-indicators/oceans.

UN News,
news.un.org/en/story/2017/06/558822-feature-climate-change-and-worlds-oceans.

N

National forests

FIELDS OF STUDY
American History; Biology; Botany; Ecology; Energy; Economics; Ecosystems; Environment; Environmentalism; Forestry; Horticulture; Life Sciences; Preservation; Public Policy; Renewable Resources; Wildlife Conservation

SUMMARY
The U.S. national forest system was one of the first and most successful resource management and conservation programs initiated by the federal government. Similar programs have been developed in most of the forested nations of the world to manage the forests' economic and ecological functions.

PRINCIPAL TERMS
- **American Association for the Advancement of Science**: the largest international nonprofit organization with the stated goals of promoting cooperation among scientists, defending scientific freedom, encouraging scientific responsibility, and supporting scientific education and science outreach for the betterment of all humanity
- **debt-for-nature swaps:** strategy for reducing foreign debt in developing nations by trading debt forgiveness or debt reduction for guarantees of environmental activities by debtor nations
- **deforestation:** clearing or thinning of forests by humans to make the land available for other uses
- **hectare:** unit of surface, or land, measure equal to 100 acres, or 10,000 square meters, equivalent to 2,471 acres
- **multiple-use management:** the management of resources, especially forests, for many purposes
- **national forests:** forestlands owned and managed by a national government
- **United States Division of Forestry:** an agency of the U.S. Department of Agriculture that administers the nation's 154 national forests and 20 national grasslands

National forests are distinctly different from national parks. National parks are aesthetically pleasing, culturally or historically significant, or ecologically important tracts of land protected in perpetuity for the public benefit. National forests are publicly owned resource management areas operated to maximize the cash value of timber and forest products, benefit the domestic economy, and ensure perpetual productivity of the land.

In the United States, the nineteenth century saw increased efforts to preserve timber resources while unrestrained national growth encouraged increased timber harvesting. In 1827 US president John Quincy Adams expressed interest in developing a plan for sustainable forestry to ensure the availability of masts for ships. The American Association for the Advancement of Science (AAAS) discussed the need for sustained-yield forestry during the 1860s, and in 1873 asked Congress to preserve and manage the nation's forests.

The first congressional actions to preserve forests focused on areas of significant natural beauty, such as Yellowstone National Park and Yosemite Valley. The Division of Forestry was established within the US Department of Agriculture (USDA) in 1876 to promote harvestable timber development. Protections for harvestable stands of timber were first approved in the Creative Act of 1891, which authorized the president of the United States to withdraw public lands previously open to preemption and homesteading rights and to establish forest reserves.

President Benjamin Harrison established America's first national forest, now known as Shoshone National Forest, in Wyoming in 1891. Confusion arose over the process for including land within the

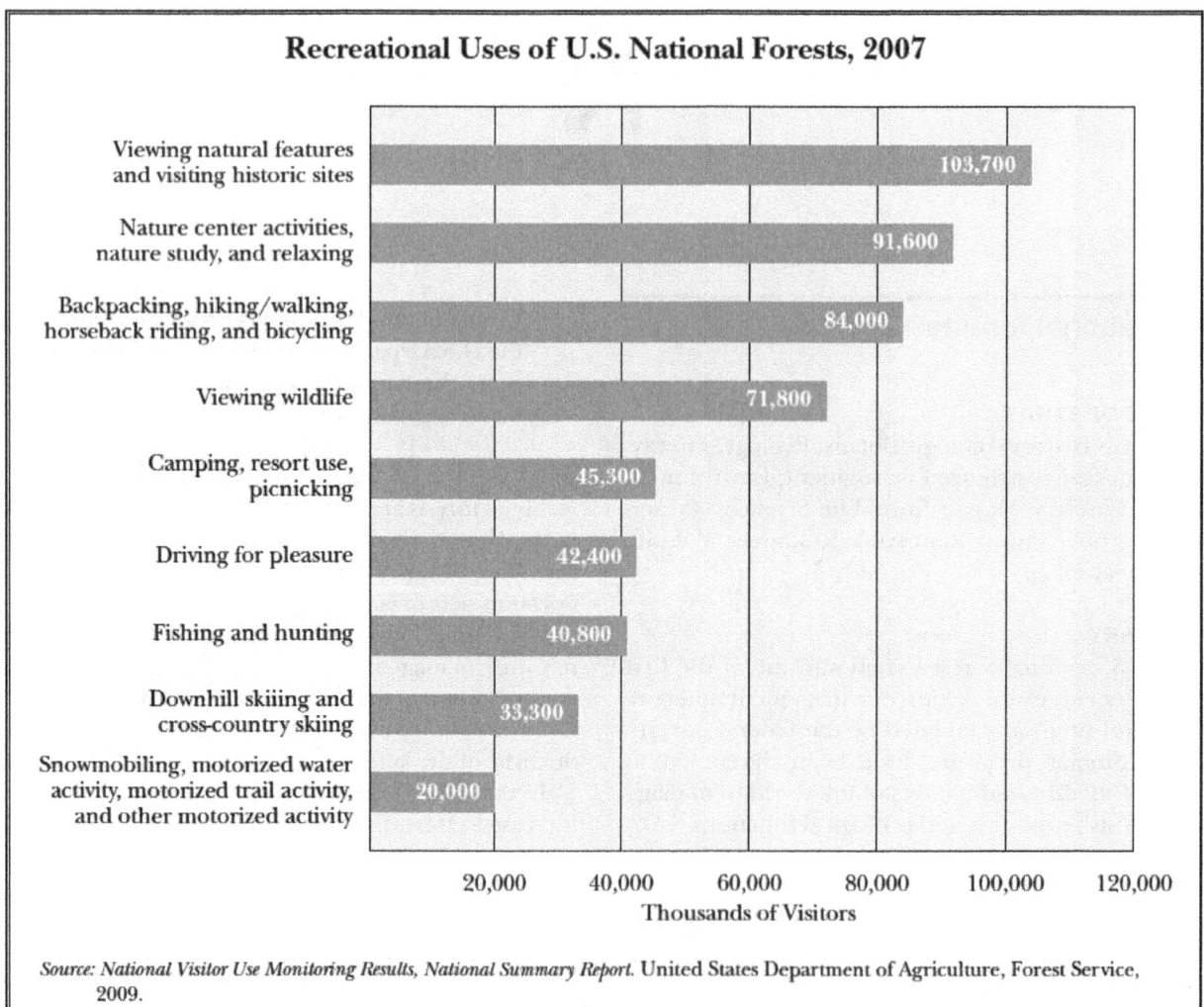

national forest system, however, and Congress suspended the law while it debated the purpose of the reserves, eventually authorizing selective cutting and marketing of timber from the existing reserves. Controversy concerning the vesting of the forest resources in the Department of the Interior or the Department of Agriculture ended in 1905 when the forest reserve system was established. President Theodore Roosevelt vested responsibility for national forest resources in the Bureau of Forestry of the Department of Agriculture under Gifford Pinchot, chief of the Division of Forestry. The bureau became known as the US Forest Service (USFS), and the reserve system became the national forest system in 1907.

Multiple Uses

From its inception, the US national forest system was intended to serve multiple purposes. The purposes of the original legislation were to improve water flows and furnish a continuous supply of timber. This evolved into the triple purposes of resource protection (especially fire protection), "wise use" of timber resources, and multiple use of system lands. The forest service lands are protected, harvested, and open to multiple other public uses. The Multiple Use-Sustained Yield Act of 1960 defined "multiple use" as a combination of outdoor recreation, fish and wildlife management, and timber production intended to meet the needs of the American people but not necessarily giving the maximum dollar value return.

The systematic management of resources for multiple uses while providing for conservation and sustainability is the goal of multiple-use management. Because some uses of natural resources are not compatible with each other and some can cause damage to the environment, multiple-use management poses several challenges.

The management of natural resources for multiple uses involves trade-offs among the current and future ecological, social, and economic use demands of citizens, governments, and private entities for goods and services. Multiple uses may take place on private and public forestlands, farmland, open lands, and along coastal resources. Forests are most often managed for multiple uses, as they provide not only wood resources but also aesthetic and environmental benefits, including regulation of climate, reduction of air pollution, provision of wildlife habitats, stabilization of soil, and preservation of water resources, some of which are used for public water supplies.

Foresters are aware of the multiple benefits that forests provide, and they manage forest resources based on the capacity of the land to accommodate multiple uses simultaneously without destroying the environmental benefits derived from forestland. Multiple-use forest management practices include sustaining the production of renewable resources such as trees and vegetation while protecting wildlife habitats, allowing recreation, and managing for fire and diseases. Private entities and individuals are also involved in multiple-use management when they engage in sustainable land-use planning that ensures development of compatible uses. Multiple-use management plans may designate areas for exclusive uses, such as timber production; dual uses, such as cultural heritage preservation and recreation; and general uses, where many compatible uses such as watershed protection, recreation, timbering, and wildlife refuges coexist simultaneously.

The National Forest Management Act of 1976 increased citizen involvement in decisions concerning forest management issues such as timber harvesting, resource conservation, and multiple uses. The 1976 act also limited the technique of timber harvesting by clear-cutting, a practice condemned by most environmentalists. In addition, the act limited logging on fragile lands and encouraged actions to maintain the diversity of plants and animals and to conserve plants, animals, soils, and watersheds in the forests. However, the act also emphasized the importance of multiple uses such as mining, oil and gas exploration, grazing, farming, hunting, recreation, and logging. The conflicts among conservation, harvesting, and multiple uses have continued to plague forest policy decision makers. Emerging forest policies in the early twenty-first century include sustained-yield forestry; substitution of alternative harvestable crops for harvestable timber, including crops such as nuts, fruits, gums, extracts, syrups, tars, and oils; and ecosystem maintenance.

The federal government owns and manages approximately 258 million hectares (640 million acres) of public land, about 28 percent of the land area of the United States. Almost 30 percent of the federal land area is part of the national forest system. In 2017, the national forest system comprised nearly 78 million hectares (193 million acres) of land in 155 national forests and 20 national grasslands located in forty-four states, Puerto Rico, and the Virgin Islands. Two-thirds of US national forests are located in the West, the Southeast, and Alaska. Twenty-three eastern states share about fifty forests. Roughly one out of every six hectares in the national forest system is part of the National Wilderness Preservation System (NWPS), and forest system lands represent about 33 percent of the NWPS.

Other National Forest Programs

Around the world, more than 140 countries have forest policy statements, and more than 150 countries have specific forest laws. Almost 75 percent of the world's forests are covered by national forest programs, and about 80 percent are under public ownership.

According to 2015 statistics from the United Nations Food and Agriculture Organization, the world's total forest area is just under 4 billion hectares (9.9 billion acres), or between 30 and 31 percent of the globe's total land area. Forests in Russia, Brazil, Canada, the United States, and China account for more than half of the world's forested area. During the 1990s global forests declined at a rate of about 16 million hectares (39.5 million acres) per year. The decline slowed to 13 million hectares (32 million acres) per year between 2000 and 2010. In addition, during this later period natural forest expansion in combination with tree-planting programs added 7 million hectares (17.3 million acres) of new forest annually. In total from 1990 to 2015, the world's forest areas declined by 3.1 percent, or 129 million hectares.

Between 1990 and 2010, the world forest area designated primarily for wood and non-wood forest products decreased, while the area designated for multiple uses increased. By 2010 about 30 percent of the world's forests were designated for productive purposes, 24 percent for multiple uses, and 8 percent for protective functions such as soil and water conservation, avalanche control, sand dune stabilization, and desertification control.

The global area of forested lands in legally protected areas such as national parks and wilderness areas increased between 1990 and 2010 by more than 94 million hectares (232 million acres). In the 2010s, legally protected areas accounted for about 13 percent of the world's total forest area. Environmental activism, ecotourism, and debt-for-nature swaps encourage the continuing development of national forest systems and other conservation measures in most nations of the world.

Gordon Neal Diem; updated by Karen N. Kähler

Further Reading

Hayes, Tanya, and Elinor Ostrom. "Conserving the World's Forests: Are Protected Areas the Only Way?" *Indiana Law Review* 38, no. 3 (2005): 595-618.

Hays, Samuel P. *The American People and the National Forests: The First Century of the U.S. Forest Service.* Pittsburgh, Pa.: University of Pittsburgh Press, 2009.

Hays, Samuel P. *Wars in the Woods: The Rise of Ecological Forestry in America.* Pittsburgh, Pa.: University of Pittsburgh Press, 2007.

Hirt, Paul W. *A Conspiracy of Optimism: Management of the National Forests Since World War Two.* Lincoln, Neb.: University of Nebraska Press, 1994.

Miller, Char and Scot J. Tilden. *America's Great National Forests, Wildernesses, and Grasslands.* New York: Rizzoli, 2016.

United Nations Food and Agriculture Organization. *Global Forest Resources Assessment, 2010.* Rome: Author, 2010.

Websites

United Nations Food and Agriculture Organization.
www.fao.org/3/a-i5588e.pdf.

Congressional Research Service, Federation of American Scientists.
fas.org/sgp/crs/misc/R42346.pdf.

■ National parks

FIELDS OF STUDY
American History; Botany; Controversies; Ecology; Ecosystems; Environment; Environmentalism; Land-use Management; Monuments; Parks; Preservation; Public Policy; Reserves; Sociology; Wildlife Conservation

SUMMARY
National park systems throughout the world provide protection for wildlife and plant life, preserve spaces for outdoor recreation, and educate people about the importance of protecting ecosystems.

PRINCIPAL TERMS

- **Dominion Forest Reserves and Parks Act:** Canada's original national parks system, established in 1911, became the National Parks Act in 1930, that regulates protection of natural areas of national significance
- **National Parks Service (or Organic Act):** established in 1916 and responsible for protecting the 35 national parks and monuments then managed by the United States Department of Interior and those yet to be established, eventually becoming an agency of the U.S. National Park Service that manages and maintains hundreds of national parks, monuments, historical sites and other designated properties of the government
- **national parks:** areas of scenic, historic, or other value that are set aside by federal governments for the preservation of animals and wildlife and for human recreation
- **United Nations Educational, Scientific, and Cultural Organization (UNESCO):** an agency of the United Nations whose declared purpose is to declared purpose is to contribute to peace and security by promoting international collaboration through educational, scientific, and cultural reforms in order to increase universal respect for justice, the rule of law, and human rights along with fundamental freedom

National parks are places where preservation for future generations must be balanced with present-day enjoyment. In the United States this balance has proved difficult to achieve, but nevertheless more than 415 park units—including national parks, historic sites, historical parks, memorials, memorial parks, battlefields, battlefield parks, battlefield sites, lakeshores, seashores, monuments, parkways, scenic trails, scenic rivers, scenic riverways, rivers, capital parks, and recreation areas—have been established across the United States. The process began when the US Congress made Yellowstone the world's first national park in 1872 (the politicians were willing to protect and preserve the geologic wonders of Yellowstone primarily because they were convinced that the lands were economically useless). Each of these park units was established to preserve and protect geologic wonders, spectacular scenery, wildlife, or an aspect of American history or culture. National parks have become important symbols of environmentalism, including occasionally conflicting forces and goals within the broad environmental movement.

Other countries have also established national parks, frequently using the United States as a model. Starting in the final decades of the twentieth century, the United Nations worked with countries to protect these areas. In 1972, on the one hundredth anniversary of the founding of Yellowstone National Park, the United Nations Educational, Scientific, and Cultural Organization (UNESCO) formed the World Heritage Committee. By 2009, 186 countries had ratified the World Heritage Convention. The committee, at its first meeting in 1977, formulated the World Heritage List, which names cultural and natural sites considered to be of "outstanding universal value." New sites have been added to the list each year. By the end of the first decade of the twenty-first century, the World Heritage List named almost nine hundred cultural and natural sites.

The list contains more cultural than natural sites, and not all the sites are parks, but many parks have received monetary support and advice from the United Nations to help ameliorate environmental problems, which abound in the United States and the rest of the world. The causes of these problems include the difficulties posed by the poaching of wildlife, exploitation of mineral deposits, and finding a balance between use and preservation.

National Coordinating Offices

Initially, national park operations in the United States were complicated because no single central federal office existed to coordinate activities. After much controversy, Congress passed the National Parks Act, or the Organic Act, of 1916. The act established a central authority called the National Park Service and stated its responsibilities, which include conserving and providing for the enjoyment of the scenery, natural and historic objects, and wildlife in the parks while leaving them unimpaired for the enjoyment of future generations.

This was not the first national office of national parks. In 1911 the Canadian parliament had passed the Dominion Forest Reserves and Parks Act, which provided for the administration of forest reserves and dominion parks and allowed dominion parks to be established from forest reserves. Thus, the Dominion Parks Service, created as a new branch in the Canadian Department of the Interior, became the first distinct bureau of national parks in the world. John Bernard Harkin served as commissioner from the service's inception in 1911 to 1936. In addition to working to separate the administration of the parks from that of the forests, Harkin emphasized resource preservation.

The numbers of national parks, as well as the numbers of visitors to the parks, continued to grow in both the United States and Canada. After World War II, as the automobile became ubiquitous, visits to the national parks increased rapidly. The facilities in place at the parks were old, however, and staffing was minimal. In 1956 the director of the US National Park Service, Conrad L. Wirth, decided that instead of asking Congress for annual appropriations, he would package the national parks' needs into a program called Mission 66. This one-billion-dollar, ten-year restoration program was designed to end in 1966, the fiftieth anniversary of the National Park Service. Canada was also influenced by Mission 66, and the Parks Policy of 1964, under the leadership of John I. Nicol, director of the Canadian National and Historic Parks Branch, emphasized the importance of protecting natural resources in the parks.

Problems of Overuse

The numbers of visitors to national parks in both countries increased steadily from the 1960s onward. With larger crowds came overuse, which created its own serious problems, including congestion, pollution, and—in some cases—destruction. Some popular parks banned automobiles from their roads, replacing them with shuttle buses. The park services of both Canada and the United States are continually faced with the problem of encouraging use while protecting valuable national resources and leaving them unimpaired for future generations.

In the United States, most of the service businesses in the national parks, such as restaurants, hotels, and souvenir shops, have traditionally been operated by private concessionaires. The operators usually sign long-term contracts under which small percentages of their profits are returned to the federal government. In the late twentieth century, as public funding for the parks declined and the numbers of visitors continued to increase, the National Park Service intensified its ties with private agencies to provide public services at national parks and other sites administered by the service. Some proposed projects became controversial and were never implemented, such as a giant theater that was to be built at Gettysburg Historic Site.

The National Park Service recognizes the importance of cooperation with the communities immediately adjacent to the parks, and partnerships and resulting plans have been developed around several national parks. Several of Canada's national parks, such as Banff and Jasper, contain towns, and the animals that live in the parks roam freely among automobiles, homes, and stores.

Management of Plants and Wildlife
National park services around the world must balance use with preservation of wildlife and plants. In general, the aim of wildlife management is to allow native species to flourish within the parks while protecting the parks from exotic, or nonnative, species. However, conflicts frequently develop regarding the remedial approaches that should be taken when exotic species are found in the parks. Some parks, such as the Shenandoah National Park and Great Smoky Mountains National Park, have been invaded by exotic insects that have destroyed whole forested areas and the component parts of the ecosystem.

One of the most controversial decisions in US National Park Service history was the decision to reintroduce the gray wolf into the Yellowstone ecosystem in 1995. Wolves in the area had been exterminated by rangers in 1915 because they were considered a menace to other native animals such as elk, deer, and mountain sheep. After many years of debate, the National Park Service was given permission and funds (approximately $6.7 million) to reintroduce wolves to the Yellowstone ecosystem, which includes the park and neighboring parts of Montana and Idaho. Many park visitors, upon being surveyed, had indicated that they wanted the wolves brought back. Local livestock owners, however, were concerned about the safety of their animals. To balance the competing needs—to serve the desires of park visitors, to restore the natural ecology of the park, to address the economic concerns of the livestock owners, and to support efforts to improve the wolf population—the Park Service developed plans to protect neighboring livestock before it undertook the careful reintroduction of wolves. In January 1995, the first set of gray wolves from Hinton, Alberta, Canada, were brought to Yellowstone. The introduction was deemed successful, and the ecology of the park returned to a more natural state as the wolf population increased, and the wolves became significant predators of elk, moose, and deer.

One problem area related to wildlife that affects almost all national parks involves the dangers that arise from human-animal interaction. It is illegal for visitors to feed any wildlife in US and Canadian national parks, but some visitors break the law. Human food is not part of the animals' natural diet, and some foods can cause digestive problems in animals and even endanger their lives. In addition, animals that come to associate humans with a ready food source can be dangerous to humans. When wild animals enter campgrounds and motor vehicles in search of something to eat, they occasionally harm humans. In extreme cases, to prevent animals from hurting people, park rangers may have to transport, or even kill, animals that have become too aggressive in searching for humans' food.

Economic Activities and Environmental Problems
The economic activities taking place outside park boundaries can cause environmental problems inside the parks. In some cases, neighboring mining or logging operations have contaminated waterways or hastened soil erosion. Sometimes compromises are reached that can reduce the impacts of such problems. For example, a planned gold mining operation

outside Yellowstone National Park was moved when the US government agreed to exchange other federal lands for the original intended site.

In other cases, distant power plants, factories, or even urban areas contribute to air pollution within park boundaries. In many parks in the United States, the combination of human-caused pollution and natural geology and meteorology has led to diminished views, more rapid weathering of natural wonders, and harm to wildlife. In South Africa, for example, the industrial district of Saldanha Bay has contributed to air and water pollution in the West Coast National Park.

National parks in parts of Africa and Central America face serious problems of wildlife poaching because of inadequate park administration and law enforcement. In areas such as Benin, disease-carrying livestock sometimes invade parks, resulting in wildlife deaths. Parts of national parks in Côte d'Ivoire and Senegal have been cultivated by farmers, but the governments of both countries have developed successful resettlement programs, thus lessening the impact on the parks.

Individual countries often look for help from the World Heritage Committee in establishing and maintaining national parks. The committee keeps a list of sites known as World Heritage in Danger, with the intent of bringing to worldwide attention the "conditions which threaten the very characteristics for which a property was inscribed on the World Heritage List." In 2018, fifty-four sites were on the World Heritage in Danger list. Sometimes, successful intervention results in a site being removed from the list, as occurred in 1998 with Plitvice Lakes National Park in Croatia. The park had been overdeveloped and overused, but after it was added to the World Heritage in Danger list in 1992, its underground water supply was protected, and a new road was built to decrease truck traffic through the park.

Margaret F. Boorstein

Further Reading

Allin, Craig W., ed. *International Handbook of National Parks and Nature Reserves.* New York: Greenwood Press, 1990.

Grusin, Richard. *Culture, Technology, and the Creation of America's National Parks.* New York: Cambridge University Press, 2004.

Heacox, Kim. *An American Idea: The Making of the National Parks.* Washington, D.C.: National Geographic Society, 2001.

Pereira, Sydney. "Trump Administration Threatens National Parks and Monuments." *Newsweek,* 5 Nov. 2017, www.newsweek.com/trump-administration-threatens-national-parks-and-monuments-701468. Accessed 22 Mar. 2018.

Ridenour, James. *The National Parks Compromised: Pork Barrel Politics and America's Treasures.* Merrillville, Ind.: ICS Books, 1994.

Runte, Alfred. *National Parks: The American Experience.* 4th ed. Lanham, Md.: Taylor Trade, 2010.

Sellars, Richard. *Preserving Nature in the National Parks: A History.* New ed. New Haven, Conn.: Yale University Press, 2009.

■ Noise pollution

FIELDS OF STUDY

Ecology; Environment; Environmentalism; Noise Control; Public Health; Public Policy; Urban Planning; Urbanization

SUMMARY

The problem of noise pollution is particularly acute because noise increases with population density; thus, a disproportionately large sector of the human population experiences the adverse effects of exposure to noise. The control of noise requires scientific, social, and sometimes political actions.

PRINCIPAL TERMS

- **Environmental Protection Agency (EPA):** an independent federal agency, created in 1970, that sets and enforces rules and standards that protect the environment and control pollution
- **noise pollution:** harmful or annoying sounds in an environment
- **Occupational Safety and Health Administration (OSHA):** an agency of the United States Department of Labor, established in 1970 assure safe and healthy conditions for working men and women by setting and enforcing standards and by providing training

Music, speech, and noise are the three basic categories of sound. Noise is simply defined as any unwanted sound. The degree to which a sound is unwanted is, however, a psychological question; the results of exposure to noise may range from moderate annoyance to hearing loss from high volume levels. Furthermore, the interpretation of what constitutes noise is subjective; both music and conversation may be regarded as noise in some places, such as in an office or a library.

With few exceptions, technological advances from the mid-twentieth century onward have resulted in a steady increase in the amount of unwanted sound. Examples include jet airplanes, automobiles, and ventilation fans. It was thought at one time that human beings should accept and tolerate the noise that went along with the benefits of many industrial advances, but problems associated with exposure to noise have manifested themselves often enough that there is now serious concern about noise pollution in the environment. Noise can be generated in a great variety of ways, but only a few prominent sources of noise emission are part of the daily lives of people in industrialized nations. Noise in the environment can be greatly reduced if these sources of noise pollution can be controlled.

Outdoor Noise
Noise from airplanes poses a major problem in urban areas. Airplanes produce noise through the efflux of jet engines and the high-pitched whine of engine fans. Since 1969 the Federal Aviation Administration (FAA) has legislated acceptable noise levels for commercial airplanes in the United States. Over the years, remarkable engineering innovations have been made in the reduction of jet noise to satisfy the relatively stringent FAA requirements. In the meantime, however, air travel has become increasingly popular. Airlines in the United States have now captured more than 80 percent of all intercity passenger traffic, and the percentage is still on the rise. It is anticipated that the number of takeoffs and landings near major cities will continue to grow throughout the early decades of the twenty-first century. Furthermore, as land prices rise, residential dwellings are encroaching on noise buffer zones near airports in greater numbers. The control of aircraft noise will certainly continue to be pressing in the future.

The deafening din of high-powered trucks and motorcycles is familiar to nearly everyone. Social surveys in cities consistently rank road traffic noise as one of the primary sources of annoyance. Cities can help to control traffic noise by rerouting heavy traffic and smoothing the flow of traffic so that vehicles avoid unnecessary starts, stops, and acceleration. Requirements that motor vehicles be maintained properly can also reduce traffic noise, as can the construction of sound-barrier walls near highways. Because most of the noise from a vehicle traveling at low speed is radiated from the exhaust system, a good muffler is very effective in controlling the acoustical emission. Given the recognition of noise pollution as a serious problem, the U.S. Environmental Protection Agency (EPA) has made several recommendations for reducing noise from motor vehicles.

Indoor Noise
A high proportion of the U.S. workforce is employed in interior environments in which the workers are subject to long periods of exposure to noise. Indoor noise also affects people living in apartments and in houses of relatively light construction. When indoor noise pollution results from sounds produced outside indoor spaces, lining the walls and ceilings of the spaces with acoustic panels and other sound-absorbing materials can be helpful in reducing the noise. In addition, the operation of most home appliances and factory machinery produces noise. Noise from washing machines, drills, and air-conditioning systems arises from friction, unbalanced rotating parts, and air turbulence created by fans. Such noise can be substantially reduced through proper lubrication, the balancing of rotating parts, and the installment of acoustic insulation. In an effort to encourage the design of quieter mechanical products, the American National Standards Institute has published guidelines for manufacturers to use in rating noise emission in their products. Adherence to the guidelines has been less than uniform, but perhaps this will change if consumers and employers express a willingness to pay higher prices for quieter appliances and machinery.

Other sources of noise pollution include some that are not even detectable by the human ear. Ultrasonic and infrasonic noises possess frequencies above and below the audible range, respectively. Such noises are emitted, for instance, through the background hum of high-voltage transmission cables. Although not technically heard, ultrasonic and infrasonic noises affect people in ways similar to audible noise.

Adverse Effects

The adverse physiological and psychological effects of noise on people have been the subject of considerable study. Researchers have found that exposure to noise interferes with people's ability to work and to sleep and also infringes on their enjoyment of recreation. Noise pollution has been associated with fatigue, loss of appetite, indigestion, irritation, and headaches.

High-intensity noise has been shown to have adverse cumulative effects on the human hearing mechanism that may produce temporary or permanent deafness. In fact, noise-induced hearing loss has been identified as a major health hazard by the U.S. Department of Labor's Occupational Safety and Health Administration. Noise pollution decreases worker efficiency and increases worker error rates.

Fai Ma

Further Reading

Berg, Richard E., and David G. Stork. *The Physics of Sound*. 3d ed. Upper Saddle River, N.J.: Prentice Hall, 2005.

Chiras, Daniel D. "Air Pollution and Noise: Living and Working in a Healthy Environment." In *Environmental Science*. Sudbury, Mass.: Jones and Bartlett, 2010.

Kotzen, Benz, and Colin English. *Environmental Noise Barriers: A Guide to Their Acoustic and Visual Design*. New York: Taylor & Francis, 2009.

Nadakavukaren, Anne. "Noise Pollution." In *Our Global Environment: A Health Perspective*. Long Grove, Ill.: Waveland Press, 2006.

Rossing, Thomas D., F. Richard Moore, and Paul A. Wheeler. *The Science of Sound*. San Francisco: Addison-Wesley, 2002.

Websites

British Library.
https://www.bl.uk/projects/save-our-sounds

Nuclear accidents

FIELDS OF STUDY
Disasters; Ecology; Environment; Environmentalism; Nuclear Power; Nuclear Weapons; Radioactive Waste

SUMMARY
Public concern about reactor safety and the risk of accidents remains an important reason the expansion of nuclear power has stalled in the United States and many other nations. Although the industry's safety record has generally been good, safety lapses and near accidents are sometimes reported, and the public has become increasingly aware that a severe accident could have widespread consequences.

PRINCIPAL TERMS
- **Chernobyl nuclear power plant:** considered the world's worst nuclear disaster to date when a sudden surge in power during a reactor-systems test in Ukraine, on April 26, 1986, resulted in an explosion and fire
- **Fukushima Daiichi nuclear power plant:** earthquake and tsunami that struck eastern Japan on March 11, 2011, caused a serious accident at the Fukushima Daiichi nuclear power plant on the northeastern coast of Japan
- **nuclear accidents:** unplanned events involving the release of radioactive materials into the environment
- **nuclear reactor:** an apparatus in which a nuclear-fission chain reaction can be initiated, sustained, and controlled, for generating heat or producing useful radiation
- **Nuclear Regulatory Commission (NRC):** an independent agency of the United States government, established in 1974, and tasked with protecting public health and safety related to nuclear energy
- **radioactivity:** the phenomenon of spontaneously emitting radiation resulting from changes in the nuclei of atoms of the element
- **Three Mile Island nuclear power plant:** the partial meltdown at Three Mile Island Unit 2 in Pennsylvania on March 28, 1978, is considered the most serious nuclear accident in U.S.

history, although it resulted in only small radioactive releases

The accident risk in nuclear power plants lies in the possibility that a malfunction caused by equipment failure, operator error, or external events will disrupt the flow of vital cooling water to the intensely hot, dangerously radioactive reactor core, a situation that could result in the fuel melting. Possible problematic external events include natural disasters such as the massive earthquake and subsequent tsunami that damaged reactors at Japan's Fukushima 1 nuclear power plant in March 2011. In the unlikely event that both the huge steel vessel holding the fuel and the massive concrete containment structure surrounding the reactor are breached, large quantities of dangerous radioactive materials could be dispersed to the environment.

The US Nuclear Regulatory Commission (NRC) is the federal agency that sets the safety requirements that plant owners must meet in the construction and operation of nuclear power reactors in the United States. NRC regulations and technical specifications cover a wide range of areas, including design requirements, safety systems, equipment quality, record keeping, and operator training.

Incidents in the United States

The worst nuclear accident to have occurred in the United States took place in 1979 at the Three Mile Island (TMI) Unit 2 plant in Pennsylvania, resulting in more than one-half of the fuel melting and the total loss of the reactor. Recurring incidents and near accidents over the years have led to some disquiet among the public and among nuclear regulators. Among these have been the 1975 fire at the Brown's Ferry plant in Alabama that burned for seven hours, during which it took plant personnel several hours to shut down the reactor; the 1983 unexpected failure of the Salem reactor in New Jersey to shut down after a safety system was activated; and the 1985 incident at the Davis-Besse plant in Ohio when multiple equipment failures caused a loss of cooling water that could have initiated a core meltdown if plant operators had not responded quickly. At some plants it has been discovered that important safety equipment has been inoperable for years. The NRC has kept several US reactors shut down for long periods until safety-related equipment problems or operator lapses could be corrected.

Proponents of nuclear energy argue that, despite recurrent mishaps, the US industry's overall safety record has been excellent. They assert that the events at TMI showed that systems designed to contain an accident worked. They claim that an accident similar to the one that occurred at Chernobyl in 1986 could not occur in the United States because the design of the older, Soviet-designed Chernobyl-style plants is flawed and permits a severe accident to occur far more easily than in US-designed plants. They note that the number of safety-related occurrences per year at U.S. reactors has declined since the TMI accident. Finally, they point to government-sponsored studies that have concluded that there is a very low probability of a severe accident occurring with significant off-site property damage and injuries to the public.

Antinuclear activists point to numerous recurring safety-related incidents and problems at US reactors, including some cases in which serious accidents were narrowly averted. They argue that the TMI incident showed that a serious accident is not as improbable as the industry claims. While critics acknowledge the design superiority of US reactors compared to the Chernobyl plant, they note that an NRC commissioner testified before Congress in 1986 that an accident at a US reactor with off-site releases equal to or worse than what occurred at Chernobyl could occur under conditions regarded as improbable but not impossible. Critics have also attacked the assumptions and methodology of the government's major nuclear accident risk studies. Finally, opponents observe that those same studies indicate that if a low-probability catastrophic accident should nonetheless occur, it could result in tens of thousands of deaths and injuries, tens of billions of dollars in property damage, and widespread, long-lived radioactive contamination.

The Price-Anderson Act limits the liability of nuclear power plant owners and equipment vendors in the event of a severe accident. This legislation was passed in 1957 after private companies made it clear that they would not participate in the development of nuclear energy without liability protection. Critics argue that the nuclear industry's continuing insistence that such protection is required is inconsistent with its denial of the possibility of a catastrophic accident.

Incidents in Other Nations

Much less information is generally available about nuclear accidents and incidents with serious safety implications in countries other than the United States because the public disclosure requirements of most nations are less stringent than US requirements. Worrisome accidents and equipment malfunctions are known to have occurred at reactors in many nations, including Japan, India, and the countries of the former Soviet Union. In Europe, grave concern remains about older Soviet-designed reactors, including some Chernobyl-style plants, still operating in the countries of Eastern Europe and the former Soviet Union. Western nations have provided some funding for safety improvements and have sought eventual permanent closure of many of these plants.

Some nuclear accidents have occurred at noncommercial facilities, some of which resulted in serious environmental consequences. In 1952 a fuel melt and explosion severely damaged an experimental Canadian reactor at Chalk River, Ontario. In 1957 an explosion at a waste site at the then-secret Chelyabinsk complex near the city of Kyshtym in Russia contaminated hundreds of square miles of the surrounding countryside. Also, in 1957, a fire occurred in England at the Windscale reactor (an atypical plant used to produce plutonium for the United Kingdom's weapons program); the surrounding countryside was contaminated, and radioactive fallout drifted into several neighboring countries.

Phillip A. Greenberg

Further Reading

Bodansky, David. *Nuclear Energy: Principles, Practices, and Prospects.* 2d ed. New York: Springer, 2004.

Cooper, John R., Keith Randle, and Ranjeet S. Sokhi. *Radioactive Releases in the Environment: Impact and Assessment.* Hoboken, N.J.: John Wiley & Sons, 2003.

Garwin, Richard L., and Georges Charpak. "Safety, Nuclear Accidents, and Industrial Hazards." In *Megawatts and Megatons: The Future of Nuclear Power and Nuclear Weapons.* Chicago: University of Chicago Press, 2001.

Savchenko, V. K. *The Ecology of the Chernobyl Catastrophe.* New York: Informa Healthcare, 1995.

Wolfson, Richard. *Nuclear Choices: A Citizen's Guide to Nuclear Technology.* Rev. ed. Cambridge, Mass.: MIT Press, 1993.

Websites

Union of Concerned Scientists.

https://www.ucsusa.org/nuclear-power/nuclear-power-accidents/history-nuclear-accidents#bf-toc-1

Nuclear power industry

FIELDS OF STUDY

Commercial Products; Ecosystems; Energy and Energy Resources; Hazardous Waste; Industries; Nuclear Power; Nuclear Weapons; Radioactive Waste; Technology and Applied Science

SUMMARY

Nuclear power has long been the most controversial source of electricity. High plant construction costs, safety concerns, waste disposal problems, and public mistrust all served to slow the growth of the nuclear power industry during the late twentieth century. In the early twenty-first century, factors such as nuclear power's potential to meet electricity demand while reducing greenhouse gas emissions have increased its appeal.

PRINCIPAL TERMS

- **breeder reactors:** a nuclear reactor that generates more fissile material than it consumes
- **nuclear fission:** subdivision of a heavy atomic nucleus, such as that , a process accompanied by the release of massive energy
- **Nuclear Industries Indemnity Act (Price-Anderson Act):** a United States federal law, first passed in 1957 and since renewed several times, which governs liability-related issues for all nonmilitary nuclear facilities constructed in the United States before 2026
- **nuclear power:** electricity generated through the harnessing of the heat energy produced by controlled nuclear fission
- **radioactivity:** the phenomenon, exhibited by and being a property of certain elements, of spontaneously emitting radiation resulting from changes in the nuclei of atoms of the element
- **uraninite:** a black octahedral mineral that consists of an oxide of uranium usually containing

- thorium, lead, and rare earth elements and is the chief ore of uranium
- **uranium:** a silvery heavy radioactive polyvalent metallic element that is found especially in uraninite and exists naturally as a mixture of mostly nonfissionable isotopes

Many of the environmental impacts of nuclear power plants are common to all large-scale electricity-generating facilities, regardless of their fuel. The most important are land use and related impacts on plants, animals, and ecosystems; nonradioactive water effluent and water quality; thermal pollution of adjacent waters; and social impacts on nearby communities. Unique to nuclear power plants is the hazardous radiation emitted by radioactive materials present in all stages of the nuclear fuel cycle. Such radiation is contained in uranium ore when it is mined and processed, in fabricated uranium reactor fuel, in spent fuel that has been fissioned, in contaminated reactor components, and in low-level radioactive waste—such as contaminated tools, protective clothing, and replaced reactor parts—generated by routine plant operation and maintenance.

Spent reactor fuel is highly radioactive and must be isolated from the environment for tens of thousands of years or more. Spent fuel can be reprocessed to separate out usable uranium and plutonium, but doing so generates large volumes of low-level waste that present disposal challenges of their own. Other challenges in waste handling and disposal arise at the end of a plant's useful life, when the contaminated reactor must be dismantled and the radioactive and nonradioactive components disposed of in a process known as decommissioning.

Under normal operating circumstances, commercial nuclear power plants release negligible radioactive emissions. The principal safety concern is that a severe accident could release large quantities of dangerous radioactive materials into the environment, as happened in 1986 at the Chernobyl nuclear plant in Ukraine. Roughly a quarter century after one of Chernobyl's reactors exploded, radiation levels remain dangerously high within the deteriorating concrete containment structure surrounding the reactor, and levels within the exclusion zone that extends in a 30-kilometer (18.6-mile) radius around the plant site continue to be higher than normal background levels.

Early History

The nuclear power industry arose out of the technology developed during World War II to produce the atomic bomb. Postwar enthusiasm for new technology in general, combined with pressure to demonstrate peaceful uses for expensive and fearsome wartime nuclear technology, led to a strong U.S. government effort beginning during the early 1950s to induce industry to develop nuclear energy. Large government subsidies and preferential treatment assisted the industry from its early days. Between fiscal year (FY) 1948 and FY2007, nuclear power received 53.5 percent of all federal energy research and development funds, totaling $85.01 billion in constant FY2008 dollars. One of the industry's unique subsidies was the passage of the 1957 Price-Anderson Act, which limits the liability of nuclear power plant owners and equipment vendors in the event of a reactor accident.

The basic design of the first U.S. nuclear power plants was adapted from early pressurized water reactor technology developed for submarines and other naval propulsion applications. Roughly two-thirds of the 104 commercial reactors in use in the United States are pressurized water reactors. In this reactor design, light (ordinary) water surrounds the nuclear fuel, which is made up of enriched uranium. The system is pressurized so that the fuel heats the water without boiling it. The resulting heat is used to boil a separate water supply, creating steam. The steam spins a turbine to generate electricity. The choice of this reactor design was largely driven by expediency and political considerations rather than an explicit effort to seek safe or reliable design features. The first large-scale commercial nuclear plant in the United States, which began operating at Shippingport, Pennsylvania, on a demonstration basis in 1957 and continued until 1982, was a conversion of a naval reactor project for which funding had been canceled. Some observers believe this early technology decision was largely responsible for the industry's problems in later years.

Another U.S. nuclear power plant that began operations in 1957, the Vallecitos plant near Pleasanton, California, was a research and development facility that became the first privately owned and operated nuclear power plant to provide significant quantities of electricity to a public utility grid. The Vallecitos facility employed a boiling-water reactor, a type of reactor in which the light water surrounding the enriched uranium fuel is converted directly into steam. The steam is piped to a turbine, which rotates to power an electric generator that produces electricity. The Vallecitos reactor was shut down in 1963. Approximately one-third of the commercial nuclear reactors operating in the United States are boiling-water reactors.

Yet another nuclear reactor technology emerged in the postwar years. Beginning in 1946, the United States developed a series of experimental prototype breeder reactors, a type of nuclear reactor in which the reaction is controlled in such a way that more fuel is produced than consumed. In 1963, the first commercial breeder reactor began low-power test operations at the Enrico Fermi Atomic Power Plant in Michigan. An accident in 1966 resulted in a partial core meltdown and caused reactor and fuel assembly damage that took almost four years to repair. Operations resumed in 1970 and continued until the decision was made in 1972 to decommission the reactor.

Late-Twentieth Century Developments
Government, industry, and public opinion were all largely positive about nuclear energy up through the late 1960s. Two-thirds of the U.S. commercial reactors operating in 2010 were issued construction permits between 1966 and 1973, a time of great optimism about the technology. During this period, the National Environmental Policy Act of 1969 was enacted, which forced prospective reactor owners to address environmental impacts in plant proposals; the environmental movement arose, beginning with the first Earth Day in 1970; and there was a widespread increase in environmental activism on the part of the public. By the mid-1970s, several widely publicized safety hearings, plant incidents, and government studies had begun to focus public and media attention on nuclear plant regulatory and safety lapses, accident risks, and the problem of nuclear waste disposal. These forces combined to create a sizable antinuclear movement in the United States, which was bolstered by the 1979 accident at the Three Mile Island plant in Pennsylvania.

Although proponents of nuclear power frequently blamed licensing interventions by antinuclear activists for numerous plant cost overruns and delays, most analyses concluded that in the majority of cases other factors, such as capital availability and shifting regulatory requirements, were primarily responsible for a slowdown in the development of nuclear power. By the late 1980s a decline in the number of plants under construction and a shift of public concern to the threat of the nuclear arms race had begun to reduce the ranks of the antinuclear power movement.

Development of breeder reactor technology in particular slowed significantly in the United States during the late twentieth century. One reason for the decline was the unique safety challenges presented by breeder reactors. Unlike light-water reactor systems, fast-neutron breeder reactors employ molten sodium as a coolant. Sodium burns when exposed to air and reacts explosively with water. Economic considerations also put breeder reactors at a disadvantage. As long as uranium supplies remained abundant, light-water reactor facilities were the more competitive option. Finally, the quantities of plutonium that could be created in breeder reactors raised concerns that the material could be used for nuclear weapons applications. The threat of nuclear proliferation led to a directive from President Jimmy Carter in 1977 that indefinitely deferred commercial reprocessing of spent nuclear fuel and plutonium recycling. This effective ban severely curbed breeder reactor progress in the United States. Congressional funding continued for a demonstration breeder plant in Oak Ridge, Tennessee, despite Carter's opposition, but in 1983 Congress cut funding for the project, which effectively ended the country's breeder reactor development for the rest of the century.

Between 1987 and 1994 nuclear power proponents won long-sought changes in regulations governing emergency planning requirements, the licensing of new reactors, siting, and reactor design certification. The essence of these changes was to facilitate the process for approving new reactors while sharply reducing opportunities for public participation in the regulatory process. Nuclear opponents adamantly opposed many of these changes, especially a 1992 congressional decision authorizing the U.S. Nuclear Regulatory Commission (NRC)—the federal agency responsible for nuclear power licensing and safety regulation—to forgo a long history of issuing separate licenses for construction and operation, each of which allowed for hearings. Instead, Congress mandated the issuance of a single combined license for both activities that permits few opportunities for safety challenges after construction has been completed.

Status and Projections: United States
The early twenty-first century has seen increasing receptivity to nuclear power in the United States. Among the factors contributing to this renewed interest are the need to meet the nation's continued growth in demand for electricity, the rising prices of fossil fuels, worries over possible interruptions in oil and gas availability, particularly from Middle Eastern sources, and concerns regarding the impact of the burning of fossil fuels on air quality and global climate. Some environmentalists tout nuclear power as a clean-air, carbon-free technology.

In the United States, nuclear power plants contributed approximately 20 percent of the electricity generated in 2009. In mid-2010 there were 104 licensed, operable reactors at sixty-five sites in thirty-one states. The newest of these received its construction license in 1973 and began its operational life in 1996. As of July, 2010, thirteen license applications were under active NRC review for up to twenty-two new reactors.

The nation's FY2011 budget authorized a total of $54.5 billion for nuclear power facilities, $36 billion of it in new loan authority established by the 2005 Energy Policy Act for DOE projects that cut greenhouse gas emissions. In February, 2010, President Barack Obama announced $8 billion in loan guarantees to go toward the construction of two new reactors at the existing Plant Vogtle nuclear power facility near Augusta, Georgia. These reactors are scheduled to come online in 2016 and 2017. If the NRC grants final approval for construction, these will be the first new nuclear reactors built in the United States in the twenty-first century.

All U.S. reactors were originally granted forty-year operating licenses. Under NRC regulations adopted in 1992, plant owners may seek twenty-year license extensions. In the case of some reactors, competitive economic pressures or the costs of expensive equipment upgrades make retirement preferable to an extended operating life; reactors may even be permanently shut down before their scheduled license expiration dates because competing power sources or needs for expensive repairs make continued operation economically undesirable. In other cases, equipment upgrades and license renewal prove more cost-effective than construction of a new power facility. (The license renewal process alone, not including equipment upgrades, costs $10 million to $15 million.) As of July 2010, the NRC had granted license renewals to fifty-nine reactors, and more renewal applications were expected.

In 2006 President George W. Bush proposed the Global Nuclear Energy Partnership (GNEP), a program for reducing the risk of nuclear weapons proliferation while minimizing reactor wastes. Spent fuel reprocessing and advanced breeder technologies were components of GNEP. In 2009, under the Obama administration, the DOE announced that it would not be pursuing domestic commercial reprocessing; however, research and development on proliferation-resistant fuel cycles and waste management would continue.

Status and Projections: World
As of November, 2010, 441 nuclear reactors were in operation in twenty-nine countries. The number of reactors in the United States (104), France (58), Japan (55), and the Russian Federation (32) accounted for more than half of the world's total operating reactors. The majority of these reactors were between twenty and thirty-nine years old. In eighteen countries, nuclear power supplied at least 20 percent of total electricity for 2009.

While most of the world's nuclear reactors are U.S.-designed pressurized water reactors and boiling-water reactors, other types are also in operation. Many of these employ graphite or heavy water (water enriched in deuterium) to moderate the nuclear reaction, which allows them to use less expensive natural (nonenriched) uranium as a fuel. The advanced gas-cooled reactors that predominate in the United Kingdom, for example, use graphite as a moderator and carbon dioxide instead of water as a coolant. In the pressurized heavy-water reactors common in Canada, heavy water serves as both coolant and moderator. The light-water graphite reactors of the Russian Federation are similar to boiling-water reactors but employ graphite moderators.

Although several countries established breeder reactor programs in the previous century, few of these programs were thriving as of 2010. The United

Kingdom had not constructed a new breeder facility since it shut down its last breeder reactor in 1994. Germany's one fast-breeder reactor permanently ended operations in 1991. France shut down Superphénix, the world's only commercial-sized breeder reactor, in 1998, after a spotty history of malfunction and repairs that kept the plant offline for more than half of its operational life. France's Phénix reactor enjoyed a more successful run, but it came to a close in 2010 with no immediate plans for another breeder reactor. In Japan, after a 1995 sodium coolant leak and fire at the Monju reactor, repairs and public controversy suspended the facility's operations until 2010. The only countries constructing new fast-breeder reactors as of August, 2010, were the Russian Federation and India, both of which were motivated to pursue the technology at least in part by concerns regarding uranium supplies.

In 2010, sixty new reactors were under construction in fifteen nations, although some of them may ultimately not be brought online. Fifty-one of them were pressurized water reactors, four were boiling-water reactors, two were pressurized heavy-water reactors, two were fast-breeder reactors, and one was a light-water graphite reactor. China, the Russian Federation, the Republic of Korea (South Korea), and India accounted for roughly 70 percent of the reactors under construction.

Waste- and Contamination-Related Issues
Spent fuel and high-level radioactive waste can pose threats to human health and the environment for many thousands of years, so they must be properly isolated and secured. Deep subterranean storage appears to be the best solution, but finding a repository site with the right geological characteristics is a technical challenge that is invariably complicated by political controversy. In the United States, Congress designated Yucca Mountain, Nevada, as the nation's sole repository in 1987, and site suitability studies were conducted at Yucca Mountain for nearly two decades. Opponents, including the state of Nevada, charged that the DOE's studies were geared more toward preparing the site for operation than for objectively assessing its suitability. The DOE submitted its license application for the repository in 2008, ten years after the original target date for opening. In early 2010, under the Obama administration, funding for the site was cut, and the DOE withdrew its license application. Lawsuits have been filed to challenge Yucca Mountain's closure, which leaves high-level waste in temporary storage at nuclear facilities around the country.

Some countries—notably France, England, Russia, Japan, and India—reprocess spent fuel from nuclear reactors. Reprocessing strips the waste of uranium and plutonium, which can be used to fuel reactors. While reprocessing results in a small reduction of high-level nuclear wastes, it generates large volumes of low-level nuclear wastes, which also require environmentally responsible disposal. Reprocessing is more costly than the single use and disposal of spent fuel. Also, because it creates stockpiles of plutonium, reprocessing has the potential to contribute to nuclear weapons proliferation and terrorism. Separating plutonium from more highly radioactive spent fuel assemblies makes it easier for it to be stolen. With this danger in mind, President Jimmy Carter issued an executive order in 1977 that indefinitely deferred U.S. reprocessing of spent nuclear reactor fuel.

Concern about nuclear power plant accidents continues to trouble the public. Precise accident probability estimates are impossible to derive, given the complexity of nuclear reactor systems. Industry proponents point to several government-sponsored studies that have concluded that the probability of a plant accident with off-site consequences is extremely low. Critics have countered that these studies were methodologically flawed, omitted important factors, and underestimated the true risks. They also emphasize the catastrophic consequences that could result if a low-probability accident should nonetheless occur.

Arguments Pro and Con
Nuclear proponents cite several points in the technology's favor: a good safety record, improved operating performance since the Three Mile Island accident, studies concluding that the risk of a severe accident is low, and the fact that nuclear power plants emit no significant amounts of common air pollutants or gases that contribute to global warming. Critics of nuclear power point to the long-standing failure of any country to establish a site for the permanent disposal of spent fuel and high-level wastes; flaws, uncertainties, and omissions in accident probability studies; the catastrophic consequences that could result from a severe reactor accident; a large number of safety-related incidents and near

accidents; and the high cost of nuclear plants compared to competing electricity-generating technologies and energy-efficiency improvements. In addition, for those countries that do not already have nuclear weapons arsenals, inherent proliferation risks are associated with nuclear power technology and fuels, which, if misused, could provide the expertise, infrastructure, and basic materials for a program to develop nuclear weapons.

High costs remain another obstacle to new plant orders. U.S. reactors completed during the 1990s cost $2 billion to $6 billion each, averaging more than $3 billion. Although proponents argue that new plant designs could be built more cheaply, the industry's history of large cost overruns discomforts prospective plant owners, capital markets, and state regulators. Many analysts believe that unpredictably high capital costs have been among the most important reason for the industry's decline. The picture has been complicated further by the deregulation of the electric utility industry that followed passage of the Public Utility Regulatory Policies Act in 1978 and the Energy Policy Act in 1992. Uncertainty surrounding the future economic and regulatory climate raises questions about the competitiveness of both new and many existing nuclear reactors.

Public opinion surveys in the United States after the events at Three Mile Island showed a growing majority opposing the construction of new nuclear power plants. Numerous influential policy assessments from the late twentieth century concluded that it was unlikely that new nuclear power plants would be built in the United States unless the key issues of waste disposal, costs, safety, and public acceptance were satisfactorily resolved. Although the situation differs from country to country, some combination of these factors served to hamper the growth of nuclear power in most countries during the late twentieth century. In the United States, additional factors that contributed to nuclear power's decline during the 1980s and 1990s included unexpectedly plentiful supplies of natural gas and slower-than-anticipated growth in electricity demand.

The public's mistrust of nuclear power is generally acknowledged as a roadblock to new plant orders. Risk perception studies have shown that the public's lack of trust is deeply rooted, widely felt, and resistant to change. Surveys assessing public support in the United States for the use of nuclear power reflected a decline between 1994, when 57 percent of those polled were in favor, and 2001, when 46 percent were in favor. Since 2001, however, support has increased, reaching 62 percent in 2010. Concerns about such environmental problems as greenhouse gas emissions and the cost and availability of fossil fuels played a role in this upswing in nuclear power's popularity. NRC officials and others have noted that, regardless of one's explanation of the public's views, public acceptance plays an important role in determining whether new plants will be built in the United States.

Phillip A. Greenberg; updated by Karen N. Kähler

Further Reading

Bodansky, David. *Nuclear Energy: Principles, Practices, and Prospects.* 2d ed. New York: Springer, 2004.

Garwin, Richard L., and Georges Charpak. *Megawatts and Megatons: The Future of Nuclear Power and Nuclear Weapons.* Chicago: University of Chicago Press, 2001.

Hagen, Ronald E., John R. Moens, and Zdenek D. Nikodem. *Impact of U.S. Nuclear Generation on Greenhouse Gas Emissions.* Washington, D.C.: Energy Information Administration, U.S. Department of Energy, 2001.

Hore-Lacy, Ian. *Nuclear Energy in the Twenty-first Century: The World Nuclear University Primer.* London: World Nuclear University Press, 2006.

Mahaffey, James. *Atomic Awakening: A New Look at the History and Future of Nuclear Power.* New York: Pegasus Books, 2009.

Murray, Raymond LeRoy. *Nuclear Energy: An Introduction to the Concepts, Systems, and Applications of Nuclear Processes.* Burlington, Vt.: Butterworth-Heinemann/Elsevier, 2009.

Wolfson, Richard. *Nuclear Choices: A Citizen's Guide to Nuclear Technology.* Rev. ed. Cambridge, Mass.: The MIT Press, 1993.

Ocean debris

FIELDS OF STUDY

Biology; Ecology; Ecosystems; Emissions; Environment; Environmentalism; Hazardous Materials; Life Sciences; Limnology; Marine Biology; Oceanography; Pollution; Toxic Waste; Zoology

SUMMARY

The presence of large amounts of litter in the world's oceans is a matter of major environmental concern. Such debris can threaten marine wildlife and ecosystems in numerous ways, such as by entangling animals, by blocking the digestive systems of animals that ingest nonfood objects, and by contaminating ocean waters with dangerous chemical compounds.

PRINCIPAL TERMS

- **biodegradable:** capable of decaying through the action of living organisms
- **Great Pacific Garbage Patch:** the largest accumulation of ocean plastic in the world, located between Hawaii and California, and currently considered to be three times the size of France
- **gyre:** a ring-like system of ocean currents rotating clockwise in the Northern Hemisphere and counterclockwise in the Southern Hemisphere
- **hydrophobic:** molecules and surfaces that repel water or liquids, such as oil, that separate from water; original meaning, "fear of water"
- **insoluable:** not capable of being dissolved in a liquid and especially water
- **North Pacific Gyre:** one of the five major oceanic gyres, the largest ecosystem on Earth covering most of the northern Pacific Ocean and comprising 20 million square kilometers (approximately 78 square miles)
- **nurdles:** pre-production plastic pellets created separately from the user plastics they are melted down to form and released into the open environment, creating pollution in the oceans and on beaches
- **ocean debris:** solid litter discarded by humans that finds its way into the oceans

Much of human-made material that is discarded after use ends up in the oceans to become marine debris. Of this debris, plastic products are among the most troublesome because they are produced in vast amounts, are lightweight and hydrophobic (water repelling), and resist degradation. In large parts of the ocean, plastic debris can be more abundant than the natural prey of marine animals. Animals that ingest these items are exposed to high levels of chemicals and may suffer blockages in their digestive systems. Large pieces of debris can entangle and choke large marine animals, including endangered turtles and marine mammals.

Various Kinds of Debris

The stability and the density of marine debris are the most important properties that affect the fates of these materials. Degradable materials such as paper and wood do not remain marine debris for long because microbes readily break down these materials. Non-soluble and nondegradable materials such as rubber, plastic, most metals, and concrete remain in water for years and therefore represent most marine debris. The density of the debris determines whether it sinks to the bottom or remains in the surface waters. Dense materials such as metals, concrete, ceramic, glass, and some plastics sink to the bottom of the sea, unless their shapes enable them to float. Materials that are less dense than seawater or equal to seawater in density, such as polystyrene and light

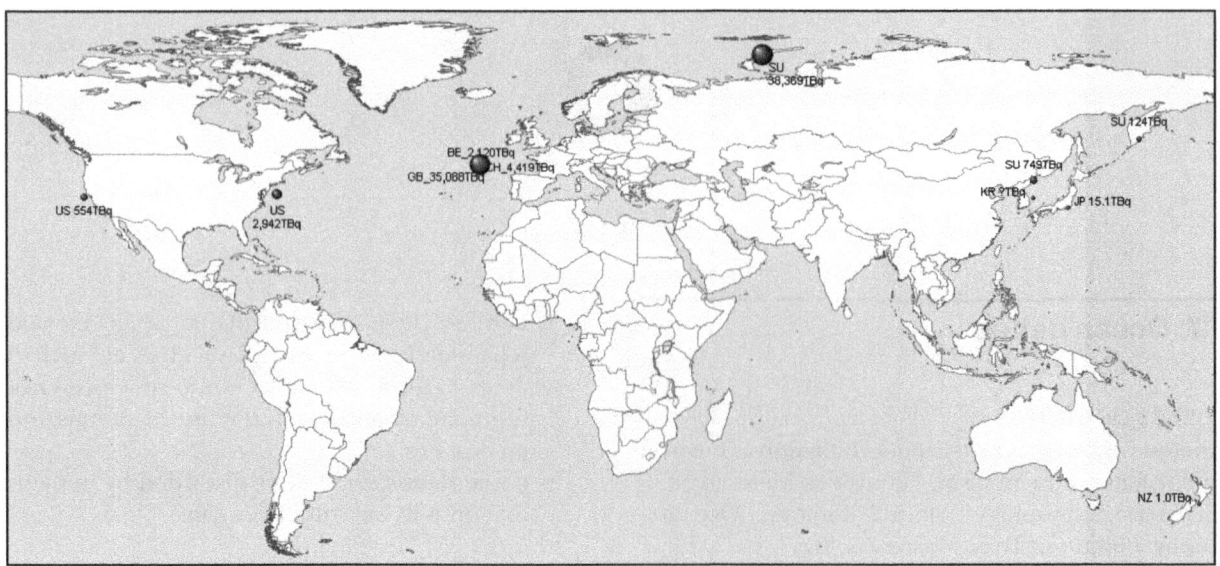

Country total at the major site. SU=Soviet Union (39,243TBq), GB=UK (35,088TBq), CH=Switzerland (4,419TBq), BE=Belgium (2,120TBq), France (354TBq), Germany (0.2TBq), Italy (0.2TBq), Netherlands (336TBq), Sweden (3.2TBq) are within GB marker, Russia (2.8TBq). By Masaqui (Own work).

plastics, drift in the surface currents of the oceans and often wash up on the shore with waves.

Large ocean gyres are vast areas where surface waters come together and down well. This water movement causes light, nondegradable debris to concentrate in these areas. The amount of debris in these waters often exceeds the amount of living organisms, a condition known elsewhere only in landfills and some urban environments.

Plastics are by far the most dangerous of all marine debris.

The Impact of Plastics

At the North Pacific Gyre, the large accumulation of debris has been dubbed the Great Pacific Garbage Patch. Because the plastic materials have been broken down to small particles that are not easily seen from satellites, the size of this debris field is difficult to measure, but it is estimated to be on the order of 1 to 10 million square kilometers (0.39 to 3.9 million square miles). In April 2010, the Associated Press reported that scientists had discovered similar plastic garbage patches in the Atlantic Ocean. Several expeditions to study the garbage patch have been undertaken, including one in 2012 by the Sea Education Association.

Every year, 8 million metric tons of plastics enter our ocean on top of the estimated 150 million metric tons that enter the oceans through rivers and circulate our marine environments (comparable to dumping one New York City garbage truck full of plastic into the ocean every minute of every day). Whether by errant plastic bags or plastic straws winding their way into gutters or large amounts of mismanaged plastic waste streaming from rapidly growing economies. The debris is so abundant that some boat captains refuse to travel in the area because of the risk that their vessels could become entangled.

More than half of this plastic is less dense than the water, meaning that it will not sink once it encounters the sea. The stronger, more buoyant plastics show resiliency in the marine environment, allowing them to be transported over extended distances. They persist at the sea surface as they make their way offshore, transported by converging currents and finally accumulating in the patch.

Effects on Marine Wildlife

Large pieces of debris provide structures and serve as habitats for marine organisms. In fact, old bridges, barges, and boats are intentionally discarded in some places to create artificial reefs for the purpose of increasing the productivity and diversity of marine life. Scientific research regarding the colonization and utilization of artificial reefs has revealed how the arrangement and shapes of debris can optimize its utility as habitat. Even large floating objects such as plastic buckets, buoys, and crates are used as habitat in much the same way as sargassum seaweed is used as natural habitat.

Most of the effects of debris on marine wildlife are negative. Marine debris can entangle, strangle, and choke some kinds of creatures. "Ghost nets" are fishing nets abandoned or lost in the sea that continue to function as they were designed to function, entangling animals. Fish, reptiles, mammals, and even birds that become entangled in ghost nets usually die from starvation, stress, injuries, or suffocation. Debris that gets wrapped around the throats of animals can cause strangulation. Ingested debris can block the digestive tracts of animals, interfering with their ability to feed, digest food, or pass waste. These animals usually die of starvation.

Predatory marine animals often consume plastic debris because it resembles their natural prey. Floating plastic bags and balloons are mistaken for jellyfish by sea turtles. Small plastic nurdles (resin pellets from which larger plastic items are made) are mistaken for fish eggs or zooplankton. Because most of the nurdles and small particles that have been found in debris samples are blue, black, white, or green, researchers have hypothesized that yellow and red plastic particles are selectively eaten by marine organisms.

Plastic has been found in more than 60 percent of all seabirds and in 100 percent of sea turtle species, that mistake plastic for food. And when animals ingest plastic, it can cause life-threatening problems, including reduced fitness, nutrient uptake and feeding efficiency—all vital for survival.

Even if ingested plastic does not kill animals by choking their digestive systems, it can still have significant sublethal effects on organisms of all size. Plastic particles concentrate hydrophobic compounds on their surfaces. These compounds interfere with the endocrine systems of animals, creating reproductive, developmental, and immunological problems for these animals. By concentrating these compounds by many orders of magnitude, even small plastic particles can negatively affect large animals by increasing their exposure to endocrine-disrupting compounds, a problem that is already serious owing to the biomagnification of compounds in long marine food webs.

In 2017, the United Nations Under-Secretary and Environment Director, Erik Solheim said, "We're facing an ocean Armageddon. ... At the current rate [of dumping], we'll end up with more plastic in the oceans than fish by the middle of the century, and ultimately that comes back to our own food chain. We need to understand that if we kill our oceans, we also kill ourselves."

Prevention and Mitigation

The 1973 International Convention for the Prevention of Pollution from Ships (along with the 1978 protocol that amended it, collectively known as MARPOL 73/78) and the Convention on the Prevention of Marine Pollution by Dumping of Wastes and Other Matter (also known as the London Convention) are international laws that aim to limit marine debris by prohibiting the disposal of certain kinds of wastes in the open oceans. Enforcing such laws has proven to be difficult, however.

Marine debris is removed in a variety of ways, from individual scuba divers gathering litter by hand to boat-mounted skimmers picking debris from surface waters and teams of people and machines picking up the trash that washes up on beaches. Given that approximately 80 percent of marine debris comes from continental sources, strainers placed on storm drains and across streams to retain litter are effective preventive measures to combat marine debris. One unique program meant to reduce the number of ghost nets in the world's oceans is Hawaii's Nets-to-Energy Program, which has collected more than 830 tons of derelict fishing nets and has used them to produce enough electricity to power three hundred homes for a year.

Greg Cronin

Further Reading

Andrady, Anthony L., ed. *Plastics and the Environment.* Hoboken, N.J.: Wiley, 2003.

Hill, Marquita K. "Solid Waste." *Understanding Environmental Pollution.* 3rd ed. New York: Cambridge University Press, 2010.

Laws, Edward A. *Aquatic Pollution: An Introductory Text.* 3rd ed. New York: Wiley, 2000.

Moore, Charles. "Trashed: Across the Pacific Ocean, Plastics, Plastics, Everywhere." *Natural History,* Nov. 2003.

Ocean Conservancy. *Trash Travels: From Our Hands to the Sea, around the Globe, and through Time.* Washington, D.C: Author, 2010.

United Nations Environment Programme. *Marine Litter: A Global Challenge.* Nairobi, Kenya: Author, 2009.

Wolf, Nancy, and Ellen Feldman. *Plastics: America's Packaging Dilemma.* Washington, D.C.: Island, 1990.

Websites

Albatross.
https://www.albatrossthefilm.com/

Chris Jordan: Photographic Arts: Midway-Message from the Gyre.
http://www.chrisjordan.com/gallery/midway/#CF000313%2018x24

The Great Pacific Garbage Patch.
https://www.theoceancleanup.com/great-pacific-garbage-patch/

One Beach Plastic: Richard and Judith Selby Lang.
https://www.beachplastic.com/

Ocean Conservancy.
https://oceanconservancy.org/trash-free-seas/plastics-in-the-ocean/

Oceans and rising temperatures

FIELDS OF STUDY

Barometry; Chemistry; Climate Change; Ecology; Ecosystems; Environment; Environmentalism; Global Warming; Oceanography

SUMMARY

Elevated land temperatures include displacement of currents and upwelling zones and increased runoff in major river systems. Effects attributable to elevated sea surface temperatures include melting sea ice in the Arctic and Antarctic, reducing habitat for plants and animals and in the tropics, where higher sea surface temperatures have degraded coral reefs, profoundly affecting many organisms.

PRINCIPAL TERMS

- **clathrates:** a compound formed by the inclusion of molecules of one kind in cavities of the crystal lattice of another
- **dead zones:** areas of deep-water oxygen depletion due to surface algal blooms or disruption of thermohaline circulation
- **El Niño–Southern Oscillation (ENSO):** periodic fluctuation of temperatures and currents in the Pacific Ocean on a four-, ten-, and ninety-year cycle
- **global warming:** long-term rise in the average temperature of the Earth's climate system, an aspect of climate change shown by temperature measurements and by multiple effects of the warming
- **gyres:** a giant circular oceanic surface current
- **primary production:** production of fixed carbon through photosynthesis
- **pteropods:** certain orders of small, thin-shelled or shell-less gastropod mollusks that swim by means of wing-like lobes on the foot
- **thermohaline circulation:** the rising and sinking of water caused by differences in water density due to differences in temperature and salinity

Until quite recently, scientists and the general public considered Earth's oceans to be impervious to anthropogenic degradation. Oceans cover 71 percent of Earth's surface and account for a little less than half of its primary production, that is, the photosynthetic conversion of carbon dioxide (CO_2) into the organic compounds that make up the bodies of living organisms.

The ocean is far from being a uniform habitat; however, except for some near-shore environments, ecological niches cover wide areas and intergrade, meaning that species can readily adapt by shifting their ranges. In consequence, environmental pressures producing elevated extinction rates on land have a less dramatic effect in the open ocean.

Nonetheless, human activity has had an adverse effect on marine life, from phytoplankton to top marine predators such as sharks. While overfishing, pollution, and damming of rivers that serve as spawning grounds for marine fish have all taken their toll, these are only indirectly related to global warming.

Present effects attributable to elevated land temperatures include displacement of currents and upwelling zones and increased runoff in major river systems. Effects attributable to elevated sea surface temperatures include melting sea ice in the Arctic and Antarctic, reducing habitat for polar bears, penguins, and the many humbler species of plants and animals that thrive at the margins of the polar ice caps. In the tropics, higher sea surface temperatures alter coral metabolism, causing corals to bleach when they lose symbiotic algae (zooxanthellae) critical to their growth. Degradation of coral reefs profoundly affects the many organisms restricted to this habitat.

As atmospheric CO_2 has continued to increase, altered seawater chemistry has become a growing concern. An estimated one-quarter of the CO_2 generated by burning fossil fuels does not remain in the atmosphere but rather is dissolved in the oceans, increasing seawater acidity. According to the US National Oceanic and Atmospheric Administration (NOAA) Pacific Marine Environmental Laboratory (PMEL), since the start of the Industrial Revolution, ocean acidity has increased by approximately 30 percent. One of the effects of ocean acidification is the stunting of shell production. Shells and similar structures found in marine organisms are formed from calcium carbonate, a process known as calcification; as the acidity of the ocean increases, the calcium carbonate in these structures becomes more vulnerable to dissolution. Coelenterates such as corals, whose skeletons are made up of aragonite (one form of calcium carbonate), are more susceptible than are mollusks, which have shells of calcite (another form of calcium carbonate), but both remain vulnerable. The threat posed to marine calcifying organisms by ocean acidification was largely theoretical until 2012, when an article published in the journal *Nature Geoscience* reported the "extensive dissolution" of the shells of pteropods found in the Antarctic Ocean.

Climate Change and Marine Life in the Geologic Record

Scientists recognize at least five major global mass extinction events, of which the Permian extinction, which took place 251 million years ago at the Permian-Triassic boundary, was the most devastating. At that time, approximately 95 percent of marine species and 70 percent of land species became extinct in three distinct pulses over a period of about eighty thousand years.

While several theories have been proposed to account for the Permian extinction, among the best supported of these theories is one that attributes it to drastic climate change triggered by massive volcanic eruptions in Siberia. According to this theory, each eruption caused brief cooling episodes due to volcanic dust and sulfuric acid aerosols in the atmosphere; once the dust and aerosols dispersed, this cooling was followed by a period of warming caused by the huge amounts of carbon dioxide, sulfur dioxide, and other greenhouse gases (GHGs) that were released along with the lava. Over the course of a million years, repeated eruptions, dwarfing anything humans have experienced in their brief tenure on Earth, eventually overwhelmed the planet's capacity to self-correct.

Another theory postulates that the ocean depths became increasingly oxygen depleted, favoring the growth of bacteria that produce hydrogen sulfide. High pressures and cold temperatures in the abyss allowed the hydrogen sulfide to build up, only to be released in a gigantic "burp" of highly toxic fumes. Still another theory points to the storage of large quantities of methane in the form of clathrates in deep-sea sediments, suggesting that this methane was abruptly released when warming raised the temperature of the deeper regions of the ocean by 5 degrees Celsius. In addition to being toxic and a powerful GHG, methane is explosive at concentrations as low as 5 percent. Both of these events—deoxygenation of the oceans and the release of trapped methane—could have similarly been triggered by the Siberian eruptions and the significant global warming that resulted.

Whatever the cause, the Permian extinction, which devastated every group of plants and animals, was extremely abrupt by geological standards. Sedimentary rocks dating from after the cataclysm are nearly bare of fossils for the first ten million years of the Triassic period.

The Present and Near Future

Unless the most carefully researched models are far off the mark, nothing resembling the devastating geochemical upheavals of the Permian-Triassic period looms in the foreseeable future, even if present levels of fossil fuel consumption persist. These models presuppose that volcanic activity will continue at levels typical of the Holocene and that no asteroids are headed in Earth's direction.

Possible changes due to increasing ocean acidity are being closely monitored, with the discovery of shell dissolution in Antarctic pteropods presenting new cause for concern. Acidity alone is not expected to reach lethal levels in the near future, but temperatures are rising rapidly, and marine organisms in the Antarctic cannot adjust their ranges southward. A wide variety of fish and birds depend on this snail for continued survival.

Polar Regions

In both the Arctic and the Antarctic, chilled surface seawater sinks, allowing nutrient-rich waters to well up from below and support high phytoplankton productivity. The polar seas teem with life. The lower surfaces of ice sheets also support dense growth of attached algae. Global warming near the North Pole causes the most productive zone to retreat northward and contract in extent. This restricts the number of both herbivores and carnivores the system can support. Most polar animals are unable to extend their ranges into temperate seas, because their unique adaptations to frigid temperatures make them poor competitors and susceptible to disease in warmer climates. The situation in the Southern Hemisphere is even more acute, as species migrating southward encounter the continental margin.

The plight of polar bears has received considerable attention. These huge carnivores prey almost entirely on seals that they hunt on sea ice. As the seals are declining in numbers and retreating farther from shore in response to the shrinking of the ice cap, bears are starving and failing to reproduce. Whale populations that had begun to recover from overexploitation by the whaling industry are also declining again as a result of low food supplies. Antarctic penguins also face declining food supplies and an influx of predators, including sharks, which are extending their ranges southward.

Coral Reefs

Reef-building corals, and the numerous species that depend on them, have a narrow temperature range for optimum growth. They are also vulnerable to changes in sea level due to either global warming or global cooling. During the last Pleistocene glaciation, the resulting drop in sea level exposed much of Australia's Great Barrier Reef, restricting this unique ecosystem to isolated pockets. A rapid rise in sea level would damage existing reefs by reducing light levels below those needed by symbiotic algae.

A 2-degree Celsius rise in surface temperature is sufficient to cause bleaching in corals as the individual polyps eject symbiotic algae. Bleaching initially causes growth to cease and eventually kills the coral colony. In recent years, there have been massive die-offs of corals. Notably, a severe bleaching event in 2016 killed an average of 67 percent of shallow-water corals in the northern part of the Great Barrier Reef over an eight- to nine-month period. A previous die-off of corals in the South Pacific was associated with a severe El Niño–Southern Oscillation (ENSO) event to which global warming may have contributed. Near-shore pollution also devastates coral reefs in populated areas.

Dead Zones

In several parts of the world, extensive areas of ocean have become depleted in oxygen, turning once-productive fisheries into wastelands. Most of these dead zones are associated with rivers that drain populated areas; one of the largest lies offshore of the mouth of the Mississippi River. This dead zone owes its existence to influxes of nutrient-laden freshwater to the Gulf of Mexico. These nutrients stimulate massive algal blooms. There is an indirect connection to global warming, in that warming generally causes increased precipitation and therefore increased runoff. Dead zones off the west coasts of the United States and South Africa result from disruption of cold currents and associated upwelling zones and thus are believed to be directly related to global warming.

Gyres

The oceans' gyres are basin-scale circulation patterns in which the net flow of water occurs in a circular pattern around the basin. Each gyre is generally made up of four distinct currents that are driven by wind stresses, and its circulation direction is governed by the Coriolis effect. The major, subtropical gyres in the Atlantic and Pacific Oceans are located between the equator and approximately 45° north latitude, while the smaller, subpolar gyres lie north of that latitude. The Antarctic Circumpolar Current is a gyre that flows continuously around Antarctica, because there are no landmasses to impede it.

Individual currents that make up a gyre have different characteristics. Because of the Coriolis effect and resulting differences in sea surface elevation, the western boundary currents (currents that flow on the western side of the ocean basins and flow northward from the equator) are narrow, are relatively deep (up to 1,200 meters), and have velocities of up to 178 kilometers per day. Examples of western boundary currents include the Gulf Stream in the Atlantic Ocean and the Kuroshio Current in the Pacific Ocean.

Eastern boundary currents have widths of up to almost 1,000 kilometers but are only 500 meters deep. Examples of eastern boundary currents include the California Current in the Pacific Ocean and the Canary Current in the Atlantic Ocean. Offshore Ekman transport associated with the eastern boundary currents leads to upwelling and high levels of productivity in these surface waters. Circulation of the gyres is completed by transverse currents that flow east and west across the ocean basins, connecting the western and eastern boundary currents.

The subtropical gyres are associated with persistent high-pressure regions in the atmosphere, which leads to a net motion of surface water to the center of the gyre—a process known as convergence. These regions of high pressure are associated with low annual rainfall totals, so the salinity of water in the center of the gyres is somewhat elevated. Elevated salinity and convergence lead to downwelling, so the central gyres are regions of low productivity.

Gyres transfer large amounts of heat from the equator to the poles. For example, the Gulf Stream carries heat north from the Caribbean Ocean, travels along the East Coast of the United States, and then curves eastward toward Europe. The heat released from the Gulf Stream may lead to warmer average temperatures in Europe than those found at similar latitudes in North America. It has been hypothesized that global warming will lead to an increased influx of freshwater to the Arctic Ocean, which could block the northward flow of the warm, salty water of the Gulf Stream. In turn, this could prevent heat transport from the equator and could lead to cooling of the Northern Hemisphere.

Productivity
While global warming due to elevated CO_2 levels can cause local drops in productivity due to drought on land and disruption of thermohaline cycles in the ocean, the long-term predicted effect of such warming on a global scale is a net increase in photosynthesis, with an upper limit that far exceeds any projections based on realistic economic indicators. In the long term, if Earth begins producing more food, both the numbers and the diversity of herbivores and predators can be expected to increase.

In the short term, however, such changes lead to the proliferation of weedy species with high reproductive rates and broad ecological ranges, loss of diversity, and generally unstable conditions. Species with specialized ecological requirements become extinct, and natural ecosystems increasingly resemble intentional agriculture or aquaculture. A glimpse of the future may be gleaned from the formerly rich fisheries off the West Coast of North America. These have been in decline for several decades, mainly because of pollution and overfishing. Warmer waters coupled with a persistent dead zone off the coast of California and Oregon have further reduced stocks of commercial and sport fishes, but they have favored proliferation of the Humboldt giant squid, an aggressive predator adapted to warm temperatures and low oxygen levels.

Rising temperatures can be expected to reduce areas of high planktonic productivity near the poles while expanding them near the equator and at continental margins, threatening polar species with starvation and extinction while increasing in numbers in warmer climates without a corresponding increase in diversity.

There may be reef-building organisms ready to replace corals should the seas become inhospitable. During the very warm late Cretaceous, rudists, a group of bivalve mollusks related to clams, were the main reef builders. Several types of algae also have limestone skeletons. If any of these groups were to replace corals, the structural integrity of reefs would be preserved, but the beauty and diversity of the ecosystem would be sadly compromised on any conceivable human time scale.

Context
The main threats to the abundance and diversity of marine life derive from Earth's human population explosion and its concomitant overexploitation and pollution of coastal waters. The exploding population is also a major factor in global warming. In terms of direct threats posed to marine life from rising temperatures, dwindling sea ice in the Arctic and especially the Antarctic is probably the most clear-cut. While anthropogenic climate change is undoubtedly a factor in coral-reef destruction and the decline of fisheries, it is likely not the only one. As scientists learn more about long-term cycles involving ENSO

and analogous oscillating pressure and current systems in other oceans, a better understanding of the relationship of current extreme events to long-term trends should emerge.

Anna M. Cruse and Martha A. Sherwood

Further Reading

Bednaršek, Nina, et al. "Extensive Dissolution of Live Pteropods in the Southern Ocean." *Nature Geoscience*, vol. 5, no. 12, 2012, pp. 881–85.

Di Lorenzo, Emanuele, et al. "North Pacific Gyre Oscillation Links Ocean Climate and Ecosystem Change." *Geophysical Research Letters* 35, no. 108607 (April 2008).

Peters, Robert L., and Thomas E. Lovejoy, editors. *Global Warming and Biological Diversity*. New Haven, Conn.: Yale University Press, 1992.

Reynolds, Colin S. *Ecology of Phytoplankton*. Cambridge, England: Cambridge University Press, 2006.

Saltzman, Barry. *Dynamical Paleoclimatology: Generalized Theory of Global Climate Change*. New York: Academic Press, 2002.

Websites

ARC Centre of Excellence for Coral Reef Studies,
www.coralcoe.org.au/media-releases/life-and-death-after-great-barrier-reef-bleaching.

BBC Earth.
www.bbc.com/earth/story/20151218-when-a-volcanic-apocalypse-nearly-killed-life-on-earth.

Pacific Marine Environmental Library: National Oceanic and Atmospheric Administration.
www.pmel.noaa.gov/co2/story/What+is+Ocean+Acidification%3F.

Oil drilling

FIELDS OF STUDY

Business; Chemistry; Commerce; Ecology; Economics; Energy and Energy Resources; Environment; Environmental Engineering; Environmentalism; Fossil Fuels; Industries; Production

SUMMARY

Careful management of oil drilling projects is crucial because the processes involved in such projects—including site preparation, equipment setup, drilling, fluid circulation, and waste disposal—all have the potential to degrade the environment.

PRINCIPAL TERMS

- **blowouts:** an uncontrollable escape of oil, gas, or water from a well
- **oil drilling:** activities involved in boring through earth and rock to tap petroleum reservoirs
- **Outer Continental Shelf Lands Act:** part of the internationally recognized continental shelf of the United States which does not fall under the jurisdictions of the individual U.S. State
- **Superfund:** established as the Comprehensive Environmental Response, Compensation, and Liability Act (CERCLA) of 1980, a a United States federal government program designed to fund the cleanup of sites contaminated with hazardous substances and pollutants.
- **Water Pollution Control Act:** the primary federal law in the United States to restore and maintain the chemical, physical, and biological integrity of the nation's waters

The process of drilling a hole through the rock layers that overlie a petroleum reservoir is simple, at least in principle. A diamond-tipped drill bit is attached at the base of a length of vertical pipe, and the pipe is rotated. The bit grinds away the rock, producing a hole the width of the drill bit and the length of the attached pipe, or drill string. As the top of the rotating drill string approaches ground level, a new piece of pipe is added, lengthening the drill string and allowing the depth of the hole to increase. The weight of the drill string is supported by a drilling rig, which helps keep the hole straight.

Meanwhile, a fluid known as drilling mud is pumped down the hole to keep the bit cool, help keep the hole from collapsing, carry rock chips (cuttings) up the hole, and prevent over-pressurized, subsurface liquid or gas from causing a blowout. Blowouts occur when subsurface zones of high pressure force the fluid in the well to flow out of the hole at a high rate of speed, in some cases blowing the drill string completely out of the hole and destroying the drill rig. Drilling mud is pumped downhole through the interior of the drill string, out through holes in the bit at the bottom of the drill string, then back to the surface between the drill string exterior and the drill hole, a space known as the well annulus. At the surface, the cuttings are collected, and the mud is recirculated downhole. Drilling mud is carefully monitored for the correct viscosity and weight. Chemicals are added, as needed, to modify these characteristics.

At intervals during drilling, the drill string is removed, and a steel liner, or casing, is cemented into place downhole. This supports the sides of the hole and isolates the annulus from the surrounding downhole environment.

If insufficient amounts of oil are found to make the well economically viable, the well is termed a "dry hole." It must be carefully cased from top to bottom and filled with cement so that the hole is sealed permanently, a procedure known as plugging and abandoning the well. If oil is found in sufficient amounts, it is pumped from the well, a process known as primary production. After an oil well's output begins to decline significantly, alternative methods of production may be applied. The term "enhanced oil recovery" refers to these more complicated—and expensive—methods for increasing an oil well's production. The most common method, saltwater injection, involves pumping saltwater brine, produced from the subsurface with the oil, back into adjacent wells, forcing more oil to the surface. Chemicals such as surfactants may also be employed to help mobilize the oil.

Environmental Concerns

The environmental concerns related to oil drilling on land are somewhat different from those related to drilling at sea. In either case, however, the effects of poor operations management can be devastating to the environment, particularly in ecologically sensitive areas. Methods of site access and preparation can have significant impacts on the ecology of land sites, especially in forest preserves, wetlands, and tundra. Soils, as well as surface-water and groundwater supplies, are at risk from oil spills and improper disposal of saltwater, drill cuttings, and drilling mud. Groundwater is also at risk from improper casing, saltwater injection, and well plugging and abandonment. Groundwater pollution may go undetected for years and can affect relatively large areas of the subsurface.

At sea, much larger areas may be at risk because of down-current transport of pollutants. Environments particularly threatened by marine drilling platforms include coral reefs, oyster banks, mangrove swamps, and tidal estuaries. Disposal of drilling mud and drill cuttings threatens the benthic (seafloor) environment surrounding the platform and the pelagic (open sea) environment down current. Improper operations can release oil and brine into the water column, and faulty casing can create oil and brine seeps on the seafloor.

In the United States, numerous federal laws regulate oil drilling, production, and transport on land and at sea. These include the Water Pollution Control Act of 1972; the Outer Continental Shelf Lands Act of 1953; the Comprehensive Environmental Response, Compensation, and Liability Act (CERCLA) of 1980 (also known as Superfund); the Marine Protection, Research, and Sanctuaries Act of 1972; and the Safe Drinking Water Act of 1974. Most U.S. states have stringent permit requirements, site inspection programs, and accidental spill reporting and response programs for drilling operations. Industry compliance with these programs is generally excellent.

Oil spills from offshore platforms represent only a small percentage of the petroleum released into the oceans each year. Natural oil seeps contribute as much as half the total petroleum found in coastal and marine environments annually, and their contribution has increased steadily as onshore and offshore oil production has reduced the pressure in hydrocarbon reservoirs. Marine drilling platforms, while perhaps eyesores above water, provide habitats for marine species that otherwise would not be present—these structures might be likened to artificial reefs. Scuba divers and fishing boat captains alike seek out platform substructures for the lush growth of invertebrates encrusting them and the many fish that this growth attracts. When offshore drilling platforms are decommissioned, debates often ensue among environmentalists as to whether the structures should be completely dismantled or allowed to remain intact so that the marine community may benefit.

Clayton D. Harris

Further Reading

Boomer, Paul M. *A Primer of Oilwell Drilling.* 7th ed. Austin: Petroleum Extension Service, University of Texas, 2008.

Camp, William G., and Thomas B. Daugherty. "Fossil Fuel Management." In *Managing Our Natural Resources.* 4th ed. Albany, N.Y.: Delmar, 2004.

Mtsiva, V. C., ed. *Oil and Natural Gas: Issues and Policies.* Hauppauge, N.Y.: Nova Science, 2003.

■ Oil shale and tar sands

FIELDS OF STUDY
Ecology; Emissions; Energy and Energy Resources; Environment; Environmentalism; Geology; Geosciences; Hazardous Materials; Industries; Pollution; Resource Management; Toxic Waste

SUMMARY
The conversion of the hydrocarbons in oil shale and tar sands into oil could provide an energy source several times the amount of the earth's current known oil reserves, but the conversion process has many environmental costs.

PRINCIPAL TERMS
- **acid mine drainage:** one of the most severe threats to water brought about by mining
- **bitumen:** any of various mixtures of hydrocarbons, such as tar, often together with nonmetallic derivatives that occur naturally or are obtained as residues after heat-refining natural substances, such as petroleum
- **oil shale:** finely grained sedimentary rock that incorporates a solid hydrocarbon substance in its structure
- **Surface Mining Control and Reclamation Act (SMCRA):** is the primary federal law, established 1977, that regulates the environmental effects of coal mining in the United States
- **tar sands:** sands permeated with a tarlike petroleum that is too thick to flow

Oil is not found in vast oil-filled underground caverns; rather, it is found in the spaces between grains that make up rock. Permeable rock has more space between grains than does impermeable rock. If the oil is thin enough, it will slowly flow through permeable rock and into the borehole of a well, from which it can be brought to the surface.

The hydrocarbons in oil shale cannot flow because they are in a solid form called kerogen. When some types of kerogen are heated by natural processes (60 to 150 degrees Celsius, or 140 to 300 degrees Fahrenheit) they slowly release crude oil and natural gas. In place of nature's slow process, oil shale may be mined and then burned as a low-grade fuel, or it can be crushed and then heated in a retort to produce crude oil and natural gas. Aboveground retorting is usually done between 300 and 520 degrees Celsius (570 and 970 degrees Fahrenheit). It is also possible to process shale underground (in situ, or in place) by fracturing the rock underground and then using the heat of an underground fire to produce crude oil, which can be pumped to the surface. In place of the fire, steam or other hot fluid can be circulated underground. After retorting, the oil is usually upgraded through a process in which it is passed over a catalyst and treated with hydrogen gas to increase the hydrogen content and remove impurities such as sulfur, nitrogen, iron, and arsenic.

Shale oil is very expensive to produce, although expected increases in production and improvements in the process are projected to bring its cost down significantly over time. The process also has environmental costs. Processing oil shale releases the carbon dioxide, which has been linked to global warming. Near-surface deposits of oil shale are open-pit mined or strip-mined, both processes that damage surrounding ecosystems; underground mining uses the less efficient room-and-pillar method. In any kind of mining of oil shale, one to five barrels of water are used for each barrel of shale oil produced; in addition, the process produces mountains of crushed rocks that must be disposed of.

The world's use of shale oil peaked in 1980, several years after the 1973 oil crisis. In 2008 Estonia was the top producer at 5,500 barrels per day, followed by Brazil at 3,100 barrels per day. It has been estimated that worldwide deposits of oil shale could produce 2.8 to 3.3 trillion barrels of shale oil. The United States has the largest share, with enough oil shale to produce 1.5 to 2.6 trillion barrels in the Green River Formation (a geologic formation located in parts of Utah, Wyoming, and Colorado). In comparison, the world's conventional oil reserves are estimated at 1.317 trillion barrels.

Different grades of oil shale yield different amounts of oil. About 16 percent of the Green River Formation yields 95 to 380 liters (25 to 100 gallons) of oil per ton of shale, about 33 percent yields 38 to 95 liters (10 to 25 U.S. gallons), and the final 51 percent yields 19 to 38 liters (5 to 10 U.S. gallons) per ton of shale. Unable to compete with cheap conventional oil, oil shale mining ceased in the United States in 1982, but in 2003 an oil shale development program was restarted.

Tar Sands

Tar sands (also called oil sands) are mixtures of sand, clay, and bitumen, a tarlike form of petroleum. The largest deposits of tar sands are in Canada and Venezuela, each country having the equivalent of the world's reserves of crude oil. Canada's production of oil from tar sands reached 1.25 million barrels per day in 2009 (at a cost of about $27 per barrel) and provided 47 percent of the nation's oil production. This allowed Canada to become the biggest supplier of products to the United States. In Canada about two tons of tar sand are processed to produce one barrel of oil. Producing oil in this way has the same environmental problems as producing oil from oil shale: Strip mining degrades ecosystems, and the process releases more carbon dioxide into the atmosphere than does the production of conventional oil, uses large amounts of water, and leaves tons of spent sand that must be disposed of.

Bitumen may be extracted in situ through the injection of steam, hot air, or solvents to allow the bitumen to flow. Alternatively, a deposit may be stripmined, and the materials then hauled to a processing facility, where the bitumen is separated from water, sand, and other waste. Like shale oil, the heavy crude extracted from bitumen must be upgraded through a process in which it is passed over a catalyst and treated with hydrogen gas to increase the hydrogen content and remove impurities such as sulfur, nitrogen, and arsenic. It can then be used as feedstock for conventional refining.

The only significant tar sands in the United States are in eastern Utah. The Utah tar sand deposit is far smaller than Canada's, and the Utah sands are hydrocarbon-wetted rather than water-wetted as the Canadian sands are, so their use would require different processing techniques.

Charles W. Rogers

Further Reading

Bartis, James T., et al. *Oil Shale Development in the United States: Prospects and Policy Issues.* Santa Monica, Calif.: RAND Corporation, 2005.

Marsden, William. *Stupid to the Last Drop: How Alberta Is Bringing Environmental Armageddon to Canada (and Doesn't Seem to Care).* Toronto: Vintage Canada, 2008.

Speight, James G. *Enhanced Recovery Methods for Heavy Oil and Tar Sands.* Houston: Gulf, 2009.

Websites

Union of Concerned Scientists.
https://www.ucsusa.org/clean-vehicles/all-about-oil/what-are-tar-sands

Native American Rights Fund.
https://www.narf.org/cases/keystone/

■ Oil spills

FIELDS OF STUDY

Disasters; Ecology; Emissions; Energy and Energy Resources; Environment; Environmentalism; Hazardous Materials; Limnology; Oceanography; Pollution; Toxic Waste

SUMMARY

Oil spills pose both short- and long-term environmental threats, including injuries and deaths among fish, birds, and other wildlife; damage to shoreline recreational areas; and pollution of water supplies.

PRINCIPAL TERMS

- **benzene:** a colorless, volatile, flammable, toxic, slightly water-soluble, liquid, aromatic compound, C_6H_6, obtained chiefly from coal tar and used to manufacture commercial and medicinal chemicals, dyes, and as a solvent
- **Kolva River:** a river in Perm Krai, Russia, 460 kilometers (290 mi) long, with a drainage basin covering 13,500 square kilometers (5,200 sq. mi)
- **Mingbulak:** a district in the Namangan Region of Uzbekistan, known primarily for its oil
- **oil spills:** accidental or intentional discharges of raw or refined petroleum products on land or at sea
- **Persian Gulf War:** (also The Gulf War), a war waged from 1990 to 1991, by coalition forces from thirty-five nations led by the United States against Iraq in response to its invasion and

annexation of Kuwait due to rising oil prices and production disputes
- **toluene:** a colorless, water-insoluble, flammable liquid, C7H8, having a benzene-like odor, obtained chiefly from coal tar and petroleum and used as a solvent for various organic compounds
- **vadose zone:** (also unsaturated zone) part of Earth that extends from the top of the ground surface to the water table
- **xylene:** any of three oily, colorless, water-insoluble, flammable, toxic, isomeric liquids, C8H10, of the benzene series, obtained mostly from coal tar and used chiefly in the manufacture of dyes.

Oil spills occur in both terrestrial and marine settings. Terrestrial spills affect land areas, including drainage courses and bodies of surface water impounded in lakes and ponds. Subsurface waters (groundwater) are also at risk from leakage of polluted water downward through the vadose zone (zone of aeration) to the water table (zone of saturation). Some crude oil is lost during exploratory drilling, workover operations, and tank storage. Mud pits utilized during rotary drilling usually contain oil recovered from the well during testing or oil derived from an oil-based mud employed in the drilling process. Oil and brine released from wells are sometimes stored in unlined ponds, and these fluids can soak into the ground and kill beneficial microbes in the soil and plant life in the immediate area. Subsurface water can also be contaminated if the oil or brine migrates downward to the groundwater table. Runoff from these areas can enter streams and ponded bodies of water and kill fish and other aquatic animals as well as reduce the natural vegetation.

The disposal of petroleum products (diesel fuel, gasoline, kerosene, jet fuel, and used motor oil) is a significant problem worldwide. Some of these products contain carcinogens such as benzene, toluene, and xylene, which require special handling. Many millions of gallons of used oil products are disposed of on land every year, and it has been estimated that more than 6.6 million tons, or about 45 million barrels, of petroleum and petroleum products enter the oceans per year. Of this total, 44 percent is land-derived. These inputs include coastal city contributions from refineries, wastewater outlets, and other sources; urban runoff, including storm drains; and river runoff.

Terrestrial Oil Spills

Terrestrial oil spills also occur during the loading, transportation, and offloading of petroleum from tank trucks and railroad tank cars. Petroleum pipelines, both buried and aboveground, are highly vulnerable to rupture from welding defects, corrosion, earthquakes, and shifting soils. Fire is a constant danger associated with such spills. During the mid-1980s a ruptured pipeline near São Paulo, Brazil, caused the deaths of more than five hundred people and resulted in the destruction by fire of twenty-five hundred homes in the town of Vila Socco. The 1992 Uzbekistan Mingbulak incident is, to date, the largest land-based oil spill in history. A blowout at a well spewed oil into the valley near the city of Fergana, caught fire and burned for two months before the well pressure subsided.

During the 1991 Persian Gulf War, large areas of Kuwait were devastated when Iraqi soldiers damaged oil pipelines and refineries and set fire to hundreds of oil wells. Pools of oil formed near the wells and infiltrated the porous and permeable soil and rock, thereby endangering the water supply of Kuwait City and other areas.

The 1994 Kolva River oil spill was caused by a breach in a corroded oil pipeline in the Russian Arctic. Oil pooled around the rupture site for eight months, contained by a dike, but the dike later gave way, spilling roughly 84 million gallons of oil into the Kolva River.

Offshore Spills

The waters of oceans and restricted seas can be polluted by oil in several ways. Oil sources include natural seeps along the ocean floor, stream runoff from the land, wastewater drainage outlets from industrial complexes, offshore drilling accidents, deliberate purging of ballast or cargo areas of ships, and accidents involving oil supertankers. The petroleum input into the marine environment from ocean-derived factors has been estimated at about 56 percent. This calculation includes transportation losses during loading and unloading (30 percent), oil spills (5 percent), offshore production losses (1 percent), atmospheric pollution (10 percent), and offshore oil seeps (10 percent).

Supertanker accidents that result in oil spills are major events. Causes include shipboard explosions, collisions with other vessels, and grounding on barriers (mostly rocks or coral reefs) because of navigation error, mechanical failure, or inclement weather. In 1978, the *Amoco Cadiz*, a large carrier stocked with nearly 69 million gallons of light crude oil, ran aground on shallow rocks off the coast of Brittany, France, where the impact slashed holes in the hull and container tanks and released the oil onto 321 kilometers (200 miles) of French coast, killing millions of invertebrates, such as mollusks and crustaceans, and an estimated 20,000 birds.

Although such oil spills are often quite dramatic, and the damage they do to the environment receives extensive news media coverage, only about 5 percent of the total input of petroleum into the oceans results from oil spills.

Oil slicks are especially problematic in semi-enclosed bodies of water, such as the Baltic and Mediterranean Seas. Because tides are minor in these seas, any oil spilled is not easily eliminated. These places are among the world's most polluted bodies of water.

Spills in the Open Ocean
A fire aboard the oil tanker *Castillo de Bellver* in August 1983 was responsible for the tanker's capsizing in the South Atlantic Ocean, roughly 70 miles from Cape Town, South Africa, releasing some 79 million gallons of crude oil. The slick caught fire, but most of the oil was caught in the Benguela Current and carried to out sea, where it dispersed, so that the environmental damage cannot be accurately measured.

When such an oil spill occurs in the open ocean, the oil floats on the surface of the water because of a density contrast between the two substances. Water has a specific gravity from 1 to 1.2, while oil is lighter, with a specific gravity from 0.7 to 1.0. The lighter components of the spilled petroleum immediately begin to evaporate; also, oil-degrading bacteria in the water begin to feed on the organic deposit and multiply. Warmer water and ambient air temperatures increase the rates of bacterial growth and evaporation. The major part of petroleum dumped into the open ocean evaporates or is reduced by bacteria. With time, the oil turns into an inert, tar-like substance that becomes extremely hard; some such residues have been found with barnacles attached to them.

In areas where the spilled oil is washed ashore and subjected to intense wave action, a gooey emulsion sometimes forms. This foamy mass, which has been described as resembling chocolate mousse, coats everything along the beach, including sand-sized particles, large boulders, aquatic animals and plants, boats, and human-made shore facilities. The polluted waters can devastate resort beaches, oyster and shrimp beds, fish hatcheries, and prime fishing grounds.

Impacts of Oil Spills
The potential damage resulting from a large oil spill in the ocean includes the loss of substantial numbers of commercial and rough fish, waterfowl, aquatic mammals, shellfish, algae, and plankton. Oil contamination reduces the amount of oxygen in the water column, and this can cause the deaths of large numbers of fish in polluted areas. Millions of fish died as a result of the 1989 *Exxon Valdez* disaster in Alaskan waters, and the Alaskan fishing industry, which is extremely important to the state's economy, was seriously threatened. Although this event is generally considered a major disaster, David McConnell reported in the *Journal of Geological Education* (1999) that, compared with other spills worldwide, the *Exxon Valdez* spill ranked only fifty-third in size. One of the largest marine oil spills in history occurred following an explosion at the *Deepwater Horizon* oil rig in April 2010 in the Gulf of Mexico, uncapping a sea-floor oil gusher that flowed for nearly three months before it was recapped. An estimated 4.9 million barrels were discharged into the gulf, causing widespread and serious damage to marine and coastal wildlife, beaches, estuaries, and wetlands along the Gulf Coast and severely affecting the local fishing and tourism industries.

When seabirds encounter spilled oil to the extent that their feathers are soaked, they lose the ability to fly and float, as well as their natural body insulation. Many such birds starve to death, die from exposure, or drown. The *Exxon Valdez* disaster resulted in the deaths of more than 200,000 seabirds. During the Persian Gulf War, more than 20,000 birds perished from oil-related causes. Oil spills also cause the deaths of aquatic mammals, which die from the loss of their food supply, exposure to cold, or poisoning when they ingest the toxic oil. Volunteer workers have saved many birds and mammals at animal centers set up after large spills; workers wash the animals with solvents to remove oil and often hold them until their normal body resilience returns. Following the *Deepwater Horizon* oil spill, dolphins and sea turtles died in record numbers, and many stillborn dolphins and stranded sea turtles were found.

Shellfish in bays and marshes can perish in oil-polluted waters from asphyxiation or toxicity. Some effects of oil pollution remain for years. Shellfish in a salt marsh in Massachusetts were quarantined for more than five years after heating oil was spilled nearby in 1969, and traces of petroleum persisted for decades.

Cleanup Procedures

Oil spills in the open ocean have been reduced due to several techniques and safety measures. Igniting the slick, if attempted shortly after a spill, can be effective; however, if the volatile components of the oil have evaporated, it is difficult to start and maintain such fires. In 1967, when the damaged supertanker *Torrey Canyon* began to founder and break up off the coast of England, Great Britain's air force dropped bombs to ignite the oil that remained in the vessel.

Another technique used to reduce the damage caused by oil slicks is the application of chemicals such as detergents to disperse the oil droplets. This procedure is not viable if the temperature of the water is too low for the chemicals to be effective. Floating booms are sometimes used to confine oil streamers or protect inlet areas. These barriers are useful if the waves are not too high. Skimmers may be used to collect floating oil during the early stages of a spill. Plant material (threshed hay, peat moss, or wood shavings) or pulverized rock (chalk or claystone) is sometimes spread over slicks to absorb the oil. Plant material can then be removed and burned or buried in sealed containers; applied rock particles absorb the oil, clump together, and sink to the ocean floor. In many parts of the oceans, oil-eating microbes (mostly bacteria) utilize petroleum as a nutrient. If these microbes are not present or exist only in small numbers in the area, engineered microbes can be introduced. Under ideal temperature and ocean chemistry conditions, the bacteria will rapidly multiply and consume the oil slick. This technique is known as bioremediation. Other factors such as wave action, timing, and scale can also affect the effectiveness of such methods, however, and the use of dispersants can kill oil-eating microbes.

In nearshore areas, beach sand and boulders can be steamed clean or washed with water or solvents. Pools of oil can be vacuumed into tank trucks and hauled away from the site. The recovered oil can sometimes be processed into useful products. Plant material can also be applied to crude oil along the shoreline. After the 1969 Santa Barbara oil spill, the beaches along the California coast were coated by approximately ten thousand barrels of gooey oil. Numerous volunteers helped spread fresh hay along the beachfront and later recovered the oil-soaked plant material.

Even cleanup efforts can damage flora and fauna. Hot water used in cleaning or walking over fragile organisms can harm them, for instance. Animals that are cleaned often die shortly thereafter, and if they do survive, such animals may not reproduce.

Human Health Effects

Human health can also be adversely affected by oil spills. Cleanup workers, animal rescuers, and nearby residents who are exposed to the volatile organic compounds and other petroleum components may exhibit respiratory, gastric, and dermatological ailments in the near term. People living near an oil spill may also have difficulty getting adequate nutrition and water or otherwise maintaining their livelihoods as drinking water, game animals, fish, or surrounding croplands become contaminated.

The long-term effects of human exposure to oil components and cleanup chemicals remain subject to debate. For one, it can be difficult to isolate a precise cause of conditions such as cancer, lung dysfunction, or developmental disorders. Nevertheless, certain petroleum constituents such as benzene have been associated with neurological affects and cancers among industry workers, but as of 2013, safety standards had not been set for people exposed unexpectedly during leaks, spills, or other accidents involving oil. Other constituent chemicals had yet to undergo scientific study.

Donald F. Reaser

Further Reading

Clifton, Adam, ed. *Oil Spills: Environmental Issues, Prevention and Ecological Impacts*. New York: Nova Science, 2014.

Davidson, Jon P., Walter E. Reed, and Paul M. Davis. *Exploring Earth: An Introduction to Physical Geology.* 2nd ed. Upper Saddle River, N.J.: Prentice Hall, 2002.

Fingas, Merv. *The Basics of Oil Spill Cleanup.* 2nd ed. Boca Raton, Fla.: CRC, 2001.

Kemp, David D. "Threats to the Availability and Quality of Water." *Exploring Environmental Issues: An Integrated Approach.* New York: Routledge, 2004.

Lehr, Jay, et al. "Oil Spills and Leaks." In *Handbook of Complex Environmental Remediation Problems.* New York: McGraw-Hill, 2002.

Montgomery, Carla W. *Environmental Geology.* 9th ed. New York: McGraw-Hill, 2010.

National Research Council. *Oil in the Sea III: Inputs, Fates, and Effects.* Washington, D.C.: National Academies Press, 2003.

National Research Council. *Oil Spill Dispersants: Efficacy and Effects.* Washington, D.C.: National Academies Press, 2005.

National Response Team. *On Scene Coordinator Report: Deepwater Horizon Oil Spill.* N.p.: n.p., 2011. PDF file.

Ott, Riki. *Not One Drop: Betrayal and Courage in the Wake of the Exxon Valdez Oil Spill.* White River Junction, Vt.: Chelsea Green, 2008.

Vegas, Jennifer. "Record Dolphin, Sea Turtle Deaths since Gulf Spill." *Discovery.* Discovery Communications, 2 Ap. 2013. Web. 2 Feb. 2015.

Walker, Jane. *Oil Spills.* New York: Gloucester, 1993.

Websites

Bloomberg
www.bloomberg.com/news/2013-06-19/what-sickens-people-in-oil-spills-and-how-badly-is-anybody-s-guess.html.

Britannica
https://www.britannica.com/list/9-of-the-biggest-oil-spills-in-history

Michel Varisco: Shifting
https://www.michelvarisco.com/project/shifting/

Smithsonian.com
www.smithsonianmag.com/science-nature/oil-spill-cleanup-illusion-180959783/.

■ Organic farming and gardening

FIELDS OF STUDY
Agriculture; Biology; Botany; Ecology; Ecosystems; Environment; Environmentalism; Gardening; Horticulture; Life Sciences; Water Resources

SUMMARY
Interest in organic farming has grown since the late twentieth century for various reasons, including concerns about the environmental effects of the chemical pesticides and fertilizers used by conventional large agricultural operations, and particularly about the possibility of negative health effects associated with ingestion of the food produced using such chemicals.

PRINCIPAL TERMS

- **chemical fertilizers:** any inorganic material of wholly or partially synthetic origin that is added to the soil to sustain plant growth, many of which contain acids, reduce the soil's beneficial organism population and interfere with plant growth
- **compost:** a mixture consisting largely of decayed organic matter used for fertilizing and conditioning soil
- **crop rotation:** the practice of growing different crops in succession on the same land chiefly to preserve the productive capacity of the soil
- **green manure:** plants, such as clover and other nitrogen-fixing plants, grown as fertilizer and plowed under to enrich the soil
- **monoculture:** the cultivation or growth of a single crop or organism especially on agricultural or forest land
- **Organic Foods Production Act (OFPA):** enacted in 1990, OFPA served to establish uniform national standards for the production and handling of foods labeled "organic"
- **organic:** not using artificial chemicals in the production of plants and animals for food
- **rotation grazing:** shifting livestock to different units of a pasture or range in regular sequence to permit recovery and growth after grazing

At the beginning of agricultural history, farmers believed that plants "ate the soil" in order to grow. During the nineteenth century, German chemist Justus von Liebig discovered that plants extract nitrogen, phosphorus, and potash from soil. His findings dramatically changed agriculture as farmers found they could grow crops in any type of soil, even sand and water solutions, if they added the right chemicals.

Diversified family farms eventually gave way to huge specialized operations; by the end of the twentieth century, less than 2 percent of the U.S. population was directly involved in crop production. Crop yields were raised with the use of chemicals, but farms and their soil and water were not being cared for as the unique ecosystems they are. The quality of the soil was ignored as chemicals were used to produce high crop yields to feed the 98 percent of the population not involved in farming. Over time, the organic quality of the soil was lost even though the chemicals remained in the soil.

Chemicals used in agriculture were also found to leach into the water supply. In 1988 the Environmental Protection Agency (EPA) found that groundwater in thirty-two U.S. states was contaminated with seventy-four different agricultural chemicals. Aside from the health consequences to humans, leaching causes once friable and fertile soils to turn hard and become nonproductive. Further, the use of chemical insecticides was proving to have toxic effects on both the foods grown and the farmworkers encountering them. According to the World Health Organization (WHO), cases of acute pesticide poisoning account for significant mortality worldwide, especially in developing countries where environmental regulations do not exist or are not enforced.

Organic farming gained popularity as some individuals attempted to return to diversified farming practices that emphasize working with nature to create a renewing, ecologically sound, and sustainable system of agriculture. Organic farmers are committed to crop-growing practices that are free of synthetic chemicals and genetic engineering. Soil must be free of chemicals for at least three years before products grown in it can be certified as organic. The organic certification requirements further state that both plants and livestock must be raised organically, without the use of chemicals, antibiotics, hormones, or synthetic feed additives. Organic farmers must comply with both organic regulations in their states and the 1990 federal Organic Foods Production Act (OFPA). In the late 1990s the U.S. Department of Agriculture, through the National Organic Standards Board, developed standards and regulations to ensure consistent national standards for organic products.

Soil Fertility

To produce chemical-free crops, organic farmers must rely on organic practices to ensure soil fertility and control unwanted plants and insects. Organic farmers build organic materials in the soil by adding green manure, compost, or animal manure.

"Green manure" is a term used for crops that are grown specifically to introduce organic matter and nutrients into the soil. These crops are raised expressly for the purpose of being plowed under rather than sold to consumers. Green manure crops protect against soil erosion, cycle nutrients from lower levels of the soil into the upper layers, suppress weeds, and keep much-needed nutrients in the soil rather than allowing them to leach out. Legumes are excellent green manure crops because they are able to extract nitrogen from the air and transfer it into the soil, leaving a supply of nitrogen for the next crop. Legumes have nitrogen nodules on their roots; when the legumes are tilled under and decompose, they add more nitrogen to the soil. As plants decay, they also make insoluble plant nutrients—such as carbon dioxide and acetic, butyric, lactic, and other organic acids—available in the soil.

Much of the organic material derived from green manure comes in the form of decaying roots. Alfalfa, one type of legume, sends its roots several feet down into the soil. When the alfalfa plants are turned, the entire root system decomposes into organic material, thus helping to improve water retention and soil quality at the same time. Some examples of legumes used for green manure are sweet clover, ladino clover, alfalfa, and trefoil. Nonlegumes or grasses used for green manure include rye, redtop, and timothy grass.

Greater soil fertility can be achieved if green manure is grazed by animals and the manure from those animals is deposited on the soil. Left on their own in a large field, cows, sheep, and horses will choose only their favorite bits of grasses to graze and thus will graze the field unevenly. Organic farmers use portable fencing to achieve strip rotational grazing, in which animals graze a strip of a field all the way down and then are moved to another strip of the field while the first pasture recovers. This process is repeated until the entire field has been grazed and fertilized. Also, when the grass is grazed, some of the roots die and rot to form a good amount of humus. This humus is the stable organic material that acts as a catalyst for allowing plants to find nutrients.

Compost can work for large or small farmers, but it is particularly useful to small farmers and gardeners who do not have enough land to grow green manure crops to plow under. Composting is a natural soil-building process that began with the first plants that existed on earth and continues as a natural process today. As falling leaves and dead plants, animals, and insects decompose into the soil, they form a rich organic layer. Farmers and gardeners can make compost by alternating layers of carbohydrate-rich plant cuttings and leaves, animal manure, and topsoil in a container or defined area and allowing the components to decay. The rich organic matter that results can then be added to the soil to be cultivated.

Animal manure is used as an organic fertilizer. Rich in nitrogen, the best animal manures come from animals that have a high amount of protein in their diets. For example, beef cattle being fattened for market have a higher level of protein in their solid waste than do dairy cattle that are producing milk for market. The application of manure to fields improves the structure of the soil, raises the organic nitrogen content, and stimulates the growth of soil bacteria and fungi necessary for healthy soil.

Crop Rotation
Planting the same crop year after year on the same piece of ground results in depleted soil. Crops such as corn, tobacco, and cotton remove nutrients, especially nitrogen, from the soil. Crop rotation involves planting different crops in the same field every other year to keep soil fertile by adding nitrogen and achieving a balanced nutrient level. Planting a winter cover crop such as rye grass helps to protect the land from erosion, and when the crop is plowed under in the spring, it creates a nutrient-rich soil for the planting of a cash crop. Crop rotation also improves the physical condition of the soil because different crops vary in root depth, are cultivated differently, and respond to either deep or shallow soil preparation.

Monoculture farming puts a large source of the same crop in easy proximity to insects that are destructive to that crop. Insect offspring can multiply out of proportion when the same crop is grown in a given field year after year. Since insects are drawn to a home area, they will not be able to proliferate and thrive if the crop is changed every other year to a crop they do not eat. This is one of the reasons organic farmers rely heavily on crop rotation as one aspect of insect control. Rotating crops can also help control weeds. Some crops and cultivation methods inadvertently allow certain weeds to thrive. Farmers using crop rotation can incorporate successor crops that eradicate those weeds. Some crops, such as potatoes and winter squash, work as "cleaning crops" because of the different styles of cultivation that are used on them.

Organic Insect & Weed Control
Advocates of organic farming methods assert that plants within a balanced ecosystem are resistant to disease and insect infestation and that plants stressed by unfavorable growing situations are much more susceptible to such problems. The whole premise of organic farming is that farmers should work with nature to help provide healthy, unstressed plants.

Rather than using chemical pesticides, which often create resistant generations of insects and thus the need for newer and stronger chemicals, organic farmers seek natural ways to diminish pest problems. They strive to maintain and replenish soil fertility to produce healthy plants that are resistant to insects. They also try to select plant species that are resistant to insects, weeds, and disease. Crop rotation, as noted above, is another method of keeping insect infestations down. Organic farmers may create barriers to insect pests by planting, in and around the crop field, rows of hedges, trees, or even plants that are not desirable to insects. They may also provide habitats for the insects' natural enemies: birds, beneficial insects, and garden snakes.

The ladybug and the praying mantis are two beneficial insects that can help rid a farm or garden of aphids, mites, mealy bugs, and grasshoppers. Trichogramma is a small wasp that will destroy moth eggs, squash borers, cankerworms, cabbage loopers, and corn earworm. Many beneficial insects are available to farmers and gardeners in large quantities through catalog or online retailers. Given that just one ladybug can consume fifty or more aphids per day and can lay up to one thousand fertile eggs, the overall cost of purchasing ladybugs is much lower than the cost of chemical insecticides.

Organic farmers and gardeners also often rely on so-called insecticide crops. Garlic planted near lettuce or peas will deter aphids. Geraniums or marigolds grown close to grapes, cantaloupes, corn, or cucumbers will deter Japanese and cucumber beetles. Herbs such as rosemary, sage, and thyme planted by cabbages will deter white butterfly pests. Potatoes will repel Mexican bean beetles if planted near beans, and tomatoes planted near asparagus will ward off asparagus beetles. Natural insecticides such as red pepper juice can be used for ant control, while a combination of garlic oil and lemon may be used against fleas, mosquito larvae, houseflies, and other insects.

Organic farming relies on the physical control of weeds, especially through the use of cutting (cultivation) or smothering (mulching and hilling). The cultivation method involves the shallow stirring of surface soil to cut off small developing weeds and prevent more from growing. This can be done with tractors, wheel hoes, tillers, or, for small gardens, hand hoes. Mulch is a soil cover that prevents weeds from getting the sunlight they need for growth. Mulching with fully biodegradable materials can help build soil fertility while controlling weeds. Plastic mulches can be used on organic farms as long as they are removed from the fields at the end of the growing or harvest season.

Organic farming offers a safe alternative to the use of synthetic chemicals in the production of food. Growing public awareness of the possible toxic effects of the traditional agricultural reliance on synthetic chemicals is reflected in a growing demand for organically produced foods, leading to significant growth in the organic food industry.

Dion Stewart and Toby Stewart

Further Reading

Coleman, Eliot. *New Organic Grower: A Master's Manual of Tools and Techniques for the Home and Market Gardener, 30th Anniversary Edition.* White River Junction, Vt.: Chelsea Green, 2018.

Duram, Leslie A. *Good Growing: Why Organic Farming Works.* Lincoln, Neb.: University of Nebraska Press, 2005.

Fortier, Jean-Martin. *The Market Gardener: A Successful Grower's Handbook for Small-Scale Organic Farming.* Gabriola Island, Canada: New Society Publishers, 2014.

Fossel, Peter V. *Organic Farming: How to Raise, Certify, and Market Organic Crops and Livestock.* Minneapolis, Minn.: Voyageur Press, 2014.

Hamilton, Geoff. *Organic Gardening.* Rev. ed., Dorling Kindersley, 2008.

Koepf, Herbert H. *The Biodynamic Farm: Agriculture in the Service of the Earth and Humanity.* New York: Anthroposophic Press, 2006.

Kristiansen, Paul, Acram Taji, and John Reganold, editors. *Organic Agriculture: A Global Perspective.* Ithaca, N.Y.: Cornell University Press, 2006.

Lampkin, Nicolas. *Organic Farming.* Rev. ed., Alexandria Bay, N.Y.: Diamond Farm Enterprises, 2002.

Paarlberg, Robert. "Organic and Local Food." *Food Politics: What Everyone Needs to Know.* New York: Oxford University Press, 2010.

Websites

US Dept. of Agriculture
www.usda.gov/wps/portal/usda/usdahome?navid=organic-agriculture.

Ozone layer

FIELDS OF STUDY
Atmospheric Chemistry; Atmospheric Science; Chemical Engineering; Earth System Modeling; Ecology; Environmental Chemistry; Environmentalism; Hazardous Materials; Meteorology; Petroleum Refining; Photochemistry; Spectroscopy; Thermodynamics; Toxicology; Waste Management

SUMMARY
Depletion of stratospheric ozone allows incoming ultraviolet radiation to reach Earth's surface, resulting in severe damage to all living organisms if ozone depletion continues. Ozone-hole formation, directly related to the use of CFCs, is evidence that human activities can significantly alter the composition of the atmosphere.

PRINCIPAL TERMS
- **chlorofluorocarbon (CFC):** a group of chemical compounds containing carbon, fluorine, and

chlorine, used in air conditioners, refrigerators, fire extinguishers, spray cans, and other applications
- **ozone layer:** a region in the lower stratosphere, centered about 25 kilometers above the surface of Earth, which contains the highest concentration of ozone found in the atmosphere
- **ozone:** the molecular form of oxygen containing three atoms of oxygen per molecule as O_3, as compared to elemental oxygen having the molecular formula O_2
- **polar vortex:** a closed atmospheric circulation pattern around the South Pole that exists during the winter and early spring; atmospheric mixing between the polar vortex and regions outside the vortex is slow
- **stratosphere:** the region of the atmosphere between 10 and 50 kilometers above the surface of Earth
- **total column abundance of ozone:** the total number of molecules of ozone above a 1-centimeter-square area of Earth's surface
- **Total Ozone Mapping Spectrometer (TOMS):** a space-based instrument for measuring the total column abundance of ozone globally
- **ultraviolet solar radiation:** electromagnetic radiation having wavelengths between 4 and 400 nanometers

Ozone, although only a minor component of the atmosphere, plays a vital role in the survival of life on Earth. Ozone molecules in the stratosphere absorb incoming high-energy ultraviolet (UV) light from the sun. Absorption of UV light in the stratospheric ozone layer, a region that contains the maximum concentration of atmospheric ozone, though only about 12 parts per million, prevents most UV light from reaching the surface of the planet. If none of the sun's ultraviolet radiation were blocked by the ozone layer, it would be difficult, if not impossible, for most forms of life, including humans, to survive on land.

The concentration of ozone in the atmosphere is highly variable, changing with altitude, geographic location, time of day, time of year, and prevailing local atmospheric conditions. Long-term fluctuations in ozone concentration are also seen, some of which are related to the solar sunspot cycle. While long-term average ozone concentrations are relatively stable, short-term fluctuations of as much as 10 percent in total column abundance of ozone as a result of the natural variability in ozone concentration are often observed.

Beginning in the early 1970s, a new and unexpected decrease in stratospheric ozone concentration was first observed. The decrease was localized near Antarctica and appears in early spring (which begins in September in the Southern Hemisphere). The initial decrease in ozone was small, but by 1980, decreases in total column abundance of ozone of as much as 30 percent were being recorded, well outside the range of variation expected as a result of random fluctuations. This seasonal depletion of stratospheric ozone in a circular region above Antarctica, which by 1990 had reached 50 percent of the total column abundance of ozone, was soon given the label "ozone hole."

Discovering the Role of CHLOROFLUOROCARBONS (CFCs)

Chlorofluorocarbons (CFCs) are organic molecules containing chlorine, fluorine, and carbon atoms. CFCs are a family of chemical compounds used in air conditions, refrigerators and aerosol spray cans. The first CFCs were discovered by Thomas Midgley, Jr., in 1928. Freon, a CFC trade name of the DuPont Corporation, was introduced as a nontoxic, nonflammable refrigerant in 1930. Because these molecules are chemically inert and easily liquefied, CFCs soon became the standard coolants in refrigerators and air conditioners. They also became widely used as propellants in aerosol spray cans. By 1968, 2.3 billion aerosol cans containing CFCs had been sold in the United States.

In 1970 the British scientist James Lovelock determined that most CFCs entering the atmosphere remained there without significant decomposition. Three years later, Frank Sherwood Rowland and Mario Molina, working at the University of California at Irvine, suggested that CFCs would eventually migrate into the stratosphere. Once there, absorption of ultraviolet light would cause CFCs to release chlorine atoms, which would then react catalytically to remove ozone. Since ozone in the stratosphere prevents high-energy ultraviolet light from reaching the surface of the earth, any decrease in ozone would lead to increased exposure to ultraviolet light on the

Earth's surface, causing higher levels of skin cancer in humans and damage to plants and animals.

Despite uncertainty as to the degree of ozone destruction CFCs would cause, in 1975 Oregon became the first U.S. state to ban CFCs in aerosol spray cans. Several other states took similar actions, and in 1977 the Food and Drug Administration (FDA) implemented a ban on the use of CFCs as aerosol propellants to be phased in over a two-year period. Continued uncertainties in predictions of ozone loss and the lack of direct evidence for ozone depletion kept most other countries from restricting the use of CFCs. While the U.S. Environmental Protection Agency (EPA) discussed instituting a total ban on CFCs, no action was taken, in part because of the difficulty in finding adequate substitutes for CFCs.

Although it was initially unclear whether the formation of the Antarctic ozone hole stemmed from natural causes or from anthropogenic effects on the environment, in 1985 a team of British scientists led by Joseph Farman announced the discovery of significant loss of ozone over the Antarctic. Beginning in the early 1970s, springtime levels of ozone had slowly decreased. By 1985 as much as 40 percent of the ozone usually present in the Antarctic stratosphere during the spring had disappeared. In addition, both the duration and the geographic extent of this ozone hole were increasing. Evidence linking formation of the ozone hole to CFCs in the atmosphere was quickly found.

Extensive field studies combined with the results of laboratory experiments and computer modeling of the atmosphere quickly led to a consistent and detailed explanation for ozone-hole formation. The formation of the ozone hole has two principal causes: chemical reactions that occur generally throughout the stratosphere, and special conditions that exist in the Antarctic region.

The discovery of the ozone hole led to further restrictions on CFCs. In 1987 an international agreement called the Montreal Protocol was reached to ban the manufacture and use of CFCs by the year 2010. In the United States, passage of the 1990 Clean Air Act amendments resulted in an accelerated timetable for restrictions on CFCs and related compounds. By the mid-1990s levels of CFCs in the atmosphere had stabilized, and CFCs are expected to disappear gradually from the atmosphere over the next century.

How CFCs Contribute to Ozone Depletion

Under normal conditions, the concentration of ozone in the stratosphere is determined by an equilibrium balance between reactions that remove ozone and those that produce ozone. The removal reactions are mainly catalytic chain reactions, in which trace atmospheric chemical species destroy ozone molecules without themselves being consumed. In such processes, it is possible for one chain component to remove many ozone molecules before being itself removed. A single chlorine atom, for example, is estimated to remove as many as 100,000 ozone molecules through chemical chain reactions before it is itself removed by forming a nonreactive species. The trace species involved in ozone removal include hydrogen oxides and nitrogen oxides, formed primarily by naturally occurring processes, and chlorine and bromine atoms and their corresponding oxides.

A major source of chlorine in the stratosphere is the decomposition of a class of compounds called chlorofluorocarbons (CFCs). Such compounds are used as refrigerants in refrigeration and air conditioning applications, and were commonly used as aerosol propellants and solvents, freely released into the atmosphere. Their use and handling is now strictly regulated. Chlorofluorocarbons are extremely stable in the lower atmosphere, with lifetimes of several decades, due to their extreme lack of chemical reactivity. The main fate of chlorofluorocarbons in the atmosphere, however, is slow migration into the stratosphere, where they absorb ultraviolet light and fragment to release chlorine atoms. The chlorine atoms produced from the breakdown of chlorofluorocarbons in the stratosphere provide an additional catalytic process by which stratospheric ozone is destroyed. A similar set of reactions involving a class of bromine-containing compounds called halons, used in some types of fire extinguishers, leads to additional ozone destruction by similar photochemical processes. By 1986, the average global loss of stratospheric ozone caused by the release of chlorofluorocarbons, halons, and related compounds into the environment was estimated to be 2 percent.

Antarctic Conditions

While the decomposition and subsequent reaction of chlorofluorocarbons, halons, and other synthetic compounds can explain the slow general decline in

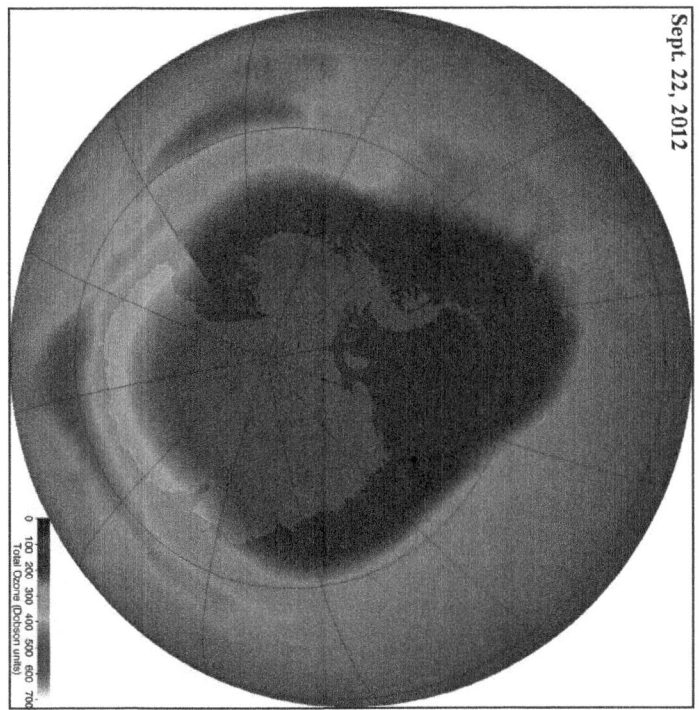

Antarctic Ozone Hole as at Sept. 22, 2012. (National Oceanic and Atmospheric Administration).

ozone concentration observed in the stratosphere, additional processes are needed to account for the more massive seasonal ozone depletion observed above Antarctica. These processes involve a set of special conditions that in combination are unique to the stratosphere above Antarctica.

During daylight hours, a portion of the chlorine present in the stratosphere is tied up in the form of reservoir species, compounds such as hydrogen chloride and chlorine nitrate that do not react with ozone. This slows the rate of removal of ozone by chlorine. Processes that directly or indirectly involve absorption of sunlight transform reservoir species and release ozone-destroying chlorine atoms. During the Antarctic winter, when sunlight is entirely absent, stratospheric chlorine is rapidly converted into reservoir species.

In the absence of additional chemical processes, the onset of spring in Antarctica and the return of sunlight convert a portion of the reservoir compounds into reactive chlorine species and reestablish the balance between ozone-producing and ozone-destroying processes. However, the extremely low temperatures occurring in the stratosphere above Antarctica during the winter months lead to the formation of polar stratospheric clouds, which, because of the extremely low concentration of water vapor in the stratosphere, do not form during other seasons or outside the polar regions of the globe.

The ice crystals that compose the clouds act as catalysts that convert reservoir species into diatomic chlorine and other gaseous chlorine compounds that, in the presence of sunlight, re-form ozone-destroying species. At the same time, nitrogen oxides in the collection of reservoir species are converted into nitric acid, which remains attached to the ice crystals. As these ice crystals are slowly removed from the stratosphere by gravity, the potential for conversion of active forms of chlorine into reservoir species is greatly reduced. Because of this, when spring arrives, large amounts of ozone-destroying chlorine species are produced by the action of sunlight, and only a small fraction of this reactive chlorine is converted into reservoir species. The increased rate of ozone removal caused by the abundance of reactive chlorine present in the stratosphere leads to ozone depletion and formation of the ozone hole.

An additional process important in formation of the ozone hole is the unique air-circulation pattern in the stratosphere above Antarctica. During the winter and early spring, a vortex of winds circulates about the South Pole. This polar vortex minimizes movement of ozone and reservoir-forming compounds from other regions of the stratosphere. As this polar vortex breaks up in midspring, ozone concentrations in the Antarctic stratosphere return to normal levels, and the ozone hole gradually disappears.

Atmospheric Ozone Study and Interpretation

Researchers utilize a great diversity of devices and techniques in the study and interpretation of atmospheric ozone. One popular technique is the use of simulation models. A good model is one that simulates the interrelationships and interactions of the various parts of the known system. The weakness of models is that, often, not enough is known to give an accurate picture of the total system or to make accurate predictions. Most modeling is done on computers. Scientists estimate how fast chemicals such as CFCs and nitrous oxide will be produced in the future and build a computer model of the way these chemicals react with ozone and with one another. From this model, it is possible to estimate future ozone levels at different altitudes and at different future dates.

Similar processes appear to be at work in the Arctic stratosphere, leading to ozone depletion, as in the Antarctic. However, the National Oceanic and Atmospheric Administration (NOAA) Aeronomy Laboratory in Boulder, Colorado, reported a discrepancy between observed ozone depletion and predicted levels, based on models that account accurately for Antarctic ozone depletion. This report suggests that some other mechanism is at work in the Arctic. Thus, while good models can be very useful in studying new data, observed discrepancies highlight the need for better system models and modeling algorithms.

There are two models favored by most scientists in this area. Some scientists put forth a chemical model that says the depletion is caused by chemical events promoted by the presence of chlorofluorocarbons created by industrial processes. Acceptance of this model was promoted by the discovery of fluorine in the stratosphere. Fluorine does not naturally occur there, but it is related to and can be formed photochemically from CFCs. The other model assumes that the ozone hole was formed by dynamic air movement and mixing. This model best fits data gathered by ozone-sensing balloons that sample altitudes up to 30 kilometers and then radio the data back to Earth. Ozone depletion is confined to the atmosphere at altitudes between 12 and 20 kilometers. While the total ozone depletion is 35 percent, different strata have shown various amounts of depletion from 70 to 90 percent. Surprisingly, about half the ozone was gone in twenty-five days. This finding does not fit the chemical model very well.

Besides ozone-sensing balloons, satellite survey data provide more direct measurements obtained over longer periods of observation time. The National Aeronautics and Space Administration (NASA) obtains measurements with its Nimbus 7 satellite. Ozone measurements made by this satellite helped to develop flight plans for the specialized aircraft NASA also deploys in ozone studies. NASA's ER-2 aircraft is a modified U-2 reconnaissance plane that carries instruments up to 20 kilometers in altitude for seven-hour flights to 80 degrees north latitude. A DC-8, operating during the same period, is able to survey the polar vortex, owing to its greater range. In addition, scientists utilize many meteorological techniques and instruments, including chemical analysis of gases by means of infrared spectroscopy, mass spectroscopy and gas spectroscopy combined, gas chromatography, and oceanographic analysis of planktonic life in the southern Atlantic, Pacific, and Indian oceans. As new research methods and techniques become available, they are also applied to this essential study.

Public Health Concerns

Stratospheric ozone provides global protection from the lethal effects of ultraviolet radiation from the sun. This ability to absorb ultraviolet radiation protects all life-forms on Earth's surface from excessive ultraviolet radiation, which destroys plant and animal cells. Currently, between 10 and 30 percent of the sun's ultraviolet B (UV-B) radiation reaches the ground surface. If ozone levels were to drop by 10 percent, the amount of UV-B radiation reaching Earth would increase by 20 percent.

Present-day UV-B levels are responsible for the fading of paints and the yellowing of window glazing and for car finishes becoming chalky. These kinds of degradation will accelerate as the ozone layer is depleted. There could also be increased smog, urban air pollution, and a worsening of the problem of acid rain in cities. In humans, UV-B causes sunburn, snow blindness, skin cancer, and cataracts, and promotes aging and wrinkling of the skin. Skin cancer is the most common form of cancer, with more than 400,000 new cases reported every year in the United States alone. The National Academy of Sciences has estimated that each 1 percent decline in ozone would increase the incidence of skin cancer by 2 percent. Therefore, a 3 percent depletion in ozone would be expected to produce some 24,000 more cases of skin cancer in the United States every year.

Ecological Concerns

Many other forms of life, from bacteria to forests and food crops, are adversely affected by excessive radiation. Ultraviolet radiation affects plant growth by slowing photosynthesis and by delaying germination in many plants, including trees and food crops. Scientists have a great concern for the organisms that live in the ocean and the effect ozone depletion may have on them. Phytoplankton, zooplankton, and krill (a shrimplike crustacean) could be greatly depleted if there were a drastic increase in ultraviolet A and B. The result would be a tremendous drop in the population of these free-floating organisms, which are extremely important because they are the foundation species of the global food chain. Phytoplankton use the energy of sunlight to convert inorganic compounds into organic plant matter. This process provides food for the next step in the food chain, the herbivorous zooplankton and krill. They, in turn, become the food for the next higher level of animals in the food chain. Initial studies of this food chain in the Antarctic suggest that elevated levels of ultraviolet radiation impair photosynthetic activity. Recent studies show that a fifteen-day exposure to UV-B levels 20 percent higher than normal can kill off all anchovy larvae down to a depth of 10 meters. There is also concern that ozone depletion may alter the food chain and even cause changes in the organisms' genetic makeup. An increase in the ultraviolet radiation is likely to lower fish catches and upset marine ecology, which has already suffered damage from human-made pollution. On a worldwide basis, fish presently provide 18 percent of all the animal protein consumed.

International Response
The United Nations Environmental Program (UNEP) is working with governments, international organizations, and industry to develop a framework within which the international community can make decisions to minimize atmospheric changes and the effects they could have on Earth. In 1977, UNEP convened a meeting of experts to draft the World Plan of Action on the Ozone Layer. The plan called for a program of research on the ozone layer and on what would happen if the layer were compromised. In addition, UNEP created a group of experts and government representatives who framed the Convention for the Protection of the Ozone Layer. This convention was adopted in Vienna in March 1985, by twenty-one nations and the European Economic Community and has subsequently been signed by many more nations. The convention pledges that the nations that sign are to protect human health and the environment from the effects of ozone depletion. Action has already been taken under the Convention to protect the ozone layer. Several countries have restricted the use of CFCs or the amounts produced. The United States banned the use of CFCs in all nonessential applications in 1978. Some countries, such as Belgium and the Nordic countries, have in effect banned CFC production altogether. The group has also worked with governments on a Protocol to the Convention that required signatory nations to limit their production of CFCs. It is the hope and aim of these nations that such international cooperation will lead to a better global environment.

In 2017 it was announced that the ozone hole was the smallest it had been since 1988. However, this was mainly due to unusually warm temperatures, not to human efforts to reduce ozone-depleting chemicals. Scientists cautioned against seeing the smaller hole as being a sign of recovery.

George K. Attwood and Jeffrey A. Joens

Further Reading
Bielle, David. "Ozone Hole Closing Up, Thanks to Global Action." *Scientific American*. Scientific American, 15 Sept. 2014. Web. 23 Mar. 2015.
Dauvergne, Peter. "Refrigerating the Ozone Layer." In *The Shadows of Consumption: Consequences for the Global Environment*. Cambridge, Mass.: The MIT Press, 2008.
Douglass, Anne, Natalya Kramarova, and Susan Strahan. *Inside the Ozone Hole*. NASA's Goddard Space Flight Center, 2015. PDF file.
Joesten, Melvin D., John L. Hogg, and Mary E. Castellion. "Chlorofluorocarbons and the Ozone Layer." In *The World of Chemistry: Essentials*. Belmont, Calif.: Thomson Brooks/Cole, 2007.
Lang, Kenneth. *Sun, Earth and Sky*. New York: Springer, 2006.
Marshal, John, and R. Alan Plumb. *Atmosphere, Ocean and Climate Dynamics: An Introductory Text*. Cambridge, Mass: Elsevier Academic Press, 2008.
Newman, Michael C., and Michael A. Unger. "Environmental Contaminants." In *Fundamentals of Ecotoxicology*. Boca Raton, Fla.: CRC Press, 2003.
Parson, Edward A. *Protecting the Ozone Layer: Science and Strategy*. New York: Oxford University Press, 2003.
Seinfeld, J. H., and S. N. Pandis. *Atmospheric Chemistry and Physics: From Air Pollution to Climate Change*. Hoboken, N.J.: Wiley, 2006.
Somerville, Richard C. J. *The Forgiving Air: Understanding Environmental Change*. Boston, Mass.: American Meteorological Society, 2008.

Sportisse, Bruno. *Fundamentals in Air Pollution: From Processes to Modelling.* New York: Springer, 2009.

Wallace, John M., and Peter V. Hobbs. *Atmospheric Science: An Introductory Survey.* Cambridge, Mass.: Academic Press, 2006.

World Meteorological Organization. *Global Atmosphere Watch: Antarctic Ozone Bulletin* 1 (August 28, 2008).

Websites

NASA, 2 Nov. 2017
www.nasa.gov/feature/goddard/2017/warm-air-helped-make-2017-ozone-hole-smallest-since-1988.

P

■ Population

FIELDS OF STUDY
Biology; Conservation Biology; Demographics; Ecology; Environment; Geriatrics; Public Health

SUMMARY
The number of a species that can survive within a given area is limited by that area's biological carrying capacity. Although populations might exceed or dip below the carrying capacity temporarily, the numbers fluctuate around a stable line of equilibrium. Environmentalists have expressed concern that with continued human population growth, humankind may one day exceed the earth's carrying capacity.

PRINCIPAL TERMS
- **carrying capacity:** the number of individuals within a given population that an environment can sustain indefinitely
- **habitat:** the natural home or environment of an animal, plant, or other organism.
- **predator:** an organism that obtains food by killing and consuming other organisms
- **prey:** an animal hunted or seized for food

In theory, all populations have the potential for exponential growth or the ability to increase indefinitely in number. In nature, however, there are not infinite numbers of every possible species. No population is able to grow exponentially for long because certain factors will cause the birthrate to decrease and the death rate to increase. Every population within a given area has a maximum number it can reach, and this population size is limited by the area's carrying capacity. Each additional individual introduced beyond the carrying capacity makes it more difficult for the existing population to survive. Therefore, at a certain point, population growth stops.

Several limiting factors can cause population growth to stop, such as food and water availability. If there is only enough food to feed a set population, any individuals that exceed that capacity will not survive. Similarly, a limited amount of habitat is available for each species within an area. If an area does not have enough nesting sites to support a large population, individuals will migrate from that area or adapt until the numbers in the area are again within its carrying capacity. Another factor in the limiting of population size is interaction with predators. If the number of prey increases dramatically in an area, more predators are likely to enter the area, which in turn leads to a reduction in the number of prey. Conversely, if the number of predators exceeds the number of prey necessary to sustain them, the excess predators will die or be forced to migrate to new areas. Disease is another factor that can limit population size: In large populations diseases tend to affect more individuals, and this can result in large reductions in numbers.

As a species, human beings are biologically limited by environment just as are other species. For centuries the carrying capacity for humankind was naturally low. High rates of disease, high infant mortality rates, and short life expectancies kept populations in check. However, in time human beings learned cultivation and domestication. The need to find food became less of a limiting factor on population growth when humans learned to grow enough food to sustain themselves. Humans also discovered ways to use fossil fuels, learned how to treat diseases and injuries, and improved hygiene, all of which led to increases in population. With the inventions of new technologies, it seemed as if the limits of biological carrying capacity no longer applied to humans. This is a controversial topic, however; many environmentalists and other observers assert that there will be a point at which the earth will no longer be able to sustain an increasing human population. They note that, in any case, continued growth in the human population will be accompanied by high costs, particularly in the numbers of other species the planet can sustain.

Kathryn A. Cochran

Further Reading

Brown, Lester R., and Hal Kane. *Full House: Reassessing the Earth's Population Carrying Capacity.* New York: W. W. Norton, 1994.

Freedman, Ronald. *Population Growth: The Vital Revolution.* New York: Routledge, 2017.

Jensen, Derrick. *Endgame.* 2 vols. New York: Seven Stories Press, 2006.

Manning, Robert E. *Parks and Carrying Capacity: Commons Without Tragedy.* Washington, D.C.: Island Press, 2007.

■ Population control and social justice

FIELDS OF STUDY
Demographics; Public Health; Environment; Politics; Gender; Feminism; Social Justice.

SUMMARY
Rapid population growth in recent decades has prompted concerns regarding overpopulation and environmental problems. A commonly-suggested solution to fears of overpopulation is to regulate global population growth through greater access to birth control. However, feminist approaches to issues of overpopulation suggest that reproductive politics must consider the social conditions, political structures, and economic systems that create environmental degradation, rather than isolating impoverished women's individual choices as the driving force behind population growth.

PRINCIPAL TERMS
- **carrying capacity:** the number of individuals within a given population that an environment can sustain indefinitely
- **overpopulation:** the unsustainability of an existing population if the number of individuals exceeds the carrying capacity of the environment
- **Global South:** economically disadvantaged countries who have been negatively impacted by capitalist globalization
- **Green Revolution:** a period from 1950 through the 1960s in which global agricultural production increased through the implementation of new technologies and techniques
- **reproductive justice:** a conceptualization of reproductive rights that expands its focus beyond an individual's choices to have children, and that challenges broader issues such as inequality, discrimination, violence, or colonialism

Since homo sapiens first emerged in Africa 200,000 years ago, human population numbers have been low, only reaching one billion individuals around the start of the nineteenth century. While the rate of population growth began to increase during the Industrial Revolution, dramatic population growth is an even more recent phenomenon. Since the 1950s, the industrialized agriculture of the Green

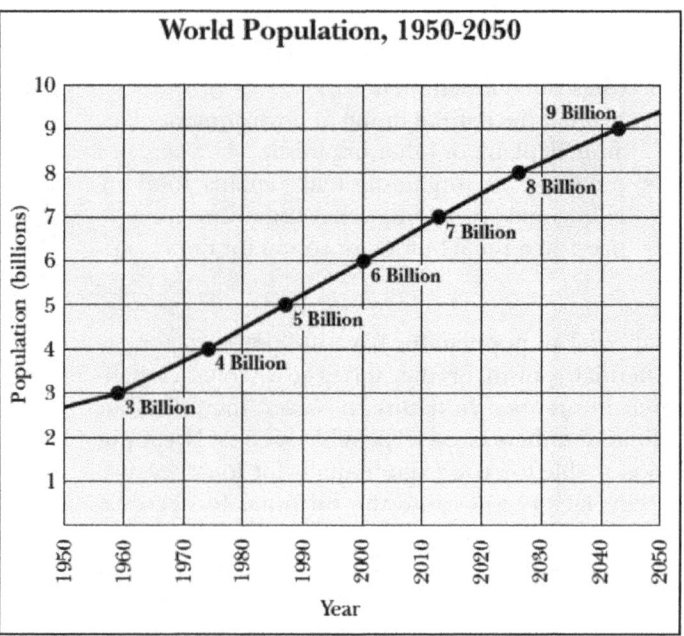

The world population is projected to grow to 8.1 billion by 2025 and 9.6 billion by 2050.

Revolution has enabled greater food production, which supported even more rapid population growth over the following decades. The global human population reached 7 billion individuals in 2011. The speed of this growth in human population has led to concerns among environmentalists that the earth will reach a limit in its capacity to support increasing numbers of humans. Environmental problems typically associated with human overpopulation include the depletion of natural resources, inadequate sources of food or water, increased chances of devastating diseases, and the mass extinction of other species due to habitat loss.

Efforts to respond to these risks of overpopulation have largely focused on limiting human reproduction in order to slow the rate of population growth. These efforts have also focused on providing contraceptives to poor women in the Global South with the assumption that a smaller global population would emit smaller amounts of greenhouse gases and mitigate the effects of anthropogenic climate change. But scholars have also argued that focusing on environmental damage as the result of a large human population obscures global inequalities in the ways natural resources are consumed, as well as how fertility rates really fit into environmental issues. The people around the world with the highest reproductive rates are not emitting the vast majority of greenhouse gases. Nations like China, the United States, India, and Russia produce the most emissions, but each of these countries have relatively low fertility rates, whereas nations with higher fertility rates are among the lowest emitters of greenhouse gases.

Fertility rates have also been declining all over the world for the past fifty years, a fact that challenges assumptions that population growth in developing countries can be slowed by distributing birth control. Some have therefore argued population growth in the Global South is a symptom rather than a cause of problems such as inequality, underdeveloped economies, political instability, and environmental destruction. For example, resource scarcity is not a product of overpopulation; rather, it arises from the unequal distribution of wealth and resources. People suffer from resource scarcity when they do not have access to land on which to grow food or money with which to purchase it. So, any policies that seek to limit reproduction in the Global South without taking into account the inequalities produced by systems of economic inequality or colonialism are ultimately harmful to the health of women and their families. Advocates of global reproductive justice claim that it is only through combatting these broader systems of injustice that social and environmental justice can be achieved.

Julia Kendrick

Further Reading

Corrêa, Sonia and Rebecca Lynn Reichmann. *Population and Reproductive Rights: Feminist Perspectives from the South.* London: Zed Books Ltd, 1994.

England, Marcia R., Maria Fannin, and Helen Hazen, editors. *Reproductive Geographies: Bodies, Places and Politics.* London: Routledge, 2019.

Hartmann, Betsy. *Reproductive Rights and Wrongs: The Global Politics of Population Control.* 3rd ed. Chicago: Haymarket Books, 2016.

Mazur, Laurie, editor. *A Pivotal Moment: Population, Justice & the Environmental Challenge.* Washington, D.C.: Island Press, 2009.

Ross, Loretta and Rickie Solinger. *Reproductive Justice: An Introduction.* Berkeley, Calif.: University of California Press, 2017.

Sasser, Jade S. *On Infertile Ground: Population Control and Women's Rights in the Era of Climate Change.* New York: New York University Press, 2018.

Recycling

FIELDS OF STUDY
Ecology; Energy and Energy Resources; Environment; Environmentalism; Industries; Land-Use Management; Mapping; Mathematics; Renewable Resources; Technology and Applied Science; Waste Management

SUMMARY
Recycling constitutes the primary component of modern waste management. It benefits the environment by reducing waste, conserving resources, saving energy, and reducing air and water pollution that can result from waste disposal methods, thus protecting human health. In addition, recycling can provide fiscal benefits, such as by expanding manufacturing jobs.

PRINCIPAL TERMS
- **compost:** a mixture of various decaying organic substances, as dead leaves or manure, used for fertilizing soil
- **electronic waste:** the disposal of broken or obsolete electronic components and materials, such as computers, televisions, cell phones, stereos, copiers and fax machines
- **Environmental Protection Agency (EPA):** an independent agency of the United States federal government, established in 1970 for environmental protection
- **landfill:** method of disposing large amounts of garbage by burying it
- **life-cycle assessment:** technique for investigating and evaluating the environmental consequences arising from the provision of a product or service
- **recycling:** processing of used materials into fresh supplies of the same materials for new products or salvaging of certain materials from more complex products

In 2009 the United States produced 251 million tons of trash, with 82 million tons (32.5 percent) of those materials recycled. In the first decade of the twenty-first century, total in the United States increased approximately 100 percent. At the national level, the U.S. Environmental Protection Agency (EPA) oversees management, including hazardous wastes and landfills, and sets recycling goals. However, no national laws for recycling exist; instead, individual state and local governments have created their own laws concerning recycling.

Several U.S. states have laws establishing deposits and refunds for beverage containers. Other states ban the deposition of recyclable materials into landfills. Some cities, including New York and Seattle, have passed laws that include fines for failing to recycle certain materials. Several organized voluntary and educational programs have been established to increase recycling where it has not been mandated by law. Recycling education is usually integrated into science or social studies classes at the elementary, middle school, and high school levels. November 15 is celebrated as America Recycles Day, which is dedicated to raising awareness about the importance of recycling and to encouraging Americans to recycle and buy recycled products.

For recycling to be economically feasible, efficiently managed, and environmentally effective, adequate recyclable materials must be available, a system must be put in place to extract those materials from the waste stream, and a facility must be available locally where the materials can be reprocessed. In addition, there must be demand for the recycled products.

Recyclable materials are generally collected in three ways: through curbside pickup at consumer homes and businesses, through consumer delivery to drop-off centers, and through consumer return through deposit or refund programs. After collection, recyclables are sent to materials recovery facilities, where they are sorted and prepared for processing into marketable items made entirely or partially of recycled content.

In 2008 the United States generated some 250 million tons of (MSW), commonly called or garbage. Organic materials such as yard trimmings, paper products, and food wastes make up more than two-thirds of human trash. This waste could become useful instead of being deposited in landfills. In composting, natural decomposers break down organic materials into fertile topsoil for plant cultivation. Humus and its nutrients in compost help regenerate and enrich soils, help remediate contaminated soils, prevent pollution, and provide economic benefits by, for example, reducing the need for fertilizers and pesticides.

Commonly Recycled Materials

Paper is the most common material in municipal solid waste, approximately 35 percent of the total. Americans recycled more than 50 percent of paper used in 2008, but that percentage could be increased. Approximately 80 percent of the paper mills in the United States are designed to depend on paper recycling; they reduce recycled paper to pulp, which is then combined with pulp from newly harvested wood. Wood fiber can be recycled only up to five times, however, since damage to the fibers with each recycling process decreases the quality of the resulting product; hence, more new fibers must be added with each cycle.

Recycled fabric items are sorted into grades; some are then processed to make industrial wiping cloths, whereas others are made into filling products or used in the manufacture of paper. Discarded clothing items often end up in landfills in the United States, but many unwanted but still wearable clothes can be recycled through charitable giving. This practice helps to reduce unwanted waste and conserves the resources needed to make new clothing, while providing clothing to persons in need.

Plastic recyclable items are sorted and sent to facilities where they are washed and ground into small flakes that are dried, melted, filtered, and formed into pellets used to manufacture new plastic products. In 2008 the U.S. MSW stream included some 13 million tons of plastics—about 12 percent, an increase from less than 1 percent in 1960. The recycling of plastics has been found to reduce energy use in the United States by 26 percent. Given that 70 percent of the plastics made in the United States are made by factories that use domestic natural gas, the energy saved is freed up for other uses, such as heating and cooling homes.

When glass bottles and jars are recycled, the containers are first smashed and the broken glass, or cullet, is examined for purity and cleaned of contaminants at a glass recycling plant. The cullet is then further crushed and placed in a melting furnace with a mix of raw materials. The resulting material is then mechanically blown or molded into new bottles, jars, and other glass items. Of all the glass recycled in the United States, 90 percent goes into the making of new containers. Glass cullet is also used for aggregate and "glassphalt" for road construction (glassphalt contains 30 percent recycled glass). Unlike paper, glass can be recycled repeatedly, as reprocessing does not affect glass's structure. Because cullet costs less than the raw materials needed to make new glass, recycling glass saves manufacturers money at the same time it conserves natural resources; it also helps glass-making furnaces last longer because it melts at a relatively low temperature. Of the 12.2 million tons of glass that were part of the U.S. MSW stream in 2008, 2.8 million tons, or 23 percent, was recovered for recycling. That figure represents a significant increase from 750,000 tons in 1980.

Iron and steel, the world's most commonly recycled ferrous metals, are generally separated from the using magnets. Recycled steel is processed at steelworks, where steel is re-melted and forged into new products. Like glass, steel can be recycled repeatedly with no reduction in quality.

Aluminum is one of the most widely recycled non-ferrous metals and one of the most efficiently processed. Approximately 40 percent of aluminum in the average aluminum can is recycled material. Aluminum beverage containers are the largest source of aluminum in the U.S. MSW stream; in 2008 they accounted for 3.4 million tons (1.3 percent) of MSW, up from only 0.4 percent in 1960. The aluminum recycling process uses 92 percent less energy than is needed to produce aluminum from bauxite ore, in part because the temperature necessary to melt recycled aluminum is 600 degrees Celsius (1,112 degrees Fahrenheit), in contrast to the 900 degrees Celsius (1,652 degrees Fahrenheit) needed to extract mined aluminum from bauxite. The processes of recycling and reuse produce no changes in aluminum, so it can be recycled repeatedly.

Electronic wastes pose problems of energy cost and of toxic substances that can be reduced by recycling. In many parts of the world the disposal of computers, televisions, mobile phones, and other electronic devices into the general waste stream is forbidden because of the toxic contents of certain components in these products. In the United States, the EPA advises consumers and local governments regarding the safe recovery and recycling of components of electronic waste. Electronic waste recycling plants process discarded electronic items by taking them apart and separating the various metal, plastic, and other components, collecting those that can be reused.

Batteries are sometimes recycled, but because of wide variations in sizes and types, recycling is not always practical. Those that cannot be recycled must be disposed of carefully, as batteries contain heavy metals that can pollute soil and water. The lead-acid batteries used in motor vehicles are easily recycled, and in the United States many local and state governments require businesses that sell such batteries to receive and recycle used batteries from consumers. The recycling rate for automotive batteries in the United States is 90 percent; new batteries generally contain up to 80 percent recycled material.

Tires are frequently recycled at an industrial level. Used tires are ground up and the resulting material is used for a variety of purposes, including insulation and road surfacing. Rubber made from old tires is used to soften the surfaces of playgrounds. Tires are also sometimes recycled into consumer products. Recycled rubber from tires is used to make items such as doormats, trash cans, and even messenger bags; also, a number of companies make shoes that have soles made of recycled tire treads.

Technologies are advancing to produce improved materials for faster recycling, such as completely packing materials. An example is a plant-based, all-natural material that breaks down into inert proteins when it meets water and is consumed by soil bacteria to produce a product used as lawn fertilizer. Scientists at Sony discovered that expanded polystyrene foam, such as that used as packing "peanuts" for cushioning fragile items inside boxes, completely dissolves at room temperature when sprayed with limonene (a natural oil extracted from the skins of citrus fruits) and can be processed for reuse.

Life-Cycle Assessment

Life-cycle assessment (LCA) is a method of considering the "cradle-to-grave" environmental impacts a product has during its entire life cycle—from raw material acquisition through production, use, consumption, and reuse or final disposal. The technique has the advantage of revealing hidden environmental impacts.

The driving force behind the first LCA was an environmental debate about waste and packaging in the United States during the late 1960s. The first LCAs compared the resource consumption and emissions of beverage cans, plastic bottles, and refillable glass bottles. During the global oil crises of 1973 and 1979 the focus of LCAs shifted to the analysis of different petroleum substitutes. With growing environmental awareness LCA became an attractive methodology for product-oriented environmental policies in several countries.

An LCA study consists of four distinct phases: goal and scope definition, inventory analysis, impact assessment, and interpretation. In the goal and scope phase the product and the purpose of the study is set. The general categories of impacts considered in an LCA are ecological consequences, human health effects, and resource use. The application of LCA is widespread and ranges from the analysis of waste management to production processes and different sorts of consumer goods. Despite the popularity of LCA and the international standardization of the process, LCA has received some criticism for a lack of consistency. Possible sources of inconsistency are variations in system boundaries and cutoff criteria, as well as the quality and availability of data. Critics have also noted that any assessment of the social implications of products is generally lacking in LCA.

Debates

Some critics of the promotion of recycling as a solution to environmental problems have suggested that the modern waste management system has fundamental flaws that call for reexamination—they have therefore added a "fourth R," rethink, to the "three

R's": reduce, reuse, and recycle. They argue that the costs of collecting and transporting recyclable materials for processing far outweigh the costs of production processes using raw materials. They also assert that more jobs are lost through the reduced collection of raw materials than are created by the recycling industry, which, additionally, pays low wages and offers poor working conditions. Critics argue further that when all processes are considered, the production of recycled products consumes more energy than would the traditional landfill disposal of the recycled materials used to make the products.

Recycling proponents counter that the benefits of recycling compensate for any higher monetary costs it may create. Landfilled wastes pollute and waterways and contribute significantly to global warming through the release of methane into the atmosphere, producing long-term financial costs of remediation. Proponents also argue that the workers that would be needed to gather amounts of equal to the amounts provided through recycling would be working in jobs, such as timber harvesting and ore mining, that have much more dangerous workplace conditions than are found in the recycling industry. Recycling proponents note also the reductions in energy needs represented by the recycling of such materials as paper and aluminum compared with the processing of raw materials. The EPA strongly supports recycling, emphasizing that by saving energy, recycling reduces emissions of carbon dioxide, a linked with global warming.

Samuel V. A. Kisseadoo and Claudia Reitinger

Further Reading

Ackerman, Frank. *Why Do We Recycle? Markets, Values, and Public Policy.* Washington, D.C.: Island Press, 1997.

Dietz, Thomas, and Paul C. Stern, eds. *Public Participation in Environmental Assessment and Decision Making.* Washington, D.C.: National Academies Press, 2008.

Loeffe, Christian V., ed. *Trends in Conservation and Recycling of Resources.* Hauppauge, N.Y.: Nova Science, 2006.

McKinney, Michael L., Robert M. Schoch, and Logan Yonavjak. "Resource Use and Management." In *Environmental Science: Systems and Solutions.* 4th ed. Sudbury, Mass.: Jones and Bartlett, 2007.

Tammemagi, Hans. *The Waste Crisis: Landfills, Incinerators, and the Search for a Sustainable Future.* New York: Oxford University Press, 1999.

Weeks, Jennifer. "Future of Recycling: Is a Zero-Waste Society Achievable?" *CQ Researcher* 17, no. 44 (December 2007): 1033-1060.

Williams, Paul T. "Waste Recycling." In *Waste Treatment and Disposal.* 2d ed. Hoboken, N.J.: John Wiley & Sons, 2005.

■ Renewable energy

FIELDS OF STUDY
Alternative Energy Sources; Ecology; Ecosystems; Energy and Energy Resources; Environment; Environmentalism; Industries; Renewable Resources; Technology and Applied Science

SUMMARY
The burning of fossil fuels such as coal, natural gas, and petroleum releases emissions that contain greenhouse gases and cause air pollution and acid rain, whereas most forms of renewable energy are nonpolluting. In addition, because the earth has a finite supply of fossil fuels, the development of renewable energy sources is important to the long-term future of humankind.

PRINCIPAL TERMS
- **renewable energy:** energy derived from natural, unlimited, and replenishable sources
- **fossil fuels:** a fuel, like coal, oil, or natural gas, formed in the earth from plant or animal remains
- **greenhouse gases:** any of the gases whose absorption of solar radiation is responsible for the greenhouse effect, including carbon dioxide, methane, ozone, and the fluorocarbons
- **biomass:** organic matter, especially plant matter, that can be converted to fuel and is therefore regarded as a potential energy source
- **photovoltaic:** of, relating to, or utilizing the generation of a voltage when radiant energy falls on the boundary between dissimilar substances

Renewable energy

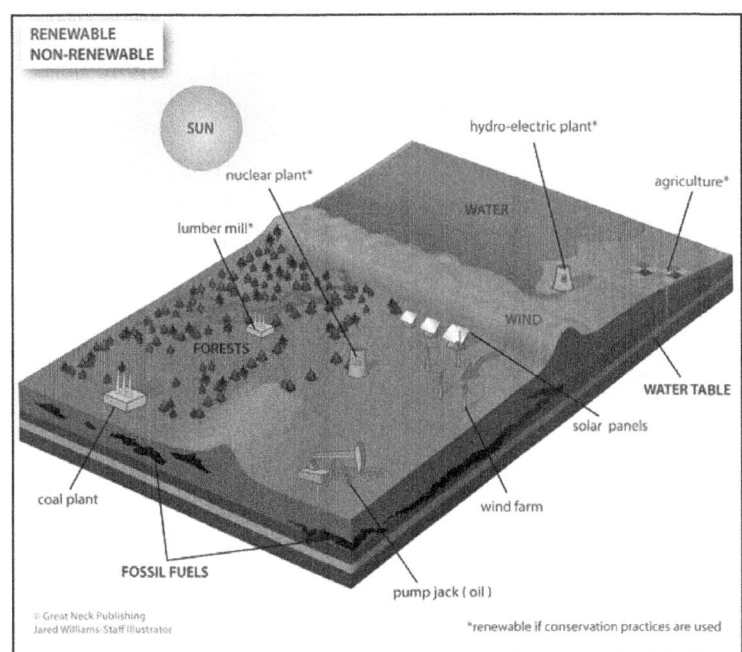

The earth has many natural resources, some are renewable and others are not. Conservation practices will ensure that resources do not become depleted.
© EBSCO

The environmental movement and the oil crises of the 1970s led to interest in the development of energy sources that would offer alternatives to the use of fossil fuels. Fossil fuels are limited resources, and the burning of fossil fuels to generate energy creates emissions of carbon dioxide, toxic chemicals, and air pollutants that harm the environment and human health. Because renewable, or clean, energy systems use natural, local sources that are inexhaustible and such systems have fewer negative impacts on human life and the environment, governments have provided increasing support for the development of renewable energy technologies.

Biomass
The oldest renewable energy source is biomass, which is organic animal and plant material and waste. Biomass resources include grass crops, trees, and agricultural, municipal, and forestry wastes. Since the discovery of fire, humans have burned biomass to release its chemical energy as heat. For example, wood has been burned to cook food and to provide heat. Biomass energy has also been used to make steam and electricity. Biomass oils can be chemically converted into liquid fuels or biodiesel, a transportation fuel. Ethanol, another transportation fuel, comes from fermented corn or sugarcane. Crops such as willow trees and switchgrass are also cultivated for biomass energy generation.

Biomass energy has many environmental benefits when compared with fossil-fuel energy. It contributes little to air pollution, as it releases 90 percent less carbon dioxide than do fossil fuels. Energy crops, such as prairie grasses, require fewer pesticides and fertilizers than do high-yield food crops such as wheat, soy beans, and corn, so they cause less water pollution. Energy crops also add nutrients to the soil. About 4 percent of the energy used in the United States is biomass energy.

Solar Energy
One of the most promising and popular kinds of renewable energy is solar energy, which uses radiant energy produced by the sun. Solar energy was used as early as the seventh century BCE, when a magnifying glass was used to concentrate sunlight to light fires. In 1767, Swiss scientist Horace-Bénédict de Saussure invented the first solar collector, a device for storing the sun's radiation and converting it into a usable form, such as by heating water to create steam. In 1891, American inventor Clarence Kemp patented the first commercial solar water heater.

Sunlight can be converted directly into electricity at the atomic level by photovoltaic (PV) cells, also called solar cells. The photovoltaic phenomenon was first noted in the eighteenth century and became more practical with the use of silicon for the cells in the twentieth century. The cells are joined together in panels, often connected together in an array. They can be placed on rooftops and connected to a grid. In the twenty-first century, solar cells are used worldwide in home and commercial electrical systems, satellites, and various consumer products.

Solar energy has numerous environmental benefits. Photovoltaics produce electricity without gaseous or liquid fuel combustion or hazardous waste by-products. Decentralized PV systems can be used to provide electricity for rural populations, saving more expensive conventional energy for industrial, commercial, and urban needs. By providing electricity for remote and rural areas, solar energy also reduces the use of disposable lead-acid cell batteries, which can contaminate water and soil if they are not disposed of properly. The use of solar energy in rural areas also reduces air pollution by decreasing the use of diesel generators and kerosene lamps.

Since solar energy depends on sunlight, the efficiency and performance of solar energy systems are affected by weather conditions and location. Nevertheless, the amount of inexhaustible solar energy that could be generated around the globe exceeds the amount needed to meet the world's energy requirements. It has been estimated that if PV systems were installed in only 4 percent of the world's deserts, they could supply enough electricity for the entire world. As solar technologies improve and costs decrease, solar energy has the potential to be the leading alternative energy source of the future.

Wind Energy
One of the fastest-growing types of renewable energy during the 1990s, wind energy has been used by humans for centuries. Windmills appear in Persian drawings from 500 C.E., and they are known to have been used throughout the Middle East and China. The English and the French built windmills during the twelfth century, and windmills were indispensable for pumping underground water in the western and Great Plains regions of the United States during the nineteenth and early twentieth centuries. These windmills converted wind into mechanical power.

The modern windmills used to convert wind energy into electricity are called wind turbines or wind generators. In 1890, Poul la Cour, a Danish inventor, built the first wind turbine to generate electricity. Another Dane, Johannes Juul, built the world's first alternating current (AC) wind turbine in 1957. In the twenty-first century, large wind plants are connected to local electric utility transmission networks to relieve congestion in existing systems and to increase reliability for consumers. Wind energy is also used on a smaller scale by home owners in what is known as distributed energy; home-based wind turbines, with batteries as backup, can lower electricity bills by up to 90 percent.

The use of wind energy has long-term environmental benefits. Unlike nuclear and fossil-fuel electricity generation plants, wind generation of electricity does not consume fuel, cause acid rain and greenhouses gases, or require waste cleanups. For example, it has been estimated that when Cape Wind, America's first offshore wind farm under development in Nantucket Sound, becomes fully operational, its 130 wind turbines will reduce greenhouse gas emissions by 734,000 tons annually.

In 2010 the annual wind energy generating capacity of the United States was more than 35,000 megawatts, enough electricity to power 9.7 million homes. This amount of electricity generated by fossil-fuel-burning plants would have released some 62 million tons of carbon dioxide; avoiding the release of so much carbon dioxide is the equivalent of keeping 10.5 million cars off the roads.

Wind energy technology has advanced to the point that wind power is affordable and can compete successfully with fossil fuels and other conventional energy generation. According to a report by the US Department of Energy, wind energy could provide 20 percent of the US electricity supply by 2030. Wind power as a commercial enterprise has been established in more than eighty countries, and between 2013 and 2014 wind energy capacity increased by 51,477 MW, with China, Germany, and the United States seeing the biggest increases.

The main disadvantage of wind energy is that it is intermittent, because wind velocities are inconsistent even in areas of strong winds. In addition, some environmentalists have objected to the establishment of wind farms because of their potential to harm wildlife and the aesthetic damage they do to natural landscapes.

Hydropower
Long before electricity was harnessed, in about 4000 BCE, ancient civilizations used hydropower, or energy from moving or flowing water, in the waterwheel, the first device employed by humans to produce mechanical energy as a substitute for animal and human labor. Running water in a stream or river moves the wooden paddles mounted around a waterwheel, and the resulting rotation in the shaft drives machinery. The earliest waterwheels were used to grind grain, and the technology went on to be used worldwide for that purpose, as well as to supply drinking water, irrigate crops, drive pumps, and power sawmills and textile mills.

In the nineteenth century, the water turbine replaced the waterwheel in mills, but then the steam engine replaced the turbine in mills. The hydraulic turbine reemerged, however, to power electric generators in the world's first hydroelectric power stations during the 1880s. By the early twentieth century, 40 percent of the US electricity supply was hydroelectric power. Modern large hydropower plants are attached to dams or reservoirs that store the water for turning the turbines and are connected to electrical grids or substations that transmit the electricity to consumers.

Hydropower is the leading renewable energy source for generating electricity. It has both negative and beneficial effects on the environment. Building dams and reservoirs changes the environment and can harm native habitats and their fish, animal, and plant life. In addition, reservoirs sometimes emit methane, a greenhouse gas. Water is a natural and inexpensive energy source, however. No fuel combustion takes place in the generation of hydropower, so the process does not pollute the air, and energy storage is clean. Hydroelectric power accounts for about 16 percent of electricity used worldwide, and 7 percent of electricity in the United States. Hydroelectric energy capacity has been growing by approximately 3.3 percent per year throughout the twenty-first century, although widespread drought in 2015 slowed growth in this area.

Geothermal Energy
Geothermal energy comes from heat produced deep inside the earth. Deep wells and pumps bring underground hot water and steam to the earth's surface to heat buildings and generate electricity. Some geothermal energy sources come to the surface naturally, including hot springs, geysers, and volcanoes. The ancient Chinese, Native Americans, and Romans used hot mineral-rich springs for bathing, heating, and cooking. Food dehydration became the major industrial use of this form of energy. In 1904, the first electricity from geothermal energy was generated in Larderello, Italy.

Although not as popular a renewable energy source as wind or solar energy, geothermal energy has significant advantages and benefits for the environment. Because the earth's heat and temperatures are basically constant, geothermal energy is reliable and inexhaustible; it is also not affected by changes in climate or weather. It is very cost-efficient as well; heat pumps can be operated at relatively low cost. The steam and water used in geothermal systems are recycled back into the earth.

Geothermal plants are environmentally friendly. Because they do not burn fuel to generate electricity, they release little or no carbon dioxide and other harmful compounds. Geothermal plants produce no noise pollution and have minimal visual impacts on the surrounding environment, because they do not occupy large surface areas.

The US Environmental Protection Agency and the Department of Energy support the use of geothermal heat pumps. The American Recovery and Reinvestment Act of 2009 (ARRA, also known generally as the Stimulus) provided for grants and tax incentives worth $400 million to the industry, which added 144 geothermal energy plants in fourteen states at the beginning of 2010. ARRA measures benefited the development of all renewable energy sources; it included a Treasury Department grant program for renewable energy developers, increased funding for research and development, and a three-year extension of the production tax credit for many renewable energy facilities. By the end of 2014, the United States had an installed geothermal energy capacity of about 3.5 GW.

It is hoped that by 2070, 60 percent of all global energy will come from renewable energy sources. The World Bank, the World Solar Decade, and the World Solar Summit have designated $2 billion for projects focused on renewable energy resources and the environment.

Alice Myers

Further Reading
Craddock, David. *Renewable Energy Made Easy: Free Energy from Solar, Wind, Hydropower, and Other Alternative Energy Sources.* Ocala, Fla.: Atlantic, 2008.
Da Rosa, Aldo Vieira. *Fundamentals of Renewable Energy Processes.* 3rd ed. Boston, Mass.: Elsevier, 2012.
Goswami, D. Yogi, and Frank Kreith, eds. *Energy Efficiency and Renewable Energy Handbook.* 2nd ed. Boca Raton, Fla.: CRC, 2016.
Jacobson, Mark Z. "How Renewable Energy Could Make Climate Treaties Moot." *Scientific American.* Nature America, 23 Nov. 2015. Web. 6 Jan. 2016.
Langwith, Jacqueline, ed. *Renewable Energy.* Detroit, Ill.: Greenhaven, 2009.
MacKay, David J. C. *Sustainable Energy—Without the Hot Air.* Cambridge, England: UIT Cambridge, 2009.

Nelson, Vaughn. *Wind Energy: Renewable Energy and the Environment.* Boca Raton, Fla.: CRC, 2009.

Peake, Stephen. *Renewable Energy: Power for a Sustainable Future.* London: Cambridge University Press, 2018.

Pimentel, David, ed. *Biofuels, Solar, and Wind as Renewable Energy Systems: Benefits and Risks.* New York: Springer, 2008.

Twidell, John, and Tony Weir. *Renewable Energy Resources.* 3rd ed. New York: Routledge, 2015.

■ Rising sea levels

FIELDS OF STUDY
Architecture; Ecology; Environment; Environmental Architecture; Landscape Ecology; Ocean and Tidal Energy Technology; Oceanography; Urban Planning and Development

SUMMARY
Rising seas associated with global warming have challenged many millions of people living close to oceans around the world. Policies and building practices have been changed in some countries to adapt to, or accommodate, these changes. Local responses to rising sea levels involving adaptations such as surrendering of land to the sea, raising dikes, and building homes designed to rise with water levels

PRICIPALS TERMS
- **bathymetric:** measurement of the depths of oceans, seas, or other large bodies of water
- **climate accommodation:** local responses to rising sea levels involving adaptions such as surrendering of land to the sea, raising dikes, and building homes designed to rise with water levels
- **climate change:** a long-term change in the earth's climate, especially due to an increase in the average atmospheric temperature
- **global warming:** an increase in the earth's average atmospheric temperature that causes corresponding changes in climate and that may result from the greenhouse effect
- **managed realignment:** a coastal management strategy allowing the shoreline to move inland instead of attempting to hold the line with structural engineering, while natural coastal habitat is enhanced seaward of a new line of defense
- **topographic:** detailed mapping or charting of the features of a relatively small area, district, or locality

Rising seas associated with global warming have challenged many millions of people living close to oceans around the world. Policies and building practices have been changed in some countries to adapt to, or accommodate, these changes. The anticipation of rising sea levels around the world as a result of global warming has led many coastal nations to develop strategies that can protect their people while making accommodations to the reality of their changed coastlines.

British Strategies
Parts of Great Britain's coastline are afflicted by the same problems as the eastern and Gulf coasts of the United States: The land is subsiding as ice melt and thermal expansion slowly raise sea levels. The United Kingdom Climate Impact Programme, a government-funded program at Oxford University, forecasts that the sea level could rise by as much as 1 meter (3.3 feet) by late in the twenty-first century. In addition to climate change, isostatic rebound (the rise of Scotland's coast following the last ice age) is contributing to subsidence of the land southward along the English coast. As the sea rises 3 millimeters (0.118 inch) per year abreast of Essex, the land itself is sinking half as rapidly, producing a net sea-level rise averaging 4.5 millimeters (0.177 inch) per year.

In a strategy officially termed "managed realignment," the British government decided to allow the sea to flood low-lying farmland rather than attempt to fend off the invading waters by building ever-higher defenses. The policy, which will eventually allow the encroaching sea to submerge several thousand hectares, has been welcomed by environmentalists. Farmers, however, have contended that the strategy is unviable and have demanded more flood defenses. The affected area of the coast ranges from the Humber estuary, around East Anglia, to the Thames estuary and west to the Solent. Strategic withdrawal also has been planned for sections of the Severn estuary. The first site surrendered to the sea was in Lincolnshire. About 81 hectares (200 acres) of farmland were flooded by seawater at Freiston Shore after diggers broke through the flood-defense banks to create a salt-marsh bird reserve.

Until around the end of the twentieth century, the Thames Barrier, built to protect London and surrounding areas from unusually high river tides and storm surges, closed an average of two or three times a year. Between November 2001, and March 2002, however, the barrier was raised twenty-three times. A British report released in September 2002, said that 152,800 square kilometers (59,000 square miles)—home to 750,000 people—in and around London are vulnerable to flooding because they are below high-tide levels, some by as much as 3.7 meters (12 feet).

Dutch Strategies

The Dutch fear that rising storm surges could inundate much of the Netherlands, large areas of which have been reclaimed from the sea. Fears have been expressed that the country's western provinces may flood. The Hague, for example, may become uninhabitable as low-lying suburbs of Amsterdam return to marshland or open water.

By 2010 the Dutch had been forced to anticipate surrendering 200,000 hectares (494,211 acres) of farmland to river floodplains and had begun a major construction program involving floating homes. Pieter van Geel, the Dutch minister of housing, spatial planning, and the environment, stated in 2004 that half of the Netherlands is below sea level, and so beyond a certain level of sea-level rise, it is not feasible for the nation to build more extensive or higher dikes in many areas. Above 2 meters (6.6 feet) of additional sea-level rise, much of the land that the Netherlands has reclaimed from the ocean over several hundred years could be lost.

During mid-2008 a Dutch governmental commission recommended that the country spend $144 billion to reinforce its sea defenses through the year 2100 as a precaution against sea-level rise. The measures proposed include widening dunes facing the North Sea and raising the height of dikes along the coastline and rivers.

In the Netherlands the threat of sea-level rise also is being met with amphibious homes. One development of forty-six homes in the town of Maasbommel, for example, features two-bedroom, two-story houses with foundations of hollow concrete attached to iron posts sunk into a lake bottom; these homes can accommodate water levels as much as 5.5 meters (18 feet) higher than the levels that existed when they were built.

New York

In 2010, New York City's Museum of Modern Art (MoMA) and P.S.1 Contemporary Art Center collaborated on a major initiative to propose solutions for the effects of climate change on New York's waterfronts, with an exhibition titled *Rising Currents: Projects for the New York Waterfront*. The exhibit presented architectural proposals emphasizing adaptive "soft" infrastructure solutions for New York and New Jersey's Upper Bay to make New York City and surrounding areas more resilient in responding to rising sea levels and ever-increasing storm surges. Proposals ranged from the creation of salt- and freshwater wetlands along the banks of the bay and an aqueous landscape—like that in Venice, Italy—to habitable piers and manmade islands, as well as a protective reef of living oysters.

Five multidisciplinary teams of New York-based architects, engineers, and landscape designers were selected to develop proposals. The exhibition was organized by Barry Bergdoll, the Philip Johnson Chief Curator of Architecture and Design at MoMA and designed to create visions for a restructured harbor city responding to the realities of ongoing rising sea level and for stimulating debate about vital issues of public concern in architecture and urban planning. The teams were required to examine and contextualize problems and concerns, display detailed presentations of topographic and bathymetric data, as well as projected flooding based on incremental sea-level rise. The exhibition's goal was to address possibilities and resilience for climate change as it intensifies.

With *Rising Currents*, New York began moving to join other major cities, such as Copenhagen, Denmark, Amsterdam, Holland, Singapore and Hong Kong, which increasingly focus on active waterfront mixed use and in exploring climate accommodation and public dialogue to mitigate crises to come.

Bruce E. Johansen, updated by Jennifer Heath

Further Reading

Archer, David. *The Long Thaw: How Humans Are Changing the Next 100,000 Years of Earth's Climate.* Princeton, N.J.: Princeton University Press, 2009.

Bergdoll, Barry, ed. *Rising Currents: Projects for New York's Waterfront.* New York: Museum of Modern Art, 2011.

Cline, William R. *The Economics of Global Warming.* Washington, D.C.: Institute for International Economics, 1992.

Dutch, Steven I., ed. *Encyclopedia of Global Warming.* 3 vols. Pasadena, Calif.: Salem Press, 2010.

Jamail, Dahr. *The End of Ice: Bearing Witness and Finding Meaning in the Path of Climate Disruption.* New York: The New Press, 2019.

Lyall, Sarah, "At Risk from Floods, but Looking Ahead with Floating Houses." *New York Times,* April 3, 2007.

McGranahan, Gordon, Deborah Balk, and Bridget Anderson. "The Rising Tide: Assessing the Risks of Climate Change and Human Settlements in Low Elevation Coastal Zones." *Environment and Urbanization* 19, no. 1 (2007): 17-37.

Orff, Kate. Toward an Urban Ecology: SCAPE / Landscape Architecture. New York: Monacelli Press, 2016.

Pilkey, Orrin H., and Rob Young. *The Rising Sea.* Washington, D.C.: Island Press, 2009.

Rosenthal, Elisabeth. "As the Climate Changes, Bits of England's Coast Crumble." *New York Times,* May 4, 2007.

Road systems

FIELDS OF STUDY
Civil Engineering; Ecology; Emissions; Environment; Environmentalism; Hazardous Materials; Hydrology; Pollution; Societies; Toxic Waste; Transportation; Urban Planning; Urbanization

SUMMARY
The construction and heavy use of highways can have severe impacts on the environment, including air and water pollution, land degradation, and loss of open space. Since motorized transportation is a common means of travel, compromise is often necessary to balance the need for additional roads and the need to protect the environment.

PRINCIPAL TERMS
- **acoustical barriers:** an exterior structure designed to protect inhabitants from noise pollution and the most effective method of mitigating roadway, railway, and industrial noise sources
- **carbon monoxide (CO):** a colorless odorless very toxic gas formed as a product of the incomplete combustion of carbon or a carbon compound
- **Highway Trust Fund:** a U.S. transportation fund receiving money from a federal fuel tax for spending on road construction and other surface transportation projects, including mass transit
- **highways and freeways:** paved thoroughfares created for largely unobstructed motor vehicle traffic
- **hydrocarbons:** an organic compound containing only carbon and hydrogen, often occurring in petroleum, natural gas, coal, and bitumens
- **hydrology:** a science dealing with the properties, distribution, and circulation of water on and below the Earth's surface and in the atmosphere
- **National System of Interstate and Defense Highways:** also known as the Interstate Highway System formed by President Dwight D. Eisenhower in 1956, is a network of controlled-access highways
- **nitrogen oxide:** any of several oxides of nitrogen mostly produced in combustion and considered to be atmospheric pollutants

The era of the freeway began during the 1950s, when lobbying groups in the United States encouraged a political vision of a nationwide high-standard, high-speed road network. Retired U.S. Army general Lucius D. Clay led a committee that studied transportation needs across the United States and advised President Dwight Eisenhower that the nation needed what came to be called the National System of Interstate and Defense Highways. The costs of creating and maintaining the more than 69,000 kilometers (43,000 miles) of interstate highways built after 1956 have been shared by federal and state governments on a 90/10 (federal/state) matching basis. The federal share comes from the Highway Trust Fund, which receives revenues from federal taxes on fuels, lubricants, vehicles, and vehicle parts. Although the interstate system accounts for only 1 percent of the total road miles in the United States, it carries 20 percent of the traffic.

Studies of the U.S. road and highway systems as a whole—including local roads and services—have found that motor vehicle user fees cover only two-thirds of public expenditures, not including the substantial nonmonetary external costs of environmental impacts. Because many of the costs of using private vehicles are hidden, many Americans perceive driving their own cars to be less expensive than using public-transit alternatives. U.S. highway statistics for 2008 indicated the existence of more than 248 million registered motor vehicles (including 137 million automobiles) and more than 4 million miles of public roads.

Traffic congestion is most severe in areas experiencing rapid growth in both total population and number of vehicles in use. In fact, rapid population growth tends to offset the beneficial impacts of remedies adopted to reduce traffic congestion. Low-density settlement generates more total automotive vehicle trip miles per day, which consume more energy and cause greater emission of pollutants. One important development that has helped to reduce congestion, and thus motor vehicle emissions, on some urban and suburban freeways since the 1970s is the so-called carpool lane—a lane reserved for the use of buses and other vehicles carrying more than one or two passengers.

Environmental Impacts

Air pollution constitutes the most serious environmental impact caused by highway transportation. Developments that may generate traffic, such as parking lots for shopping centers, may be classified as indirect sources of pollution. Internal combustion and diesel engines are the principal sources of carbon monoxides and hydrocarbons, account for nearly one-half of the nitrogen oxides, and are the chief source of particulate lead in the atmosphere. Highway emissions are directly related to traffic volume and density, vehicle type, speed, and mode (idle, acceleration, cruise, or deceleration). Increased speed produces a demand on an engine for increased power, which leads to more fuel consumed and greater emissions, but the vehicle also passes through an area more quickly. Long trips by motor vehicle and traffic congestion both increase the emissions discharged into the atmosphere.

The building of roads and highways also consumes open space, affecting plant and animal life, as well as climate and water runoff. Highways facilitate the spread of urban areas and often lead to low-density developments, which are difficult to provide with services. Highways that connect developed areas usually follow valleys and other areas with flat terrain, and consequently highways are often built near streams, lakes, and wetlands. Until very late in the twentieth century, hydrologic features that blocked proposed roads were seen primarily as obstacles to be bridged, filled, or moved at lowest cost. Laws that were put in place to protect endangered animal and plant species changed this approach, requiring highway developers sometimes to employ routes or construction procedures that are more expensive than those they would have used in the past.

Among the subtler and probably more serious impacts of road construction are changes in local hydrologic patterns, such as changes in the water table that affect vegetation. Erosion and sedimentation are also associated with road construction activities. Another environmental concern related to roads and highways involves the runoff from street surfaces into waterways; such runoff can include salt (spread on roads to melt ice in winter) and petroleum products that contaminate the soil.

Highway noise has also been identified as an environmental problem. Noise—excessive or unwanted sound—has been found to be more than simply an annoyance; in the extreme, the sleep disturbances and other negative impacts of noise can be considered a danger to public health. Highway noise is troublesome to control, but some attenuation can be achieved through proper planning. Buffer zones and acoustical barriers, modifications of highway alignment, and traffic management measures can reduce the exposure to noise for those who live or work near highways. The U.S. Federal Highway Administration specifies a methodology that should be used for the evaluation of traffic noise and abutting land use and provides guidance on when noise abatement should be considered.

In balancing transportation needs and environmental goals, engineers must distinguish between mobility (simple movement) and access (the ability to reach destinations and desired services). Increasing mobility through more roads, more vehicles, and more traffic may actually reduce access over the long term. For example, older neighborhoods tend to have stores, schools, and transit services within walking distance of residential areas, whereas newer auto-dependent neighborhoods tend to be lower in density with few local services and often no pedestrian facilities, so more trips require driving. Hence mobility increases but access declines.

Stephen B. Dobrow

Further Reading

Goddard, Stephen B. *Getting There: The Epic Struggle Between Road and Rail in the American Century.* New York: Basic Books, 1994.

Gonzalez, George A. *The Politics of Air Pollution: Urban Growth, Ecological Modernization, and Symbolic Inclusion.* Albany, N.Y.: State University of New York Press, 2005.

Hester, R. E., and R. M. Harrison, eds. *Transport and the Environment.* Cambridge, England: Royal Society of Chemistry, 2004.

Lay, M. G. *Ways of the World: A History of the World's Roads and of the Vehicles That Used Them.* New Brunswick, N.J.: Rutgers University Press, 1992.

Lewis, Tom. *Divided Highways: Building the Interstate Highways, Transforming American Life.* New York: Viking Press, 1997.

S

■ Severe and anomalous weather

FIELDS OF STUDY
Atmospheric Science; Climate Modeling; Earth System Modeling; Ecology; Electromagnetism; Environmental Sciences; Environmental Studies; Fluid Dynamics; Heat Transfer; Hydroclimatology; Hydrology; Hydrometeorology; Meteorology; Oceanography; Physical Geography; Spectroscopy; Thermodynamics

SUMMARY
Severe storms and meteorological events in and around the United States in the last few decades have led to greater concern about climate change and sustainable development. Highly publicized tornadoes, nor'easters and other snow and ice storms, hurricanes, and cyclones have cost billions in monetary damage and have led to significant human casualties.

PRINCIPAL TERMS
- **condensation:** the process by which water, or any other substance, changes from a vapor state to a liquid state, releasing heat into the surrounding air; this process is the opposite of evaporation, which requires the input of heat
- **El Niño:** meteorological condition in which the waters of the eastern, tropical Pacific Ocean are warmed by the atmosphere
- **Fujita scale:** scale that rates the severity of tornadoes based on the amount of destruction they cause
- **La Niña:** meteorological condition in which the waters of the eastern, tropical Pacific Ocean are cooled by a lack of radiation from the atmosphere
- **nor'easter:** severe storm in which storm fronts combine off the Atlantic seaboard, resulting in a circulating, high-precipitation storm
- **Saffir-Simpson scale:** system used to categorize the strength of a hurricane or typhoon based on wind speed
- **tropical storm:** a thunderstorm with cyclonic winds circulating at speeds of 64 to 118 kilometers per hour
- **vortex:** a mass of air, water, or other fluid that spins about a central axis, capable of reaching high velocities

A severe storm is a violent weather phenomenon that has a specific structure, often associated with heavy precipitation and air circulating in a cyclonic or anti-cyclonic manner. Wings affected by a storm are often of high velocity, a factor used to differentiate among storm stages.

Severe storms and meteorological events in and around the United States and worldwide in last few decades have led to greater concern about climate change and sustainable development. Highly publicized tornadoes, nor'easters and other snow and ice storms, hurricanes and cyclones have cost billions in monetary damage and have led to significant human casualties. Scientists, emergency professional and political leaders are working to understand these phenomena and how they affect developed areas. Such studies could help introduce systems and protocols that warn citizens and protect against weather-related disasters.

Tornadoes
An unpredictable and incredibly destructive weather phenomenon is the tornado, which is a vortex of violently circulating winds associated with severe storms. Tornadoes are formed when systems of warm and cold air intersect to form powerful storm clouds known as mesocyclones. The mix of warm and cold air, resulting in strong updrafts and downdrafts, cause the air near the surface to spin.

Scientists are learning more and more about how tornadoes form and move. Such knowledge has helped meteorologists and emergency personnel to alert at-risk residents, giving them time to take shelter before the arrival of a tornado. However, science has not reached a point at which tornadoes can be accurately predicted.

In 1971, University of Chicago researcher Tetsuya Fujita developed what would be called the Fujita scale, which assesses the strength of a tornado based predominantly on the amount of damage it causes. An EF-0 tornado (with winds ranging between 64 and 116 kilometers, or 40 and 72 miles, per hour), for example, may strip shingles from a house, knock over television antennae, and cause tree damage. Meanwhile, an EF-5 tornado, the most powerful type on record, can contain sustained winds between 400 and 500 km (about 250 to 310 mi) per hour or more, destroying homes and large structures and causing wide swaths of damage.

In recent years several severe tornadoes (categorized EF-3 or higher) have touched down in the United States. For example, on May 22, 2011, a seemingly unremarkable storm system formed over southeastern Kansas. This storm quickly intensified into a supercell (a thunderstorm that contains a mesocyclone), generating small tornadoes. A massive, wedge-shaped funnel cloud (a cone-shaped, circulating cloud emanating downward from a mesocyclone that, when it reaches the ground, becomes a tornado) dropped from the storm and headed from the southwest in a northeasterly direction—the general trajectory on which tornadoes travel—to the city of Joplin, Missouri. At its apex, the EF-5 wedge (which actually contained multiple vortices) was approximately 1 km (0.62 mi wide), cutting a swath of devastation through the southern area of the city of 49,000 people. About 160 people were killed and another 1,150 were injured by the twister, which caused approximately $3 billion in damage; almost seven thousand homes were destroyed.

The Joplin tornado was preceded by another major storm, an EF-4 tornado that leveled a large portion of Tuscaloosa, Alabama, in April 2011. Nearly three hundred people were killed, thirty-two of whom were in Tuscaloosa, where the tornado cut a swath nearly 1.5 km (1 mi) wide in some areas.

The Joplin and Tuscaloosa tornadoes were among the 1,691 confirmed tornadoes that touched down in 2011, 59 of which were deemed "killer tornadoes" by the National Oceanic and Atmospheric Administration (NOAA). Since the start of the twenty-first century, this total number has been exceeded twice only: The year 2004 had about 1,800 tornadoes in the United States and 2008 had about 1,700 tornadoes.

Hurricanes and Typhoons

Hurricanes and typhoons are some of the most destructive natural forces on Earth. Like tornadoes, these massive storms (which are both identified scientifically as cyclones) feature high winds but also include extremely large amounts of precipitation. This combination of wind and water is potentially devastating for any area in its path.

Cyclones are tropical storms, forming over the warm waters of the Atlantic and Pacific oceans near the Earth's equator. The warm, moist air in these areas rises upward, creating areas of low pressure beneath the developing system. Air from nearby systems with higher pressure rushes into these lower-pressure areas. This air is quickly warmed and sent into the growing clouds (formed as the rising, moist air cools off). This constant updraft causes the clouds to start spinning counterclockwise if the system is above the equator and clockwise if below the equator.

The spinning eventually creates an eye in the center of the storm; the eye is a region in which the sky is clear, and winds are calm. Outside the eye, winds continue to build as more air rushes into the low-pressure system through the eye and along the ocean surface. Once winds in the storm reach speeds of 63 km (39 mi) per hour, the system is classified as a tropical storm. At 120 km (74 mi) per hour, the system becomes a cyclone, or hurricane in the Atlantic and eastern Pacific oceans and typhoon in the central and western Pacific.

There are five categories in which hurricanes and typhoons are classified. According to the Saffir-Simpson scale, the weakest of these storms, category 1, has sustained winds of between 120 and 153 km (74 and 95 mi) per hour, while the strongest type, category 5, has sustained winds of more than 250 km (155 mi) per hour. At the beginning of the annual Atlantic and Pacific cyclone seasons (June 1 through November 30 and April through December, respectively), a list of names (developed by the World Meteorological Organization) is available for each storm once it reaches tropical storm status.

Hurricane Andrew, which reached Florida and Louisiana in 1992, was at its strongest a category 4 storm. Hurricane Katrina, which reached the Gulf region in 2005, was a category 5, as was Hurricane Camille in 1969. (This storm's final wind speed could not be determined because the storm had destroyed all available measuring devices.)

Because they are oceanic storms, hurricanes and typhoons will quickly weaken once they reach land. However, even a weak hurricane can cause massive damage, generating a storm surge (high water caused by a cyclone's wind and low pressure) and triggering coastal flooding. Such a storm also can produce several inches of rain in a short of time, causing coastal and inland flooding.

Using computer modeling, remote sensor technologies, and other meteorological tools, scientists work to predict the number of cyclones that will occur in a given season. Prediction is important, not only because it will help scientists, public safety officials, and private citizens take stock of emergency measures but also because it could help researchers develop technologies that could weaken such storms before they strike land. This latter endeavor has received much attention in recent years, although most scientists agree that prediction is a theoretical and not a practical field of science.

Severe Winter Storms

Some forms of severe and anomalous weather occur in the winter, producing heavy snow, ice, and high winds. Blizzards, for example, are massive winter storms in which sustained winds of 56 km (35 mi) per hour or higher are coupled with snowfall.

The more common type of condition associated with blizzards is heavy snowfall of 30 centimeters (1 foot) or more. However, ground blizzards, in contrast, do not produce as much snow; instead, they have strong winds that kick up snow that is already on the ground. Whether the snow falls from the sky or is blown from the ground, one of the most common features of a blizzard, in addition to high winds, is extremely poor visibility of 0.5 km (0.25 mi) or less. The severe winds and poor visibility characteristic of a blizzard make these storms extremely dangerous.

Some unusual winter weather patterns do not produce a large amount of snow; rather, they produce rapid changes in temperature, which cause rain that already has fallen to quickly freeze. Ice storms occur when warm, moist air typically driven from the south meets colder weather patterns. Such systems may produce freezing rain, which is rain that freezes once it contacts a colder surface. One of the best-known examples of a severe storm in the last two decades is a system that led to the accumulation of nearly 7 cm (3 in) of solid ice in northern New York and New England and parts of Canada in 1998. That storm snapped trees and telephone wires, caused power outages for millions of people, and caused more than $3 billion in damage. It also led to the death of nearly forty people.

Another type of severe weather system, the nor'easter, continues to garner study, especially because it affects more people and has a wider geographic effect, particularly in the northeastern United States, than both tornadoes and hurricanes. A nor'easter is a strong storm that involves the interplay of cold air from Canada with the warm air of the Atlantic Ocean. The two fronts create a slow-moving, counterclockwise storm that features high winds and heavy precipitation.

Nor'easters are frequently known as winter storms, although they occur year-round. One of the most famous examples of a nor'easter is the so-called perfect storm of October 1991, which caused damage from as far south as Florida and as far north as Maine. This storm was so powerful that a hurricane formed inside the larger nor'easter. This hurricane was never named, becoming one of only eight unnamed cyclones since the naming practice was introduced in the 1950s; a more pressing need was to track the major storm and the devastation it caused.

Climate Change

The high number of devastating tornadoes, nor'easters, and hurricanes since the early 1990s may be attributed to a number of atypical (although not unnatural) factors. For example, two critical phenomena, El Niño and La Niña, are well-known contributors to weather patterns.

El Niño, a warming trend in the eastern tropical Pacific, is known to contribute to the creation of storms with heavy precipitation. La Niña, in contrast, is a period marked by cooler water temperatures that brings colder, drier air along the jet stream (the band of air currents that proceeds from the west to the east) and causes periods of cooler air in the United States.

El Niño and La Niña are cyclical events caused by the interaction of the atmosphere and the surface of the ocean in the tropical Pacific. However, a growing school of scientific thought, concerned with global warming and climate change, argues that the long-time emission of greenhouse gases into the atmosphere has caused the atmosphere to increase in temperature. Such changes are theorized to foster El Niño and La Niña conditions more frequently than in previous centuries. Such shifts could lead to more severe droughts and to severe hurricanes, tornado-producing storms, flooding, and blizzards. Many scientists attribute the occurrence of high-profile and devastating storms in recent decades to this trend. Although data are not complete (tornadoes, for example, are too difficult to predict and model), researchers continue to seek direct connections between global warming and severe weather.

Michael P. Auerbach

Further Reading

Ahrens, C. Donald and Henson, R. *Meteorology Today.* 11th ed. Boston, Mass.: Cengage Learning, 2016.

Bluestein, Howard B. *Tornado Alley: Monster Storms of the Great Plains.* New York: Oxford University Press, 2006.

Burt, Christopher C. *Extreme Weather: A Guide and Record Book.* New York: Norton, 2007.

D'Aleo, Joseph S., and Pamela G. Grube. *The Oryx Resource Guide to El Niño and La Niña.* Westport, Conn.: Greenwood Press, 2002.

Dunlop, Storm. *The Weather Identification Handbook.* Lyons, 2003.

Grazulis, Thomas P. *The Tornado: Nature's Ultimate Windstorm.* Norman, Okla.: University of Oklahoma Press, 2003.

Holton, James R., and Gregory J. Hakim. *An Introduction to Dynamic Meteorology.* New York: Academic, 2013.

Mogil, H. Michael. *Extreme Weather: Understanding the Science of Hurricanes, Tornadoes, Floods, Heat Waves, Snow Storms, Global Warming, and Other Atmospheric Disturbances.* New York: Black Dog & Leventhal, 2007.

_____. *Extreme Weather: Understanding the Science of Hurricanes, Tornadoes, Floods, Heat Waves, Snow Storms, Global Warming, and Other Atmospheric Disturbances.* New York: Black Dog & Leventhal, 2007.

Mooney, Chris. *Storm World: Hurricanes, Politics, and the Battle over Global Warming.* New York Harcourt, 2007.

Repetto, Robert, and Robert Easton. "Climate Change and Damage from Extreme Weather Events." *Environment* 52, no. 2 (2010): 22-33.

Schneider, Stephen Henry, and Michael D. Mastrandrea, eds. *Encyclopedia of Climate and Weather.* New York: Oxford University Press, 2011.

■ Smog and heavy metals

FIELDS OF STUDY

Biology; Chemicals; Chemistry; Compounds; Ecology; Elements; Emissions; Environmentalism; Hazardous Materials; Medicine; Pollution; Public Health; Substances; Toxic Waste

SUMMARY

Severe smog episodes have been responsible for many deaths and widespread illness in cities around the world. Growing recognition of the detrimental health effects of smog have led many governments to pass laws designed to reduce chemical pollutants in the air. Heavy metals released into the environment by mining, industry, and disposal of wastes can poison living organisms by interfering with metabolic functions. Such poisoning can cause impaired nervous system functioning, birth defects, and death.

PRINCIPAL TERMS

- **fog:** a cloud that touches ground
- **fossil fuels:** any combustible organic material, as oil, coal, or natural gas, derived from the remains of former life
- **heavy metals:** dense metallic chemical elements
- **photochemical smog:** air pollution containing ozone and other reactive chemical compounds formed by the action of sunlight on nitrogen oxides and hydrocarbons, especially those in automobile exhaust
- **smog:** air pollution resulting from the combination of smoke with fog or from sunlight acting on unburned hydrocarbons emitted from automobiles
- **sulfurous smog:** also called "London smog," resulting from a high concentration of sulfur oxides in the air, caused using sulfur-bearing fossil fuels, particularly coal and aggravated by

dampness and a high concentration of suspended particulate matter in the air
- **temperature inversion:** a reversal of the normal behavior of temperature in the troposphere (the region of the atmosphere nearest the Earth's surface), in which a layer of cool air at the surface is overlaid by a layer of warmer air

Originally a blend of the words "smoke" and "fog," the term "smog" was coined to describe the severe air pollution that results when smoke from factories combines with fog during a temperature inversion. As one ascends upward from the earth's surface, the air temperature drops by about 3 degrees Celsius (5.5 degrees Fahrenheit) every 300 meters (1,000 feet). Temperature inversions occur when this normal condition is reversed so that a blanket of warm air is sandwiched between two cooler layers. A temperature inversion restricts the normal rise of surface air to the cooler upper layers, in effect placing a lid over a region. When the air above a city cannot rise, the air currents that carry pollutants away from their sources stagnate, causing pollution levels to increase drastically. A combination of severe air pollution, prolonged temperature inversion, and moisture-laden air may result in what has been termed "killer fog."

Killer Fogs

Several acute episodes of killer fog occurred during the twentieth century. One was in the Meuse Valley of Belgium. During the first week of December 1930, a thick fog and stagnant air from a temperature inversion-concentrated pollutant spewing forth from a variety of factories in this heavily industrialized river valley. After three days of such abnormal conditions, thousands of residents became ill with nausea, shortness of breath, and coughing. Approximately sixty people died, primarily elderly people and persons with chronic heart and lung diseases. The detrimental effects on health were later attributed to sulfur oxide gases emitted by combusting fossil fuels; the gases were concentrated to lethal levels by the abnormal weather. The presence of coal soot, combined with moisture from the fog, exacerbated the effect.

A second episode occurred in Donora, Pennsylvania, during the last week of October 1948. Donora is situated in a highly industrialized river valley south of Pittsburgh. A five-day temperature inversion with fog concentrated the gaseous effluents from steel mills with the sulfur oxides released by burning fossil fuels. Severe respiratory tract infections began to occur, especially in the elderly, and 50 percent of the population became ill. Twenty people died, a tenfold increase in the normal death rate.

A third major episode occurred in London, England, in early December 1952. At that time, many residents burned soft coal in open grates to heat their homes. When a strong temperature inversion and fog enveloped the city for five consecutive days, Londoners began complaining of respiratory ailments. By the time the inversion had lifted, four thousand excess deaths had been recorded. In this case it was not only the elderly who were affected— deaths occurred in all age categories. During the next decade London experienced two additional episodes: one in 1956, which claimed the lives of one thousand people, and one in 1962, which caused seven hundred deaths. The decline in mortality rates resulted from the restriction of the use of soft coal, with its high sulfur content, as a source of fuel. Sulfur oxide compounds are responsible for causing lung problems during such episodes; therefore, the term "killer fog" has come to be replaced by the more accurate "sulfurous smog."

Photochemical Smog

Photochemical smog, first noticed in the Los Angeles basin in the late 1940s, has been an increasingly serious problem in cities around the world. Moisture is not part of the equation in this type of air pollution, and smoke-belching factories dumping tons of sulfur oxide compounds into the atmosphere are not required. Rather, photochemical smog results when unburned hydrocarbon fuel, emitted in automobile exhaust, is acted upon by sunlight. The Los Angeles basin, hemmed in by mountains to the east and ocean to the west, has a high density of automotive traffic and plenty of sunshine. Varying driving conditions mean that gasoline is never completely consumed by automobile engines; instead, it is often changed into other highly reactive substances. Sunlight acts as an energy catalyst that changes these compounds into the variety of powerful oxidizing agents that constitute photochemical smog. This type of smog has a faint bluish-brown tint and typically contains several powerful eye irritants. The chemical reactions also produce aldehydes, a class of organic chemical best typified by an unpleasant odor.

The complicated chemistry of photochemical smog also produces ozone, which is extremely reactive; it damages plants and irritates human lungs. Because ozone production is stimulated by sunlight and high temperatures, it becomes a particularly pernicious problem during the summer, especially during morning rush hours. Under temperature inversion conditions, the ozone created in photochemical smog can increase to dangerous levels. Ozone is highly toxic. It irritates the eyes, causes chest irritation and coughing, exacerbates asthma, and damages the lungs.

Photochemical smog and ozone are now common ingredients in urban air. Although acute episodes of ozone-induced mortality are rare, concerns have grown about the detrimental long-term consequences of the brief but repetitive exposures to ozone consistently inflicted on commuters. It appears no curtailment of the problem will be possible in urban areas in the United States without significant changes in transportation systems, strict limits on growth, and radical alterations in lifestyle, including automobile use.

Heavy Metals

The chemical elements lead, mercury, cadmium, and thallium are located together at the central bottom portion of the periodic table. Although mercury, thallium, and cadmium are rare, lead is an abundant element on earth. All are dense, soft metals (mercury is a liquid at room temperature) that have a high affinity for chemically bonding with the element sulfur. They occur on or near the earth's surface as sulfur-containing minerals (sulfides) that are insoluble in water. This lack of water-solubility kept the heavy metals isolated from life-forms as they were evolving prior to the onset of the technological era, at which point humans began to mine and purify these useful elements. Once the heavy metals and their ions became more abundant in the environment, human beings—particularly mine workers and workers in certain industries—were at risk for toxic exposure to elements their bodies were not well equipped to process.

The ions of heavy metals resemble beneficial lighter metals such as zinc, calcium, magnesium, and iron in terms of their diameters and charges. This enables heavy metals to substitute for the beneficial elements and thus reside in the body over time. These lighter metals are much more abundant in the environment, and so life evolved with them, even employing them in critical roles.

The strong ability of heavy metals to bond to sulfur is significant because body biochemicals known as metalloproteins are composed of lighter metals such as zinc bonded to sulfur. (Sulfur is a normal constituent element of proteins.) In cases of heavy metal intoxication, defective metalloproteins are biosynthesized, employing an incorrect toxic metal that is strongly bound to sulfur. Although the bonding ability and even the size and charge of the heavy metal are appropriate for fitting into the structure of the biochemical, the biochemical fails to function as required, leaving the organism deficient in some vital way.

The degree of poisoning depends on the level and duration of exposure to the toxic element. Organisms have limited detoxification defenses, including proteins called metallothioneins, which scavenge metal atoms by virtue of having many sulfur atoms in their structures. Medical intervention is possible in certain cases; for example, chelating agents are drugs designed to scavenge metal atoms and make them easier to eliminate from the body in the urine.

Wendy Halpin Hallows and George R. Plitnik

Further Reading

Elsom, Derek M. *Smog Alert: Managing Urban Air Quality*. London: Earthscan Publications, 1996.

Grant, Wyn. *Autos, Smog, and Pollution Control: The Politics of Air Quality Management in California*. Brookfield, Vt.: Edward Elgar, 1995.

Hinrichs, Roger A., and Merlin Kleinbach. *Energy: Its Use and the Environment*. Belmont, Calif.: Thomson Brooks/Cole, 2006.

Jacobs, Chip, and William J. Kelly. *Smogtown: The Lung-Burning History of Pollution in Los Angeles*. Woodstock, N.Y.: Overlook Press, 2008.

Kaim, Wolfgang, and Brigitte Schwederski. *Bioinorganic Chemistry*. Hoboken, N.J.: Wiley, 2006.

Vallero, Daniel. *Fundamentals of Air Pollution*. Boston: Elsevier, 2008.

Species loss and the Sixth Extinction

FIELDS OF STUDY
Animals; Biology; Botany; Ecology; Ecosystems; Environment; Environmentalism; Forestry; Forests; Horticulture; Life Sciences; Marine Biology; Plants; Vegetation; Zoology

SUMMARY
In general, the term "extinction" refers to global extinction, the elimination of a species from the earth, but species can also be said to be extinct locally or regionally. By some estimates, before humans arrived on the scene animal and plant species disappeared from the planet at an average rate of about one per year. According to the International Union for Conservation of Nature and Natural Resources (IUCN), the modern extinction rate is anywhere from one thousand to ten thousand or more times higher than the expected natural rate.

PRINCIPAL TERMS
- **acid rain:** a broad term including any form of precipitation with acidic components, such as sulfuric or nitric acid that fall to the ground from the atmosphere in wet or dry forms, such as rain, snow, fog, hail or dust that is acidic
- **extirpation:** to remove or destroy totally; do away with; exterminate.
- **extinction:** the complete local, regional, or global die-off of a species
- **Industrial Revolution:** a process—which began in 18th-century Britain and spread to other parts of the world—of changing from an agrarian and handicraft economy to one dominated by industry and machine manufacturing.
- **Mesozoic:** an era occurring between 230 and 65 million years ago, characterized by the appearance of flowering plants and by the appearance and extinction of dinosaurs
- **minimum viable population (MVP):** ecological threshold that specifies the smallest number of individuals in a species or population capable of persisting at a specific statistical probability level for a predetermined amount of time
- **ocean acidification:** term describing significant changes to the chemistry of the ocean when carbon dioxide gas (or CO_2) is absorbed and reacts with seawater to produce acid
- **speciation:** the creation of new species

In general, the term "extinction" refers to global extinction, the elimination of a species from the earth, but species can also be said to be extinct locally or regionally. The term "extirpation" is sometimes used to describe local or regional extinction. By some estimates, historical baseline rates of extinction for animal and plant species average at about one per year. According to the International Union for Conservation of Nature and Natural Resources (IUCN), the modern extinction rate is anywhere from one thousand to ten thousand or more times higher than the estimated baseline rate.

The fossil record indicates that throughout the history of life on the earth, extinctions have resulted primarily from the planet's changing surface and climate. The rates of speciation (the creation of new species) and extinction have remained constant over millions of years, and both creation and extinction are elements of the planet's ever-evolving biological composition. However, five mass extinctions have occurred over the course of Earth's history. The most famous, if not the largest, extinction event occurred 65 million years ago at the end of the Mesozoic era, when a large number of land and marine animals, including the dinosaurs, suddenly disappeared from the fossil record. Some scientists credit this extinction to a large meteor that hit the earth and subsequently caused climate changes that devastated some life-forms. Other scientists credit the demise of dinosaurs to the evolutionary superiority of mammals. The current accelerated loss of species is sometimes called the "sixth extinction crisis." It is the only mass extinction that is attributable to a single species: human beings.

Causes of Recent Extinctions
The capacity of humans to alter their environments is not new. Through building tools and trading materials— as well as developing linguistic skills and intricate cultures— humans have increased their ability to shape the ecology of the landscapes they have occupied. There is evidence that hunter-gatherer societies around 14,000 years ago played a key role in the extinction of about half of earth's megafauna, such as wooly mammoths and the *Megatherium*, or giant

ground sloths. As human populations have continued to grow, people have cleared forests and land at increasing rates to build communities and support agriculture. This process disrupts the habitats that are home to microbes, plants, and animals, which all have unique ecological relationships to one another.

Habitat destruction is one of the most prominent causes of extinctions in recent history. With the destruction of one species, there is the potential for the secondary extinction of other species that are dependent on the first species for reproduction or food. For instance, a bird species may disperse seeds by eating the fruit of a plant and expelling the seeds in its droppings as it travels through its home range. If the bird dies off, the plant loses an important means of propagation. If a prey species becomes extinct, predators may not be able to find enough food to survive. If a major predator becomes extinct, there may be a population explosion among prey species, followed by a catastrophic depletion of food sources for those species and a subsequent die-off.

Habitat fragmentation can also lead to extinction. Highway construction and other development can divide sections of a habitat, preventing migration between the two fragments. This can cause a species to die out if neither area can support a viable population of the species. This is also true of habitats that are decreasing in size. In the United States at the close of the twentieth century, 99.8 percent of the nation's tallgrass prairies had been destroyed, 50 percent of the wetlands were gone, and 98 percent of virgin and old-growth forests had been cut. At least five hundred native species had become extinct, with tremendous losses in the populations of animals such as wolves, black bears, bison, and cougars.

Deforestation is another threat to species' survival. Tropical rain forests are home to anywhere from one-half to two-thirds of the world's plants and animals. By some calculations, tropical rain forests lose approximately four to six thousand or more species per year because of deforestation. Many of the lost species are unique to the tropical forests. This destruction has also contributed to a serious decline in the numbers of many migratory songbirds that winter in Central and South America. The songbirds' northern summer habitats are becoming fragmented as well, causing more losses in their populations. In addition, feral and domestic cats kill billions of songbirds, as well as other wildlife every year billions of birds are killed by cats every year in the United States alone. Cats have caused the extinction of at least thirty-three endemic wildlife species on islands worldwide.

Even when a species is not completely obliterated by development, extinction can occur if the population size becomes too small to recover. Such a population is said to fall into an extinction vortex. This problem may occur if there are too few females left in the population to breed, or, if the habitat is too fragmented, if individuals are not able to locate partners with which to mate. If enough mates cannot be found, genetic inbreeding can destroy the viability of the species. For instance, a 2009 study of moss carder bumblebees living on the Hebridean islands off the west coast of Scotland found that many generations of inbreeding had left the insects more susceptible to parasitic disease than their mainland counterparts. Inbreeding among these bumblebees had also resulted in an increase in infertile males.

Small populations that are vulnerable to environmental fluctuations may also fall into an extinction vortex. For example, extremely harsh winter weather can wipe out an entire species that has already been reduced to a small population. Larger groups, by contrast, are better able to survive adversity. The chance of being destroyed by environmental fluctuations increases exponentially with decreasing population size.

The smallest population of a species that is able to stay above the extinction vortex is often called the minimum viable population (MVP). If a population declines below this size, it is usually only a matter of time before breeding problems and climatic fluctuations will destroy the whole population. Likewise, if a habitat is reduced to a point where it is unable to support the MVP in an adverse year, the population will vanish.

Pollution, Overuse, and Overhunting

Pollution can kill many plants and animals and at the same time alter and destroy habitats. Acid rain and air pollution are detrimental to forests and forest animals. Sediment and excessive nutrients, such as nitrous oxide from fertilized fields, that run off into lakes, rivers, and bays often have adverse effects on aquatic life. Pesticides that persist in the environment, such as Dichloro-diphenyl-trichloroethane (DDT), have caused large losses and near extinction among some birds, notably meat- and fish-eating species such as bald eagles, ospreys, peregrine falcons, and brown pelicans. These species were particularly susceptible to DDT because the effects of chemicals are amplified as they become increasingly concentrated in the fatty tissue of organisms at higher trophic levels in the food chain.

Exotic species are animals or plants that have been introduced into an area to which they are not native. The introduction of exotic species contributed to approximately 58 percent of extinctions worldwide between 1500 and 2015 CE. Because the new exotic species may not have any natural predators or competitors in its new habitat, it can dominate the new ecosystem and reduce the populations of many native species. Islands are particularly vulnerable to invasive exotic species. The brown tree snake, a mildly venomous reptile, was introduced to the Pacific island of Guam during World War II, probably as an accidental stowaway on a military cargo ship from Papua New Guinea. By the end of the century, the snakes had decimated most of Guam's native species of birds and small mammals and reptiles. Among the island's native fauna, nine of eighteen bird species, two of three bat species, and four of ten lizard species had been extirpated, and more had become uncommon or rare. Because the snake has no natural predator on Guam, and overall lizard populations have remained high enough to provide the snakes with an ample food supply, in some areas the snake has maintained population densities of nearly thirteen thousand individuals per square mile.

Overhunting can also decrease species populations to dangerously low numbers. During the nineteenth century whales were harvested at a rate of fewer than one hundred every three years. With the development of faster boats and more efficient weapons, however, by 1933 the whaling industry was killing thirty thousand whales per year; by 1967 that number had more than doubled. It is interesting to note that 2.5 million barrels of whale oil were harvested in 1933, as opposed to only 1.5 million barrels in 1967. This is because the larger whales, the blues and fins, had been hunted to the brink of extinction by 1967. An international whaling moratorium observed by many nations, not all, since 1986 has allowed the partial recovery of some species. However, many may lack the genetic diversity to thrive in their pollution-stressed environment.

In the United States, overhunting has led to the extinction or near extinction of many species, such as the American bison. Early in the nineteenth century the passenger pigeon was one of the most abundant birds on earth. Alexander Wilson, a renowned ornithologist, is said to have observed a flock of passenger pigeons that took several hours to fly by him. He estimated that the flock, which appeared to be approximately 1.6 kilometers (1 mile) wide and 386 kilometers (240 miles) long, was composed of two billion birds. The passenger pigeon became extinct in 1914, largely because of overhunting by market hunters. The hunters used nets, guns, and even dynamite to trap the birds, which were viewed as a culinary delicacy.

Many species are under government and international protection but are still hunted because the economic incentive of selling skins, horns, or bone outweighs the risk of a small fine or short prison sentence. In the early twenty-first century, a Bengal tiger skin can fetch $35,000 on the black market, and rhinoceros horns used in traditional medicines and aphrodisiacs can sell for as much as $1,200 per kilogram. A poacher can earn millions of dollars annually.

Pest Control, Monocultures, and Wild Plants
Species can become endangered or even extinct if they must compete with the human population for food. Farmers in Africa have killed elephants to prevent the elephants from eating or trampling food crops. In the United States, the Carolina parakeet was exterminated by farmers in 1914 because it fed on fruit crops, and 98 percent of the nation's prairie dog population has been exterminated with poisons so that horses and cattle do not break their legs stepping into the rodents' burrows. In turn, the prairie dog's primary predator, the black-footed ferret, has come close to extinction with the massive reduction of the population of its food source.

Extinctions and species losses are critical problems for the human population because human life is dependent on the biodiversity of species. One area in which this is apparent is agriculture. Of the world's estimated 200,000 to 300,000 species of edible plants, humans generally eat about 200 species and farmers have developed high-yielding monocultures of only a few dozen species, each with a minimum of genetic variation. These monocultures lack the vigor of wild plants, which are constantly developing new ways to adapt to adverse conditions and fend off the animals and microorganisms that attack them. When a monoculture crop fails because of disease or other problems, plant breeders must go back to wild species to find the traits that their crop needs to thrive and breed these characteristics into the crop species. If wild species are not protected, their gene pools will dwindle, and the species will eventually disappear.

Wild plants, especially in the tropical rain forests, have important uses as medicine. In 1960 the rosy periwinkle, a shrub that grows in Madagascar, was found to contain two chemicals that revolutionized the treatment of childhood leukemia and Hodgkin's disease. The use of these drugs brought about a 95 percent remission rate in children who previously had little chance of surviving leukemia. Paclitaxel (originally known as taxol), a drug extracted from the bark of the Pacific yew tree, has been valuable in the treatment of Kaposi's sarcoma and breast, ovarian, and small-cell cancers. A plant related to the periwinkle, rauwolfia, provides an alkaloid that has been widely used in the control of high blood pressure. Digitalis, which comes from the foxglove plant, is a highly effective drug used for centuries in the treatment of various heart conditions. Of the world's estimated 400,000 plant species, only a few thousand have been studied for medicinal use. It is believed that the world's tropical forests may be home to thousands of plants that have cancer-fighting properties. At least 25 percent of the active ingredients in the anticancer drugs in use come from organisms that are native to rain forests.

Global Climate Change

Five mass extinctions have occurred periodically throughout earth's history and are frequently attributed to fluctuations in global climatic conditions. The drastic increase in extinctions in recent years has alerted scientists to what has been called a sixth mass extinction, the only one to be brought about by the actions of any single species. Human-driven global climate change combines with other human pressures that lead to extinctions, creating a powerful force with the capacity to reshape earth's ecology for millennia to come.

Since the Industrial Revolution, humans have burned enough fossil fuels and cut down enough forests to add about 550 billion metric tons of carbon to the atmosphere, and each year, another 9 billion tons is added to this sum. The atmosphere works as part of a system that functions on a planetary scale: vegetation, soils, and the ocean all work to take in gasses from the atmosphere and process them. Changing the balance of this system can lead to extinctions in several ways.

Ocean acidification is one such consequence of global climate change that creates inhospitable conditions for numerous aquatic species. Under normal conditions, the ocean takes in an amount of carbon, dissolves it, and releases a roughly equivalent amount of gas. When more carbon is present in the atmosphere, more carbon enters the ocean than can be dissolved. This results in a decrease in pH levels in the ocean, or an increase in acidity. By the beginning of the twenty-first century, pH levels of the oceans' surface had decreased from an average of 8.2 to 8.1. This amounts to a thirty percent increase in acidity. While some species do thrive in an increasingly acidic environment, many, such as clownfish or Pacific oysters, do not. Experiments have shown that many aquatic species disappear around pH 7.8, which oceans are expected to reach by the year 2100 if current trends continue.

Global increases in temperature are another outcome of climate change that can drastically alter species' chances of survival. The current rates of greenhouse gas accumulation in the atmosphere is eventually expected to lead to an average global temperature rise of 3.5 to 7 degrees Fahrenheit. Global warming can have such ecosystem-transforming effects as melting ice caps, rising sea levels, and flooding islands and coastal areas, but rising temperatures can also make tropical ecosystems uninhabitable for many species that live there in the most ecologically diverse region on earth. Scientists can only speculate about whether these species have the capabilities to survive an increasingly warmer world, as temperatures as high as today's have not existed on earth since approximately fifteen million years ago.

Changing the chemistry of the atmosphere and oceans will undoubtedly continue to influence extinction rates in this era of global climate change, and this new reality has led some humans to ask what this means for our own survival as a species. It is possible that humans will also find themselves vulnerable to inhospitable conditions, as people continue to rely on earth's biological and chemical composition for food, water, and the very air we breathe. Some scientists argue that by disrupting earth's systems, we are jeopardizing our own future as a species. Others suggest that humans are capable of immense creativity, and while it may be human actions that has brought about this crisis, it will be human innovation that solves it.

*Toby Stewart and Dion Stewart;
updated by Karen N. Kähler and Julia Kendrick*

Further Reading

Barrow, Mark V., Jr. *Nature's Ghosts: Confronting Extinction from the Age of Jefferson to the Age of Ecology.* Chicago: University of Chicago Press, 2009.

Bellard, Céline, et al. "Biology of Extinction: Alien Species as a Driver of Recent Extinctions." *Biology Letter*, vol. 12, 2016, doi: 10.1098/rsbl.2015.0623. Accessed 26 Mar. 2018.

Ehrlich, Paul R., and Anne H. Ehrlich. *Extinction: The Causes and Consequences of the Disappearance of Species.* New York: Ballantine Books, 1981.

Goudie, Andrew. *The Human Impact on the Natural Environment: Past, Present, and Future.* 6th ed. Malden, Mass.: Blackwell, 2006.

Kaufman, Les, and Kenneth Mallory, eds. *The Last Extinction.* 2d ed. Cambridge, Mass.: The MIT Press, 1993.

Kolbert, Elizabeth. *The Sixth Extinction: An Unnatural History.* London: Bloomsbury, 2014.

McGavin, George. *Endangered: Wildlife on the Brink of Extinction.* Buffalo, N.Y.: Firefly Books, 2006.

Novacek, Michael J. *Terra: Our 100-Million-Year-Old Ecosystem—and the Threats That Now Put It at Risk.* New York: Farrar, Straus and Giroux, 2007.

Raup, David M. *Extinction: Bad Genes or Bad Luck?* New York: W. W. Norton, 1991.

Ward, Peter D. *Under a Green Sky: Global Warming, the Mass Extinctions of the Past, and What They Mean for Our Future.* New York: Smithsonian Books/Collins, 2007.

Warren, John. *The Nature of Crops: How We Came to Eat the Plants We Do.* CABI, 2015.

Wilson, Edward O. *The Diversity of Life.* New ed. New York: W. W. Norton, 1999.

Tidal energy

FIELDS OF STUDY
Biology; Conservation Biology; Ecosystems; Ecology; Energy and Energy Resources; Environment; Environmentalism; Fisheries; Life Sciences; Limnology; Oceanography; Public Policy; Renewable Resources

SUMMARY
The energy generated during the rise and fall of the tides may be cleanly and safely converted into electrical power, but large-scale tidal power installations can have severe consequences for the environment, including decimation of fisheries; destruction of the feeding grounds of migrating birds; damage to shellfish populations; interference with ship travel, port facilities, and recreational boating; and disruption of the tidal cycle over a wide area.

PRINCIPAL TERMS
- **tidal energy:** power generated during the rise and fall of the tides
- **tides:** he periodic rise and fall of the waters of the ocean and its inlets, produced by the attraction of the moon and sun, and occurring about every 12 hours
- **megawatts:** a unit of power; one megawatt is a million watts

Tidal power can be an important source of local electricity generation because such projects produce energy that is free, clean, and renewable; they produce neither air pollution nor thermal pollution, and they do not consume exhaustible natural resources. Only a limited number of places in the world offer the potential for such power installations, however, because a vertical tidal rise of 5 meters (16.4 feet) or more is required. Installations must also be near major population centers so that transmission requirements are minimized, and a natural bay or river estuary is typically required to store a large amount of water with a minimum of expense for dam construction. The seawater impounded behind the dam at high tide produces a hydrostatic head so that electricity is generated as the water passes through the dam's turbines when sea level falls. If the turbines in the dam are reversible, power can be generated on both incoming and outgoing tides.

Tidal power plants have been constructed on the Rance River near St. Malo, France (240 megawatts of power), on the Annapolis River in Nova Scotia, Canada (20 megawatts), on the Yalu River in the People's Republic of China (3.2 megawatts), in Kislaya Guba, Russia (1.7 megawatts), and on Strangford Lough in Northern Ireland (1.2 megawatts). In the first decades of the twenty-first century, South Korea constructed a plant on Sihwa Lake (254 megawatts) and made plans to build several more around the country. The Rance River plant, which has been in continuous operation since November 1966, was for many years the world's largest tidal power installation. It bridges the estuary with a dam nearly 0.8 kilometer (0.5 mile) long and provides power for 300,000 people. In the early 2010s, plans were drawn up for a new lagoon-style installation to be built offshore in Wales's Swansea Bay.

The environmental impacts of the Rance River dam have generally been limited to the modification of fish species distributions, the disappearance of some sandbanks, and the creation of high-speed currents near the sluices and the powerhouse. Tidal patterns have also changed, with the maximum average rise reduced from about 13.4 meters to 12.8 meters (44 feet to 42 feet) and a corresponding increase in the height of the mean low-tide level.

The environmental impacts of the smaller Annapolis River plant in Nova Scotia, which became operational in 1984, reportedly have included the generation of silt, which destroyed clam beds in the basin behind the dam, and increased erosion of the river's banks. The Nova Scotia Power Corporation reached a settlement with one nearby landowner whose house suffered a cracked foundation and shifted toward the river as a result of erosion.

Several tidal power projects were proposed for the United States during the early and mid-twentieth century but were never built because of environmental concerns. A proposed tidal power plant on the upper Saint John River in Maine was halted, for example, because damming the river would have destroyed a unique stand of a rare wildflower. The flower was later found growing elsewhere. Objections cited for other projects included possible effects on historic and archaeological sites, as well as presumed economic and social impacts on Native American communities such as the Passamaquoddy.

Shortly after the dramatic jump in world oil prices during the 1970s, the Tidal Power Corporation, a venture owned by the Nova Scotia government, proposed building a major tidal power project in the Bay of Fundy, which lies between Nova Scotia and New Brunswick in eastern Canada. This plant would have been the world's largest tidal power installation, producing 4,560 megawatts of power—nearly twenty times the output of the Rance River plant and more than three times the output of Hoover Dam on the Colorado River in the United States. A major feature of the project was to be a dam 8.5 kilometers (5.3 miles) long across the Bay of Fundy, which has the largest tidal range in the world, averaging more than 15 meters (50 feet). The enormous scope of the project forced scientists to pay close attention to its anticipated environmental consequences, and these appeared to be so severe that the project was never begun.

Disrupted bird migrations were predicted after the dam's completion because of the submersion of tidal mudflats where large numbers of semipalmated sandpipers and other shorebirds annually gorge on mud shrimp before beginning their fall migrations to wintering grounds in South America and the Caribbean. Damage to fish stocks was also predicted because of repeated passage of the fish through the dam's turbines as the tides rose and fell. Particularly affected would have been the American shad, a member of the herring family, which migrates to the Bay of Fundy each year from as far away as Florida in order to fatten itself on mysid shrimp living on the tidal mudflats. Oceanographers also used computer modeling to show that dam construction would alter tidal patterns over a broad area, resulting in tidal levels 10 percent higher and lower as far south as Cape Cod, Massachusetts, 400 kilometers (250 miles) away. They predicted that these tidal changes would flood coastal lands and threaten roads, bridges, waterfront homes, water wells, sewage systems, salt marsh areas, harbors, and docking areas along the entire coast.

Donald W. Lovejoy

Further Reading

Charlier, R. H., and C. W. Finkl. *Ocean Energy: Tide and Tidal Power.* London: Springer, 2009.

Cruz, João, ed. *Ocean Wave Energy: Current Status and Future Perspectives.* New York: Springer, 2008.

Fairley, Peter. "Tidal Power Makes a Surprising Comeback." *IEEE Spectrum,* June 19, 2013.

Hardisty, Jack. *The Analysis of Tidal Stream Power.* New York: John Wiley & Sons, 2009.

McKinney, Michael L., Robert M. Schoch, and Logan Yonavjak. *Environmental Science: Systems and Solutions.* 4th ed. Sudbury, Mass.: Jones and Bartlett, 2007.

Peppas, Lynne. *Ocean, Tidal, and Wave Energy: Power from the Sea.* New York: Crabtree, 2008.

■ Tragedy of the commons

FIELDS OF STUDY

Advocacy; American and European History; Commerce; Ecology; Economics; Environment; Environmentalism; Ethics; Evolutionary Biology; Industrial Revolution; Philosophy and History of Science; Public Policy

SUMMARY

Environmentalists, conservationists, and others concerned with the depletion of the world's shared natural resources have developed various approaches to averting the situation known as the tragedy of the commons.

PRINCIPAL TERMS

- **anti-environmentalism**: philosophy that human beings' immediate economic and lifestyle needs are more important than concerns about the

- **fates of other specifies and the general environment**
- **commons:** a resource to which all persons have open, free, and unrestrained access
- **evolutionary biology:** branches of biology that deal with the processes of change in populations of organisms, especially taxonomy, paleontology, ethology, population genetics, and ecology
- **privatization:** to change from public to private control or ownership
- **tragedy of the commons:** individuals acting independently and solely in their own self-interest, collide with the interests and needs of the larger community, resulting in the depletion of resources against the long-term interests of both individuals and the group

Although the concept of the tragedy of the commons can be traced to Aristotle, the modern application of the concept is closely associated with ecologist Garrett Hardin, who published an article about it in 1968. The commons can be defined as any resource to which all persons have open, free, and unrestrained access. Examples of such shared resources include the atmosphere, rain forests, outer space, oceans, fisheries, and public land. The tragedy of the commons is a situation in which an individual makes a rational decision concerning a shared resource based on his or her own best interest. For example, if farmers share a plot of land where they graze their cattle, individual farmers could increase their own individual profits by grazing additional animals. They would gain the sole benefit of each additional animal in their herds and only bear a fraction of the cost. When all the farmers sharing the land choose to increase their individual profits in this way, however, they harm the common resource by overgrazing the land and compacting the soil.

Although the tragedy of the commons is a rational choice for individuals, it can be overcome. Sometimes averting the tragedy can be as simple as educating individuals about their behavior and appealing to their consciences. Some commentators place faith in technology to overcome the tragedy, asserting that scientific advancements can repair or counterbalance any potential harm to the commons. Another way to prevent people from overusing common resources is to create economic incentives: When individuals must pay taxes or fees to use a common resource, they may be inclined to reduce their use of that resource. For some commons, such as the seas, limits can be placed on the numbers or amounts of given resources, such as fish, that can be harvested within a given period. The use of taxes, fees, and quotas to avert the tragedy of the commons requires that someone or something, often a government, exists that will monitor the compliance of individuals.

Another way to prevent the tragedy of the commons from occurring is to privatize the resource, so that it is no longer a commons. If a farmer owns the land on which he grazes cattle, he still must make choices about the number of cattle that should be grazed in areas. If the number of cattle is increased beyond a certain point, the carrying capacity of the land, then the land will be harmed. Each additional animal represents a benefit to the farmer, but because the farmer owns the land, he bears the entire cost as well. Rational individuals will not compromise their own private resources. Though initially an ecological concept, the tragedy of the commons has also been applied to evolutionary biology and the Internet, specifically to what is termed "knowledge commons," a space where a group of people create and control information.

Kathryn A. Cochran

Further Reading

Easton, Thomas A., ed. *Environmental Studies*. 4th ed. New York: McGraw, 2012.

Hanley, Nick, and Colin J. Roberts, eds. *Issues in Environmental Economics*. Malden: Blackwell, 2002.

Hardin, Garrett. "The Tragedy of the Commons." *Science* 162 (1968): 1243–48.

Ohler, Adrienne M., and Sherrilyn M. Billger. "Does Environmental Concern Change the Tragedy of the Commons? Factors Affecting Energy Saving Behaviors and Electricity Usage." *Ecological Economics* (2014): 1. *Academic OneFile*. Web. 29 Jan. 2015.

Stavins, Robert N., ed. *Economics of the Environment: Selected Readings*. New York: Norton, 2000.

U

Urban rain gardens and greenbelts

FIELDS OF STUDY
Biology; Botany; Ecology; Ecosystems; Environment; Environmentalism; Gardening; Horticulture; Hydrology; Life Sciences; Reserves; Urban Planning; Water Resources; Zoology

SUMMARY
Rain gardens provide various environmental benefits, including the filtering and neutralization of water pollutants, the reduction of stormwater flooding, and the creation of small islands of natural habitat in urban areas. Greenbelts provide numerous environmental, social and economic benefits to the areas that surround them.

PRINCIPAL TERMS
- **aquifer:** any geological formation containing or conducting ground water
- **compost:** a mixture of various decaying organic substances used for fertilizing soil
- **greenbelts:** tracts of open space preserved to control urban growth patterns
- **habitat:** place that is natural for the life and growth of an organism
- **rain gardens:** garden areas designed to capture, slow, and filter rainwater runoff
- **topsoil:** the fertile, upper part of the soil
- **urbanization:** an increase in a population in cities and towns versus rural areas

The basic idea behind rain gardens is the replication, on a small scale, of the natural conditions that existed before urbanization. In forests and on farmland, rainwater soaks into the soil and percolates slowly through it. Water that is not recycled by immediate evaporation or used by plants eventually works its way through the soil to end up replenishing underground aquifers. The average suburban lawn is relatively impervious to water infiltration by this process, because the grass has shallow roots and many soils, especially clay, are not porous enough to drain well. A well-sited rain garden planted with native shrubs, perennials, and other hardy plants will allow at least 30 percent more water from rainstorms or snowmelt to seep into the ground than a lawn does.

Although the basic idea is surprisingly simple, rain gardens to counter environmental damage were not widely recognized as a workable concept until the early 1990's. Larry Coffman, the environmental official in charge of creating a plan for handling stormwater in Prince George's County, Maryland, coined the term. The use of rain gardens in some large-scale projects, such as one in Maplewood, Minnesota, in 1996, helped to popularize the practice. Maplewood officials, led by landscape architect Joan I. Nassauer, sponsored the building of rainwater-gathering garden strips along the edges of suburban streets.

Most rain-garden projects in the United States have been small in scale, created by individual home owners or by communities around schools and parks with drainage problems, rather than large initiatives by developers or municipalities. The rain garden's acceptance as a tool to combat water pollution has been rapid, however, and relatively conflict-free. European countries, with their traditions of centralized planning, have been more innovative in using rain gardens than have North American nations.

Environmental Benefits
As cities grow, their land area becomes almost entirely covered with the hard surfaces of buildings, streets, and sidewalk pavements. In order to move rainwater quickly out of the way, most modern cities have built gutters and storm sewers that ultimately drain rainfall into rivers and lakes. In the journey from the city, the water picks up many toxic

substances from the urban landscape through which it flows. Oil products, organic wastes, pesticides, and other residues of industrial processes are all carried along in storm runoff. The U.S. Environmental Protection Agency estimates that 70 percent of all water pollution is the result of rainwater runoff. Such water is seldom even treated before it is discharged into lakes and rivers, and any attempts at treatment are likely to be prohibitively expensive.

When stormwater runoff is received by rain gardens, most of the pollutants are filtered out. The soil is enriched by the movement of nitrogen, phosphorus, and other compounds through the plants' root systems and soil processes. As an added benefit, rain gardens' comparatively slow absorption rate helps reduce downstream erosion from the rapid runoff that can come with heavy storms. The existence of garden spaces in urban areas also helps in a small way to combat the heat island effect, which results when heat is absorbed by and then radiates off the unrelieved hard surfaces of buildings and pavements.

So long as those who create rain gardens follow a few basic rules—such as not putting a rain garden so close to a house that its foundations are undermined by excess water—there is virtually no downside to this conservation concept. Rain gardens bring a touch of nature to urban and suburban residents while helping to purify water resources.

Practical Considerations
Rain gardens do not necessarily look different from purely ornamental gardens, but the creation of rain gardens requires special attention to location and soil preparation. Because water should flow into a rain garden from impermeable surfaces such as roofs and driveways as well as from lawn areas subject to flooding, the rain garden needs to be somewhat lower than the surrounding ground. Occasionally a natural depression can be used, but normally some digging—to a depth of about 20 to 25 centimeters (8 or 10 inches)—is necessary. The digging also enables the replacement of the original soil with "rain-garden soil," an optimal mix for root establishment and permeability. A mixture of half sand and 20 to 30 percent each of topsoil and compost is usually suggested.

A downspout with a drainpipe or shallow troughs usually needs to be installed so that water is directed from roof or driveway to the rain garden. At the upper or higher end of the garden, a border of grass can keep water from entering the garden too fast on stormy days. A berm or even a low wall can keep it from overflowing at the downslope border. With all this, the object is not to create a pond or small swamp. An effective rain garden will drain rain within forty-eight hours after a heavy rainfall.

Effective rain gardens contain a variety of plants that can thrive under both wet and dry conditions. Enough perennials and ornamental foliage fit this description that rain gardens have the potential to be pleasing landscape design elements. Once they are well established, rain gardens often become places where birds and other wildlife shelter, providing additional touches of nature in the city.

Greenbelts
Throughout the twentieth century and into the twenty-first, developed nations have urbanized at ever-increasing rates. As a result of the automobile, the United States and Canada have been subjected to uncontrolled growth, resulting in the phenomenon referred to as urban sprawl. As development moves outward from a central city, prime agricultural and forested lands are converted to more intensive uses, resulting in a significant loss of wildlife and plant habitats. This decrease in natural areas also leads to a subsequent degradation of air and water quality.

The concept of creating greenbelts, or greenways, developed as a grassroots response to address these problems. With limited public funds for open-space preservation, greenbelt proponents have focused attention on "leftover" or abandoned lands. These parcels are often found along ridgelines and streams, areas that are too steep or too wet for development. Abandoned railroad and utility rights-of-way have become important potential resources as well. All of these areas have common physical characteristics: They are long, thin tracts of land that relate to the topography, threading through land more suitable for development.

Greenbelts, as linear open-space corridors, provide several important benefits. First, they enable urban areas to retain their biodiversity. This is important for maintaining plant and animal habitats, as well as for establishing sources of protection for air and water quality. The natural corridors provide migration routes for species interchange. This movement of plant and wildlife along natural pathways is particularly significant, since it may determine the ability of some species to survive in these areas. Second, the retention of undeveloped, vegetated lands allows surface water to be returned naturally to the water table, minimizing surface runoff, erosion, and subsequent stream sedimentation.

Greenbelts offer many recreational opportunities as well. Most greenbelts include systems of trails that may link larger, more intensive recreational facilities or provide people with access to natural amenities from urban areas. By connecting different sorts of facilities, creating a system or network of urban parks, greenbelts increase the aggregate benefit to the community. Because of the linear nature of greenbelts, they have more edge area than do other kinds of parks or open spaces. This characteristic maximizes the available open space and provides potential access to greater numbers of people.

The economic benefits of greenbelts are also significant. As leftover or derelict lands, suitable parcels may be purchased relatively inexpensively; thus, minimum expenditure is often required for the development of a greenbelt system. In addition, the aesthetic improvement of the green edge provided by a greenbelt often enhances the value of adjacent properties.

Emily Alward and Steven B. McBride

Further Reading

Amati, Marco, ed. *Urban Green Belts in the Twenty-first Century.* Burlington, Vt.: Ashgate, 2008.

Dunnett, Nigel, and Andy Clayden. *Rain Gardens: Managing Water Sustainably in the Garden and Designed Landscape.* Portland, Oreg.: Timber Press, 2007.

Hough, Michael. *Cities and Natural Process: A Basis for Sustainability.* 2d ed. New York: Routledge, 2004.

Kinkade-Levario, Heather. *Design for Water: Rainwater Harvesting, Stormwater Catchment, and Alternate Water Re-Use.* Gabriola Island, B.C.: New Society, 2007.

Woelfle-Erskine, Cleo, Laura Allen, and July Oskar Cole, eds. *Dam Nation: Dispatches from the Water Underground.* New York: Soft Skull Press, 2007.

■ Urban sprawl and urban ecology

FIELDS OF STUDY
Biology; Ecology; Ecosystems; Environment; Environmentalism; Life Sciences; Peoples, Societies; Urban Planning; Urbanization

SUMMARY
When urban settlements become overcrowded, some individuals, businesses, and industries migrate to the fringes of the urban area, where population concentrations are less dense but urban services remain available. The concentration and eventual diffusion of urbanization affect existing land uses, the physical environment, the survival of plant and animal species, and the aesthetics of the landscape. Since the beginnings of the modern environmental movement in the 1970s, the science of urban ecology has gradually become accepted as an indispensable part of the urban planning process.

PRINCIPAL TERMS
- **biodiversity:** the variety of life found in a place on Earth or the total variety of life on Earth
- **habitat:** the natural environment of an organism; place that is natural for the life and growth of an organism
- **urban ecology:** subfield within the science of ecology that is concerned with the interactions among human beings, plants, and animals in urban and metropolitan areas
- **urban sprawl:** the uncontrolled and unregulated development that occurs outside the administrative boundaries of the zoning and land-use authority of municipalities and outside the conscious and deliberate direction of those authorities
- **urbanization:** an increase in a population in cities and towns versus rural areas
- **zoning:** of or relating to the division of an area so as to restrict the number and types of buildings and their uses

Urban growth and development are usually controlled by two forces. First, municipal government authorities regulate growth through urban planning, zoning, and a variety of land-use ordinances. Second,

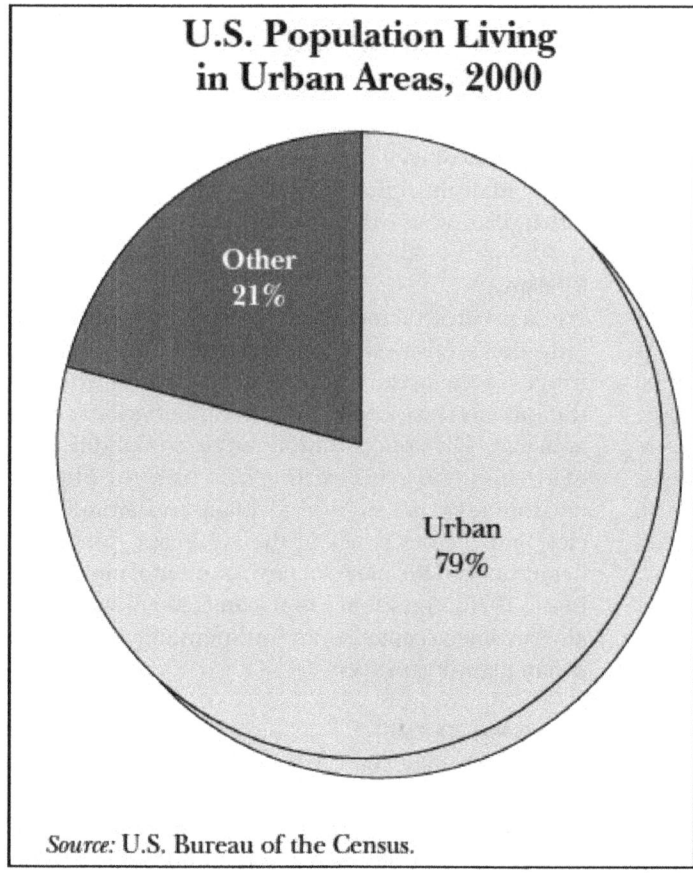

Source: U.S. Bureau of the Census.

social and economic forces combine to encourage outward growth of the urban area in one of three consistent patterns, described as concentric circle, sector, and multinuclear patterns. Urban sprawl, which is neither controlled nor regulated, often involves some break in this historic pattern. The unplanned nature of urban sprawl often results in a mix of incompatible land uses placed adjacent to one another on the same developed physical area.

The challenges presented by urbanization and urban sprawl are likely to intensify during the decades to come. In mid-2009 the world population officially became more urbanized than rural—of the earth's 6.83 billion people, 50.1 percent were found to be living in urban areas; 75 percent of the inhabitants of the more developed regions lived in urban areas, while such areas accounted for only 45 percent of the people living in less developed regions. According to United Nations projections for the year 2050, the world's population growth will continue to be concentrated within urban areas; of that increase, most is expected to be within the cities and towns of developing nations. The United Nations estimated in 2014 that by 2050 the number of people inhabiting urban areas would be roughly 6.3 billion, constituting 66 percent of the global population.

Problems Caused by Sprawl

Sprawl results in discontinuous leapfrog or checkerboard patterns of development and strip development along transportation corridors, with skipped areas remaining undeveloped. This creates inefficiencies in providing urban services to the sprawl area. Sprawl also results in less-than-maximum utilization of existing developed land in the urban center. Instead of converting existing developed land to new uses, developers establish new developments on less expensive land in suburban areas adjacent to the existing urban area.

Urban sprawl generally increases the total amount of land affected by human activity, the amount of natural areas converted to recreational uses, the amount of wasteland, and the residential, industrial, institutional, and infrastructure uses of land. Wasteland is land disturbed by humans to such an extent that natural uses cannot be restored and future development on the land is restricted. Wastelands include soil borrow pits (areas where soil has been dug out to be taken to other locations), quarries, debris landfills, and construction material storage sites. Urban sprawl generally decreases agricultural uses and timber uses of land and decreases the amount of natural barren or rocky lands, river and stream floodplain areas, and native timber- and grasslands.

Sprawl is encouraged by a variety of social and economic forces. First, it is often a consequence of increasing heterogeneity of the urban population and is a means to limit or escape from interactions between dissimilar social, economic, and ethnic groups. Second, sprawl allows individuals and businesses to escape from the negative effects caused by concentrations of undesirable land uses—such as industrial, commercial, and low-income residential zones—and relocate to suburban areas with more pristine and affordable real estate. Third, sprawl is encouraged by landowners in suburban areas seeking to maximize profits from investments in land. Finally, sprawl permits individuals and businesses to escape taxes and regulations imposed in core urban areas.

Some researchers contend that the sprawl problem is overstated. In the United States, for example, urban land in 2007 represented less than 3 percent of the nation's total land area and housed more than three-quarters of the population. By contrast, about 5 percent of the nation's land is protected from development under the 1964 Wilderness Act.

Environmental Impacts

The first significant impact of urban sprawl on the environment is the abnormal greening of the physical area. In most cases, total greening is reduced through destruction of forests, grasslands, and floodplains to make way for development. This decreases the number and variety of species able to occupy the space and creates microclimate effects, such as regional warming. However, in some cases total greening is increased through irrigation and the introduction of cultivated lawns, orchards, and other plantings in areas that are naturally arid or barren. This results in the introduction of new species into the environment, increased pollen counts, and a variety of microclimate effects, such as increases in regional humidity. In either case, the pre-sprawl natural ecology is dramatically changed, with resulting negative impacts on plant and animal species displaced by or unable to adapt to the new environment.

In the course of sprawl development, existing natural ecosystems are destroyed while pre-sprawl plants and animals are killed, displaced, or replaced. Most pre-sprawl wildlife retreats in the face of development. However, some species can adapt to the sprawl environment and find the mix of land uses, the dispersion of human activities, and the residue of human activity conducive to their survival. Among the wildlife that benefits are scavengers, such as pigeons, rats, raccoons, and opossums; vermin hunters, such as falcons, foxes, and coyotes; and foragers, such as deer, squirrels, and rabbits. Benefiting plant species include all those that homestead on disturbed

The front of the Riverside Park branch of the Urban Ecology Center. By Mark F. Heffron

soil or wasteland or that thrive in the open sunlight of most new suburban developments.

Sprawl reduces the amount of productive agricultural land and economically important forests. Productive agricultural land and forests are lost either by development of the land for residential, industrial, or institutional uses or by the transformation of productive farms and forests into hobby farms and parklands in an effort to preserve green space without preserving the productive purpose of the agricultural lands or forestlands.

Finally, sprawl increases the total amount of land surface that is impervious to rainfall penetration. Roadways, parking lots, slab building foundations, and other paved areas increase rainwater runoff and the possibility of flooding, soil erosion, and ecosystem destruction.

Impacts on Air and Water Resources

Other consequences of urban sprawl include increases in noise, light pollution, land area devoted to highways and roads, and public-utility impacts as land is cleared for underground water, sewer, and utility pipes and for aboveground utility cables.

The impact of sprawl on air and water resources is both negative and positive. The increase in human population that accompanies sprawl increases the concentration of significant amounts of unnatural substances in the soil, water, and air and also produces abnormally high concentrations of natural

305

substances at levels that may cause undesirable health effects, corrosion, and ecological change. However, studies also indicate that the population dispersal associated with sprawl actually reduces air pollution by dispersing both the mobile and stationary sources of pollutants. Increases in air pollution from automobiles associated with sprawl may be less than the air pollution produced by traffic gridlock, mass-transit buses, and trains in denser urban areas.

Subsurface water supplies and surface watercourses are less affected by sprawl than by denser patterns of development. Denser urban development increases the demands on water resources, runoff and the possibility of flooding, and the likelihood that watercourses will be channeled and hardened by concrete and other construction materials. Denser urban development, and the increase in paved surfaces that comes with it, also make it more difficult for subsurface water to replenish itself.

The Theory of Urban Ecology

In the 1920s sociologists Robert E. Park and Ernest Watson Burgess developed a theory of urban ecology that postulates that cities are environments similar to those found in nature, regulated by principles analogous to those that govern nature and natural evolution. This theory views the overall structure of cities as the consequence of the struggle for limited urban land, which influences cities to evolve gradually into five concentric rings. The term "urban ecology" also refers generally to that part of the science of ecology that studies the interactions among human beings, plants, and animals within urban and metropolitan areas, as well as the effects that urban development has on natural ecosystems and on biodiversity in these areas.

During most of the twentieth century, as cities sprawled outside their limits and within them, the environment was not a central theme in urban planning, and most planners did not develop urban biodiversity strategies. Since the 1970s, however, protection of the environment has gradually become a key component of urban planning.

Urban green spaces and green corridors are important elements in the urban planning paradigm, known as the new urbanism, that developed in part as a reaction against modern urbanism. Urban ecology also gradually became part of the core disciplines that constitute the multidisciplinary field of urban planning.

Importance of Biodiversity

High levels of biodiversity (genetic, species, and habitat diversity) are important not only for plant and animal species but also for the quality of life of human beings. The higher the level of biodiversity in an ecosystem, the better able the ecosystem is to face instability, whether of natural or anthropogenic (human-caused) origin.

Regarding biodiversity, human-nature interactions in urban and metropolitan areas face two challenges. First, the rapid and increasing pressure that will result from foreseeable demographic growth is likely to have substantial negative impacts on biodiversity. In 2013 the United Nations estimated that in 2050 the world population will reach 9.6 billion, most living in urban areas; this trend will substantially increase the human pressure on natural ecosystems. Second, global climate change, although still uncertain in its extent, requires adaptation and mitigation measures in urban and metropolitan areas that will increase the resilience of urban ecosystems. On the other side, changes in land cover affect the stocks of carbon, and through that effect these changes are also expected to have impacts on climate. For this reason, changes in land cover owing to anthropogenic factors constitute another facet of urban ecology that needs to become a central issue in urban planning.

In addition to conducting biodiversity inventories (such as inventories of native plants in risk of extinction in specific urban areas), urban ecologists examine the impacts of urbanization on native species of plants and animals (both vertebrates and invertebrates) and on wildlife in general (including reptiles, aquatic vegetation, and native fish). They also evaluate the effects of streets, roads, and other urban infrastructures on plant and animal species.

The study of the urban ecologist ends, in most cases, in policy recommendations in the form of an urban biodiversity strategy, as part of a wider urban planning process. These policy recommendations address different types of actions, both direct and indirect. An urban biodiversity strategy may suggest, for example, that a city or metropolitan area create habitats or restore natural habitats within the area, clean waste from green spaces and green corridors, connect fragmented and isolated pieces of urban green spaces so that green corridors and networks are created, develop wildlife habitats in the backyard gardens of private homes, plant native species as part of the regeneration of brownfields, collect native species in places that are going to be developed and transplant them to other parts of the area, improve the water quality in rivers, or reduce soil erosion in particular areas. Indirect actions recommended might include the establishment of environmental education programs designed to change social values related to natural ecosystems and to increase public awareness about the environment.

Gordon Neal Diem and Carlos Nunes Silva; updated by Karen N. Kähler

Further Reading

"Growth in Urban Population Outpaces Rest of Nation, Census Bureau Reports." *United States Census Bureau.* United States Census Bureau, 26 Mar. 2012. Web. 3 Feb. 2015.

"Measuring Sprawl 2014." *Smart Growth America.* Smart Growth America, 2014. Web. 3 Feb. 2015.

Frumkin, Howard, Lawrence D. Frank, and Richard Jackson. *Urban Sprawl and Public Health: Designing, Planning, and Building for Healthy Communities.* Washington, D.C.: Island, 2004.

Gillham, Oliver. *The Limitless City: A Primer on the Urban Sprawl Debate.* Washington, D.C.: Island, 2002.

Goudie, Andrew. *The Human Impact on the Natural Environment: Past, Present, and Future.* Malden, Mass.: Blackwell, 2006.

Marzluff, John M., et al., eds. *Urban Ecology: An International Perspective on the Interaction Between Humans and Nature.* New York: Springer, 2008.

McDonnell, Mark J., Amy K. Hahs, and Jürgen Breuste, eds. *Ecology of Cities and Towns: A Comparative Approach.* New York: Cambridge University Press, 2009.

Miller, Debra A., ed. *Urban Sprawl.* Detroit, Ill.: Greenhaven, 2008. Print.

Pugh, Cedric, ed. *Sustainability, the Environment, and Urbanization.* 1996. Reprint. Sterling, Va.: Earthscan, 2002.

Raven, Peter H., and Linda R. Berg. *Environment.* New York: John Wiley & Sons, 2006.

Selman, Paul. *Local Sustainability: Managing and Planning Ecologically Sound Places.* London: Paul Chapman, 1996.

Soule, David C., ed. *Urban Sprawl: A Comprehensive Reference Guide.* Westport, Conn: Greenwood Press, 2006.

Squires, Gregory D., ed. *Urban Sprawl: Causes, Consequences, and Policy Responses.* Washington, D.C.: Urban Institute, 2002.

W

■ Water pollution

FIELDS OF STUDY
Ecology; Environment; Environmental Microbiology; Hydrology; Population; Public Heath; Sanitation; Waste Management; Water Pollution Control

SUMMARY
In many parts of the world human beings lack access to safe drinking water. Millions of people die each year because of a lack of drinking water or because their water supplies are contaminated owing to unsanitary conditions. And while many must rely on bottled water, in the United States, landfills are overwhelmed with plastic water bottles.

PRINCIPAL TERMS
- **contaminants:** unwanted constituents, harmful substances or impurities in a material, physical body, natural environment, workplace
- **arsenic:** a grayish-white element having a metallic luster, vaporizing when heated, and forming poisonous compounds
- **radon:** a chemically inert, radioactive gaseous element produced by the decay of radium, whose: emissions produced by outgassing of rock or brick and other substances are a health hazard
- **World Health Organization (WHO):** a specialized agency of the United Nations, established in 1948, concerned with international public health
- **United Nations Children's Fund (UNICEF):** an agency of the UN, established in 1946, responsible for programs to aid education and the health of children and mothers in developing countries
- **chlorination:** the process of adding highly toxic chlorine or chlorine compounds to water in order to kill certain bacteria and other microbes to prevent the spread of waterborne diseases

In many parts of the world human beings lack access to safe drinking water. Millions of people die each year because of a lack of drinking water or because their water supplies are contaminated owing to unsanitary conditions. Even in developed countries where water supplies are enough, drinking water may contain dangerous contaminants, including such toxic chemicals as lead and arsenic, harmful microorganisms, and even radioactive compounds such as radon. In 2010, the United Nations declared clean water a universal human right, but as of 2017, according to a report by the World Health Organization and United Nations Children's Fund, 2.1 billion people still lacked access to safe drinking water at home and 4.5 billion did not have safely managed water sanitation services.

In June 2018, the United Nations released a report emphasizing the dire state of the world's safe water supply. The agency identified several elements that have continued to thwart the goal of ensuring access to safe water and sanitation for all people in the world by 2030. According to the report, urbanization, pollution, climate change, and population growth are just some of the factors causing the limitation of access to safe water due to environmental degradation and strain on natural resources such as water. The United Nations subsequently estimated that, based on trends at that point, between 4.8 and 5.7 billion people could be living in areas that are water-scarce for at least one month per year by 2050.

Quality, Accessibility, and Regulation
Most problems with water quality involve contamination with disease-causing (pathogenic) microorganisms. Safe, clean drinking water is free of all pathogenic microbes. Some dangerous microorganisms

can be transmitted by water, including viruses (for example, hepatitis A), bacteria (*Salmonella* and *Vibrio cholera*), protozoa (*Giardia*, *Cryptosporidium*, and *Entamoeba*), and parasitic worms, or helminths (*Ascaris lumbricoides*). Pollution and microbial contamination make the waters of most rivers, lakes, and ponds (surface water) unfit for human consumption without prior purifying treatment. The most common method of treating surface water to produce safe drinking water is disinfection by chlorination. Chloramines and ultraviolet light are also used for treatment.

Enough good-quality drinking water exists to satisfy the needs of all human beings on the planet, but the water is not distributed in such a way that it reaches populations in all parts of the world. Population growth also has impacts on drinking-water resources, even in regions with large supplies of water. In many developing countries people have no choice but to use water that is polluted with various wastes—including human sewage, animal excrement, and a variety of pathogenic microorganisms—because no water treatment systems are in place.

In many locations, drinking water is obtained from groundwater sources. Groundwater—water found beneath the ground surface in soil and rock spaces—can generally be used for drinking with minimal treatment because it has already been purified by passing through soil. Wells must be dug to reach this water. A 2014 US Geological Survey report showed that in 2010, an estimated 63 percent of US public supply water (which includes drinking water for most households) came from rivers and lakes and about 37 percent from groundwater sources. According to the US Geological Survey, over 20 percent of total national water withdrawals in the United States came from groundwater sources in 2015. In many parts of the world, groundwater usage is increasingly outpacing the rate of groundwater replacement.

Several countries have introduced regulations intended to ensure the good quality of drinking water. In the United States, the Environmental Protection Agency (EPA) sets standards for drinking water under the 1974 Safe Drinking Water Act. These standards cover some ninety-four possible contaminants, including biological and chemical substances. For example, the EPA sets the acceptable number of microorganisms per 1 milliliter (0.034 ounce) of water at fewer than 10, and water that is considered drinkable can contain no coliform bacteria. However, many pharmaceutical contaminants are not covered by the EPA's regulations.

Oil-drilling processes, such as hydraulic fracturing (fracking), intensify pressure and further drying up already-strained systems, speeding up the rise in temperatures and further reducing the availability of clean water. Huge amounts of wastewater from fracking are laced with carcinogens and contaminate the aquifers.

Desalination and Conservation
The need to obtain more drinking water in some regions has led to increased use of desalination technologies as well as increasing emphasis on the conservation of existing water supplies. The desalination of ocean water produces billions of gallons of drinking water per day around the world, but this process can have some negative environmental impacts, such as thermal pollution and damage to shoreline ecosystems.

Water conservation is the most cost-effective way to reduce demand for drinking water. Both local and national governments in many nations have introduced regulations and programs aimed at encouraging water conservation on the level of individual households; among the areas targeted for reduced usage of drinking-water supplies have been lawn maintenance and toilet flushing. The practice of rainwater harvesting is encouraged in some areas, and some governments have instituted strict compulsory water metering to raise awareness of the need to conserve water.

Another option for increasing supplies of drinking water is the restoration of municipal and industrial wastewater to drinkable quality. Using technologies such as membrane bioreactor treatment (which involves treating wastewater with certain types of microorganisms and then putting it through microfiltration, followed by disinfection with ultraviolet light), communities can turn their own wastewater into drinking water.

Plastic Water Bottle Pollution
Americans buy 29 billion water bottles a year, yet water bottles do not degrade. Landfills in the United States are overflowing with an estimated two million tons of discarded water bottles. Plastics are produced

with fossil fuels, which only makes the bottles an environmental hazard, but also an enormous waste of valuable resources It takes at least a thousand years for these bottles to degrade.

In Flint, Michigan, a new water system introduced in 2014, proved to contain enormous amounts of lead, creating a health crisis of unprecedented proportions. One hundred thousand Flint residents could rely only on water from bottles or filters during this years-long contamination emergency. As a result, millions of bottles have been assigned to landfills.

Sergei A. Markov

Further Reading

Gardner, Lynn. *Toilets, Taps & Trash: From recycled sewage masquerading as "potable water" to landfills filled with plastic water bottles masquerading as "recycling" - we've got BIG problems.* Kindle Edition. Seattle, Wash.: Amazon Digital Services LLC, 2019.

Hammer, Mark J. *Water and Wastewater Technology.* London: Pearson, 2004.

Hanna-Atisha, Mona. *What the Eyes Don't See: A Story of Crisis, Resistance and Hope in An American City.* London: Oneworld Publications, 2018.

Glennon, Robert. *Unquenchable: America's Water Crisis and What to do About it.* Washington, D.C.: Island Press: 2010.

Shannon, Mark A., et al. "Science and Technology for Water Purification in the Coming Decades." *Nature*, vol. 452, 2008, pp. 301–10.

Saltzman, James. *Drinking Water: A History.* New York: Harry H. Abrams, 2017.

Sigee, David. *Freshwater Microbiology.* Hoboken, N. J.: Wiley, 2005.

Wright, Richard T. *Environmental Science: Toward a Sustainable Future.* 10th ed., Pearson, 2008.

WEBSITES

Centers for Disease Control and Prevention. www.cdc.gov/healthywater/global/index.html

Sustainable Development Goals: 17 Goals to Transform Our World, United Nations, www.un.org/sustainabledevelopment/water-and-sanitation.

The Guardian, www.theguardian.com/environment/2018/mar/19/water-shortages-could-affect-5bn-people-by-2050-un-report-warns.

■ Watersheds

FIELDS OF STUDY

Agriculture; Animal Husbandry; Biology; Ecology; Ecosystems; Environment; Environmentalism; Erosion Control; Forestry; Geography; Geology; Hydrology; Life Sciences; Marine Biology; Water Resources; Zoology

SUMMARY

Human activity can cause unanticipated changes in watersheds, affecting the hydrologic balance. Careful management of watersheds is important because land use alters the balance between storage and dispersal of precipitation, in many cases increasing erosion, stream sedimentation, and flooding hazards.

PRINCIPAL TERMS

- **basin:** a large or small depression in the surface of the land or in the ocean floor
- **erosion:** the process by which the surface of the earth is worn away by the action of water, glaciers, winds, waves or human activity
- **habitat:** place that is natural for the life and growth of an organism
- **hydrologic balancing:** the process of optimizing the distribution of water in a building's hydronic heating or cooling system by equalizing the system pressure so it provides the intended indoor climate at optimum energy efficiency and minimal operating cost
- **sedimentation:** the tendency for particles in suspension to settle out of the fluid in which they are entrained and come to rest against a barrier
- **watershed management:** policies governing the use of land areas bounded by drainage divides within which precipitation drains to watercourses or bodies of water
- **watershed:** a region or area bounded peripherally by a divide and draining ultimately to a watercourse or body of water
- **wetlands:** land that has a wet and spongy soil, as a marsh, swamp, or bog

Characteristics of Selected Major Drainage Basins

River	Continent	Outflow	Length	Area	Average Annual Suspended Load
Amazon	South America	180.0	6,300	5,800	360
Congo	Africa	39.0	4,700	3,700	—
Yangtze	Asia	22.0	5,800	1,900	500
Mississippi	North America	18.0	6,000	3,300	296
Irawaddy	Asia	14.0	2,300	430	300
Brahmaputra	Asia	12.0	2,900	670	730
Ganges	Asia	12.0	2,500	960	1,450
Mekong	Asia	11.0	4,200	800	170
Nile	Africa	2.8	6,700	3,000	110
Colorado	North America	0.2	2,300	640	140
Ching	Asia	0.06	320	57	410

Note: Rivers are ordered by outflow; outflow is multiplied by 1,000 cumecs (cubic meters of water per second); length is measured in kilometers; area is measured in square kilometers multiplied by 1,000; average annual suspended load is measured in millions of metric tons.

Watersheds are defined at many scales: The Mississippi River watershed contains the Ohio River watershed, which in turn contains smaller watersheds. A fundamental part of the hydrologic cycle, the watershed collects and stores precipitation in soils, lakes, wetlands, or aquifers and disperses water by evaporation, plant transpiration, surface runoff, springs, and base flow to streams. Watersheds of different geographic regions have distinctive characteristics based on climate, topography, and soil type; therefore, the natural variability among watersheds is predictably large. In arid regions, precipitation occurs as intense, infrequent storms, with most of the water rapidly running off and eroding soil with little protective vegetation. Watersheds in humid areas are characterized by frequent, usually gentle rain that replenishes aquifers and sustains streams, springs, and wetlands.

Ecologically, the watershed provides habitat and nutrients for plants and animals, including humans. Land use can disrupt a watershed's ecology by disturbing habitat and nutrient cycling through soil loss and removal of native vegetation. The role of the watershed in environmental problems such as flooding, erosion, sedimentation, and ecological disruption has led to increased emphasis on the watershed as the basic unit for environmental management, rather than political units such as states or counties.

The 1954 Watershed Protection and Flood Prevention Act authorized the secretary of the U.S. Department of Agriculture (USDA) to manage watersheds in cooperation with states and local organizations, such as soil and water conservation districts. The driving idea behind the act is that floods are better controlled through management of runoff upstream in the watershed than through downstream engineering projects. The act requires local interests to contribute up to 50 percent of the costs to ensure local support for watershed projects. In contrast, Army Corps of Engineers flood-control projects originally were funded entirely by the federal government. The Watershed Protection and Flood Prevention Act is generally administered through the USDA's Natural Resources Conservation Service, formerly the Soil Conservation Service.

Recognizing the need for basin-wide planning, the federal government created the Water Resources Council through the 1965 Water Resources Planning Act. This council created river basin planning commissions but fell into disfavor and lost funding because the river basins were too large for effective planning.

A major step in watershed management was taken with the 1972 Clean Water Act. With this act, land management began to include water-quality control. Nonpoint sources of pollution were targeted, among them agriculture, forestry, mining, and waste disposal. Most states passed laws directing the use of certain widely accepted methods of preventing soil and water problems (known as best management practices, or BMPs) to protect or rehabilitate watershed functions. The Clean Water Act provided for regulation of land use, initially through incentives. The 1985 Food Security Act provided incentives for landowners to control erosion on highly erodible croplands. The act's "swampbuster" provisions directed protection of existing wetlands and provided incentives for wetlands restoration. The 1986 amendments to the 1974 Safe Drinking Water Act encouraged public suppliers of drinking water to protect wellheads. The 1987 amendments to the Clean Water Act encouraged states to address nonpoint source pollution.

State regulations on nonpoint source pollution range from voluntary compliance with BMPs to strict enforcement of BMPs with fines for noncompliance. In general, however, such regulations have become increasingly detailed and comprehensive. The concept of the total maximum daily load (TMDL) permissible for nonpoint source pollutants has been introduced but determining TMDL is costly and difficult because appropriate loads vary with land use and with watershed.

Over time, Americans' perspectives on land and water management have broadened. Whereas management initially focused on single farms or individual fields, the watershed view has come to be widely accepted. This change has been influenced in part by concerns about the greenhouse effect, in which atmosphere-biosphere-hydrosphere-terrasphere interactions are critical. The term "ecosystem management" may better reflect watershed management focus in the future.

Watersheds are managed for a spectrum of land uses, including water supply, settlement, grazing, crop production, forestry, and recreation. Management focuses on water, sediment, and wastes. Water management generally seeks to reduce runoff; exceptions are landfills and mine spoils in which infiltration is minimized. Sediment management seeks to prevent soil erosion or to trap eroded sediment. Waste management seeks to distribute the waste load properly and to prevent it from reaching water. The appropriate strategies for achieving these management goals vary from problem to problem. For example, in forestry one strategy might include revegetating logged areas, diverting water from logging roads, and closing logging roads after use. The overall approach to watershed management is to identify the problem and its source and then select and implement BMPs. While the law requires that BMPs be considered, no definitive catalog of such practices exists. Many state agencies have written and assembled their own collections of BMPs for various land uses, which are available to the public. Public education and public participation in decision making have played increasingly important roles in sustainable watershed management in the United States.

Mary W. Stoertz

Further Reading

Black, Peter E. *Watershed Hydrology*. 2d ed. Chelsea, Mich.: Ann Arbor Press, 1996.

France, Robert L., ed. *Facilitating Watershed Management: Fostering Awareness and Stewardship*. Lanham, Md.: Rowman & Littlefield, 2005.

Hartel, Diana. *Watershed Redemption: A Journey in Time on Five US Watersheds*. Arroyo Seco, N.M.: Madrona Arts Press, 2018.

Heathcote, Isobel W. *Integrated Watershed Management: Principles and Practice*. 2d ed. Hoboken, N.J.: John Wiley & Sons, 2009.

McPhee, John. *Basin and Range*. New York: Farrar, Straus and Giroux, 1982.

Newson, Malcolm. *Land, Water, and Development: Sustainable and Adaptive Management of Rivers*. 3d ed. New York: Routledge, 2009.

Satterlund, Donald R., and Paul W. Adams. *Wildland Watershed Management*. 2d ed. New York: John Wiley & Sons, 1992.

■ Wetlands

FIELDS OF STUDY

Biology; Ecology; Ecosystems; Environment; Environmentalism; Life Sciences; Preservation; Public Policy; Wilderness Preservation

SUMMARY

Wetlands are widely considered to be among the world's most important ecosystems because of their high biodiversity and productivity. They also perform important functions related to the maintenance of surface and groundwater quality and quantity, prevention of saltwater intrusion, control of coastal erosion, and regulation of climate.

PRINCIPLE TERMS
- **coterminous:** having the same or coincident boundaries
- **hectares:** a unit of surface, or land, equivalent to 2.471 acres
- **hydric soil:** a soil that formed under conditions of saturation, flooding, or ponding long enough during the growing season to develop anaerobic conditions
- **hydrophyte:** a plant that grows in water or very moist ground; an aquatic plan
- **United States Army Corps of Engineers:** a federal agency under the Department of Defense, made up of some 37,000 civilian and military personnel and the world's largest public engineering, design, and construction management agencies
- **wetlands:** transitional areas between terrestrial and aquatic ecosystems that exhibit characteristics of both

Wetlands are often distinguished by having three major components: water, hydrophytic vegetation, and hydric soils. All have water present for at least part of the year, though the depth and duration of flooding vary considerably. Some wetlands have water-saturated soil, whereas others are characterized by permanent flooding. At least periodically, wetlands support a predominance of hydrophytic vegetation—that is, plant life adapted to thrive in saturated soil conditions. Wetlands are also characterized by having undrained, or hydric, soils. These are soils in which a condition (an absence of free oxygen) has developed because of long periods of saturation, flooding, or ponding during the growing season.

No single, formal definition of wetlands has been established, because no single description is appropriate for all the diverse wetland types that exist over a large geographic scale with diverse climatic conditions. Dozens of definitions have been written, however, for specific reasons by specific interest groups and various regulatory agencies. The problem of defining wetlands is one of critical consequence to those persons who are subject to restrictions and limitations placed on them by various national, regional, state, and local laws concerning wetlands. Inconsistent definitions place a severe burden on private landowners who may be subject to such laws and do not have adequate technical or legal knowledge about wetlands.

In the United States, a regulatory definition of the term "wetland" has been developed so that the Army Corps of Engineers and the Environmental Protection Agency (EPA) can administer the permitting of dredging and filling of wetlands as prescribed in Section 404 of the Clean Water Act. For this purpose, wetlands are

> those areas that are inundated or saturated by surface or ground water at a frequency and duration enough to support, and that under normal circumstances do support, a prevalence of vegetation typically adapted for life in saturated soil conditions. Wetlands generally include swamps, marshes, bogs, and similar areas.

Thus, jurisdictional wetlands in the United States—those that are subject to section 404 permitting—must possess all three key characteristics: hydrology, hydrophytic vegetation, and hydric soils. However, because many wetlands are not permanently wet and because water may not be seen during a single site visit, positive indicators must be found, which must be supported by wetland vegetation and soils. These strict requirements for wetland identification have caused many wetlands to fall into uncertain categories. For example, some wetlands can have the appropriate hydrology but fail to develop the appropriate wetland soils and vegetation; often two of the three characteristics can be confirmed but not the third. These situations continue to cause confusion among governmental agencies and private landowners.

Another significant issue concerning wetlands is that they form ecotones (ecological transition zones) between upland and aquatic ecosystems. Thus, even if an area is identified as a wetland, determination of its exact boundaries may be extremely difficult because of the gradual, perhaps imperceptible, changes in soil and vegetation characteristics. The problem of identifying areas as wetlands and defining the boundaries of those wetlands is known as wetland delineation. The ability to perform delineations is acquired only through extensive training, especially in the areas of soils and botany.

Functions and Value

Not all wetlands perform the same set of functions or perform their functions at the same rate or efficiency. Often the size of a wetland and its location in the ecosystem determine its functions. Wetlands generally have extremely high biodiversity and rates of productivity. They provide food, shelter, and water for various invertebrates and vertebrates, many of which may be endangered or threatened. According to the U.S. Department of Agriculture's Natural Resources Conservation Service, nearly 5,000 plant species, 190 amphibian species, one-third of all bird species, and all wild ducks and geese depend on the nation's wetlands. Endangered species in the United States that have wetland habitats include the bald eagle, red wolf, whooping crane, fat mucket mussel, and swamp rose. It has been estimated that wetlands provide essential for 40 percent of the nation's endangered and 60 percent of its threatened species.

Another function associated with wetlands is the maintenance of the quantity and quality of both and groundwater. Many wetlands serve to recharge aquifers. Wetlands also accumulate sediments, nutrients, and many forms of water pollutants from their watersheds. By removing these materials, wetlands serve to clean the water.

Wetlands are sometimes referred to as nature's sponges because of their ability to ameliorate the effects of stormwater and reduce floodwater damage. Stormwater enters wetlands and spreads out over large areas and then is slowly released. Increased property damage from flooding has been shown to occur following the destruction of wetlands. The Army Corps of Engineers noted in a 1976 report that if the Charles River wetlands near Boston, Massachusetts, were destroyed, flood damage in the would increase by as much as $17 million annually. Other wetland functions are prevention of into and surface-water supplies, protection against coastal from storms, and regional and global climate stabilization. Concerns regarding have sparked interest in the ability of wetlands to function as carbon reservoirs.

A wetland value is any product, characteristic, or function of a wetland that has worth or is beneficial to the or to people. Wetland products such as timber, fiber, food, and fish have commercial value and are easily measured. Some other values of wetlands, however, are not as easily quantified. For example, wetlands may have sociocultural significance and provide sites for recreation, research, and education. Further, it is impossible to assign a dollar value to the fact that wetlands provide habitat for a high percentage of endangered and around the world.

The value placed on a wetland's functions is, in many cases, the most important factor that determines whether the wetland is preserved or converted to some other use. As society's needs and perceptions change over time, the value assigned to wetland functions also changes. The values associated with wetlands are often in conflict because of the large number of functions these ecosystems can perform. For example, if the water level in a wetland is raised, waterfowl production may increase while timber production decreases. Managers of wetlands may thus make decisions that are popular with one user group but unpopular with other user groups.

Loss and Degradation

"Wetland loss" refers to a decrease in wetland area caused by the conversion of wetland to non-wetland. "Wetland degradation" refers to the impairment of one or more wetland functions because of human activity. In most cases, wetland loss is difficult or impossible to reverse because of the complexity of wetland structure and function. Wetland degradation, in contrast, is more easily reversed through a variety of applied science and conservation tools. Wetland creation—the formation of wetlands in formerly non-wetland areas—has become an increasingly common strategy for combating wetland loss.

Wetlands are found on every continent. Even beneath Antarctica's mantle of ice, there are wetlands that support life. The exact size of global wetland areas is difficult to assess because of differences in wetland definitions and lack of documentation in many countries, but wetlands have been estimated to cover about 6 percent of the land area of the earth. The largest wetland areas are found in tropical, subtropical, and boreal regions. Since the beginning of the twentieth century, the world has lost more than 50 percent of its wetland area.

A 1990 report issued by the U.S. Fish and Wildlife Service estimated that the area of wetlands in the United States decreased from about 158 million hectares (391 million acres) during the 1780s to about 111 million hectares (274 million acres) in the mid-1980s—a 30 percent overall loss of wetland area. Within the lower forty-eight states, the estimated loss was 53 percent. The Fish and Wildlife Service has been the nation's wetlands trends since the 1950s. During the period from the mid-1950s to the mid-1970s an estimated 185,400 hectares (458,000 acres) of wetlands in the coterminous United States were lost every year. From the mid-1970s to the mid-1980s, by which time the value of wetlands had begun to be recognized, the estimated rate of loss decreased to 117,400 hectares (290,000 acres) per year. From 1986 to 1997 the loss rate dropped by 80 percent, to 23,700 hectares (58,500 acres) annually. Between 1998 and 2004 net wetlands gains in the coterminous United States surpassed net losses for the first time since the survey began, thanks to wetlands creation and restoration efforts. The net gain for the period was 77,598 wetland hectares (191,750 acres), or an average net gain of 12,950 hectares (32,000 acres) per year. As of 2004, there were an estimated 43.6 million hectares (107.7 million acres) of wetlands in the coterminous United States.

The underlying causes of wetland loss and degradation are numerous. These include poverty and economic inequality; pressures from growth, immigration, and mass tourism; social and political conflicts; high demand for wetland resources such as timber; drainage; diking and damming; air and water pollution; introduction of exotic species; natural events such as hurricanes; and economic policies. In the United States, approximately 80 percent of wetland losses from the mid-1950s through the mid-1970s were the result of agricultural practices. Since the 1970s growing awareness of the importance of wetland functions has slowed the destruction of wetlands, but wetland loss and, more frequently, degradation continue, caused by development and other stressors.

Conservation and Protection
Wetlands in the United States are protected through regulation, economic programs, and acquisitions. At the federal level, a confusing mix of programs and legislation simultaneously encourages and discourages wetland conservation. As early as 1903, President Theodore Roosevelt recognized that wetland loss had become significant. By executive order, he established Pelican Island in Florida as the nation's first wildlife refuge. The federal government also protects wetlands through several laws, including the Clean Water Act of 1972, which created a plan to control the release of dredged or scrap materials into wetlands and other waters of the United States. The Army Corps of Engineers and EPA share responsibility for implementing the program. The "swampbuster" program is part of the Food Security Act of 1985 and 1990. It seeks to remove federal incentives for the agricultural conversion of wetlands to nonwetlands. In conjunction with this act, the 1990 Farm Bill created a voluntary Wetland Reserve Program, which provides financial incentives to farmers to restore and protect wetlands through the use of long-term easements.

The North American Waterfowl Management Program represents another milestone in the conservation of important wetland habitat. This plan was signed between Canada and the United States in 1986 to restore declining waterfowl populations through habitat acquisition, development of economic incentives to change land-use practices, and improvement of water management. Mexico became a signatory to the agreement in 1994. At the global level, the most significant wetland conservation work has resulted from the Convention on Wetlands of International Importance, or Ramsar Convention, in 1971. This global treaty provides a framework for the international protection and wise use of wetlands.

U.S. presidents became active in wetland protection during the 1970s. President Jimmy Carter signed two executive orders that provide guidance for wetland and management and protection of these areas by federal agencies. President George H. W. Bush extended these efforts to recommend that the United States establish a national goal of "no net loss" of wetlands. This policy became a major force for wetland conservation in the United States.

Despite this activity, several difficulties remain. Wetlands, because of the complexity of their values and functions, continue to be managed for a variety of purposes, many of them conflicting. The policy of no net loss of wetlands applies to the loss of wetland acreage only, not to wetland functions, values, or quality. President Bill Clinton took a compromise position in wetlands protection by reaffirming the "no net loss" policy and supporting the Wetland Reserve Program; however, he created section 404 exemptions for 21.4 million hectares (53 million acres) of previously converted wetlands and for small plots of land owned by families who wanted to build single-family houses.

On Earth Day 2004, President George W. Bush announced an initiative to achieve an overall increase in the quantity and quality of U.S. wetlands. While there was a net gain in wetland area between 1998 and 2004, there were still wetland losses; urban and rural development was responsible for about 61 percent of the net freshwater wetlands losses during this period. Two notable U.S. Supreme Court cases during the Bush administration, *Solid Waste Agency of Northern Cook County v. Army Corps of Engineers* (2001) and *Rapanos v. United States* (2006, consolidated with *Carabell v. Corps of Engineers*), found that the Clean Water Act did not protect the isolated wetlands addressed in each case from being developed.

Roy Darville; updated by Karen N. Kähler

Further Reading

Batzer, Darold P., and Rebecca R. Sharitz, eds. *Ecology of Freshwater and Estuarine Wetlands*. Berkeley: University of California Press, 2006.

Fowler, Theda Braddock, and Lisa Berntsen. *Wetlands: An Introduction to Ecology, the Law, and Permitting*. 2d ed. Lanham, Md.: Government Institutes, 2007.

Keddy, Paul A. *Wetland Ecology: Principles and Conservation*. 2d ed. New York: Cambridge University Press, 2010.

Mitsch, William J., and James G. Gosselink. *Wetlands*. 4th ed. Hoboken, N.J.: John Wiley & Sons, 2007.

Spray, Sharon L., and Karen L. McGlothlin, eds. *Wetlands*. Lanham, Md.: Rowman & Littlefield, 2004.

Tiner, Ralph W. *Wetland Indicators: A Guide to Wetland Identification, Delineation, Classification, and Mapping*. Boca Raton, Fla.: CRC Press, 1999.

Wilderness areas

FIELDS OF STUDY

Animal Studies; Botany; Conservation Biology; Controversies; Ecology; Ecosystems; Environment; Environmentalism; Forestry; Geography; Land-use Management; Life Sciences; Public Policy; Wildlife Management

SUMMARY

The designation of large areas of land as protected wilderness areas is the subject of ongoing debate in the United States, with preservationists asserting that more areas need to be protected and critics arguing that the natural resources found in these areas should be available for use.

PRINCIPAL TERMS

- **national forests:** forestlands owned and managed by a national government
- **national parks:** areas of scenic, historic, or other value that are set aside by federal governments for the preservation of animals and wildlife and for human recreation
- **National Wilderness Preservation Systems (NWPS):** a Univted States agency that manages and protects wilderness areas
- **wilderness areas:** natural, undeveloped and protected areas set aside by a national government for the preservation of wildlife
- **wilderness:** a wild and uncultivated region, as of forest or desert, uninhabited or inhabited only by wild animals

Preserving areas of unspoiled nature is a relatively new idea, and in the United States, this idea began to make sense to many Americans only when the seemingly inexhaustible wilderness of North America had been, in fact, nearly exhausted. In 1924, at the urging of US Forest Service (USFS) employee and influential conservationist Aldo Leopold, 305,500 hectares (755,000 acres) of the Gila National Forest in New Mexico were set aside as the first federally protected wilderness, the Gila Wilderness Area.

As the system of wilderness areas (sometimes known as primitive areas) grew, environmentalists became concerned about inconsistent management and the fact that these areas were protected only by agency policy and not by law. They began lobbying for federal legislation that would designate and protect wilderness areas throughout the United States. The concept of preserving wilderness was strongly opposed, however, by many of those who made their livings by using natural resources; these included people involved in ranching and those in the timber and mining industries. They saw protected wilderness lands, which often had great economic value, as being "locked up" for the pleasure of a few.

Legislation

On September 3, 1964, after eight years of debate and compromise, President Lyndon B. Johnson signed the Wilderness Act, creating the National

Wilderness Preservation System (NWPS), which consisted of fifty-four areas totaling 3.6 million hectares (9 million acres). The act states:

> A wilderness, in contrast with those areas where man and his own works dominate the landscape, is hereby recognized as an area where the earth and its community of life are untrammeled by man, where man himself is a visitor who does not remain.

The act defines the mechanism for adding more areas to the system in the future. To be considered, an area must be at least 2,023 hectares (5,000 acres) "or of manageable size." This is a far cry from the early days of wilderness advocacy, when the minimum size was thought to be 202,000 hectares (500,000 acres), or, as Aldo Leopold put it, "large enough to absorb a two-week pack trip." Designated wildernesses become part of the NWPS. All roads, structures, and other installations are prohibited in designated wilderness, as is the use of motorized equipment or any mechanical transport. These areas of wild nature have been, and continue to be, the focus of intense controversy regarding their designation and management.

A significant addition to the NWPS came in 1975 with the passage of the Eastern Wilderness Act. The lack of pure, untouched wilderness in the eastern states led to the loosening of the strict standards of the original act to allow the inclusion of ecologically significant areas that show more impact from human activities than would originally have been permitted. In this way, sixteen areas totaling 83,770 hectares (207,000 acres), from 8,900-hectare (22,000-acre) Bradwell Bay in Florida to the 5,670-hectare (14,000-acre) Lye Brook Wilderness in Vermont, were added to the system. By the late 1990s, the wilderness system encompassed more than 650 areas, ranging in size from the 2-hectare (5-acre) Oregon Islands Wilderness to the 3.6 million hectares (9 million acres) of the Wrangell-St. Elias Wilderness in Alaska, for a total of more than 40 million hectares (100 million acres). By 2018, in part as a result of the passage of the Omnibus Public Land Management Act of 2009, the number of wilderness areas had grown to 765, with a total of more than 44.1 million hectares (109 million acres). Of this total, approximately 52 percent was in the state of Alaska.

Debates and Controversies

Although the amount of protected land may seem quite large, preservationists point out that only about 5 percent of the landscape of the United States is protected in its natural state. Some large areas continue to be fought over, such as the fragile Arctic coastal plain of Alaska, home of vast caribou herds and underlain by large oil deposits. Idaho, which is among the top three states in the lower forty-eight in terms of wilderness land area (exceeded only by California and Arizona), still has millions of hectares of undeveloped roadless land that many believe should be protected. Wilderness advocates also point out that many wilderness areas, as well as national parks and other protected lands, have illogical political boundaries, unrecognized by grizzly bears and other important wildlife species. They argue that areas between and adjacent to designated wilderness areas should often be protected as well, to create units based on natural, ecological boundaries.

After a wilderness area is designated, the focus shifts to the maintenance of its desired qualities, leading to the paradox of

Bisti Wilderness Area in New Mexico. By By Larry Lamsa.

"wilderness management." Although recreation is only one of the stated uses of wilderness—the others being scenic, scientific, educational, conservation, and historic—agency efforts and budgets are based primarily on the need to manage the vast numbers of human visitors. One of the stated purposes of preserving wilderness areas is to provide for "primitive and unconfined recreation," but another consideration is the protection of the resource itself. At what point do the camping and trail restrictions, quotas, and permits needed to protect the resource impinge on the unconfined recreation of the visitor?

Another issue related to wilderness areas is that of wildfire suppression. It is now understood that fire is an important component of most ecosystems, but past policies of fire suppression have left unnatural fuel conditions in many areas. Should managers allow wildfires to burn, even though these fires are likely be larger and more destructive than natural, periodic fires of the past? Other major controversies center on the reintroduction of wildlife species (especially predators such as the wolf and the grizzly bear) to wilderness areas, the disposition of long-standing mining and drilling claims, and the flying of aircraft over, or even into, remote wilderness areas.

Joseph W. Hinton

Further Reading

Allin, Craig W. *The Politics of Wilderness Preservation.* 1982. Reprint. Fairbanks: University of Alaska Press, 2008.

Brinkley, David. *The Wilderness Warrior: Theodore Roosevelt and the Crusade for America.* New York: Harper Perennial, 2010.

Dawson, Chad P., and John C. Hendee. *Wilderness Management: Stewardship and Protection of Resources and Values.* 4th ed. Boulder, Colo.: WILD Foundation, 2009.

Frome, Michael. *Battle for the Wilderness.* Rev. ed. Salt Lake City: University of Utah Press, 1997.

Muir, John. *Wilderness Essays.* Layton, Utah: Gibbs Smith, 2012.

Nash, Roderick. *Wilderness and the American Mind.* 4th ed. New Haven, Conn.: Yale University Press, 2001.

Scott, Doug. *The Enduring Wilderness: Protecting Our Natural Heritage Through the Wilderness Act.* Golden, Colo.: Fulcrum, 2004.

Thoreau, Henry David and Eliot Porter. *In Wildness is the Preservation of the World.* Pasadena, Calif.: AMMO Books, 2012

Wildlife refuges

FIELDS OF STUDY
Conservation Biology; Ecology; Ecosystems; Environment; Environmentalism; Geography; Life Sciences; Marine Biology; Monuments; Parks; Public Policy; Reserves; Sociology; Zoology

SUMMARY
The U.S. National Wildlife Refuge System has endured congressional debate and public scrutiny involving environmental issues related to the societal, governmental, and commercial use of designated sanctuaries, culminating in the 1997 National Wildlife Refuge System Improvement Act and the transformation of America's refuges into multiple-use systems.

PRINCIPAL TERMS
- **Land and Water Conservation Fund:** a federal program supports the protection of federal public lands and waters, including national parks, forests, wildlife refuges, and recreation areas, and voluntary conservation on private land
- **National Audubon Society:** a United States non-profit environmental organization dedicated to conservation, incorporated in 1905, which uses science, education and grassroots advocacy to advance its mission in honor of John James Audubon, a Franco-American ornithologist and naturalist
- **Nature Conservancy**: a charitable environmental organization, founded in 1951, whose mission is to "conserve the lands and waters on which all life depends," with non-confrontational, pragmatic solutions while working with indigenous communities, businesses, governments, multilateral institutions, and other non-profits.
- **wildlife refuges:** regions of land or water set aside by governments or private organizations to protect and preserve one or more species of wildlife

Prior to 1900, the U.S. federal government aggressively raised much-needed revenue and rewarded growing commerce by selling or giving away nearly

405 million hectares (1 billion acres) of land to states, homesteaders, veterans, railroads, and businesses. President Theodore Roosevelt initiated the protection of habitat for wildlife in 1903 when he set aside Pelican Island, a 1.2-hectare (3-acre) ecosystem of barren sand and scrub in Florida's Indian River, as a federal reservation to protect birds from hunters supplying plumes to the fashion industry. Inspired while camping in California's Yosemite Valley with naturalist John Muir, Roosevelt established more than fifty wildlife refuges, five national parks, and eighteen national monuments, such as the Grand Canyon. He also greatly increased the area of lands designated as national forests before leaving his second term in office.

To preserve additional lands "for our children and their children's children forever, with their majestic beauty all unmarred," Roosevelt guaranteed land for future refuges by separating other federal public domain regions such as national forests and rangelands from the control of commercial interests. More than 90 percent of the refuge land area in existence in the United States in the early years of the twenty-first century resulted from Roosevelt's foresight, which enabled the National Wildlife Refuge System to grow larger than the national park system and entail nearly 4 percent of the surface area of the United States.

With vital assistance by private individuals and organizations such as the Nature Conservancy and the National Audubon Society, wildlife refuges have been established for waterfowl, big game, small resident game, and colonial nongame birds. Wildfowl refuges, easily the most plentiful, are geographically patterned to supply breeding, wintering, resting, and feeding areas along the four major North American migration flyways. The sportsmen who were essential in establishing many national refuges ensured that hunting would be permitted on most sanctuary lands, with trapping allowed on many. Although the entire National Wildlife Refuge System logs substantial numbers of hunting visits annually, visitors who are interested in wildlife education and photography outnumber hunters and anglers by more than four to one.

The National Wildlife Refuge System—overseen by the U.S. Fish and Wildlife Service, which is part of the Department of the Interior—is the most comprehensive nature protection network in the world. The entire system includes more than 60.7 million hectares (150 million acres) inside refuge boundaries, encompassing 553 national wildlife refuges and other units as well as 38 wetland management districts. Nearly all of the refuges in the system are open to the public. Endangered and threatened species are supported on almost 60 refuges created specifically for that purpose, and many urban refuges have been established near large cities. At least one wildlife refuge has been established in each of the fifty states, and several are found in overseas possessions from Puerto Rico to American Samoa in the South Pacific. The two largest refuges are in Alaska: The Arctic National Wildlife Refuge and the Yukon Delta National Wildlife Refuge are both more than 7.7 million hectares (19 million acres) in size. The smallest refuge in the system is the Mille Lacs National Wildlife Refuge in Minnesota, which is just 0.24 hectare (0.60 acre) in size.

Environmental Management Issues

Creating and maintaining a refuge to provide food, cover, and protection from human development for wildlife is considerably more difficult than simply sequestering an area and allowing nature to run its course; continual management of resources is imperative to keep delicate ecosystems in balance. The many tasks that humans must perform on refuge lands include fixing broken floodgates and cleaning clogged ditches, seeding wildlife foods, plowing and burning areas that have been overrun with unwanted vegetation, and closing off areas from the public during sensitive periods for animals, such as mating and birthing seasons. In addition, refuge managers must often battle for their lands' shares of rapidly declining water supplies.

Although many refuges include areas just as spectacular as those within the national park system, the National Wildlife Refuge System as a whole was for many years not well utilized by the public. The U.S. Congress, observing this lack of public use, leased some of these public lands for commercial purposes such as grazing, farming, oil drilling, mining, logging of timber, military maneuvers, and motorized recreation. However, as public use of refuges increased during the 1980s and 1990s, Congress added eighty new refuges to the system, creating such a backlog of environmental preservation issues that the legislators then began to contemplate selling some areas to pay

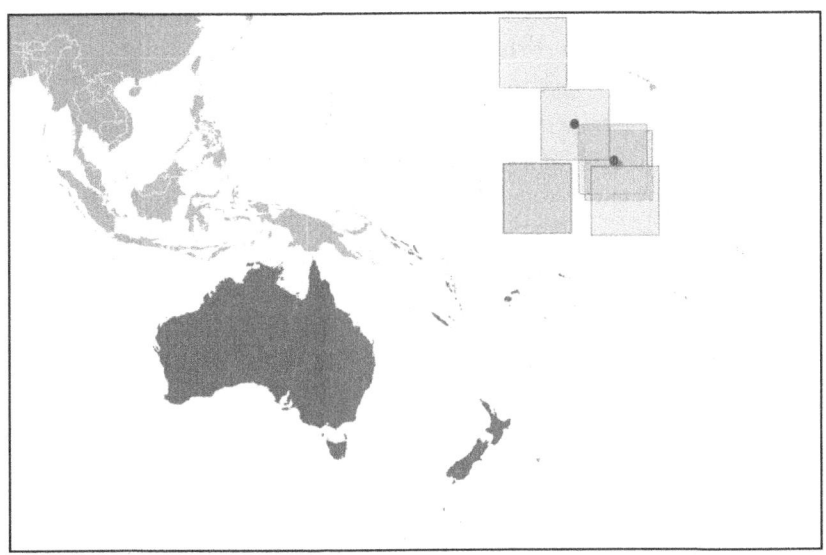

The United States Pacific Island Wildlife Refuges are highlighted in this regional map. (Courtesy of the CIA World Fact Book)

for maintenance. With minimal budgets, refuge managers are charged with making certain that all activities that take place within their refuges are compatible with wildlife while still allowing potentially destructive activities such as off-road driving and motorcycling, powerboating, and commercial fishing. The maintenance of biodiversity is also an important goal of managers; the wildlife refuge system provides homes for about 700 species of birds, 220 species of mammals, 250 species of reptiles and amphibians, and 200 species of fish, in addition to innumerable species of plants.

Refuge areas owned by private corporations have often sacrificed key habitat for short-term economic gain with little regard for long-term environmental and social consequences. However, business executives have realized that environmental issues are of genuine concern to most Americans. In response to increased environmental awareness and pressure from consumers, employees, and stockholders, many large businesses have implemented stewardship strategies that protect natural resources, enhance wildlife habitat, and provide for public enjoyment of their underdeveloped land.

Key Legislative Actions

Following Theodore Roosevelt's initial work, new additions to the refuge system came slowly until the Dust Bowl years of the 1930's, when migratory bird populations became depleted. Congress then passed the 1934 Migratory Bird Hunting and Conservation Stamp Act, widely known as the Duck Stamp Act, which added a conservation fee to the price paid for every waterfowl license purchased by a hunter; the revenue collected in this way enabled the Fish and Wildlife Service to acquire wetlands along major bird migration flyways. Additional moneys to purchase refuge lands came from the Land and Water Conservation Fund set up in the 1960s to increase public space for outdoor recreation, which generated considerable revenues from offshore oil drilling leases. Passage of the 1980 Alaska National Interest Lands Conservation Act (ANILCA) enabled both the refuge system and the national park system to double in size. Although 96 percent of all refuge units are outside Alaska, Alaska contains about 83 percent of National Wildlife Refuge System lands.

The 1973 Endangered Species Act spurred managers of refuges to make more concessions for certain species of flora and fauna. At the beginning of the twenty-first century, refuges in the U.S. system harbored about 280 threatened or endangered species. Surveys in the late 1980s by the Fish and Wildlife Service and the General Accounting Office revealed that more than 60 percent of refuges were permitting activities known to be harmful to wildlife. The most harmful practices, such as military activities and drilling, were not under the control of the Fish and Wildlife Service. The high-profile activist group Defenders of Wildlife organized a citizens' commission in 1992 that confirmed that the National Wildlife Refuge System was "falling far short of meeting the urgent habitat needs of the nation's wildlife" and was suffering from "chronic fiscal starvation and administrative neglect."

The National Wildlife Refuge System Improvement Act, signed into law by President Bill Clinton on October 9, 1997, dramatically shifted the priorities of the refuge system from its original sole purpose of protecting wildlife to the formation of a multiple-use system. The legislation redefined the system's mission regarding the conservation of habitat for fish, wildlife, and plants; designated priority public uses such as hunting, fishing, wildlife observation and photography, and environmental education and interpretation; and required that the environmental health of the refuge system be maintained. This monumental bill gave hunting, fishing, commercial trapping, and recreation equal status in the refuges with the conservation of plants, birds, and animals. It also limited new or secondary refuge use to activities compatible with wildlife protection and made legislative changes more difficult for future congressional cycles. The principle of multiple use, however, continues to allow and may result in increased mining, drilling, grazing, logging, and motorized recreation on refuge lands, in addition to increased military training, including bombing and tank and troop exercises. Upon signing the bill, Clinton stated that he "hoped and trusted that the process by which this bill was enacted will serve as a model for future congressional action on other environmental issues," with the future of the National Wildlife Refuge System to be shaped by the future of other bills such as the Clean Water Act, the Wetlands Protection Act, and the Endangered Species Act.

Daniel G. Graetzer

Further Reading

Butcher, Russell D. *America's National Wildlife Refuges: A Complete Guide.* 2d ed. Lanham, Md.: Taylor Trade, 2008.

Dolin, Eric Jay. *Smithsonian Book of National Wildlife Refuges.* Washington, D.C.: Smithsonian Institution Press, 2003.

Nelson, Lisa. "Wildlife Policy." In *Western Public Lands and Environmental Politics,* edited by Charles Davis. 2d ed. Boulder, Colo.: Westview Press, 2001.

Patent, Dorothy Hinshaw. *Places of Refuge: Our National Wildlife Refuge System.* Boston: Houghton Mifflin, 1992.

Riley, Laura, and William Riley. *Guide to the National Wildlife Refuges.* Rev. ed. New York: Macmillan, 1992.

■ Wind energy

FIELDS OF STUDY
Ecology; Ecosystems; Energy and Energy Resources; Energy Efficiency; Environment; Environmental Engineering; Environmentalism; Renewable Resources; Technology and Applied Science

SUMMARY
Wind energy is one of several nonpolluting, renewable types of energy that are considered potential candidates to replace fossil fuels, which are finite in supply and produce by-products that are harmful to the environment. It has been predicted that wind and solar industries will be significant sources of new manufacturing jobs in the twenty-first century.

PRINCIPAL TERMS
- **American Wind Energy Association:** a Washington, D.C.-based trade association formed in 1974, representing wind power project developers, equipment suppliers, service providers, parts manufacturers, utilities, researchers, and others involved in the wind industry and which promotes wind energy as a clean source of electricity for consumers in the United States and abroad
- **fossil fuels:** any combustible organic material, as oil, coal, or natural gas, derived from the remains of former life
- **renewable energy:** energy derived from natural, unlimited, and replenishable sources
- **smog:** smoke or other atmospheric pollutants combined with fog in an unhealthy or irritating mixture
- **wind energy:** energy harnessed from moving air to produce mechanical or electrical power
- **wind turbine:** a tall structure with blades that are blown round by the wind and produce power to make electricity
- **windmills:** any of various machines for grinding, pumping, etc., driven by the force of the wind acting upon several vanes or sails

Human harnessing of wind energy goes back thousands of years. Historically, people used sails to harness wind energy to propel ships long before the invention of the steam engine. Wind energy has also long been used to drive windmills to grind grain, pump water for irrigation, and keep lands from being flooded with seawater. At the dawn of the twentieth century, however, as fossil fuels became cheap and widely accessible and as the usage and applications of electricity became widespread, windmills began to be neglected except by a few interested researchers and users.

Rapid increases in the prices of fossil fuels during the 1970's brought a resurgence of interest in wind energy as an alternative source of power. This led to the progressive evolution of windmills into wind turbines—wind-driven machines connected to electrical generators to produce electricity. By the beginning of the twenty-first century, the combination of practical experience, advances in technology, and scientific research had led to the sophisticated wind turbines that dot the landscapes of many countries, including Germany, Denmark, the Netherlands, South Africa, and the United States.

How Wind Turbines Work

A turbine is a machine that converts the energy that is stored in a moving fluid (such as air, water, or steam) into another form of energy (such as electricity or mechanical work). Wind turbines catch energy from the wind by using blades that are shaped like propellers. The blades are attached to a shaft and are tilted in such a way that the force of the wind on them attempts to lift each blade. The lift is only partially complete, because the shaft begins to turn before the blade rises very high above its original station. This lift effect holds true for each blade, and it is repeated over and over again. The net result is that the shaft rotates continuously as long as the speed of the wind remains above a certain threshold. The assembly consisting of the blades and the shaft to which they are attached is part of what is called the rotor.

During the early days, windmills had six or more blades. It is now known that, by carefully shaping the blades, one can use fewer of them and capture much more energy than windmills did. Thus, in modern times, turbines are equipped with only two or three blades. The wind turbine assembly is mounted onto a tall tower. As a general rule, the taller the tower, the better. This is because the higher above the ground the turbine is located, the less the wind that reaches it is disturbed or reduced by what is on the ground and by surrounding objects such as trees and buildings.

Applications and Systems

Wind energy applications can be divided into three types: stand-alone wind turbines, distributed energy systems, and turbines that are connected to utility power grids. Stand-alone systems are generally used by home owners and by small business owners—such as ranchers, farmers, and owners of small retail stores—seeking to reduce the size of their electric bills. Others use them for communications and for pumping water. Distributed energy systems are various small power generation technologies that can be grouped and combined for the purpose of improving or expanding the operation and delivery of electrical energy.

For wind turbines to be connected to a local power grid, large numbers of them, generating many megawatts of power, are needed to make the required costs of construction, operation, and maintenance worthwhile. Such arrangements are called wind farms or wind plants. Several providers of electrical power, in the United States and other nations, use wind farms to supply power to their customers.

Benefits

Research by the American Wind Energy Association indicates that two main categories of benefits are associated with the adoption of wind energy: First, the production of electricity using wind energy reduces environmental risks while enhancing health benefits; and second, the installation of wind farms spurs economic development in the areas, usually rural, where they are located.

The generation of electricity using wind energy produces little air pollution. It is estimated that extensive use of this technology could reduce total U.S. emissions of carbon dioxide, a associated with global warming, by one-third. This corresponds to a reduction of 4 percent at the world level. Survey data show that forty-six of fifty states in the United States have wind resources that could be developed; thus, the potential for growth in this area is very great. For example, if ten of the windiest states in the United States were to develop 10 percent of their wind energy potential, the result would offset the carbon dioxide emissions from all U.S. power plants that burn coal.

The American Wind Energy Association estimated in 2010 that every megawatt of electrical power produced from wind energy generated $1 million in economic development. The association found that when wind energy is adopted by rural communities, local farmers and other landowners receive steady income through the lease of their land and the payment of royalties. Furthermore, the advent of wind energy operations brings new jobs to the communities where they are located. Some of these jobs are directly related to the installation and maintenance of wind turbines, whereas others come from such activities as road and building construction and from the transportation, hospitality, and food services needed as the local economy changes.

Josué Njock Libii

Further Reading

Burton, Tony, et al. *Wind Energy Handbook.* New York: John Wiley & Sons, 2001.

Eggleston, David M., and Forrest S. Stoddard. *Wind Turbine Engineering Design.* New York: Van Nostrand Reinhold, 1987.

Gipe, Paul. *Wind Power: Renewable Energy for Home, Farm, and Business.* 2d ed. White River Junction, Vt.: Charles Green, 2004.

Hansen, Martin O. L. *Aerodynamics of Wind Turbines.* 2d ed. Sterling, Va.: Earthscan, 2008.

McKinney, Michael L., Robert M. Schoch, and Logan Yonavjak. "Renewable and Alternative Energy Sources." In *Environmental Science: Systems and Solutions.* 4th ed. Sudbury, Mass.: Jones and Bartlett, 2007.

Appendixes

U.S. federal laws concerning the environment

Anh Tran

ALASKA NATIONAL INTEREST LANDS CONSERVATION ACT (1980)
Designated certain public lands in Alaska as units of the national park and national forest systems, the national wildlife refuge system, the National Wild and Scenic Rivers System, and the National Wilderness Preservation System. Provided comprehensive management guidance for all public lands in Alaska, including provisions regarding wilderness; subsistence; transportation and utility corridors; oil and gas leasing; mining; public access; hunting, trapping, and fishing; and implementation of the Alaska Native Claims Settlement Act (1971).

ANTIQUITIES ACT (ACT FOR THE PRESERVATION OF AMERICAN ANTIQUITIES; 1906)
Authorized permits for legitimate archaeological investigations and penalties for taking or destroying antiquities. Authorized presidents to protect all forms of American historical sites (natural, scientific, and archaeological) by proclaiming them to be national monuments.

CLEAN AIR ACT (1963)
Regulated air emissions from area, stationary, and mobile sources. Amendments in 1970 authorized the Environmental Protection Agency to create national air-quality standards to protect health and the environment, and required states to prepare and submit plans to implement clean air standards. Amendments in 1977 extended the deadline for areas that had not reached compliance levels by 1975. Amendments in 1990 addressed such issues as acid rain, ozone depletion, and air toxins.

CLEAN WATER ACT (1977)
An amendment to the Federal Water Pollution Control Act of 1972; prohibited discharge of any pollutant from a source point into navigable waters of the United States unless a special permit had been obtained from the Environmental Protection Agency. Amendment in 1987 (the Water Quality Act) included provisions for toxic pollutants, citizen suits, and funding of sewage treatment plants.

COASTAL ZONE MANAGEMENT ACT (1972)
Provided for management of the nation's coastal resources, including the Great Lakes, and balanced economic developments with environmental conservation. Encouraged states and Native American tribal governments to preserve, protect, develop, and restore or enhance valuable national coastal resources. Amendments in 1990 called on states and tribes to develop and implement coastal nonpoint pollution control programs.

EMERGENCY PLANNING AND COMMUNITY RIGHT-TO-KNOW ACT (1986)
Provided assistance to local communities in protecting the environment and public health and safety from chemical hazards. Required each state to create a State Response Commission, to divide itself into districts, and to appoint an Emergency Planning Committee for each district. Required both commissions to provide the community with information on chemical hazards that might affect the public, and required the dissemination of procedures to be followed in the event of emergency hazardous situations.

ENDANGERED SPECIES ACT (1973)
Repealed the Endangered Species Conservation Act (1969), which had amended the Endangered Species Preservation Act (1966). Implemented the Convention on International Trade in Endangered Species of Wild Fauna and Flora in 1973 and the 1940 Convention on Nature Protection and Wildlife Preservation in the Western Hemisphere. Provided for the conservation of ecosystems on which threatened and endangered species of fish, wildlife, and plants depend and required the U.S. Fish and Wildlife Service to designate which plants and animals were threatened or endangered. Prohibited activities that could have adverse effects on endangered or threatened species and their habitats.

ENERGY POLICY AND CONSERVATION ACT (1975)
Enacted to help cut the amount of energy consumed by various industrial and consumer products. Introduced Corporate Average Fuel Economy (CAFE)

standards for automobile manufacturers, extended oil price controls to 1979, and created the Strategic Petroleum Reserve. Amended in part by the Alternative Fuels Act (1988), which encouraged the development, production, and demonstration of alternative motor fuels and vehicles that could run on such fuels.

FEDERAL INSECTICIDE, FUNGICIDE, AND RODENTICIDE ACT (1947)
Amendments in 1972 prohibited the sale, distribution, or use of pesticides that might adversely affect threatened or endangered species. Required users of pesticides to register when they purchase pesticides and to take and pass a certification examination in order to apply pesticides. Required that all pesticides used in the United States be approved and licensed by the Environmental Protection Agency.

FEDERAL LAND POLICY AND MANAGEMENT ACT (1976)
Guided the Bureau of Land Management in the management, protection, development, and enhancement of public lands. Required the agency to manage lands for multiple uses and for sustained yield of resources for both present and future generations.

FISH AND WILDLIFE ACT (1956)
Directed the secretary of the interior to develop the policies and procedures necessary for carrying out fish and wildlife laws and to research and report on fish and wildlife matters. Authorized the administrator of the Environmental Protection Agency to undertake studies on the effects of insecticides, herbicides, fungicides, and other pesticides on fish and wildlife resources to determine the amounts, percentages, and formulations of chemicals injurious to fish and wildlife and thus the amounts, percentages, and formulations that could be used without losses of fish and wildlife from spraying, dusting, or other treatments using these chemicals.

FLOOD CONTROL ACT (1944)
Limited the authorization and construction of navigation, flood control, and other water projects to those having significant benefits for navigation and that could be operated consistent with other river uses.

GENERAL MINING ACT (1872)
Authorized and governed prospecting and mining for economic minerals on public lands. Amendments and new acts that superseded the 1872 provisions established regulations on the removal and use of resources such as oil and natural gas and also provided protection for national parks and other historic sites.

MARINE MAMMAL PROTECTION ACT (1972)
Enacted in partial response to growing concerns that certain marine mammals were in danger of extinction or depletion as a result of human activities. Prohibited, with certain specified exceptions, the act of hunting, killing, capture, and harassment of mammals in U.S. waters and by American citizens on the high seas, and the importation of marine mammals and their products into the United States.

MARINE PROTECTION, RESEARCH, AND SANCTUARIES ACT (OCEAN DUMPING ACT; 1972)
Prohibited all municipal sewage sludge and industrial waste dumping into the ocean after December 31, 1991.

MIGRATORY BIRD TREATY ACT (1918)
Implemented the 1916 convention between the United States and Great Britain and incorporated the provisions in the 1913 Migratory Bird Act (also known as the Weeks-McLean Act). Made it unlawful to pursue, hunt, take, capture, kill, or sell more than eight hundred species of birds that migrate between the United States and Canada. Scope of the act's protection expanded after similar conventions were signed between the United States and Mexico, and between Japan and the Soviet Union.

NATIONAL ENVIRONMENTAL POLICY ACT (1970)
Required federal agencies to take environmental factors into consideration before undertaking any major action such as construction of new highways, airports, or military complexes. Required the government to disclose the probable environmental effects of all projects by completing environmental assessments and environmental impact statements. Involved federal courts in environmental questions, expanded judicial review into agency decisions, and gave Congress additional power over matters concerning the environment.

National Forest Management Act (1976)
An amendment to the Forest and Rangeland and Renewable Resources Planning Act (1974). Required the secretary of agriculture to assess forestlands, develop a management program based on multiple-use, sustained-yield principles, and implement a resource management plan for each unit of the national forest system.

National Landscape Conservation System Act (2009)
Unified individual units as a public lands system to ensure that the conservation system was appropriately managed, funded, and protected for future generations. Included in the Omnibus Public Land Management Act (2009), which also added newly designated sites to the system, including a national monument and three national conservation areas.

National Trails System Act (1968)
Authorized a national system of trails and defined four categories of national trails. Provided for outdoor recreation opportunities and promoted the preservation of access to the wildnerness areas and historic resources of the nation.

National Wildlife Refuge System Administration Act (1966)
Provided guidelines and directives for administration and management of all areas in the system, including wildlife refuges, areas for the protection and conservation of fish and wildlife threatened with extinction, game ranges, wildlife management areas, and waterfowl production areas. Amended by the National Wildlife Refuge System Improvement Act (1997).

Occupational Safety and Health Act (1970)
Required employers to provide workers with safe workplaces, addressing such issues as the workplace use of toxic and hazardous substances. Required that state workplace safety and health acts meet or exceed federal requirements.

Oil Pollution Act (1990)
Regulated oil storage facilities and required oil-carrying vessels to submit plans for response in the case of large discharges. Required the development of area contingency plans to prepare and plan for oil spill response on a regional scale.

Organic Foods Production Act (1990)
Required the U.S. Department of Agriculture to develop national standards for organically produced agricultural products. Required producers to be in full compliance with the resulting standards by October 20, 2002, to be allowed to use the word "organic" in marketing.

Pollution Prevention Act (1990)
Focused on source reduction of pollution through requiring cost-effective changes in production, operation, and raw-material use by both private industry and the government. Included provisions regarding recycling and sustainable agricultural practices that increase efficiency in the use of energy, water, and other natural resources.

Reclamation Act (Newlands Act; 1902)
Established the Reclamation Fund and provided for the construction of irrigation projects in the arid lands of the American West. The newly irrigated land would be sold and money put into a revolving fund for similar projects.

Resource Conservation and Recovery Act (1976)
Contained provisions for the control of the generation, transportation, treatment, storage, and disposal of hazardous waste and the management of nonhazardous solid wastes. Amendments in 1984 established the national hazardous waste management program and required the Environmental Protection Agency to identify hazardous waste characteristics and list specific substances as hazardous wastes. Amendments in 1986 addressed problems related to underground storage of petroleum and other hazardous substances.

Safe Drinking Water Act (1974)
Addressed issues relating to the quality and safety of drinking water. Authorized the Environmental Protection Agency to establish purity standards for both underground and surface sources of water for human use.

Small Business Liability Relief and Brownfields Revitalization Act (2002)
Provided funds to assess and clean up brownfields (property affected by hazardous substances) and

provided funds to enhance brownfield response programs of states and Native American tribal governments.

Superfund (Comprehensive Environmental Response, Compensation, and Liability Act; 1980)
Addressed the handling of hazardous waste sites, accidents, spills, and other emergency releases of pollutants or contaminants. Authorized the Environmental Protection Agency (EPA) to locate the parties responsible for any release and enforce their cooperation in the cleanup. Required the EPA to do cleanup if the releasing parties could not be found or refused to cooperate, but allowed the EPA to recover the costs of the action from those involved. Directed the EPA to revise its Hazard Ranking System and take into account degree of risk, human health, and the environment when placing uncontrolled waste sites on the National Priorities List. Reauthorized by the Superfund Amendments and Reauthorization Act (1986).

Surface Mining Control and Reclamation Act (1977)
Established mandatory uniform standards for surface mining and required minimized adverse impacts on fish, wildlife, and related environmental values. Created a fund for reclaiming and restoring land and water resources adversely affected by coal-mining practices.

Taylor Grazing Act (1934)
Enacted to stop injury to rangelands caused by overgrazing; to provide for the lands' orderly use, improvement, and development; and to stabilize the livestock industry dependent on the public rangelands. Authorized the secretary of the interior to establish grazing districts on public lands and to develop regulations necessary to administer the districts.

Toxic Substances Control Act (1976)
Provided for the testing, regulation, and screening of all chemicals produced in or imported into the United States before they reach the consumer marketplace. Required the tracking of chemicals that pose health or environmental hazards and provided for the implementation of cleanup procedures in the case of contamination by toxic materials.

Uranium Mill Tailings Radiation Control Act (1978)
Gave the U.S. Department of Energy the responsibility of stabilizing, disposing, and controlling uranium mill tailings and other radiation-contaminated material at twenty-four uranium mill processing sites located across ten states and at more than five thousand associated properties.

Wild and Scenic Rivers Act (1968)
Created the National Wild and Scenic Rivers System to preserve select rivers with outstanding scenic, recreational, geologic, fish and wildlife, historic, cultural, or other important values in free-flowing conditions for the benefit of present and future generations.

Wilderness Act (1964)
Established the National Wilderness Preservation System and specified criteria for inclusion in the system. Made eligible every roadless area of 2,023 hectares (5,000 acres) or more, every roadless island within the national wildlife refuge and national park systems, and national forestlands.

Major world national parks and protected areas

Narayanan M. Komerath and Padma P. Komerath

This article lists a selection of the world's most important protected areas and national parks, noting the year of establishment and approximate area of each. The list is organized alphabetically by continent and by country within continents. Many of the sites mentioned here are designated by the United Nations Educational, Scientific, and Cultural Organization (UNESCO) as World Heritage Sites; more information on these protected areas is available at the official UNESCO Web site devoted to the World Heritage List: http://whc.unesco.org/en/list.

Many of the sites in the United States fall under the jurisdiction of the US National Park Service (NPS). A complete listing of that agency's managed areas can be found on its official website: https://www.nps.gov/findapark/advanced-search.htm.

Africa

Cameroon
Dja Faunal Reserve
   Year established: 1950:
   Area: 5,260 square kilometers (2,031 square miles)

Central African Republic
Manovo-Gounda St. Floris National Park (World Heritage Site)
   Year established: 1988:
   Area: 17,400 square kilometers (6,718 square miles)

Côte d'Ivoire
Comoé National Park (World Heritage Site)
   Year established: 1983:
   Area: 11,493 square kilometers (4,437 square miles)

Democratic Republic of the Congo
Garamba National Park
   Year established: 1938:
   Area: 4,920 square kilometers (1,900 square miles)

Kahuzi-Biéga National Park
   Year established: 1970:
   Area: 6,000 square kilometers (2,317 square miles)

Okapi Wildlife Reserve
   Year established: 1992:
   Area: 13,726 square kilometers (5,300 square miles)

Salonga National Park
   Year established: 1970:
   Area: 36,000 square kilometers (13,900 square miles)

Virunga National Park
   Year established: 1925:
   Area: 7,800 square kilometers (3,012 square miles)

Ethiopia
Simien National Park
   Year established: 1969:
   Area: 220 square kilometers (85 square miles)

Kenya
Lake Turkana National Parks (World Heritage Site)
   Year established: 1997:
   Area: 1,615 square kilometers (624 square miles)

Mount Kenya National Park/Natural Forest (World Heritage Site)
   Year established: 1997:
   Area: 1,420 square kilometers (548 square miles)

Madagascar
Rainforests of the Atsinanana (World Heritage Site)
   Year established: 2007:
   Area: 4,797 square kilometers (1,852 square miles)

Namibia
Etosha National Park
   Year established: 1975:
   Area: 22,270 square kilometers (8,598 square miles)

Niger
Air and Ténéré Natural Reserves (World Heritage Site)
   Year established: 1991:
   Area: 77,360 square kilometers (29,869 square miles)

Senegal
Niokolo-Koba National Park
   Year established: 1954:
   Area: 9,130 square kilometers (3,525 square miles)

South Africa
Cape Floral Region Protected Areas (World Heritage Site)
   Year established: 2004:
   Area: 5,530 square kilometers (2,135 square miles)

Golden Gate Highlands National Park
   Year established: 1963:
   Area: 340 square kilometers (131 square miles)

iSimangaliso Wetland Park (World Heritage Site)
   Year established: 1999:
   Area: 3,280 square kilometers (1,266 square miles)

Kruger National Park
   Year established: 1926:
   Area: 18,989 square kilometers (7,332 square miles)

Vredefort Dome (World Heritage Site)
   Year established: 2005:
   Area: 300 square kilometers (116 square miles)

Tanzania
Kilimanjaro National Park
   Year established: 1973:
   Area: 753 square kilometers (291 square miles)

Ngorongoro Conservation Area
   Year established: 1959:
   Area: 8,288 square kilometers (3,200 square miles)

Selous Game Reserve
   Year established: 1922:
   Area: 44,800 square kilometers (17,297 square miles)

Serengeti National Park
   Year established: 1951:
   Area: 14,760 square kilometers (5,700 square miles)

Uganda
Bwindi Impenetrable National Park
   Year established: 1991:
   Area: 331 square kilometers (128 square miles)

Rwenzori Mountains National Park
   Year established: 1991:
   Area: 998 square kilometers (385 square miles)

Zambia
Mosi-oa-Tunya/Victoria Falls (World Heritage Site)
   Year established: 1989:
   Area: 88 square kilometers (34 square miles)

Zimbabwe
Mana Pools National Park, Sapi and Chewore Safari Areas (World Heritage Site)
   Year established: 1984:
   Area: 6,766 square kilometers (2,612 square miles)

ASIA

China
Huanglong Scenic and Historic Interest Area (World Heritage Site)
   Year established: 1992:
   Area: 600 square kilometers (232 square miles)

Sichuan Giant Panda Sanctuaries-Wolong, Mt. Siguniang and Jiajin Mountains (World Heritage Site)
   Year established: 2006:
   Area: 9,245 square kilometers (3,570 square miles)

South China Karst (World Heritage Site)
  Year established: 2007:
  Area: 47,600 square kilometers (18,378 square miles)

Three Parallel Rivers of Yunnan Protected Areas (World Heritage Site)
  Year established: 2003:
  Area: 17,000 square kilometers (6,564 square miles)

Wulingyuan Scenic and Historic Interest Area (World Heritage Site)
  Year established: 1992:
  Area: 264 square kilometers (102 square miles)

India
Great Nicobar Biosphere Reserve
  Year established: 1989:
  Area: 885 square kilometers (342 square miles)

Kaziranga National Park
  Year established: 1974:
  Area: 430 square kilometers (166 square miles)

Manas Wildlife Sanctuary (World Heritage Site)
  Year established: 1985:
  Area: 500 square kilometers (193 square miles)

Nanda Devi and Valley of Flowers National Parks (World Heritage Site)
  Year established: 1988:
  Area: 718 square kilometers (277 square miles)

Sundarbans National Park
  Year established: 1984:
  Area: 1,330 square kilometers (514 square miles)

Indonesia
Komodo National Park
  Year established: 1980:
  Area: 1,733 square kilometers (669 square miles)

Lorentz National Park
  Year established: 1997:
  Area: 25,056 square kilometers (9,674 square miles)

Tropical Rainforest Heritage of Sumatra (World Heritage Site)
  Year established: 2004:
  Area: 25,951 square kilometers (10,020 square miles)

Ujung Kulon National Park
  Year established: 1980:
  Area: 1,206 square kilometers (466 square miles)

Japan
Shirakami-Sanchi (World Heritage Site)
  Year established: 1993:
  Area: 169 square kilometers (65 square miles)

Kazakhstan
Saryarka-Steppe and Lakes of Northern Kazakhstan (World Heritage Site)
  Year established: 2008:
  Area: 4,503 square kilometers (1,739 square miles)

Malaysia
Gunung Mulu National Park
  Year established: 1974:
  Area: 529 square kilometers (204 square miles)

Kinabalu National Park
  Year established: 1964:
  Area: 754 square kilometers (291 square miles)

Mongolia
Uvs Nuur Basin (World Heritage Site)
  Year established: 2003:
  Area: 8,981 square kilometers (3,468 square miles)

Nepal
Chitwan National Park
  Year established: 1973:
  Area: 932 square kilometers (360 square miles)

Sagarmatha National Park
  Year established: 1976:
  Area: 1,148 square kilometers (443 square miles)

Philippines
Tubbataha Reefs Natural Park (World Heritage Site)
  Year established: 1993:
  Area: 1,300 square kilometers (502 square miles)

Republic of Korea (South Korea)
Jeju Volcanic Island and Lava Tubes (World Heritage Site)
    Year established: 2007:
    Area: 95 square kilometers (37 square miles)

Russia
Central Sikhote-Alin (World Heritage Site)
    Year established: 2001:
    Area: 15,539 square kilometers (6,000 square miles)

Golden Mountains of Altai (World Heritage Site)
    Year established: 1998:
    Area: 16,115 square kilometers (6,222 square miles)

Lake Baikal (World Heritage Site)
    Year established: 1996:
    Area: 88,000 square kilometers (33,977 square miles)

Natural System of Wrangel Island Reserve (World Heritage Site)
    Year established: 2004:
    Area: 9,163 square kilometers (3,538 square miles)

Volcanoes of Kamchatka (World Heritage Site)
    Year established: 1996:
    Area: 38,302 square kilometers (14,788 square miles)

Thailand
Thungyai-Huai Kha Khaeng Wildlife Sanctuaries (World Heritage Site)
    Year established: 1991:
    Area: 5,775) square kilometers (2,230 square miles)

Vietnam
Phong Nha-Ke Bang National Park
    Year established: 2001:
    Area: 858 square kilometers (331 square miles)

AUSTRALIA

Gondwana Rainforests (World Heritage Site)
    Year established: 1986:
    Area: 3,700 square kilometers (1,429 square miles)

Great Barrier Reef (World Heritage Site)
    Year established: 1981:
    Area: 348,700 square kilometers (134,634 square miles)

Greater Blue Mountains Area (World Heritage Site)
    Year established: 2000:
    Area: 10,326 square kilometers (3,987 square miles)

Heard and McDonald Islands (World Heritage Site)
    Year established: 1997:
    Area: 386 square kilometers (149 square miles)

Purnululu National Park
    Year established: 1987:
    Area: 2,397 square kilometers (925 square miles)

Shark Bay, Western Australia (World Heritage Site)
    Year established: 1991:
    Area: 21,973 square kilometers (8,484 square miles)

Tasmanian Wilderness (World Heritage Site)
    Year established: 1982:
    Area: 14,075 square kilometers (5,434 square miles)

Willandra Lakes Region (World Heritage Site)
    Year established: 1981:
    Area: 2,400 square kilometers (927 square miles)

EUROPE

Belarus
Belovezhskaya Pushcha/Białowieża Forest (World Heritage Site)
    Year established: 1979:
    Area: 927 square kilometers (358 square miles)

Finland
High Coast/Kvarken Archipelago (World Heritage Site)
   Year established: 2000:
   Area: 1,944 square kilometers (751 square miles)

Germany
Wadden Sea (World Heritage Site)
   Year established: 2009:
   Area: 10,000 square kilometers (3,861 square miles)

Greenland
Ilulissat Icefjord (World Heritage Site)
   Year established: 2004:
   Area: 4,024 square kilometers (1,554 square miles)

Iceland
Surtsey (World Heritage Site)
   Year established: 2008:
   Area: 34 square kilometers (13 square miles)

Italy
Aeolian Islands (World Heritage Site)
   Year established: 2000:
   Area: 12 square kilometers (5 square miles)

Dolomites mountains range (World Heritage Site)
   Year established: 2009:
   Area: 1,419 square kilometers (548 square miles)

Norway
West Norwegian Fjords-Geirangerfjord and Nærøyfjord (World Heritage Site)
   Year established: 2005:
   Area: 1,227 square kilometers (474 square miles)

Romania
Danube Delta (World Heritage Site)
   Year established: 1991:
   Area: 3,124 square kilometers (1,206 square miles)

Spain
Doñana National Park
   Year established: 1969:
   Area: 543 square kilometers (210 square miles)

Garajonay National Park
   Year established: 1981:
   Area: 40 square kilometers (15 square miles)

Teide National Park
   Year established: 1954:
   Area: 190 square kilometers (73 square miles)

Sweden
Laponian Area (World Heritage Site)
   Year established: 1996:
   Area: 9,400 square kilometers (3,629 square miles)

United Kingdom
Dorset and East Devon Coast (World Heritage Site)
   Year established: 2001:
   Area: 25 square kilometers (10 square miles)

English Lake District (World Heritage Site)
   Year established: 2017:
   Area: 2,292 square kilometers (884 square miles)

Giant's Causeway and Causeway Coast (World Heritage Site)
   Year established: 1986:
   Area: 0.7 square kilometer (0.27 square mile)

NORTH AMERICA—CANADA

Alberta
Wood Buffalo National Park
   Year established: 1922:
   Area: 44,807 square kilometers (17,300 square miles)

Manitoba
Wapusk National Park
   Year established: 1996:
   Area: 11,475 square kilometers (4,431 square miles)

Northwest Territories
Tuktut Nogait National Park
   Year established: 1996:
   Area: 16,340 square kilometers (6,309 square miles)

Nunavut

Auyuittuq National Park
   Year established: 1976:
   Area: 19,089 square kilometers (7,370 square miles)

Quttinirpaaq National Park
   Year established: 1986:
   Area: 37,775 square kilometers (14,585 square miles)

Yukon Territory

Ivvavik National Park
   Year established: 1984:
   Area: 10,168 square kilometers (3,926 square miles)

Kluane National Park and Reserve
   Year established: 1976:
   Area: 22,013 square kilometers (8,499 square miles)

NORTH AMERICA—MEXICO

Calakmul Biosphere Reserve
   Year established: 1989:
   Area: 7,284 square kilometers (2,812 square miles)

El Vizcaíno Biosphere Reserve
   Year established: 1988:
   Area: 143,600 square kilometers (55,444 square miles)

Isla del Golfo Special Biosphere Reserve
   Year established: 1978:
   Area: 1,517 square kilometers (586 square miles)

Islands and Protected Areas of the Gulf of California (World Heritage Site)
   Year established: 2005:
   Area: 7,046 square kilometers (2,720 square miles)

Montes Azules Biosphere Reserve
   Year established: 1971:
   Area: 3,351 square kilometers (1,294 square miles)

Sierra de San Pedro Mártir National Park
   Year established: 1947:
   Area: 757 square kilometers (292 square miles)

NORTH AMERICA—UNITED STATES

Alaska

Arctic National Wildlife Refuge
   Year established: 1960:
   Area: 78,000 square kilometers (30,116 square miles)

Denali National Park and Preserve
   Year established: 1917:
   Area: 24,585 square kilometers (9,492 square miles)

Gates of the Arctic National Park and Preserve
   Year established: 1980:
   Area: 39,460 square kilometers (15,236 square miles)

Glacier Bay National Park and Preserve
   Year established: 1980:
   Area: 13,287 square kilometers (5,130 square miles)

Katmai National Park and Preserve
   Year established: 1980:
   Area: 19,120 square kilometers (7,382 square miles)

Kenai Fjords National Park
   Year established: 1980:
   Area: 4,600 square kilometers (1,776 square miles)

Kobuk Valley National Park
   Year established: 1980:
   Area: 6,757 square kilometers (2,609 square miles)

Lake Clark National Park and Preserve
   Year established: 1980:
   Area: 16,308 square kilometers (6,297 square miles)

Wrangell-St. Elias National Park and Preserve
    Year established: 1980:
    Area: 53,321 square kilometers (20,587 square miles)

Yukon Delta National Wildlife Refuge
    Year established: 1909:
    Area: 77,500 square kilometers (29,923 square miles)

Arizona
Coronado National Forest
    Year established: 1941:
    Area: 7,203 square kilometers (2,781 square miles)

Glen Canyon National Recreation Area
    Year established: 1972:
    Area: 4,856 square kilometers (1,875 square miles)

Grand Canyon National Park
    Year established: 1919:
    Area: 4,950 square kilometers (1,911 square miles)

Petrified Forest National Park
    Year established: 1962:
    Area: 380 square kilometers (147 square miles)

Saguaro National Park
    Year established: 1994:
    Area: 370 square kilometers (143 square miles)

Arkansas
Hot Springs National Park
    Year established: 1921:
    Area: 23 square kilometers (9 square miles)

Ozark-St. Francis National Forest, Arkansas
    Year established: 1908:
    Area: 4,694 square kilometers (1,812 square miles)

California
Channel Islands National Park
    Year established: 1980:
    Area: 1,010 square kilometers (390 square miles)

Death Valley National Park
    Year established: 1933:
    Area: 13,638 square kilometers (5,266 square miles)

Joshua Tree National Park
    Year established: 1994:
    Area: 3,196 square kilometers (1,234 square miles)

Kings Canyon National Park
    Year established: 1940:
    Area: 1,873 square kilometers (723 square miles)

Lassen National Forest
    Year established: 1905:
    Area: 4,300 square kilometers (1,700 square miles)

Lassen Volcanic National Park
    Year established: 1916:
    Area: 429 square kilometers (166 square miles)

Mojave National Preserve
    Year established: 1994:
    Area: 6,475 square kilometers (2,500 square miles)

Point Reyes National Seashore
    Year established: 1962:
    Area: 288 square kilometers (111 square miles)

Redwood National and State Parks
    Year established: 1968:
    Area: 456 square kilometers (178 square miles)

Santa Monica National Recreation Area
    Year established: 1978:
    Area: 620 square kilometers (239 square miles)

Sequoia National Park
    Year established: 1890:
    Area: 1,635 square kilometers (631 square miles)

Yosemite National Park
    Year established: 1890:
    Area: 3,081 square kilometers (1,190 square miles)

Colorado

Black Canyon of the Gunnison National Park
   Year established: 1999:
   Area: 124 square kilometers (48 square miles)

Dinosaur National Monument
   Year established: 1915:
   Area: 853 square kilometers (329 square miles)

Great Sand Dunes National Park and Preserve
   Year established: 1932:
   Area: 170 square kilometers (66 square miles)

Mesa Verde National Park
   Year established: 1906:
   Area: 211 square kilometers (81 square miles)

Rocky Mountain National Park
   Year established: 1915:
   Area: 1,076 square kilometers (415 square miles)

Florida

Big Cypress National Preserve
   Year established: 1974:
   Area: 2,916 square kilometers (1,126 square miles)

Biscayne National Park
   Year established: 1980:
   Area: 700 square kilometers (270 square miles)

Dry Tortugas National Park
   Year established: 1935:
   Area: 249 square kilometers (96 square miles)

Everglades National Park
   Year established: 1947:
   Area: 6,050 square kilometers (2,336 square miles)

Georgia

Chattahoochee-Oconee National Forest
   Years established: 1936 (Chattahoochee), 1959 (Oconee):
   Area: 3,504 square kilometers (1,353 square miles)

Okefenokee National Wildlife Refuge
   Year established: 1937:
   Area: 1,627 square kilometers (628 square miles)

Hawaii

Haleakalā National Park
   Year established: 1961:
   Area: 116 square kilometers (45 square miles)

Hawaii Volcanoes National Park
   Year established: 1916:
   Area: 931 square kilometers (359 square miles)

Idaho

Craters of the Moon National Monument and Preserve
   Year established: 1924:
   Area: 2,893 square kilometers (1,117 square miles)

Kentucky

Big South Fork National River and Recreation Area
   Year established: 1974:
   Area: 507 square kilometers (196 square miles)

Mammoth Cave National Park
Year established: 1941:
Area: 214 square kilometers (83 square miles)

Maine

Acadia National Park
   Year established: 1929:
   Area: 186 square kilometers (72 square miles)

Massachusetts

Cape Cod National Seashore
   Year established: 1961:
   Area: 176 square kilometers (68 square miles)

Michigan

Isle Royale National Park
   Year established: 1940:
   Area: 2,314 square kilometers (893 square miles)

Minnesota

Mississippi National River and Recreation Area
   Year established: 1988:
   Area: 218 square kilometers (84 square miles)

Voyageurs National Park
   Year established: 1975:
   Area: 882 square kilometers (341 square miles)

Mississippi
Gulf Islands National Seashore
    Year established: 1971:
    Area: 549 square kilometers (212 square miles)

Montana
Bighorn Canyon National Recreation Area
    Year established: 1966:
    Area: 487 square kilometers (188 square miles)

Glacier National Park
    Year established: 1910:
    Area: 4,101 square kilometers (1,583 square miles)

Nevada
Great Basin National Park
    Year established: 1986:
    Area: 310 square kilometers (120 square miles)

New Mexico
Carlsbad Caverns National Park
    Year established: 1930:
    Area: 186 square kilometers (72 square miles)

White Sands National Monument
    Year established: 1933:
    Area: 581 square kilometers (224 square miles)

North Dakota
Theodore Roosevelt National Park
    Year established: 1978:
    Area: 285 square kilometers (110 square miles)

Ohio
Cuyahoga Valley National Park
    Year established: 2000:
    Area: 136 square kilometers (52 square miles)

Oregon
Crater Lake National Park
    Year established: 1902:
    Area: 732 square kilometers (283 square miles)

South Carolina
Congaree National Park
    Year established: 2003:
    Area: 85 square kilometers (33 square miles)

South Dakota
Badlands National Park
    Year established: 1978:
    Area: 970 square kilometers (375 square miles)

Wind Cave National Park
    Year established: 1903:
    Area: 115 square kilometers (44 square miles)

Tennessee
Great Smoky Mountains National Park
    Year established: 1934:
    Area: 2,109 square kilometers (814 square miles)

Texas
Big Bend National Park
    Year established: 1944:
    Area: 3,242 square kilometers (1,252 square miles)

Guadalupe Mountains National Park
    Year established: 1972:
    Area: 350 square kilometers (135 square miles)

Utah
Arches National Park
    Year established: 1971:
    Area: 310 square kilometers (120 square miles)

Bears Ears National Monument
    Year established: 2016:
    Area: originally 5,470 square kilometers (2,112 square miles), reduced to 816 square kilometers (315 square miles) in 2017

Bryce Canyon National Park
    Year established: 1928:
    Area: 145 square kilometers (56 square kilometers)

Canyonlands National Park
    Year established: 1964:
    Area: 1,366 square kilometers (527 square miles)

Capitol Reef National Park
    Year established: 1971:
    Area: 979 square kilometers (378 square miles)

Zion National Park
   Year established: 1909:
   Area: 579 square kilometers (224 square miles)

Virginia
Shenandoah National Park
   Year established: 1935:
   Area: 796 square kilometers (307 square miles)

Washington
Mount Rainier National Park
   Year established: 1899:
   Area: 942 square kilometers (364 square miles)

North Cascades National Park
   Year established: 1968:
   Area: 2,769 square kilometers (1,069 square miles)

Olympic National Park
   Year established: 1938:
   Area: 3,584 square kilometers (1,384 square miles)

Wyoming
Grand Teton National Park
   Year established: 1929:
   Area: 1,242 square kilometers (480 square miles)

Yellowstone National Park
   Year established: 1872:
   Area: 8,990 square kilometers (3,470 square miles)

NORTH AMERICA—UNITED STATES TERRITORIES

American Samoa
National Park of American Samoa
   Year established: 1988:
   Area: 42 square kilometers (16 square miles)

Virgin Islands
Virgin Islands National Park
   Year established: 1956:
   Area: 59 square kilometers (23 square miles)

SOUTH AMERICA AND CENTRAL AMERICA

Argentina
Ischigualasto/Talampaya Natural Parks (World Heritage Site)
   Year established: 2000:
   Area: 2,750 square kilometers (1,062 square miles)

Los Glaciares National Park
   Year established: 1981:
   Area: 4,459 square miles (1,722 square miles)

Valdes Peninsula (World Heritage Site)
   Year established: 1999:
   Area: 3,600 square kilometers (1,390 square miles)

Bolivia
Noel Kempff Mercado National Park
   Year established: 2000:
   Area: 15,230 square kilometers (5,880 square miles)

Brazil
Central Amazon Conservation Complex (World Heritage Site)
   Year established: 2000:
   Area: 60,000 square kilometers (23,166 square miles)

Cerrado Protected Areas: Chapada dos Veadeiros and Emas National Parks (World Heritage Site)
   Year established: 2001:
   Area: 65,500 square kilometers (25,290 square miles)

Iguazu National Park
   Year established: 1939:
   Area: 1,700 square kilometers (656 square miles)

Pantanal Conservation Area
   Year established: 2000:
   Area: 769 square kilometers (297 square miles)

Chile
Torres del Paine National Park
   Year established: 1959:
   Area: 2,400 square kilometers (927 square miles)

Costa Rica
Cocos Island National Park
 Year established: 1997:
 Area: 65,500 square kilometers (25,290 square miles)

Talamanca Range-La Amistad Reserves/La Amistad National Park (World Heritage Site)
 Year established: 1983:
 Area: 5,678 square kilometers (2,192 square miles)

Ecuador
Cotopaxi National Park
 Year established: 1975:
 Area: 334 square kilometers (129 square miles)

Galápagos Islands (World Heritage Site)
 Year established: 1978:
 Area: 140,665 square kilometers (54,311 square miles)

Peru
Manú National Park (World Heritage Site)
 Year established: 1987:
 Area: 15,328 square kilometers (5,918 square miles)

Suriname
Central Suriname Nature Reserve
 Year established: 2000:
 Area: 16,000 square kilometers (6,178 square miles)

Venezuela
Canaima National Park (World Heritage Site)
 Year established: 1994:
 Area: 30,000 square kilometers (11,583 square miles)

BIBLIOGRAPHY
"Find a Park." National Park Service, US Dept. of the Interior, www.nps.gov/findapark/advanced-search.htm. Accessed 23 Mar. 2018.

"World Heritage List." World Heritage Convention, United Nations Educational, Scientific and Cultural Organization (UNESCO), whc.unesco.org/en/list. Accessed 23 Mar. 2018.

■ Environmental organizations

Laurence W. Mazzeno

From the late twentieth century onward, especially since the 1970's, worldwide awareness of environmental issues has been heightened by the work of hundreds of nongovernmental organizations established to promote research, influence public policy, and encourage citizen engagement in solving problems associated with the earth's environment and ecosystems. Some of the most important of these organizations are briefly described below; asterisks on entries indicate organizations that are profiled in more depth in individual essays in this encyclopedia.

African Conservation Foundation (ACF)
Year founded: 1999.
http://www.africanconservation.org
ACF focuses on protecting and conserving African wildlife by seeking to find workable approaches to managing natural resources that offer a balance between needs for development and conservation initiatives. The organization provides training, support, and assistance to groups that share similar aims. It also sponsors research and conservation projects in Africa and engages in fund-raising to promote awareness and support of these efforts. ACF has offices in several African countries and in the United Kingdom.

American Farmland Trust (AFT)
Year founded: 1980.
http://www.farmland.org
AFT was founded by farmers and ranchers in the United States to help preserve farms and ranches, promote a healthy and sustainable environment, and build communities. It engages in lobbying activities and produces publications outlining problems associated with the loss of farm and ranch lands and suggesting solutions for retaining these resources. AFT has been influential in the passage of several state and federal laws governing farm and ranch preservation, including special provisions in the 1996, 2002, and 2008 farm bills passed by the U.S. Congress.

Antinea Foundation
Year founded: 2007.
http://www.antinea-foundation.org
The Antinea Foundation, based in Switzerland, supports research, education, and public awareness programs aimed at promoting conservation of the earth's oceans and marine ecosystems. The current organization was formed from a merger of two Swiss groups, Association Pacifique and Association Antinea. The centerpiece of the foundation's initiatives is a ten-year voyage of exploration and scientific research conducted aboard the Antinea Foundation's ship Fleur de Passion, a converted German warship.

Association for Environment Conscious Building (AECB)
Year founded: 1989.
http://www.aecb.net
Operating in the United Kingdom, AECB promotes building projects that respect and help preserve the environment. Its membership includes builders, designers, housing professionals, and government officials. The organization conducts seminars on green building practices, develops building standards and codes that enhance conservation and reduce harmful construction practices, and lobbies for implementation of rigorous requirements to reduce the impacts of new construction and renovation on the environment, especially standards that reduce harmful carbon emissions.

Association of Environmental and Resource Economists (AERE)
Year founded: 1979.
http://www.aere.org
AERE is an association of academics, professionals from government agencies and private research organizations, and representatives of consulting firms who are committed to promoting the study of environmental and natural resource economics. The group sponsors workshops, conferences, and symposia as a means of sharing information and stimulating further investigation of the economic ramifications of environmental problems. A sister organization, the European

Association of Environmental and Resource Economists, was established in 1990 and works toward the same ends.

Association of Environmental Professionals (AEP)
Year founded: 1975.
http://www.califaep.org

AEP draws its members from the fields of environmental planning, natural resources management, and environmental science. It promotes awareness of environmental issues, serves as a watchdog on governmental policies relating to the improvement or degradation of the environment, monitors the impacts of current legislation, and lobbies for changes to laws and policies to promote sustainability. The organization is also actively involved in helping members improve their skills as environmental activists and natural resource managers.

Australian Conservation Foundation
Year founded: 1966.
http://www.acfonline.org.au

The Australian Conservation Foundation is a nonprofit organization focused on protecting the natural resources of Australia. The group's chief interests are research, policy development, education, and advocacy. Over four decades it has made significant contributions to conservation efforts targeted at the Great Barrier Reef as well as several endangered rivers and rain forests. It has campaigned against hazardous mining practices and encouraged environmentally responsible farming and land management. Since 1990 it has sought more active involvement with businesses to promote environmentally responsible development.

Canadian Parks and Wilderness Society (CPAWS)
Year founded: 1963.
http://www.cpaws.org

CPAWS (also known as Société pour la nature et les parcs du Canada, or SNAP) is actively engaged in efforts to preserve the Canadian wilderness. The group monitors government and private activity in national and province parks, advocates for restricted development in natural environments, and promotes maintenance of healthy natural ecosystems. Annually it honors Canadians who have made notable contributions to conservation.

Center for Health, Environment, and Justice (CHEJ)
Year founded: 1981.
http://www.chej.org

Formerly known as Citizens Clearinghouse for Hazardous Waste, CHEJ is a grassroots organization that engages citizens in campaigns to promote healthy communities through elimination of pollutants and creation of sustainable urban ecosystems.

Center for International Environmental Law (CIEL)
Year founded: 1989.
http://www.ciel.org

CIEL is a nonprofit organization with offices in the United States and Switzerland. The organization uses international courts to regulate activities that might prove harmful to the environment. CIEL offers legal counsel, conducts policy research, engages in advocacy initiatives, and assists in building other groups' capacities to make meaningful contributions to environmental causes. It also sponsors educational activities, most notably a curriculum in international environmental law at the American University in Washington, D.C.

Ceres
Year founded: 1989.
http://www.ceres.org

Ceres is a network of investors, environmental organizations, and other public interest groups working to address a variety of sustainability issues, including matters such as global climate change.

Climate Project
Year founded: 2006.
http://www.theclimateprojectus.org

Founded by former U.S. vice president and 2007 Nobel Peace Prize recipient Al Gore, the Climate Project is designed to bring worldwide awareness to the growing problems associated with global warming and climate change. The organization sponsors lectures and seminars at which carefully trained presenters stress the urgency for action to reverse the damages of global warming. By 2009 the group had trained more than two thousand people to help deliver its message. The Climate Project has been the target of criticism from some observers who have expressed concerns that the organization presents an alarmist message that is not always rooted in sound science.

Comité de Liaison Energies Renouvelables (CLER)
Year founded: 1984.
http://www.cler.org/info
CLER is an advocacy and educational association made up of professionals from industry, commerce, architecture, building trades groups, and the academic community working collaboratively to promote sustainable energy for France. Members of CLER engage in educational programs to promote awareness of energy issues and garner support for projects that enhance the development of renewable energy resources.

Conservation International
Year founded: 1987.
http://www.conservation.org
Based in the United States, Conservation International has been active in promoting projects to preserve the world's biodiversity both on land and in marine environments. Working in nearly four dozen countries, the group has been active in sponsoring explorations of remote regions, where dozens of previously unknown species of animals and plants have been identified. The organization has engaged in lobbying efforts aimed at restricting activities harmful to natural ecosystems, but it has also worked to find solutions for allowing humans to live harmoniously within the natural environment.

Conservation Law Foundation (CLF)
Year founded: 1966.
http://www.clf.org
CLF is a regional environmental advocacy group focused on issues affecting the northeastern United States. It has lobbied successfully to halt development that would have degraded several traditionally important natural areas in the region, such as New Hampshire's White Mountains; has become involved in activities to clean up Atlantic coastal areas, several rivers, and Boston Harbor; has advocated for mass transportation as a means of reducing energy pollution; has fought against the proliferation of nuclear power; and has initiated collaborative efforts to combat global warming by rewarding companies that reduce emissions.

Earth First!
Year founded: 1980.
http://www.earthfirst.org
Earth First! is a radical advocacy group with branches in nearly twenty countries. It promotes direct action, including civil disobedience, to protect the environment from commercial efforts at development.

Earth Policy Institute (EPI)
Year founded: 2001.
http://www.earth-policy.org
EPI, an advocacy group based in the United States, is primarily interested in raising public awareness about dangers facing the world's population, among them environmental problems such as global warming and the loss of plant and animal species. Established to promote the ideas of environmentalist and activist Lester Brown, EPI publishes books and reports that lay out a vision of a sustainable environment, document current problems, and track efforts by governments and private groups in meeting attainable systemic changes.

Earthwatch Institute
Year founded: 1971.
http://www.earthwatch.org
The Earthwatch Institute is an international organization that engages people in activities that help them understand what is required to create a sustainable natural environment. Operating from offices in the United States, the United Kingdom, Australia, and Japan, the Institute sponsors research projects around the globe in which volunteers are encouraged to join professionals on expeditions where they can gather data on rain-forest and marine ecology, wildlife conservation, and archaeology.

Energy Action Coalition
Year founded: 2004.
http://energyactioncoalition.org
Energy Action Coalition is a network of more than fifty U.S. and Canadian environmental groups committed to raising awareness among college students regarding environmental problems facing the planet. Headquartered in the United States, the organization lobbies for changes to environmental law and

policies and engages communities in efforts to deal with challenges posed by the deteriorating condition of the earth's natural resources. The coalition encourages student involvement in numerous conferences, seminars, and lobbying efforts to get federal governments to reverse long-standing economic and environmental policies thought to favor development over conservation and sustainability of ecosystems.

Environmental and Energy Study Institute (EESI)
Year founded: 1984.
http://www.eesi.org

EESI grew out of a program established by the U.S. Congress in 1975 to gather and disseminate information about environment and energy issues as a means of assisting lawmakers in developing sound policies. The group became an independent nonprofit in 1984 but continued to keep its focus on education and data collection. EESI sponsors congressional briefings, meetings, and seminars and issues publications that address the topics of global warming and air pollution. It endorses policies that promote energy security and rural economic development and encourages increased use of renewable energy sources and improved energy efficiency. The organization also lobbies for the protection of areas such as the nation's Arctic and coastal regions.

Environmental Defense Fund (EDF)
Year founded: 1967.
http://www.edf.org

EDF has been one of the most effective and also one of the most controversial environmental advocacy organizations in the United States. Growing from a grassroots movement to save endangered raptors in New York, EDF emerged as a national force lobbying for laws and policies that promote species conservation, clean water, and sustainable ecosystems. It claims to have been the driving force behind the U.S. ban on the use of the pesticide dichloro-diphenyl-trichloroethane (DDT) and dangerous chemical compounds including chlorofluorocarbons (CFCs), passage of the Safe Drinking Water Act of 1974 and the 1990 amendments to the Clean Air Act, establishment of the Northwestern Hawaiian Islands Coral Reef Ecosystem Reserve, and creation of the U.S. Climate Action Partnership. Critics have charged, however, that EDF has often exercised little concern for human communities in its zeal to enforce tighter restrictions on the use of chemicals or improve natural ecosystems.

Environmental Design Research Association (EDRA)
Year founded: 1968.
http://www.edra.org

EDRA was founded in the United States to bring together design professionals, social scientists, facilities managers, and others interested in creating buildable spaces compatible with and respectful of the natural environment. Its members meet regularly to share scholarship and best practices, and EDRA publishes annual proceedings of the group's meetings. EDRA has also established a number of awards to recognize individuals, organizations, and projects that demonstrate commitment to best practices in environmental management.

Environmental Foundation for Africa (EFA)
Year founded: 1992.
http://www.efasl.org.uk

Although founded in the United Kingdom, EFA is based in West Africa and has as its principal focus the protection and improvement of that region. EFA works locally to sponsor community involvement in projects to improve environmental quality. The organization promotes awareness of environmental problems, provides technical assistance in creating solutions to those problems, serves as an advocacy group in lobbying for funding and policy changes that will help preserve or improve the environment of the region, and organizes networks of environmental organizations, government agencies, and private funding groups to bring about change that will improve the lives of the people of West Africa while respecting the natural world in which they live.

Environmental Investigation Agency (EIA)
Year founded: 1984.
http://www.eia-international.org

Based in the United Kingdom and the United States, EIA is an activist organization that funds undercover investigations to identify violations of environmental law and expose animal cruelty. EIA has been successful in focusing a spotlight on illegal logging activities in Southeast Asia, securing a ban on the illegal trade in tiger parts (particularly in China and India), lobbying for a cessation of commercial whaling, and revealing the extent of illegal trading in hazardous chemicals. Using evidence gathered in its investigations, EIA has lobbied effectively for stricter laws governing environmental issues in numerous countries.

Environmental Protection UK (EPUK)
Year founded: 1898.
http://www.environmental-protection.org.uk
Originally the Coal Smoke Abatement Society and known for years as the National Society for Clean Air and Environmental Protection, EPUK is a nongovernmental organization dedicated to improving air quality in the United Kingdom. It has also been active in efforts to reduce noise pollution and has campaigned to protect and rehabilitate land areas. EPUK lobbies for stronger environmental standards, particularly in the area of air quality, and provides assistance to local governments in achieving those standards.

European Environmental Bureau (EEB)
Year founded: 1974.
http://www.eeb.org
EEB is a federation made up of more than 140 environmental groups from all member countries of the European Union (EU). Organized principally to provide a unified and strong voice for representing the interests of environmentalists, EEB provides information on environmental issues to its members and to officials of individual European governments and the EU Parliament and its subsidiaries. It also represents member organizations in lobbying before these government bodies on pending legislation or on issues that EEB believes require government action to protect or rehabilitate the region's natural resources.

Friends of the Earth International (FOEI)
Year founded: 1969.
http://www.foei.org
FOEI is an international network of environmental organizations with affiliates in seventy-seven countries. The organization is dedicated to challenging environmentally unsound government policies and to promoting initiatives that will sustain and improve ecosystems worldwide.

Global Witness
Year founded: 1993.
http://www.globalwitness.org
Global Witness is an international nonprofit organization whose aim is to investigate and expose criminal activities and irresponsible behaviors that damage natural resources and have negative impacts on the quality of life for people in impoverished areas of the globe. Operating out of offices in the United States and the United Kingdom, Global Witness uses information gathered in its investigations to lobby governments for changes to policies that harm the environment and perpetuate poverty.

Green Belt Movement
Year founded: 1977.
http://www.greenbeltmovement.org
The Green Belt Movement, established in Kenya by activist Wangari Maathai, is a grassroots organization aimed at getting people, especially women, directly involved in conservation efforts. It has sponsored the planting of trees across the country as a means of replenishing resources used for subsistence by Kenya's rural population. During the 1980's the program expanded to other countries in Africa, and Green Belt officials began actively engaging in efforts to protest irresponsible development in the region.

Greenpeace
Year founded: 1971.
http://www.greenpeace.org/international
Greenpeace is an international organization with affiliates in more than forty countries. It supports research and encourages lobbying efforts to identify and promote sustainable ecosystems, but it has also engaged in direct action to stop activities considered harmful to the environment.

Groupe Energies Renouvelables, Environnement, et Solidarités (GERES)
Year founded: 1976.
http://www.geres.eu
GERES is a French nongovernmental organization working in France, Africa, and Asia to provide technical expertise to improve environmental conservation efforts, mitigate the effects of climate change, and improve lives through sustainable development

activities. It promotes access to and efficient use of energy, develops plans for environmentally friendly waste management, and works to combat climate change. Its members work through local partnerships to facilitate environmental management and development of resources for indigenous populations.

International Network for Sustainable Energy (INFORSE)
Year founded: 1992.
http://www.inforse.org

Headquartered in Denmark, INFORSE is a network of environmental organizations working to promote sustainable energy, protect the environment, and decrease poverty. An outgrowth of the 1992 Earth Summit, INFORSE works through regional offices in Asia, Africa, Europe, and the Americas to help develop programs aimed at transitioning the world's economies to 100 percent renewable energy sources by the year 2050. INFORSE has been active in creating public awareness of energy issues and in collaborative efforts with other organizations to establish and enforce standards relating to energy production and use.

International Union for Conservation of Nature (IUCN)
Year founded: 1948.
http://www.iucn.org

IUCN is dedicated to promoting the conservation of natural resources; its principal focus has been on identifying endangered species and promoting efforts to prevent extinctions.

Izaak Walton League
Year founded: 1922.
http://www.iwla.org

The Izaak Walton League was founded by American sportsmen to promote protection of natural resources, especially rivers and lakes. The group has lobbied for programs to preserve and rejuvenate America's rivers, lakes, wetlands, and wilderness areas. It was instrumental in the creation of the Upper Mississippi River National Wildlife and Fish Refuge in 1924 and the passage of the Clean Water Act of 1972. Over the years the organization has also lobbied against illegal logging and for legislation to protect endangered species.

League of Conservation Voters (LCV)
Year founded: 1969.
http://www.lcv.org

LCV is an educational and advocacy group in the United States that lobbies to elect candidates to office who are likely to support environmental issues.

National Audubon Society
Year founded: 1905.
http://www.audubon.org

The National Audubon Society is a conservancy group noted for its advocacy of programs aimed at the preservation of bird species and habitats. It has been active in supporting bans on harmful chemicals and in creating wildlife sanctuaries.

National Council for Science and the Environment (NCSE)
Year founded: 1990.
http://ncseonline.org

Originally known as the Committee for the National Institute for the Environment, NCSE is a U.S.-based nonprofit organization that works to assist policy makers responsible for making decisions about the environment by providing scientific data that they can use in forming judgments. NCSE supports research, disseminates information, and operates public education and outreach programs to communicate to the public accurate, scientifically based information about the environment.

National Wildlife Federation (NWF)
Year founded: 1936.
http://www.nwf.org

Originally known as the General Wildlife Federation, NWF is the largest environmental educational and advocacy group in the United States. The federation's principal aims are to connect individuals with the natural world, protect and restore critical wildlife habitats, and work toward reversing trends in global warming. NWF members come from a wide variety of interests—sports enthusiasts, nature lovers, environmentalists, and others—but work collaboratively to find ways to balance the needs of human communities with those of the natural world. The group sponsors a number of educational activities, information programs, and conferences and has also partnered with other conservation organizations on a number of important environmental projects.

Natural Resources Defense Council (NRDC)
Year founded: 1970.
http://www.nrdc.org
NRDC is an advocacy group that has operated principally in the United States but has also branched out to other countries; it engages in lobbying activities and occasionally takes legal action to promote sound environmental policy or seek the prohibition of activities it considers detrimental to the environment.

Naturfreunde International (NFI)
Year founded: 1895.
http://www.nfi.at
NFI (also known as Friends of Nature) is an international organization founded as an offshoot of the Social Democratic movement in Europe to promote appreciation for and responsible use of the region's natural resources. NFI initially encouraged recreation and tourism as a means of stimulating people to become familiar with the natural world; over time the organization has become effective and insistent in lobbying for responsible conservation and sustainable development, especially for regions that cross political boundaries.

Naturschutzbund Deutschland
Year founded: 1899.
http://www.nabu.de
Naturschutzbund Deutschland (also known as the Nature and Biodiversity Conservation Union) is one of Germany's oldest and most widely respected conservation groups, a private nonprofit that works on conservation projects both within Germany and outside the country's borders. The group publishes periodic reports to inform the public about environmental issues, conducts education programs, and works closely with governmental agencies in crafting laws and policies affecting the environment.

Ocean Conservancy
Year founded: 1972.
http://www.oceanconservancy.org
Known as the Center for Marine Conservation until 2001, the Ocean Conservancy is a nonprofit organization based in the United States that is interested in preserving and improving the world's marine resources. The group works to promote healthy and diverse ocean ecosystems and opposes practices that it considers to be threats to both marine and human life. The Ocean Conservancy has been active in efforts to restore sustainable American fisheries, protect wildlife from human activities, and encourage government reforms that can bring about improved stewardship for the oceans.

Organisationen for Vedvarende Energi (OVE)
Year founded: 1975.
http://www.ove.org
OVE (also known as the Danish Organization for Renewable Energy) is an association of individuals, business groups, and educational institutions working to promote renewable energy for Denmark. Often associated with the antinuclear movement, OVE has initiated grassroots campaigns to lobby the Danish government to restrict the use of nonrenewable energy sources and support the development of energy sources that can be replenished. The organization conducts informational campaigns and has become involved in cooperative efforts with other environmental groups to tackle issues such as climate change and global warming.

People for the Ethical Treatment of Animals (PETA)
Year founded: 1980.
http://www.peta.org
PETA is the largest animal rights group in the world; it campaigns against various forms of animal cruelty, including exploitation of animals in medical research, entertainment, and the use of animals as food for humans.

Rainforest Action Network (RAN)
Year founded: 1985.
http://ran.org
RAN is an environmental group concerned principally with issues surrounding the sustainability of the world's forests; it has engaged in campaigns to pressure corporations to refrain from activities that would deplete habitat or permanently alter the condition of forest terrains worldwide.

Sierra Club
Year founded: 1892.
http://www.sierraclub.org
The Sierra Club promotes responsible use of the earth's ecosystems and engages in educational initiatives and lobbying activities to promote conservation and responsible use of natural resources.

Stockholm Environment Institute (SEI)
  Year founded: 1989.
  http://www.sei-international.org
Although established by the government of Sweden, SEI is an independent organization performing research and developing policies that promote sustainable environments. It has offices in six countries in addition to Sweden, giving it an international reach. SEI researchers focus on overarching issues such as climate change, energy systems, ecosystem vulnerability, and governance issues, as well as specific matters such as water resources and air pollution. The group has influenced governmental policies through its work on sustainability modeling and vulnerability assessments.

Tellus Institute
  Year founded: 1976.
  http://www.tellus.org
The Tellus Institute, headquartered in the United States, has focused on scientific investigation to advance what it sees as a necessary transition to a sustainable, equitable, and humane global civilization. With funding from a variety of sources, including governments, nongovernmental agencies, and corporations, Tellus Institute researchers have produced reports on topics such as water quality, energy issues, requirements for sustainable communities, corporate social responsibility, and climate change. The institute has consistently sought to promote what it calls the "Great Transition," a paradigm shift in cultural values away from materialism and consumerism to a greater sense of global citizenship in which individual fulfillment can be achieved in societies that provide sufficiently for all their members.

Union of Concerned Scientists (UCS)
  Year founded: 1969.
  http://www.ucsusa.org
UCS, which was founded by researchers to investigate the scientific ramifications of government policies regarding the environment, promotes research and lobbying on a wide range of issues, including energy policy, climate change, and developments in technology, especially military technology, that may potentially affect the environment.

U.S. Climate Action Partnership (USCAP)
  Year founded: 2007.
  http://www.us-cap.org
USCAP is an umbrella organization uniting efforts of businesses and environmental groups to encourage government action to reduce greenhouse gas emissions.

Wetlands International
  Year founded: 1954.
  http://www.wetlands.org
Wetlands International, a conservation group headquartered in the Netherlands, began as the International Wildfowl Inquiry, dedicated to the protection of waterfowl. It gradually broadened its focus to include protection of wetlands and changed its name to International Waterfowl and Wetlands Research Bureau (IWRB). In 1991 this group merged with the Asian Wetland Bureau and Wetlands for the Americas and assumed its current name. Wetlands International works to protect and restore the world's wetlands through scientific investigation, educational programs, and advocacy initiatives to influence government policy regarding these areas.

Wilderness Society
  Year founded: 1935.
  http://wilderness.org
The Wilderness Society is an American group devoted to the preservation and responsible use of the nation's wilderness areas; it monitors policies and activities related to government-owned areas such as national forests and areas controlled by the Bureau of Land Management.

World Business Council for Sustainable Development (WBCSD)
  Year founded: 1995.
  http://www.wbcsd.org
The WBCSD, formed by a merger of the Business Council for Sustainable Development and the World Industry Council for the Environment, is an association of more than two hundred companies interested in the relationship between business and the environment. Headquartered in Switzerland with offices in the United States and Belgium, the WBCSD promotes ecofriendly business practices and lobbies gov-

ernments for policy changes that will help bring about both sustainable environments and economies. Group members operate on the philosophy that responsible policies regarding the environment are ultimately good for business as well.

World Resources Institute (WRI)
Year founded: 1982.
http://www.wri.org
WRI is a think tank based in Washington, D.C., whose members are concerned with protecting the earth's ecosystems and simultaneously enhancing people's lives. The group unites scientists and business leaders to conduct research, provide information to the public, and promote environmentally responsible development that improves communities worldwide.

Worldwatch Institute
Year founded: 1974.
http://www.worldwatch.org
The Worldwatch Institute is a research organization that collects and disseminates data on issues such as climate change, degradation of natural resources, and population growth in order to promote sustainable development.

World Wide Fund for Nature (WWF)
Year founded: 1961.
http://wwf.panda.org
Originally known as the World Wildlife Fund, WWF is an international organization that encourages conservation of natural resources (both animal and plant) and fosters sustainable development; it has actively promoted the harmonious relationship of humans with natural ecosystems worldwide.

BIBLIOGRAPHY

Cohen, Steven, et al. *Understanding Environmental Policy.* 2nd ed. New York: Columbia UP, 2014. Print.

Jordan, Andrew. *Environmental Policy in the European Union: Actors, Institutions and Processes.* Sterling: Earthscan, 2012. Print.

Mellino, Cole. "'How to Change the World' Traces the Birth of Greenpeace." *EcoWatch*. EcoWatch, 30 Jan. 2015. Web. 31 Jan. 2015.

Spapens, Toine, et al. *Environmental Crime and Its Victims: Perspectives within Green Criminology.* Burlington: Ashgate, 2014. Print.

Taylor, Dorceta E. "The State of Diversity in Environmental Organizations." Univ. of Michigan School of Natural Resources & Management. (2014). Pdf.

■ Directory of US National Parks

List of U.S. National Parks, by state.

ALASKA

Denali National Park and Preserve
P.O. Box 9, Denali Park, AK 99755
  http://www.nps.gov/dena

Gates of the Arctic National Park and Preserve
P.O. Box 30, Bettles, AK 99726
  http://www.nps.gov/gaar

Glacier Bay National Park and Preserve
P.O. Box 140, Gustavus, AK 99826
  http://www.nps.gov/glba

Katmai National Park and Preserve
P.O. Box 7, King Salmon, AK 99613
  http://www.nps.gov/katm

Kenai Fjords National Park
P.O. Box 1727, Seward, AK 99664
  http://www.nps.gov/kefj

Kobuk Valley National Park
P.O. Box 1029, Kotzebue, AK 99752
  http://www.nps.gov/kova

Lake Clark National Park and Preserve
240 West Fifth Avenue, Suite 236, Anchorage, AK 99501
  http://www.nps.gov/lacl

Wrangell-St. Elias National Park and Preserve
P.O. Box 439, Copper Center, AK 99588
  http://www.nps.gov/wrst

ARIZONA

Grand Canyon National Park
P.O. Box 129, Grand Canyon, AZ 86023
  http://www.nps.gov/grca

Petrified Forest National Park
P.O. Box 2217, Petrified Forest, AZ 86028
  http://www.nps.gov/pefo

Saguaro National Park
3693 South Old Spanish Trail, Tucson, AZ 85730
  http://www.nps.gov/sagu

ARKANSAS

Hot Springs National Park
101 Reserve Street, Hot Springs, AR 71901
  http://www.nps.gov/hosp

CALIFORNIA

Channel Islands National Park
1901 Spinnaker Drive, Ventura, CA 93001
  http://www.nps.gov/chis

Death Valley National Park
P.O. Box 579, Death Valley, CA 92328
  http://www.nps.gov/deva

Joshua Tree National Park
74485 National Park Drive , Twentynine Palms, CA 92277
  http://www.nps.gov/jotr

Kings Canyon National Park
47050 Generals Highway, Three Rivers, CA 93271
  http://www.nps.gov/seki

Lassen Volcanic National Park
P.O. Box 100, Mineral, CA 96063
  http://www.nps.gov/lavo

Redwood National and State Parks
1111 Second Street, Crescent City, CA 95531
  http://www.nps.gov/redw

Sequoia National Park
47050 Generals Highway, Three Rivers, CA 93271
  http://www.nps.gov/seki

Yosemite National Park
P.O. Box 577, Yosemite National Park, CA 95389
  http://www.nps.gov/yose

Colorado

Black Canyon of the Gunnison National Park
102 Elk Creek, Gunnison, CO 81230
http://www.nps.gov/blca

Great Sand Dunes National Park and Preserve
11500 Highway 150, Mosca, CO 81146
http://www.nps.gov/grsa

Mesa Verde National Park
P.O. Box 8, Mesa Verde, CO 81330
http://www.nps.gov/meve

Rocky Mountain National Park
1000 Highway 36, Estes Park, CO 80517
http://www.nps.gov/romo

Florida

Biscayne National Park
9700 SW 328 Street, Homestead, FL 33033
http://www.nps.gov/bisc

Dry Tortugas National Park
P.O. Box 6208, Key West, FL 33041
http://www.nps.gov/drto

Everglades National Park
40001 State Road 9336, Homestead, FL 33034
http://www.nps.gov/ever

Hawaii

Haleakalā National Park
P.O. Box 369, Makawao, HI 96768
http://www.nps.gov/hale

Hawaii Volcanoes National Park
P.O. Box 52, Hawaii National Park, HI 96718
http://www.nps.gov/havo

Kentucky

Mammoth Cave National Park
1 Mammoth Cave Parkway, P.O. Box 7, Mammoth Cave, KY 42259
http://www.nps.gov/maca

Maine

Acadia National Park
P.O. Box 177, Bar Harbor, ME 04609
http://www.nps.gov/acad

Michigan

Isle Royale National Park
800 East Lakeshore Drive, Houghton, MI 49931
http://www.nps.gov/isro

Minnesota

Voyageurs National Park
3131 Highway 53, International Falls, MN 56649
http://www.nps.gov/voya

Montana

Glacier National Park
P.O. Box 128, West Glacier, MT 59936
http://www.nps.gov/glac

Nevada

Great Basin National Park
100 Great Basin National Park, Baker, NV 89311
http://www.nps.gov/grba

New Mexico

Carlsbad Caverns National Park
3225 National Parks Highway, Carlsbad, NM 88220
http://www.nps.gov/cave

North Dakota

Theodore Roosevelt National Park
P.O. Box 7, Medora, ND 58645
http://www.nps.gov/thro

Ohio

Cuyahoga Valley National Park
15610 Vaughn Road, Brecksville, OH 44141
http://www.nps.gov/cuva

Oregon

Crater Lake National Park
P.O. Box 7, Crater Lake, OR 97604
http://www.nps.gov/crla

South Carolina

Congaree National Park
100 National Park Road, Hopkins, SC 29061
http://www.nps.gov/cong

South Dakota

Badlands National Park
25216 Ben Reifel Road, P.O. Box 6, Interior, SD 57750
http://www.nps.gov/badl

Wind Cave National Park
26611 U.S. Highway 385, Hot Springs, SD 57747
http://www.nps.gov/wica

TENNESSEE
Great Smoky Mountains National Park
107 Park Headquarters Road, Gatlinburg, TN 37738
http://www.nps.gov/grsm

TEXAS
Big Bend National Park
P.O. Box 129, Big Bend National Park, TX 79834
http://www.nps.gov/bibe

Guadalupe Mountains National Park
400 Pine Canyon Road, Salt Flat, TX 79847
http://www.nps.gov/gumo

Arches National Park
P.O. Box 907, Moab, UT 84532
http://www.nps.gov/arch

Bryce Canyon National Park
P.O. Box 640201, Bryce Canyon, UT 84717
http://www.nps.gov/brca

Canyonlands National Park
2282 SW Resource Boulevard, Moab, UT 84532
http://www.nps.gov/cany

Capitol Reef National Park
HC 70 Box 15, Torrey, UT 84775
http://www.nps.gov/care

Zion National Park
Springdale, UT 84767
http://www.nps.gov/zion

VIRGINIA
Shenandoah National Park
3655 U.S. Highway 211 East, Luray, VA 22835
http://www.nps.gov/shen

WASHINGTON
Mount Rainier National Park
55210 238th Avenue East, Ashford, WA 98304
http://www.nps.gov/mora

North Cascades National Park
810 State Route 20, Sedro-Woolley, WA 98284
http://www.nps.gov/noca

Olympic National Park
600 East Park Avenue, Port Angeles, WA 98362
http://www.nps.gov/olym

WYOMING
Grand Teton National Park
P.O. Drawer 170, Moose, WY 83012
http://www.nps.gov/grte

Yellowstone National Park
P.O. Box 168, Yellowstone National Park, WY 82190
http://www.nps.gov/yell

NATIONAL PARK SERVICE REGIONAL OFFICES
Alaska Area Region
Regional Director, National Park Service, 240 West Fifth Avenue, Anchorage, AK
99501
(907) 644-3510

Intermountain Region
Regional Director, National Park Service, 12795 Alameda Parkway, Denver, CO
80225
(303) 969-2500

Midwest Region
Regional Director, National Park Service, 601 Riverfront Drive, Omaha, NE
68102
(402) 661-1736

National Capital Region
Regional Director, National Park Service, 1100 Ohio Drive SW, Washington, DC
20242
(202) 619-7222

Northeast Region
Regional Director, National Park Service, U.S. Custom House, 200 Chestnut Street,
Fifth Floor, Philadelphia, PA 19106
(215) 597-7013

Pacific West Region
Regional Director, National Park Service, One Jackson Center, 1111 Jackson Street, Suite 700, Oakland, CA 94607
(510) 817-1300

Southeast Region
Regional Director, National Park Service, 100 Alabama Street SW, 1924 Building, Atlanta, GA 30303
(404) 507-5600

Subject index

Symbols

2016 presidential election 28

A

acid rain 27, 84, 175, 176, 215, 216, 266, 276, 278, 291
adaptation 99, 207, 209, 306
aerobic 73
aerosols 185, 226, 249
Africa
 deforestation 112
agriculture
 slash-and-burn[slash and burn] 5, 7, 21, 112
air pollution 6, 82, 83, 84, 85, 86, 94, 198, 200, 206, 224, 231, 235, 266, 276, 277, 278, 288, 289, 292, 297, 306, 323
alloys 78
also established national parks 233
aluminum recycling 274
Amazon River basin 20
ambient air 75, 94, 257
anaerobic 61, 62, 63, 80, 314
Andes 20, 21, 69, 70
Antarctica 181, 182, 185, 250, 263, 265, 315
antelope 154, 190
anthracite 175
anthropogenic 52, 53, 76, 90, 91, 92, 98, 118, 124, 183, 184, 209, 210, 212, 227, 248, 251, 264, 271, 306
antienvironmentalism 25
antinuclear movement 241
aquifers 29, 62, 75, 116, 196, 301, 310, 312, 315
arms race 241
Atlantic Multidecadal Oscillation 184
atomic bomb 240
Audubon, John James 319
Australia 24, 45, 49, 65, 98, 101, 115, 143, 144, 190, 194, 222, 226, 250
aviation 236

B

bacteria 7, 8, 19, 30, 55, 57, 58, 59, 61, 62, 63, 73, 80, 179, 180, 205, 226, 249, 257, 258, 261, 266, 275, 309, 310
bats 38, 49

batteries
 lead-acid, 15, 275, 278
Bay of Fundy 298
Bernard, John 233
biodegradable 15, 50, 245, 262
biodiversity 2, 17, 19, 20, 24, 25, 27, 43, 112, 113, 135, 139, 147, 153, 154, 155, 156, 157, 160, 169, 171, 174, 186, 197, 198, 199, 211, 227, 293, 302, 303, 306, 321
biodiversity
 wetlands, 43
biogeography 41, 42, 49, 50, 142
biological phenomena diversity 42, 43
biological pump 74
biomass 9, 66, 67, 206, 227, 276, 277
Biophilia 44
bioremediation 61, 62, 63, 258
biosphere 46, 54, 64, 66, 67, 68, 73, 127, 141, 142, 206, 209, 313
biota 18, 41, 43, 48
birds 2, 38, 47, 48, 50, 51, 53, 65, 66, 69, 70, 89, 102, 125, 153, 155, 166, 171, 190, 224, 247, 250, 255, 257, 258, 261, 292, 293, 297, 302, 320, 321, 322
bituminous coal 175
Bonn Convention 156
Brazil 17, 18, 19, 20, 21, 22, 39, 45, 48, 57, 99, 100, 129, 156, 194, 216, 231, 254, 256
breeder reactor 241, 242, 243
Bush, George H. W. 84, 99, 129, 216, 316
Bush, George W. 28, 85, 86, 242, 316
by-product 65, 73

C

California 9, 22, 24, 36, 43, 61, 63, 68, 85, 95, 107, 114, 121, 161, 180, 206, 207, 214, 215, 227, 240, 245, 251, 258, 263, 271, 290, 317, 318, 320
Canada 43, 68, 69, 70, 98, 134, 141, 143, 149, 155, 194, 202, 204, 216, 222, 231, 232, 233, 234, 242, 255, 262, 287, 297, 298, 302, 316
capital markets 244
carbon cycle 64, 66, 68, 73, 74
carbon dioxide 2, 10, 11, 12, 13, 14, 16, 17, 19, 22, 64, 92, 93, 95, 96, 97, 109, 114, 115, 130, 134, 140, 175, 177, 184, 185, 186, 192, 194, 210, 225, 226, 242, 276, 277, 278, 279, 291, 323
carbon dioxide
 climate change, 92

carbon dioxide
    global warming, 184
carbon monoxide 75, 86, 175, 176, 282
carbon tax 2, 75, 99
carcinogens 78, 79
Carson, Rachel 26, 34, 127, 214
Carson, Rachel
    Silent Spring, 26, 34, 214
Carter, Jimmy 26, 84, 241, 243, 316
catalyst 254, 255, 261, 289
cement 36, 75, 203, 253
Central America
    deforestation, 112
Chernobyl 215, 237, 238, 239, 240
cholera 95, 310
chosen 46, 107, 128, 129
Civilian Conservation Corps 214
clean energy 204
climate change 10, 12, 91, 92, 93, 95, 96, 97, 98, 100, 114, 119, 161, 175, 181, 183, 186, 205, 207, 208, 209, 210, 212, 225, 226, 248, 249, 252, 267, 271, 282, 287, 288, 294
climate change skeptics 13, 28, 96, 97
climate change
    skeptics, 95
climatology 10, 90, 93, 96, 100, 120, 124, 157, 161, 174, 181, 183, 225
Clinton, Bill 162, 193, 316, 321
coal 1, 10, 11, 13, 15, 19, 23, 27, 73, 74, 75, 78, 80, 84, 85, 96, 119, 174, 175, 176, 177, 203, 207, 208, 254, 255, 256, 276, 282, 288, 289, 322, 323
coast ranges 280
Colorado 8, 95, 107, 125, 128, 164, 254, 266, 298
communication 141
compost 259, 260, 261, 273, 274, 301, 302
Comprehensive Environmental Response, Compensation, and Liability Act (CERCLA) 25, 26, 71, 199, 202, 252, 253
computer modeling
    climate, 264
concentration 12, 51, 57, 66, 73, 75, 92, 97, 114, 115, 119, 226, 263, 264, 265, 288, 289, 303, 305
conflicts of interest 27
Conservation 4, 17, 19, 22, 23, 24, 34, 35, 37, 41, 42, 43, 70, 100, 105, 108, 111, 125, 148, 152, 153, 154, 156, 157, 192, 212, 213, 214, 215, 216, 217, 221, 223, 227, 229, 232, 269, 315, 316, 317, 319, 321
Conservation biology 44
Consumerism 60, 172

cooling system 311
coral 42, 54, 100, 101, 102, 117, 145, 169, 171, 197, 198, 227, 248, 249, 250, 251, 253, 257
crude oil 63, 254, 255, 256, 257, 258

D

dam 106, 107, 108, 156, 297, 298
dams
    hydroelectric, 21
decomposition 13, 17, 21, 61, 62, 74, 174, 175, 263, 264
deforestation 2, 5, 12, 17, 18, 19, 20, 21, 23, 42, 43, 65, 68, 70, 80, 81, 111, 112, 113, 114, 118, 124, 129, 135, 136, 145, 174, 197, 198, 207, 208, 215, 229, 292
dengue fever 95
density 15, 49, 100, 109, 197, 220, 221, 235, 245, 248, 257, 283, 284, 289
deregulation 187, 244
desalination 95, 114, 115, 310
desertification 118, 120, 121, 124, 184, 190, 191, 232
detergents 21, 134, 258
deuterium 242
Diamond, Jared 42
dichloro-diphenyl-trichloroethane (DDT) 62
discharge 29, 87, 88
disease-carrying mosquito 94
disease vectors 93, 95
distillation 114, 115
drilling mud 252, 253
drought 6, 17, 18, 19, 21, 93, 114, 120, 121, 122, 123, 124, 125, 126, 166, 185, 190, 251, 279
dysentery 95

E

Earth Day 26, 28, 83, 127, 128, 129, 130, 143, 192, 193, 216, 241, 316
Earth Summit 45, 57, 156, 157, 216
Echo Park Dam opposition 107
ecology 19, 25, 33, 44, 46, 60, 102, 103, 110, 111, 112, 141, 142, 143, 144, 145, 146, 159, 160, 189, 227, 234, 253, 267, 291, 294, 299, 303, 305, 306, 312
economic growth 85, 139, 147, 152, 186, 211, 221
ecosystem 2, 19, 21, 37, 38, 41, 42, 44, 46, 49, 59, 60, 69, 81, 101, 102, 138, 139, 142, 143, 148, 152, 153, 154, 165, 166, 169, 170, 171, 190, 206, 221, 231, 234, 245, 250, 251, 261, 293, 294, 305, 306, 313, 315, 320
ecosystems
    diversity, 19

ecosystems
 wetlands, 313
Ehrlich, Paul R. 1, 3, 127, 152, 214
electric power 106
electricity
 hydropower, 279
electronic waste
 recycling, 151
elephants 33, 293
elk 66, 190, 213, 234
El Niño 18, 100, 101, 102, 183, 184, 248, 250, 285, 287, 288
Emerson, Ralph Waldo 213
emission 76, 77, 83, 84, 92, 93, 98, 140, 186, 211, 236, 283, 288
endangered species 43, 107, 152, 154, 155, 156, 157, 191, 205, 213, 321
Endangered Species Act, 152, 155, 215, 321, 322
energy conservation 26, 185, 193
energy efficiency 193, 194, 311
energy
 renewable, 276
England 32, 57, 76, 82, 96, 102, 115, 135, 139, 149, 151, 154, 177, 192, 193, 197, 214, 220, 239, 243, 252, 258, 271, 279, 282, 284, 287, 289
environmentalism 25, 26, 27, 28, 29, 96, 127, 135, 233, 298
environmental movement 26, 27, 34, 59, 60, 83, 119, 127, 128, 129, 162, 233, 241, 277, 303
environmental organizations 27, 111, 130, 193
Environmental Protection Agency 12, 13, 25, 26, 63, 71, 82, 83, 87, 88, 94, 96, 151, 157, 162, 167, 180, 193, 194, 199, 201, 203, 204, 215, 225, 227, 235, 236, 260, 264, 273, 279, 302, 310, 314
Environmental Protection Agency (EPA) 12, 25, 26, 63, 71, 82, 83, 87, 88, 94, 151, 162, 180, 193, 199, 201, 203, 215, 225, 235, 236, 260, 264, 273, 310, 314
environmental sustainability 140, 141, 147
erosion 7, 21, 35, 36, 37, 79, 80, 81, 108, 112, 118, 123, 124, 165, 171, 181, 190, 191, 198, 199, 210, 215, 220, 234, 260, 261, 298, 302, 303, 305, 307, 311, 312, 313, 314
ethanol 9, 13, 14, 15, 17, 18, 62, 185
experiments 92, 184, 264
explosion at the *Deepwater Horizon* oil rig 257
extinction 6, 18, 22, 24, 33, 42, 44, 48, 49, 50, 58, 70, 113, 139, 152, 153, 154, 155, 156, 170, 213, 226, 248, 249, 251, 271, 291, 292, 293, 294, 306

extinctions 44, 197, 198, 291, 292, 293, 294
Exxon Valdez disaster 257

F

famine 95, 113, 120
Federation 215, 232, 242, 243
feedback processes 97
Fermi, Enrico 241
fill 88, 108, 135, 166, 207, 208
financial incentives 316
fish 7, 20, 21, 22, 37, 48, 51, 55, 67, 101, 102, 107, 115, 144, 156, 166, 169, 170, 171, 172, 173, 175, 215, 216, 230, 256, 257, 258, 267, 279, 293, 297, 298, 299, 322
fisheries 20, 101, 144, 145, 169, 170, 171, 172, 227, 250, 251, 297, 299
fission 2, 3, 237, 239
floodplain 22, 108, 173, 174, 304
floods
 deforestation, 113
Florida 24, 36, 115, 154, 162, 214, 222, 286, 287, 298, 316, 318, 320
Food Security Act (1985) 313, 316
forestry 111, 113, 122, 210, 229, 231, 277, 312, 313
forests
 deforestation, 112
forests
 management, 206, 231
forests
 old-growth [old growth], 113, 215, 292
fossil fuel 9, 13, 14, 18, 74, 75, 92, 249
fossil fuels 2, 3, 9, 11, 12, 13, 14, 15, 16, 26, 66, 75, 76, 77, 96, 97, 102, 115, 134, 175, 176, 182, 185, 186, 193, 207, 208, 209, 213, 216, 242, 244, 249, 269, 276, 277, 278, 288, 289, 294, 311, 322, 323
France 12, 129, 149, 194, 242, 243, 245, 246, 257, 297, 313
free energy 75
Friends of the Earth International 193
Fukushima 1 238

G

gasification 177
gasohol 14
genetic diversity 41, 42, 43, 171, 293
genetic resources 45, 46, 157
geoengineering 75

geothermal energy 279
Germany 12, 19, 70, 129, 134, 216, 243, 246, 278, 323
Gettysburg Historic Site 234
giant 20, 53, 234, 248, 251, 291
glass recycling 274
Glen Canyon Dam 107
global climate 17, 23, 24, 73, 76, 81, 97, 111, 145, 183, 192, 210, 242, 294, 306, 315
global climate change 23, 24, 76, 81, 97, 145, 192, 294, 306
global warming 4, 10, 11, 12, 13, 17, 40, 41, 43, 74, 76, 90, 91, 93, 95, 96, 98, 99, 101, 102, 188, 192, 207, 208, 209, 210, 211, 212, 215, 225, 226, 227, 243, 248, 249, 250, 251, 254, 276, 280, 288, 323
global warming
    carbon cycle, 74
global warming
    climatology, 183
global warming
    skeptics, 96
Gore, Al 212
gray wolf 234
Great Barrier Reef 101, 102, 226, 250
Great Smoky Mountains National Park 234
greenhouse effect 10, 13, 17, 18, 25, 40, 73, 80, 93, 96, 184, 276, 280, 313
greenhouse gas 10, 12, 13, 73, 74, 76, 77, 80, 81, 82, 92, 98, 99, 108, 140, 175, 176, 183, 184, 185, 192, 209, 239, 242, 244, 278, 279, 294
greenhouse gas emissions 10, 12, 13, 74, 77, 80, 81, 98, 99, 175, 185, 209, 239, 242, 244, 278
greenhouse gases 10, 11, 12, 13, 14, 16, 21, 73, 74, 76, 79, 80, 82, 86, 93, 96, 97, 98, 102, 114, 130, 140, 175, 184, 185, 225, 226, 249, 271, 276, 288
greenhouse gases
    carbon dioxide, 96
Greenpeace 188, 193, 216
Green River Formation 254
groundwater 7, 8, 29, 30, 62, 65, 79, 80, 121, 123, 194, 195, 196, 198, 200, 201, 203, 253, 256, 260, 310, 313, 315
guinea worms 95

H

habitat destruction 26, 155, 156, 175, 197, 198, 199
Hansen, James E. 185, 225
heat island 91, 302
heatstroke 93
heavy water 242
Hetch Hetchy Dam 107
hole in the ozone layer 27
Hoover Dam 108, 298
human nature 27
hydrocarbons
    oil shale, 254
hydroelectric power 18, 105, 279
hydrogen peroxide 62
hydrology 60, 174, 282, 314
hypothermia 93

I

ice ages 91, 92
impoundment 105
incineration 26, 78, 199, 202
India 5, 7, 8, 12, 37, 58, 65, 114, 124, 182, 194, 204, 239, 243, 271
indicator 205, 206
Indonesia 7, 49, 113, 123
Industrial Revolution 6, 98, 106, 133, 186, 207, 208, 209, 210, 213, 215, 226, 249, 270, 291, 294, 298
influenza 94, 148
insects 7, 8, 18, 24, 33, 37, 38, 45, 48, 51, 58, 65, 66, 95, 156, 214, 234, 260, 261, 262, 292
intensive agriculture 39
Intergovernmental Panel on Climate Change 12, 19, 91, 92, 93, 96, 98, 161, 181, 183, 207, 208, 209, 212
invertebrates 48, 58, 190, 253, 257, 306, 315
irrigation 8, 9, 103, 105, 106, 107, 108, 109, 114, 118, 124, 126, 214, 215, 305, 323
island biogeography theory 42

J

Japan 69, 98, 115, 134, 193, 214, 222, 237, 238, 239, 242, 243

K

karst 43
Kyoto Protocol (1997) 10, 12, 76, 77, 98, 99, 100, 175, 176, 177, 212

L

land clearance 6
land management 191, 221, 312

League of Nations 187
Leopold, Aldo 132, 142, 146, 317, 318
Linnaeus, Carolus 33, 43, 142
liquefaction 16
lithium 15
logging
    deforestation, 112
Lomborg, Bjørn 94
Louisiana 63, 158, 161, 162, 174, 199, 286

M

Maine 72, 287, 298
malaria 95, 214
malnutrition 39, 95
Man and Nature (Marsh) 213
Mann, Michael E. 184, 185
marsh 11, 35, 80, 258, 281, 298, 311
Marsh, George Perkins 213
Massachusetts 220, 258, 298, 315
Medieval Warm Period 95, 184, 185
mercury poisoning 214
methane 11, 12, 13, 14, 62, 73, 80, 81, 82, 86, 93, 96,
    108, 175, 176, 184, 186, 203, 204, 211, 226, 249, 276,
    279
Michigan 90, 144, 241, 311
Migratory Bird Treaty Act 216
mineral 4, 22, 75, 78, 81, 114, 117, 192, 203, 233, 239,
    279
mitigation 28, 86, 93, 207, 306
model 3, 4, 82, 84, 86, 92, 107, 110, 127, 137, 138, 141,
    186, 187, 188, 233, 265, 266, 288, 322
monitoring 67, 89, 145, 194, 205, 206
Muir, John 111, 213, 320
municipal solid waste 274

N

National Aeronautics and Space Administration 145,
    185, 225, 266
National Audubon Society 154, 215, 319, 320
national parks 26, 27, 68, 148, 154, 214, 216, 229, 232,
    233, 234, 235, 317, 318, 319, 320
National Park Service 216, 232, 233, 234
natural gas 10, 11, 13, 15, 16, 19, 69, 73, 75, 80, 96, 119,
    174, 176, 198, 203, 204, 207, 244, 254, 274, 276, 282,
    288, 322
Nature Conservancy 154, 193, 215, 319, 320

nature reserves 154
Nebraska 8, 10, 28, 120, 144, 232, 262
Netherlands 77, 163, 222, 246, 281, 323
Nevada 107, 121, 164, 243
nitric acid 265, 291
North American forests 69
North Sea 281
Noss, Reed F. 43
nuclear energy 237, 238, 240, 241
nuclear fission 2, 239
nuclear fuel 240, 241
nuclear power 79, 215, 237, 238, 239, 240, 241, 242,
    243, 244
nuclear power industry 239, 240
nuclear power plants 238, 240, 242, 243, 244
nuclear proliferation 135, 136, 241
nuclear reaction 242
nuclear reactor 237, 239, 241, 243
Nuclear Regulatory Commission (NRC) 237, 238, 241
nuclear technology 240
nuclear waste disposal 3, 241
nuclear weapons
    antinuclear movement, 241
nutrient 2, 21, 53, 61, 80, 101, 108, 171, 247, 250, 258,
    261, 312

O

Oak Ridge, Tennessee 241
Obama, Barack 28, 86, 242
ocean acidification 226, 249, 291
oceans
    biological pump, 74
offshore drilling 214, 253, 256
oil drilling in the Arctic 28
oil shale 254, 255
oil spills 63, 87, 145, 176, 177, 253, 255, 256, 257, 258
Oregon 43, 60, 207, 221, 222, 251, 264, 318
organic 9, 11, 13, 15, 22, 30, 54, 57, 59, 61, 62, 63, 64,
    70, 96, 108, 109, 131, 133, 134, 139, 145, 174, 175,
    179, 194, 195, 199, 202, 203, 248, 256, 257, 263, 267,
    288, 290, 301, 302, 322
organic matter 13, 54, 57, 59, 73, 74, 75, 108, 116, 145,
    203, 259, 260, 261, 276
overfishing 101, 166, 169, 171, 172, 227, 248, 251
oxides 11, 84, 85, 264, 265, 282, 283, 288, 289

P

Pacific Decadal Oscillation 183, 184
paleoclimatology 252
particulate matter 11, 289
Pennsylvania 29, 204, 237, 238, 240, 241, 289
petroleum 13, 15, 16, 63, 71, 114, 115, 119, 175, 202, 203, 252, 253, 254, 255, 256, 257, 258, 275, 276, 282, 283
petroleum pipelines 256
photosynthesis 17, 19, 20, 54, 65, 73, 100, 248, 251, 266
photovoltaic (PV) cells 277
phytoplankton 101, 226, 227, 248, 250
Pinchot, Gifford 214, 230
plague 33, 95, 170, 231
plastics
   recycling, 274
poaching 148, 154, 155, 233, 235
pollen 37, 39, 152, 305
pollutant 11, 14, 51, 85, 177, 206, 289
pollution control 214
polystyrene 245, 275
Population Bomb, The (Ehrlich) 214
population growth 3, 19, 46, 90, 139, 166, 208, 269, 270, 271, 283, 304, 309
Population issues
   Ehrlich, Paul R., 3
power plants 11, 74, 84, 85, 86, 94, 175, 176, 177, 235, 238, 240, 242, 243, 244, 297, 323
power plants
   hydroelectric, 105
precipitation 21, 29, 49, 53, 64, 65, 66, 70, 80, 90, 91, 92, 94, 115, 116, 117, 120, 121, 122, 123, 164, 175, 250, 285, 286, 287, 291, 311, 312
preservation 42, 43, 64, 67, 68, 127, 132, 147, 149, 152, 158, 161, 162, 212, 213, 215, 216, 217, 219, 231, 232, 233, 234, 302, 317, 320
Price-Anderson Act 238, 239, 240
principal 18, 240, 264, 283
public opinion 127, 215, 241
public utility 241

R

radiation
   solar, 13, 79, 80, 93, 276
radioactive fallout 239
radioisotopes 79

ragweed 95
rainfall patterns 53, 94
rain forests 2, 5, 42, 44, 53, 66, 67, 112, 197, 292, 294, 299
Ramsar Convention on Wetlands of International Importance (1971) 316
Reagan, Ronald 26, 84, 89, 129, 130
reclamation 36, 176
recycling 26, 60, 61, 75, 135, 138, 139, 141, 151, 194, 241, 273, 274, 275, 276, 311
reefs 42, 54, 67, 100, 101, 117, 169, 171, 197, 198, 246, 248, 249, 250, 251, 253, 257
reforestation 74, 77, 113
release 11, 12, 14, 16, 18, 51, 66, 73, 74, 88, 102, 108, 114, 175, 180, 200, 204, 208, 211, 226, 237, 239, 240, 249, 253, 254, 263, 264, 265, 276, 277, 278, 279, 316
renewable energy 3, 9, 98, 194, 276, 277, 278, 279, 322
Republic 60, 120, 243, 297
reservoir 106, 107, 108, 123, 252, 265
resource depletion 213
result of human activities 93, 184
Rio de Janeiro, Brazil 45, 57, 156, 216
risk 146, 148, 152, 153, 163, 180, 190, 198, 200, 204, 227, 237, 238, 242, 243, 246, 253, 256, 286, 290, 293, 306
river basin 107, 312
Rocky Mountain spotted fever 95
Roosevelt, Franklin D. 213, 214
Roosevelt, Theodore 154, 213, 214, 230, 316, 319, 320, 321
rubber 15, 17, 245, 275
runoff 7, 8, 17, 54, 70, 89, 109, 122, 123, 165, 166, 174, 248, 250, 256, 283, 301, 302, 303, 305, 306, 312, 313
Russia 48, 65, 69, 98, 231, 239, 243, 246, 255, 271, 297
Russian Federation 242, 243

S

Safe Drinking Water Act of 1974 253
Sagebrush Rebellion 26, 27
saltwater intrusion 314
Sand County Almanac, A (Leopold) 146
Santa Barbara oil spill 258
schistosomiasis 95, 107
Scotland 182, 280, 292
scrap 220, 316
sea level 35, 36, 183, 184, 186, 211, 225, 226, 250, 280, 281, 297

secretariat 210
sediments 7, 20, 29, 91, 108, 118, 166, 171, 174, 226, 249, 315
severe storms 35, 125, 285
sewage 87, 89, 109, 166, 196, 215, 298, 310, 311
shale oil 254, 255
Sierra Club 85, 107, 111, 137, 154, 189, 193, 213, 215
Silent Spring 26, 34, 52, 214
silt 7, 20, 21, 29, 108, 165, 173, 203, 298
Singer, S. Fred 184
sinking 100, 174, 248, 280
slash-and-burn agriculture[slash and burn agriculture] 5, 7, 21, 112, 113
slurry 62
soil erosion 124, 125, 126, 198, 234, 260, 305, 307, 313
solar 9, 13, 26, 64, 65, 74, 79, 80, 92, 93, 94, 114, 138, 184, 185, 198, 209, 263, 276, 277, 278, 279, 322
Southern Oscillation 18, 100, 248, 250
South Korea 243, 297
species diversity 6, 18, 41, 42, 47
specific gravity 257
spent fuel 3, 240, 243
sprawl 8, 71, 72, 198, 219, 221, 302, 303, 304, 305, 306
storm surges 35, 281
stratosphere 184, 263, 264, 265, 266
suburbs 94, 281
sulfate aerosols 185
sulfur 11, 12, 18, 27, 64, 82, 84, 85, 86, 175, 176, 226, 249, 254, 255, 288, 289, 290
sulfur dioxide 11, 18, 27, 84, 85, 175, 226, 249
sulfuric acid 226, 249
sunlight 10, 11, 37, 66, 100, 101, 112, 116, 166, 191, 262, 265, 267, 277, 278, 288, 289, 290, 305
Superfund 25, 26, 71, 199, 202, 252, 253
supertanker *Torrey Canyon* 258
supply 5, 6, 13, 16, 19, 21, 37, 38, 51, 65, 66, 81, 95, 106, 107, 115, 117, 118, 121, 122, 123, 124, 135, 170, 174, 182, 190, 195, 198, 201, 230, 235, 240, 256, 257, 260, 276, 278, 279, 293, 309, 310, 313, 320, 322, 323
surface water 30, 63, 65, 80, 87, 88, 116, 195, 200, 251, 256, 303, 310
sustainable development 3, 68, 77, 149, 155, 161, 166, 172, 186, 207, 209, 211, 285
sustainable forestry 113, 229

T

Tanzania 45
tar sands 254, 255
taxation 75
Tellico Dam 107, 156
Tennessee Valley Authority (TVA) 105, 107, 214
Thailand 129, 174
thermal pollution 43, 145, 240, 297, 310
threatened species 23, 27, 38, 44, 45, 155, 315, 320
Three Mile Island 237, 238, 241, 243, 244
Tidal power 297
tradable pollution permits 28
trash 54, 129, 247, 248, 273, 274, 275
troposphere 289
tundra 42, 49, 52, 66, 67, 68, 69, 143, 189, 253
typhoid fever 95
typhus 95

U

United Nations Conference on the Human Environment 216
United Nations Convention on Biological Diversity (CBD) 45
United Nations Educational, Scientific, and Cultural Organization (UNESCO) 67, 102, 232, 233
upwelling 74, 109, 226, 248, 250, 251
urban heat islands 94
urban sprawl 71, 72, 198, 219, 221, 302, 303, 304, 305
US Department of Energy 278
Utah 111, 254, 255, 319
Uzbekistan 45, 255, 256

V

viscosity 253
volcanic eruptions 92, 123, 184, 226, 249

W

Washington 27, 28, 39, 44, 46, 50, 60, 63, 68, 70, 74, 86, 90, 93, 97, 100, 102, 115, 124, 128, 129, 130, 139, 144, 146, 149, 157, 172, 177, 186, 194, 196, 199, 201, 207, 214, 223, 224, 235, 244, 248, 259, 270, 271, 276, 282, 307, 311, 322
waste 2, 3, 9, 26, 51, 61, 62, 63, 74, 78, 79, 80, 81, 134, 138, 139, 141, 149, 150, 151, 161, 162, 166, 171, 172, 192, 255, 261, 273, 274, 275, 277, 278, 306, 311, 312, 313
waste management 61, 242, 273, 275
waste stream 202, 273, 275
water pollution 24, 43, 80, 87, 89, 90, 130, 133, 161, 166, 175, 192, 200, 224, 235, 273, 277, 282, 301, 302, 316

Water Pollution Control Act 87, 215, 252, 253
water quality 23, 80, 87, 88, 144, 145, 166, 167, 194, 203, 240, 302, 307, 309
water resources 203, 231, 302, 305, 306, 310
watershed 2, 21, 29, 60, 68, 105, 123, 166, 167, 173, 174, 231, 311, 312, 313
water table 29, 196, 256, 283, 303
water vapor 11, 12, 14, 64, 75, 121, 265
Watt, James 27
weathering 235
welfare 1, 2, 3, 12, 60, 86, 127, 137, 139, 143
wetlands 27, 43, 87, 89, 90, 173, 174, 191, 198, 206, 214, 253, 257, 281, 283, 292, 311, 312, 313, 314, 315, 316, 317, 321
wildlife management 144, 154, 230, 234
Wilson, Edward O. 2, 44
wind 7, 9, 35, 36, 47, 69, 70, 74, 80, 92, 116, 117, 118, 121, 123, 124, 126, 163, 164, 165, 193, 250, 278, 279, 285, 286, 287, 322, 323, 324
wind energy 74, 278, 322, 323, 324
Wind power 9, 278
wise-use movement 27, 222
World Heritage Convention 233
World War II 143, 147, 187, 188, 214, 220, 233, 240, 293
Wyoming 8, 125, 164, 190, 213, 229, 254

Y

yellow fever 95
Yellowstone National Park 190, 229, 233, 235
Yosemite National Park 229, 320

Z

zooplankton 50, 51, 91, 227, 247, 267